Epidemics and Society

THE OPEN YALE COURSES SERIES is designed to bring the depth and breadth of a Yale education to a wide variety of readers. Based on Yale's Open Yale Courses program (http://oyc.yale.edu), these books bring outstanding lectures by Yale faculty to the curious reader, whether student or adult. Covering a wide variety of topics across disciplines in the social sciences, physical sciences, and humanities, Open Yale Courses books offer accessible introductions at affordable prices.

The production of Open Yale Courses for the internet was made possible by a grant from the William and Flora Hewlett Foundation.

BOOKS IN THE OPEN YALE COURSES SERIES

Paul H. Fry, *Theory of Literature*
Roberto González Echevarría, *Cervantes' "Don Quixote"*
Christine Hayes, *Introduction to the Bible*
Shelly Kagan, *Death*
Dale B. Martin, *New Testament History and Literature*
Giuseppe Mazzotta, *Reading Dante*
R. Shankar, *Fundamentals of Physics I: Mechanics, Relativity, and Thermodynamics,* Expanded Edition
R. Shankar, *Fundamentals of Physics II: Electromagnetism, Optics, and Quantum Mechanics*
Ian Shapiro, *The Moral Foundations of Politics*
Steven B. Smith, *Political Philosophy*
Frank M. Snowden, *Epidemics and Society: From the Black Death to the Present*

Epidemics and Society

From the Black Death to the Present

With a New Preface

FRANK M. SNOWDEN

Yale
UNIVERSITY PRESS
New Haven and London

Published with assistance from the foundation established in memory
of Amasa Stone Mather of the Class of 1907, Yale College.

Yale University Press books may be purchased in quantity for educational,
business, or promotional use. For information, please e-mail
sales.press@yale.edu (U.S. office) or sales@yaleup.co.uk (U.K. office).

Set in Minion type by Westchester Publishing Services.
Printed in the United States of America.

Library of Congress Control Number: 2019933691
ISBN 978-0-300-19221-6 (hardcover : alk. paper)
ISBN 978-0-300-25639-0 (paperback)

A catalogue record for this book is available from the British Library.

For Claire and Jessica

Contents

Preface to the Paperback Edition

The coronavirus, a severe acute respiratory syndrome, has unleashed a pandemic since the original publication of *Epidemics and Society*. Coronavirus disease (COVID-19) is still too new and too poorly understood to allow us to assess its ultimate impact, but its broad contours have become sufficiently clear, and several of its features relate closely to the themes of this book.

Like all pandemics, COVID-19 is not an accidental or random event. Epidemics afflict societies through the specific vulnerabilities people have created by their relationships with the environment, other species, and each other. Microbes that ignite pandemics are those whose evolution has adapted them to fill the ecological niches that we have prepared. COVID-19 flared up and spread because it is suited to the society we have made. A world with nearly eight billion people, the majority of whom live in densely crowded cities and all linked by rapid air travel, creates innumerable opportunities for pulmonary viruses. At the same time, demographic increase and frenetic urbanization lead to the invasion and destruction of animal habitat, altering the relationship of humans to the animal world. Particularly relevant is the multiplication of contacts with bats, which are a natural reservoir of innumerable viruses capable of crossing the species barrier and spilling over to humans.

Such spillovers occur with growing frequency—usually without widespread consequences. But contingent circumstances may favor transmission from a first human host to others, as Ebola demonstrated in December 2013. At that time, a small child played in the hollow of a tree near the garden of his home in Guinea. There rampant deforestation had led fruit bats—evicted from the demolished canopy of the nearby woods—to cluster by the thousands. The misfortune of the four-year-old boy was to inhale viruses shed in the dejecta of the displaced bats. All of the subsequent victims of Ebola throughout the West African epidemic of 2014–2016 were linked in an unbroken chain of transmission from this first, or "index," case.

This sequence of events, transposed to an urban context, probably recurred in December 2019 at a bushmeat "wet market" in Wuhan, China. There a combination of factors transformed a warren of closely packed stalls lacking refrigeration and narrow, unhygienic passageways into a giant petri dish. The major conditions facilitating the interspecies exchange of microbes involved the close proximity of various species of caged domestic and wild animals, including bats; the mixing of their feces and blood after butchering; and the contamination of produce and throngs of shoppers. In this setting, "patient zero" was likely a daily customer who contracted the novel coronavirus and transmitted it to close contacts. Community spread was rapid because humanity lacks herd immunity against so newly emergent a pathogen—that is, we lack the protection that resists disease spread when large enough numbers of the population are immune (such as through vaccination) and can thereby break the chain of transmission.

Of all the issues raised by COVID-19, the most important is preparedness. Nobel laureate Joshua Lederberg famously argued that, in the contest between humans and microbes, the only defense humans possess is their wits. One could add to Lederberg's formulation our capacity to collaborate—if we so choose. Unfortunately, when COVID-19 appeared, it found a world that was not mobilized to confront a challenge long foreseen. Since World War II we have lived in an age of ever-increasing numbers of emerging diseases. Already in 2008 researchers identified 335 human diseases that had emerged between 1960 and 2004, most of them of animal origin. Their names now run the gamut from A to Z—from Avian flu to Zika, and scientists caution that far more potentially dangerous pathogens exist than have so far been discovered. Particularly since the outburst of H5N1 influenza in 1997, the public health community has consistently sounded the alarm. Its message is that future outbreaks are inevitable—particularly of pulmonary viral diseases, to which our society is highly vulnerable. The question is not whether but when. According to the virologist Brian Bird, "We live in an era now of chronic emergency." In addition, virologists stress that there is every reason to expect a disastrous pandemic in the near future rivaling the 1918 outbreak known as the "Spanish influenza." Synthesizing the scientific literature in 2012, David Quammen, in his book *Spillover*, forecast the "next human pandemic."

As predicted, challenges have occurred that can be regarded as dress rehearsals urgently demanding our intelligence to organize and fund a coherent response. Between 2003 and 2016 these included outbreaks of avian flu, SARS (severe acute respiratory syndrome), MERS (Middle East respiratory syndrome), Marburg, and Ebola.

Unfortunately, along with outbreaks, a recurring pattern of societal amnesia has prevailed. Each microbial challenge has been followed by a period of frenetic activity at every level, internationally and nationally, but has concluded with a lapse into forgetfulness. The interval between the SARS crisis of 2003 and the Ebola epidemic is illustrative. Immediately after the SARS experience, the World Health Organization (WHO) produced a *Global Influenza Preparedness Plan* (2005) to establish guidelines for country-by-country efforts; revised the International Health Regulations to include emerging disease threats as notifiable events; and devised its own rapid response capabilities. In the same year the US government issued a *National Strategy for Pandemic Influenza* and allocated funding for that purpose. Similar plans were drawn up by the Department of Defense, the Veterans Administration, the fifty states, and a series of major companies in the private sector.

But as the emergency receded and fear subsided, citizens and governments reverted to business as usual. Funds pledged to emergency response through WHO and the Centers for Disease Control (CDC) in the United States and its sister agencies abroad, and through health departments, governments, and private laboratories, were slashed. Agencies charged with coordinating the response at international, federal, and state levels were disbanded and their leaders removed.

Predictably, this pattern reasserted itself after the Ebola emergency in West Africa. In 2018, on the very day that a new Ebola epidemic began in the Democratic Republic of the Congo, President Trump dismissed the chair of the National Health Security Council and dissolved the team. As the director-general of WHO noted, the world has alternated between feast and famine concerning epidemic disease, content to imagine that periodic lurches into sudden improvisation and declarations of good intentions will carry the day. In this instance, WHO is especially important, because it is the agency charged with coordinating the international response to health emergencies. WHO appointed a commission in 2018 to assess global preparedness for the next microbial challenge after the relaxation of efforts since SARS. The report, issued in 2019, considered the world and its individual countries as comprehensively unprepared for the long-anticipated challenge. The poignant title of the report was *A World at Risk*.

When COVID-19 began its global spread, it met with success in part because the sentinels had stood down and the world slept. Here the stance taken by the United States is critical: it is the remaining superpower and economic giant; it provides the crucial funds for the efforts of the WHO; and the CDC is the agency that sets the gold standard for international response. Despite the repeated warnings since 1997, an important driver of present sufferings is the position of the American president, who, as the disease spiraled out of control on three continents, asked, "Who would have thought?" A more fitting question is whether, after COVID-19 abates, the world will return to complacency or decide upon a sustainable, long-term assessment of the likely challenges and organize the means to meet them. Scientific research, enhanced healthcare infrastructures, close international collaboration, health education, protection of biodiversity, and ample funding will all need to be deployed around the globe if we are to secure our civilization.

Preface

This book began as an undergraduate lecture course at Yale University. Its original purpose was to respond to concerns at the time about emerging diseases such as severe acute respiratory syndrome (SARS), avian flu, and Ebola that were not met within the established courses on offer to undergraduates at Yale. Specialist classes for scientists in the graduate school and for medical students at the School of Medicine dealt, of course, with these diseases from scientific and public health perspectives. Their purpose, however, was not to consider epidemics in their societal context and in their relationship to politics, the arts, and historical change. More broadly, it also became apparent that the study of the history and impact of epidemic diseases was an underdeveloped subject in the undergraduate curriculum of US universities in general. The course, therefore, was my attempt to meet what seemed a significant need for the discussion, from an interdisciplinary perspective, of the ways that infectious diseases have played a substantial role in shaping human societies and continue to pose a threat to their survival.

In transforming the course into a book, I have maintained many of the original intentions underlying the class but with the intention of reaching a wider but similar audience. The goal, in other words, is not to reach specialists in the relevant fields, but rather to encourage discussion among general readers and students with an interest in the history of epidemic diseases and a concern about our preparedness as a society to meet new microbial challenges.

That goal shapes the way in which the book is organized and written. As in the original lectures, my aim here is to preserve the accessibility of the material by not assuming prior knowledge of history or epidemiology. I have attempted to provide a self-contained discussion of the subject matter for anyone concerned with the issues the book considers. The book could serve as reading material within the context of a college course for students interested in the intersection of the humanities and the sciences. For that reason I have explained the relevant scientific terminology, provided a bibliography of additional readings for those interested and for those who wish to explore the sources of the opinions expressed, and limited the scholarly apparatus of the notes to indicate the sources only for direct quotations. My primary aim is not to furnish an original contribution to the subject but rather to place existing knowledge in a broad context of interpretation.

On the other hand, this book is not a textbook. I do not attempt here to provide a comprehensive synthesis of the material in the field but rather to focus selectively on major issues and on those epidemics that have had the deepest and most lasting impact on society. Furthermore, unlike a textbook, this book includes chapters that are primarily based on original source materials, particularly where I felt that my views differed from the conventional wisdom or where it seemed useful to fill gaps in the existing literature. The various chapters convey what is, I hope, the informed view of a single scholar who conducts research in the field and has had the good fortune of having the opportunity to learn from the comments and questions of those interested and thoughtful general readers who are Yale's undergraduates.

Acknowledgments

Since this work originated as a lecture course at Yale University, I extend my thanks to all of the students who took the class over a period of seven years and whose questions and suggestions helped me to clarify ideas and explanations. I am also deeply grateful to Professor Ingrid Walsoe-Engel, who co-taught the class with me in its first year and helped me to plan its early content; to Dr. Lee Sateline and Dr. John Boos, who audited the course and made helpful contributions; to my wife, Margaret, who made shrewd editorial comments and suggestions and gave a lecture on AIDS in the United States; to my teaching assistants, who provided helpful ideas; and to the editors at Yale University Press for their patience and wise advice. I am solely responsible for the text that appears here, and for the errors that it might contain.

Abbreviations

AIDS	acquired immune deficiency syndrome
ANC	African National Congress Party
BCG	bacillus Calmette-Guérin
CDC	Centers for Disease Control and Prevention
CFR	case fatality rate
DDT	dichlorodiphenyltrichloroethane
ERLAAS	Ente Regionale per la Lotta Anti-Anofelica in Sardegna (Regional Agency for the Anti-Anopheline Battle in Sardinia)
GPEI	Global Polio Eradication Initiative
HIV	human immunodeficiency virus
IHR	International Health Regulations
IOM	Institute of Medicine
IPV	inactivated polio vaccine
NGO	nongovernmental organization
OPV	oral polio vaccine
SARS	severe acute respiratory syndrome
UNAIDS	Joint United Nations Programme on HIV/AIDS
UNRRA	United Nations Relief and Rehabilitation Administration
WHO	World Health Organization

Epidemics and Society

CHAPTER 1

Introduction

This book began as an undergraduate lecture course at Yale University during the immediate aftermath of a series of public health emergencies—first SARS, followed by avian flu and then Ebola—that erupted at the outset of the twenty-first century and raised troubling issues regarding the unexpected vulnerability of modern society to sudden outbreaks of infectious diseases. I developed the course using my perspective as a historian to address concerns about how vulnerable society is to diseases and pandemics and applying my expertise in the history of medicine and my own research on cholera and malaria. My goal was to think about and explore, together with my students, a subject that was unfamiliar to them and was largely unrepresented in undergraduate syllabi—but that was suddenly claiming their attention.

Epidemics and Society emerged from the final iteration of the course lectures after several annual rounds of revision in the light of reflection and classroom discussions. It is not a monograph intended for scholars of medical history or public health practitioners. Several of the chapters are based principally on my own research with original sources, but that is almost coincidental. The chief goals are not to present new information but to contextualize known material, to draw general conclusions from it, and to make the subject matter accessible to the general reader. Because Yale University also offered the course online, the shape of the book further reflects the feedback of many who viewed the course on the web and shared their observations and suggestions. Although I have never met these people in person, I am grateful for their comments as well as for those of my students in the classroom.

One of the overall themes structuring *Epidemics and Society* is an intellectual hypothesis to be tested through the examination of widely dissimilar diseases in different societies over time. This hypothesis is that epidemics are not an esoteric subfield for the interested specialist but instead are a major part of the "big picture" of historical change and development. Infectious diseases, in other words, are as important to understanding societal development as economic crises, wars, revolutions, and demographic change. To examine this idea, I consider the impact of epidemics not only on the lives of individual men and women, but also on religion, the arts, the rise of modern medicine and public health, and intellectual history.

I include only a subset of high-impact infectious diseases that have affected, or threatened to infect, western Europe and North America. I exclude chronic diseases such as cancer, heart disease, diabetes, asthma, and obesity. I also do not consider occupational diseases such as miners' lung or asbestosis and silicosis, or lead poisoning; or genetic disorders such as hemophilia, sickle cell anemia, and cystic fibrosis. Also lying beyond the reach of this work is a whole range of tropical diseases that have not had a major effect on the industrial West, such as sleeping sickness (trypanosomiasis), Chagas disease, and guinea worm disease. All of these other categories of disease are important, and they merit careful study on their own behalf; but it would be unmanageable to deal with them all, and lumping them together would be to lose all intellectual coherence and rationale. Therefore, my sole focus in this book is on epidemic diseases.

The reason for this is threefold. First, epidemic diseases make sense historically as a distinctive category for analysis. Experienced differently from chronic diseases, they gave rise to characteristic fears and anxieties. To have severe heart disease, for instance, can be a frightening, even lethal, experience; however, it is qualitatively different from being diagnosed with HIV/AIDS or being stricken with smallpox, polio, or Asiatic cholera. Correspondingly, major chronic diseases such as cancer have a devastating effect on health-care systems, on the economy, and on the lives of millions of people. But unlike some epidemic diseases, heart disease and cancer do not give rise to scapegoating, mass hysteria, and outbursts of religiosity, nor are they extensively treated in literature and art. Diseases, in other words, are not simply interchangeable causes of morbidity and mortality. Epidemic diseases have left a particular legacy in their wake. Their singularity merits attention.

A second reason for concentrating on epidemic diseases is historical. Since our interest here is history, it is important to stress that, throughout human history until the twentieth century, infectious diseases have been far more devastating than other categories of illness. Indeed globally they remain leading causes of suffering and death. One of the goals of *Epidemics and Society* is to explain this feature of the history of human disease.

Finally, and perhaps most compellingly, epidemic diseases merit attention because their history is far from over. Emerging diseases such as SARS, Ebola, and Zika have provided a reminder of this ongoing susceptibility; we live with the persisting ravages of HIV/AIDS, and some older diseases once thought to be eradicable—dengue, malaria, and tuberculosis—have instead reemerged as major threats. Even the industrial West is still at risk, and climate change enhances the potential for future disasters. This microbial threat is real. How severe is it? What are our defenses? What factors promote vulnerability? How prepared are we to confront the challenge? How the global community deals with these issues may well be an important factor in determining the survival of our society, and perhaps even of our species.

Geographically, the focus in the book is primarily the industrial world of Europe and North America. A major reason is manageability. A work that was systematically and appropriately global would need to be several times longer and to include an array of additional diseases that are mainly afflictions of the tropical world. On the other hand, there are numerous occasions when dealing with the late twentieth and early twenty-first centuries when it will be important to look farther afield. It would be perverse, for example, to discuss HIV/AIDS, the campaign to eradicate polio, the third pandemic of bubonic plague, modern cholera, or Ebola without considering their places of origin, their epicenters, and the nations where these afflictions still cause an immense burden of suffering and loss. We are inescapably part of a global world in which microbes—and the insects that transmit them—refuse to recognize political borders, and we need to take that carefully into account. For that reason I include chapters that focus on South Africa, West Africa, India, Haiti, and Peru.

Chronologically, I start with everyone's idea of the worst-case scenario in terms of epidemic disease—bubonic plague (which peaked in Europe in the fourteenth century)—and end with the recent threat of Ebola today. By linking history with the events of today's newspapers and considering those events in the light of historical experience, I hope to help readers acquire the tools necessary to confront today's public health events in a more informed and productive manner.

What are my criteria for selecting the diseases to examine? Four are most important. First, I pay careful attention to those epidemics that have had the largest social, scientific, and cultural resonance. For that reason, it is critical, for example, to include tuberculosis but omit typhoid fever, for the sake of brevity.

Second, I consider diseases that encouraged the development of major public health strategies to contain them. A central concern of *Epidemics and Society* is to examine not just epidemics but also the strategies society deployed at various times to combat, prevent, cure, and even eradicate them. I therefore give a privileged place to those diseases that stimulated various styles of organized societal responses. Often the attempts failed, but the concepts guiding them still underlie public health measures to confront microbial assaults.

Third, it is important to stress biological diversity. Some major epidemic diseases are bacterial; others are viral or parasitic. They are various in their mode of transmission—the air, sexual contact, contaminated food and water, excrement, or vectors such as mosquitoes, lice, and fleas. I draw examples from each of these categories.

Finally, although the social resonance of infectious diseases cannot be correlated in any simple manner with the mortality that they cause, it is important to consider the major killers of each of the centuries under consideration. To understand the relationship of early modern societies to mortality, it is clearly essential to discuss bubonic plague, just as no study of disease in the twentieth and twenty-first centuries could fail to give a central place to HIV/AIDS.

Taking these criteria into consideration, I give careful attention to plague, cholera, smallpox, tuberculosis, polio, typhus, dysentery, yellow fever, HIV/AIDS, and Ebola. This list is by no means either canonical or all-inclusive. One could, for instance, make an excellent case for the inclusion of typhoid fever, influenza, and syphilis. The selection is simply representative: it makes no attempt to be comprehensive. The only claim here is that, for the period and places discussed, these diseases are the minimum that the historian needs to consider, and perhaps the most that can be accommodated in a single volume.

Epidemics and Society is a history book, not a biology text. On the other hand, epidemics are inescapably biological events. Therefore, in the analysis of each disease readers will need some understanding of its origins or etiology, its mode of transmission, and its course in the human body. Without some understanding of their medical and biological bases, diseases are incomprehensible. Furthermore, one of the major issues to examine is the manner in which major epidemics often induced important changes in medical philosophy. But the biology will remain in the background: the primary concern is the impact of disease on society, history, and culture.

The purpose here is to do far more than to examine a series of ghastly biological disasters. The book's emphasis is on a series of long-term developments. Of these, the most important are the following:

- *Public health strategies:* These strategies include vaccination, quarantine and sanitary cordons, urban sanitation, sanatoria, and "magic bullets" such as quinine, mercury, penicillin, and streptomycin. There is also the policy of concealment as a means to deny the presence of a disease, as China did at the onset of the SARS outbreak and as other national governments and municipalities have done over a long history.
- *Intellectual history:* Epidemic diseases have played a leading role in the development of the modern biomedical paradigm of disease, the germ theory, and disciplines such as tropical medicine. Furthermore, medical ideas were often held not only for scientific reasons, but also because of the kind of society they promote, or the power they convey to nations and to strategically placed elites within them.
- *Spontaneous public responses:* Under certain circumstances the passage of epidemic disease through communities has triggered large-scale and revealing responses among those at risk. These responses include stigmatization and scapegoating, flight and mass hysteria, riots, and upsurges in religiosity. Such events provide important lenses through which to examine the affected societies and the way in which they are constructed—the relations of human beings to one another, the moral priorities of political and religious leaders, the relationship of human beings to both the natural and the built environments, and the severely compromised standards of living that were ignored in more settled times.
- *War and disease:* Armed conflict in an era of "total war," which began with mass conscription during the French revolutionary and Napoleonic eras, involved the clash of military forces of unprecedented size, even of entire peoples. Warfare on such a scale created fertile conditions for epidemic diseases such as typhus, dysentery, typhoid fever, malaria, and syphilis to flourish. These afflictions often affected both military and

> civilian populations far from the clash of armies. In its turn, disease frequently had a decisive impact on the course of military campaigns and therefore on international politics and the fate of political regimes.
>
> To illustrate the relationship between war and epidemics, I examine two examples of military conflict in two different hemispheres during the Napoleonic period. The first is the great military force that Napoleon Bonaparte sent to the Caribbean colony of Saint-Domingue in 1802–1803 to restore slavery and impose French rule. A virulent epidemic of yellow fever, however, destroyed Napoleon's army and led to a cascade of consequences, including Haitian independence and the Louisiana Purchase.
>
> The second case is the campaign of 1812 during which the French emperor invaded Russia with the largest military force ever assembled to that point. This titanic clash in eastern Europe provides an opportunity to consider the impact of the two classic epidemic diseases of warfare—dysentery and typhus. Together, these diseases annihilated the Grande Armée and played an important part in destroying the emperor himself and transforming the geopolitical balance of power.

Evaluating the interaction of epidemics and society in the past provides background necessary to confront the questions raised by the general public during the recent challenges of SARS, avian flu, and Ebola. What have we learned as a people from the experience of the past four centuries of recurring deadly epidemics? In 1969 the US Surgeon General experienced a premature surge of optimism in the power of science and public health to combat microbes, declaring the end of the era of infectious diseases. During the same era of exuberant hubris, international public health authorities announced that it would be possible by the end of the twentieth century to eradicate one microbial threat after another, beginning with malaria and smallpox. In this triumphant climate, schools of medicine such as Yale and Harvard closed their departments of infectious diseases. Societies, especially in the developed world, were thought to be on the verge of becoming invulnerable to new plagues.

Unfortunately, this expectation has proved to be spectacularly misplaced. Well into the twenty-first century smallpox remains the only disease to have been successfully eradicated. Worldwide, infectious diseases remain leading causes of death and serious impediments to economic growth and political stability. Newly emerging diseases such as Ebola, Lassa fever, West Nile virus, avian flu, Zika, and dengue present new challenges, while familiar afflictions such as tuberculosis and malaria have reemerged, often in menacing drug-resistant forms. Public health authorities have particularly targeted the persisting threat of a devastating new pandemic of influenza such as the "Spanish lady" that swept the world with such ferocity in 1918 and 1919.

Indeed, many of the central features of a global modern society continue to render the world acutely vulnerable to the challenge of pandemic disease. The experiences of SARS and Ebola—the two major "dress rehearsals" of the new century—serve as sobering reminders that our public health and biomedicine defenses are porous. Prominent features of modernity—population growth, climate change, rapid means of transportation, the proliferation of megacities with inadequate urban infrastructures, warfare, persistent poverty, and widening social inequalities—maintain the risk. Unfortunately, not one of these factors seems likely to abate in the near future.

A final important theme of *Epidemics and Society* is that epidemic diseases are not random events that afflict societies capriciously and without warning. On the contrary, every society produces its own specific vulnerabilities. To study them is to understand that society's structure, its standard of living, and its political priorities. Epidemic diseases, in that sense, have always been signifiers, and the challenge of medical history is to decipher the meanings embedded in them.

This book has chapters of two distinct, though overlapping types—those concerned with thematic issues, and those devoted to individual epidemic diseases. Each is intended to be self-contained and can be read on its own, but the thematic chapters provide the context within which individual epidemics were experienced. An example is bubonic plague. In understanding European reactions to plague in the seventeenth century, it is helpful to consider the reigning medical doctrine of the period—humoralism—as it had been inherited from Hippocrates and Galen. Humoral theory, which was the first embodiment of what might be termed "scientific medicine," was the dominant medical paradigm for disease, and it provided the framework within which the visitations of bubonic plague were interpreted and experienced by physicians, politicians, and educated laymen.

For this reason, Chapter 2 analyzes the legacy of two of the most influential figures in the history of medicine, both of whom were Greeks: Hippocrates, of the fifth century BCE, and Galen, of the second century CE. Analyzing their philosophy of medicine illuminates the intellectual shock of those who lived through visitations of pestilence. Plague years were not only times of death and suffering, but also of intellectual disorientation. Bubonic plague undermined the foundations of the contemporary understandings of disease, leaving people bewildered and frightened. Ravages of pestilence, therefore, were biological events that presented intellectual and spiritual challenges.

With the background of humoral theory, bubonic plague is the first case study of a specific epidemic disease (Chapters 3–5). This is because plague is nearly everyone's candidate as the worst-case scenario for epidemic disease. The word "plague" is virtually synonymous with terror. It was an extraordinarily rapid and excruciating killer, whose symptoms were dehumanizing. Furthermore, in the absence of an effective treatment, a substantial majority of its sufferers died, ensuring that contemporaries feared the destruction of entire populations of major cities such as London and Paris. Here was the foundation for the horrifying cliché about plague—that too few people survived to bury the dead.

The discussion of plague, as for the other epidemics covered in this book, begins with an examination of its effects on the individual human body and then turns to its impact on society at large. The clinical manifestations of each disease are essential to the task of decoding the social responses of populations to medical crises—in the case of plague, flight, witch hunts, the cult of saints, and violence.

At the same time, however, plague also led to the first strategies of public health to combat epidemic disaster—measures that were often draconian in direct proportion to the magnitude of the perceived threat. These strategies included the establishment of boards of health with almost unlimited powers during emergencies, quarantine and the forced confinement of sufferers, lines of troops and naval blockades known as sanitary cordons intended to isolate entire cities or even countries, and pesthouses opened to take in the sick and the dying.

The other diseases considered throughout this book will be addressed in the same manner: they will be set in their intellectual contexts, followed by discussions of their etiology and clinical manifestations, of their social and cultural effects, and of the medical and public health measures deployed to contain them. The goal is to help readers understand the great variety of individual and societal responses to epidemics and introduce them to the study of the medical, social, and intellectual history of epidemic disease.

CHAPTER 2

Humoral Medicine

The Legacy of Hippocrates and Galen

An important aim of this book is to examine the meaning of "scientific medicine" in a variety of incarnations. But the place to begin is antiquity, which produced the first embodiment of a rational medicine that lasted from the fifth century BCE until the end of the eighteenth century as the dominant—though not exclusive—medical paradigm. It originated in Greece and was associated with Hippocrates (ca. 460 to ca. 377 BCE), the so-called father of medicine, whose collection of some sixty works, known as the Hippocratic corpus and almost certainly written by multiple hands, announced a radically new idea of medicine.

Of these works, some are especially well-known, such as the "Hippocratic Oath"; *On the Sacred Disease; On Human Nature; Epidemics;* and *Airs, Waters, Places.* A first aspect of the works is their variety. The corpus includes a collection of aphorisms, case histories, lecture notes, memoranda, and writings on every aspect of the medicine that was practiced at the time—such as surgery, obstetrics, diet, the environment, and therapeutics. But all of the Hippocratic writings stress a central claim: that disease is a purely naturalistic event that can be explained only by secular causes and that can be treated only by rational means. Hippocrates espoused a philosophy of medicine that resolutely insisted on viewing both the macrocosm of the universe and the microcosm of the body as governed exclusively by natural law.

Hippocrates rejected an alternative view of disease that preceded him, that continued alongside his practice, and that has persisted to our own day. This is the supernatural interpretation, which takes two dominant forms: the divine and the demonic.

Divine Interpretations of Disease

The divine theory of disease asserts that illness is a punishment sent by an angry god as chastisement for disobedience or sin. Four examples of the divine interpretation from four different eras illustrate its enormous influence on Western culture.

The Bible

The book of Genesis tells the story of the first humans, Adam and Eve, who were immortal beings inhabiting a garden free of disease, suffering, and the need for work. Everything changed when they yielded to the blandishments of the serpent. In disobedience to God's command, they tasted the forbidden apple from the tree of the knowledge of good and evil. This sin marked humanity's fall from grace and innocence. Angered by their disobedience, God expelled Adam and Eve from the garden of Eden forever

Figure 2.1. In the book of Genesis, God expels Adam and Eve from the garden of Eden and decrees they will suffer disease as part of their punishment for eating the forbidden fruit. Michelangelo, *The Fall of Man and Expulsion from the Garden* (1509–1510). Sistine Chapel, Vatican City.

and as punishment ordained that they suffer disease, hard work, pain during childbirth, and finally death. Diseases, in other words, were the "wages of sin" (fig. 2.1).

With specific reference to epidemic disease, the book of Exodus provides an interpretation fully in line with Genesis. Long after the Fall, God's chosen people, the Israelites, lived in bondage in Egypt. Through Moses and Aaron, God told Pharaoh to free his people, but Pharaoh refused. In response, God visited a series of terrible plagues upon the Egyptians. Plagues, in other words, were divine punishment for defying the will of God.

A further important biblical expression of this view of epidemic disease is Psalm 91, which again expresses the idea of pestilence as humanity's chastisement by an irate divinity. This psalm is especially significant historically because it became the great plague text that was read out from Christian pulpits across Europe during epidemic visitations. It simultaneously explained the catastrophe and provided hope:

> Thou shalt not be afraid for the terror by night; nor for the arrow that flieth by day;
> Nor for the pestilence that walketh in darkness; nor for the destruction that wasteth at noonday.
> A thousand shall fall at thy side, and ten thousand at thy right hand; but it shall not come nigh thee.
> Only with thine eyes shalt thou behold and see the reward of the wicked.
> Because thou hast made the Lord, which is my refuge, even the most High, thy habitation;
> There shall no evil befall thee, neither shall any plague come nigh thy dwelling.
> For he shall give his angels charge over thee, to keep thee in all thy ways. (Ps 91:5–11 King James Version)

The message is crystal clear: if you renounce sin and trust in the Lord, you need have no fear of plague, which afflicts only the wicked.

Homer's Iliad

Another expression of the divine interpretation of disease in Western culture is the opening scene of Homer's epic poem *The Iliad,* which narrates the climax of the Trojan War. The poem begins with the anger of Achilles, the greatest of the Greek warriors, whose concubine had been taken by the Greek king Agamemnon. Enraged, Achilles withdrew from combat and sulked in his tent. His friend, a priest of Apollo, tried to intervene by beseeching Agamemnon to right the wrong and return the woman. But Agamemnon rebuffed Apollo's priest and then mocked and threatened him. What followed is a terrifying plague scene. As the early lines of the poem announce, the priest withdrew from the Greek commander and then prayed to Apollo for revenge:

> "Hear me," he cried, "O god of the silver bow. . . . If I have ever decked your temple with garlands, or burned your thigh-bones in fat of bulls or goats, grant my prayer, and let your arrows avenge these my tears upon the Danaans."
>
> Thus did he pray, and Apollo heard his prayer. He came down furious from the summits of Olympus, with his bow and his quiver upon his shoulder, and the arrows rattled on his back with the rage that trembled within him. He sat himself down away from the ships with a face as dark as night, and his silver bow rang death as he shot his arrows in the midst of them. First he smote their mules and their hounds, but presently he aimed his shafts at the people themselves, and all day long the pyres of the dead were burning.[1]

Thus the god Apollo scourged the Greeks with plague for refusing to heed his priest.

The Perfectionists

A third example is more recent—that of the nineteenth-century divinity student John Humphrey Noyes (1811–1886). While attending the Yale Theological Seminary in the 1830s, Noyes reasoned that if diseases are the wages of sin, then logically there is a practical remedy. He and a group of friends agreed that they could reclaim immortality and immunity to disease by renouncing all sin. Appropriately calling themselves the "Perfectionists," they founded a sinless community—first in Putney, Vermont, and then at Oneida in New York State. Their quest for immortality was prominent in the history of American utopian communities, along with their unusual social practices, including what they called "complex marriage" and mutual criticism, in which they vigilantly watched each other and provided earnest rebukes for moral backsliding.

The Oneida Community was founded in 1848 in accordance with socialistic principles. By the 1890s, however, it had declined and reorganized as a joint-stock company, which still produces pottery and silverware but no longer makes special claims of moral rectitude. Despite their hopes, none of the founding members attained immortality, and the last original member died in 1950. Perhaps they lowered their standards, or perhaps the concept was misguided from the outset.

Noyes's Oneida experiment was clearly based on the idea of disease as divine punishment for transgression. His view of pestilence implies a law-governed cosmos. Noyes was logically coherent in arguing that disease exists for an intelligible reason and that there is therefore a corresponding therapeutics, which he believed was repentance and right behavior.

The Jerry Falwell Phenomenon

An even more recent version of the divine interpretation of disease is the example of Jerry Falwell. This Southern Baptist evangelical preacher from Virginia was a pioneer of the megachurch phenomenon and the founder of the Moral Majority movement. Falwell greeted the onset of the HIV/AIDS epidemic with the tirade that the epidemic was God's punishment for the sin of homosexuality. He declared, however, that it was not merely homosexuals who were being punished by an irate God, but rather a whole society that had sinned by tolerating them in its midst. In his most famous hate-filled words, he opined that "AIDS it not just God's punishment of homosexuals, it is God's punishment for the society that tolerates homosexuals."[2]

Demonic Views of Disease

The divine interpretation of disease is magical but obedient to a logic flowing from its supernatural premises. There is, however, a variant of the supernatural view of disease that is more capricious, arbitrary, and unpredictable. This view is what some have called the "demonic theory of disease." This idea postulates that the world is populated by powerful, arbitrary, and evil spirits that cause disease through the exercise of their malign influence. The spirits may be evil people, such as witches or poisoners; the disembodied spirits of the dead when they return to haunt the living; superhuman beings; or even the devil himself. We will see this view—that epidemic diseases are not natural or logical events but rather diabolical plots—recurring throughout the book. In the seventeenth century on both sides of the Atlantic the idea of such occult crimes committed by witches famously gave rise to scapegoating and witch-hunting to track down and punish the guilty. In Massachusetts this view triumphed in the 1690s at Salem, as Arthur Miller vividly recounts in his play *The Crucible.* In Europe, too, the demonic idea was clearly expressed by Martin Luther in the 1500s, who declared, "I should have no compassion on the witches; I would burn all of them."[3]

As a variant, a person could be deemed to be innocent but temporarily "possessed" by an evil spirit. In this case, the therapeutic indication was to cast out the devil through exorcism. Healers pursuing this logic invoked magic and incantations, relying on concoctions, chants, sacred rites, and spells. In European history, an illustration of this idea was the concept of the king's touch to cure disease. King Charles II of England, for example, administered the touch to nearly one hundred thousand people during the mid-1600s. Less illustrious healers chanted, recommended sacrifices, or furnished magic charms to ward off malevolent spirits. Alternatively, they suggested taking flight when disease struck a community or seeking such influential allies as the Virgin Mary or Christian saints.

The Hippocratic Breakthrough

The Hippocratic breakthrough in fifth-century Greece contrasts with the two supernatural interpretations of disease—the divine and the demonic. This naturalistic, secular view flourished in the century of Pericles (ca. 495–429 BCE). One can find it in the famous account of the Peloponnesian War by Thucydides where he recounts the Plague of Athens, which recent DNA research suggests was caused by typhoid fever. Thucydides describes the epidemic as a purely natural event in which the occult, the supernatural, and the divine play no part.

Most dramatic is the discussion by Hippocrates of what he terms the "sacred disease," which is probably what modern physicians diagnose as epilepsy—a condition that, perhaps more than any other, resembles possession by a demon. But Hippocrates stresses that even this so-called sacred disease has purely naturalistic causes:

> [This disease] appears to me to be nowise more divine nor more sacred than other diseases, but has a natural cause . . .

> like other afflictions. Men regard its nature and cause as divine, from ignorance and wonder. And this notion of its divinity is kept up by their inability to comprehend it and the simplicity of the mode by which it is cured. . . . The quotidian, tertian, and quartan fevers, seem to me no less sacred and divine in their origin than this disease, although they are not reckoned so wonderful. . . .
>
> They who first referred this malady to the gods appear to me to have been just such persons as the conjurors, purificators, mountebanks, and charlatans now are, who give themselves out for being excessively religious, and as knowing more than other people. Such persons, then, using the divinity as a pretext and screen of their own inability . . . to afford any assistance, have given out that the disease is sacred. Adding suitable reasons for this opinion they have instituted a mode of treatment which is safe for themselves, namely by applying purifications and incantations, and enforcing abstinence from baths and many articles of food which are unwholesome to men in diseases. . . . And they forbid to have a black robe because black is expressive of death; and to sleep on a goat's skin or to wear it, and to put one foot upon another, or one hand upon another; for all these things are held to be hindrances to the cure. All these they enjoin with reference to its divinity, as if possessed of more knowledge, and announcing beforehand other causes so that if the person should recover theirs would be the honor and credit; and if he should die, they would have a certain defense, as if the gods, and not they, were to blame.[4]

This was a monumentally important conceptual breakthrough—the cognitive foundation on which a scientific medicine was built. Under its naturalistic influence, healers abandoned incantations, spells, and sacrifices; they eliminated exorcism; and they renounced the appeasement of the gods. This momentous step in intellectual history and consciousness was memorably assessed in the 1940s by the epidemiologist Charles-Edward Winslow: "If disease is postulated as caused by gods, or demons, then scientific progress is impossible. If it is attributed to a hypothetical humor, the theory can be tested and improved. The conception of natural causation was the essential first step. It marks incomparably the most epochal advance in the intellectual history of mankind."[5]

Winslow's statement, which clearly expresses a traditional view of Hippocrates, undoubtedly contains a measure of hyperbole. The Hippocratic corpus is far from unified, and there were differences of opinion among the various authors of the sixty surviving writings. Furthermore, scholars have greatly diversified Winslow's picture of a single, triumphant rational medicine. Hippocratic physicians—known as *iatroi*, or doctor-healers—were one among numerous types of practitioners competing in the medical marketplace. A variety of medical sects propounded alternative doctrines, and some healers practiced their craft without any underlying philosophy. As historian Vivian Nutton put it, in ancient Greece you had "the wound-surgeon, the bone-setter, the herbalist, the midwife, the gymnast, the woman-doctor, the exorcist."[6]

Ancient Greece was thus a place of competing claims and approaches to health and disease, and patients had a choice among numerous alternatives. The Hippocratic corpus may be read, therefore, not as the authoritative expression of a medical consensus, but rather as the voice of one style of medicine stating its position and its claim for attention—and customers. It is no accident that it includes advice on the need to win the patient's confidence and to expose charlatans who practice with no doctrine to guide them. Nevertheless, while ignoring a larger context, Winslow—in accord with the traditional view of ancient medicine that he expresses—makes an important and incisive point with regard to Hippocratic medical science. Furthermore, whatever the nature of its many competitors, Hippocratic medicine—thanks in part to the posthumous and enthusiastic endorsement of the Greek physician Galen (see more below)—became a dominant influence on later medical thinking.

Galen's "fatal embrace" of Hippocratic doctrine, as the historian Nutton describes his influence, distorted Hippocratic teachings in a variety of ways. Galen isolated Hippocrates from his context, oversimplified the complexity and contradictions of Greek physicians' practice, and overstated the role of theory in their work. On the other hand, as Nutton demonstrates, Galen's authority played a critical part in establishing his version of Hippocratic thought as preeminent for centuries in Byzantium, across the Islamic world, and later in the Latin West.

A question arises almost inevitably: Why was there a breakthrough to a secular and naturalistic medicine in fifth-century Greece? A major part of the answer rests on imponderables such as the inspiration of both Hippocrates and his associates. Individual and contingent factors are always important in the chain of historical causation. Clearly, however, other factors were also at work. One was the absence of a priestly bureaucracy with the power to define and punish heretics. Others included the decentralization of the Greek city-states, the legacy of Greek natural philosophy and especially the influence of Aristotle, and a dominant culture of individualism.

It is also important to remember the position of Hippocratic physicians in Greek society and their particular niche in the marketplace. Although they were known to treat the poor and slaves, their medical attentions were by and large unavailable to the masses. Their primary clientele consisted of educated and prosperous elites, both in Greece and later in Rome. "Humoralism" was a medical philosophy built on the shared educational background of both patient and doctor. They spoke the same language of natural philosophy, and the remedies that the physician proposed, such as special diets or rest, were ones that the comfortable and well-educated could understand and afford to implement.

Humoral Medical Philosophy

A foundational assumption of the medical philosophy of humoralism is that the macrocosm of the universe and the microcosm of the body correspond to one another. Both are composed of the same elements; both are subject to the same naturalistic laws; and a disorder that occurs in one is followed by disease in the other.

According to Aristotle and Greek natural philosophy, the macrocosm consists of four elements: earth, water, air, and fire. Each element in turn is associated with two of the four "qualities" of dry or wet, hot or cold. Thus enamored of the number four, Aristotelian natural philosophers in later centuries added four seasons, four winds, four cardinal directions, and—for the Christians among them—four evangelists.

In terms of medicine, the microcosm of the body recapitulates the essential features of the macrocosm. It too is composed of liquid equivalents of the four elements, called "humors": black bile (earth), phlegm (water), blood (air), and yellow bile (fire). Like the elements, so too each humor has qualities of hot, cold, wet, and dry. All four humors flow in the body through the veins, and each is associated with its corresponding qualities and functions:

1. *Blood* is wet and hot, and it nourishes the flesh, provides heat, and transports the other humors throughout the body. It is produced by the liver more or less copiously according to age and season.

2. *Phlegm* is cold and wet. It nourishes the brain and tempers the heat of the blood. Its slipperiness also lubricates the joints and thus enables the body to move.
3. *Yellow bile,* or choler, is the fiery, dry humor. It accumulates in the gallbladder and facilitates the expulsive movements of the gut.
4. *Black bile* is dry and cold. It promotes appetite and nourishes the bones and the spleen.

The varying proportions of these humors in the body also define the four temperaments and determine whether a person is melancholic, phlegmatic, choleric, or sanguine. They also correspond to the "four ages of man" (childhood, youth, maturity, and old age) and four principal organs (the spleen, the brain, the gallbladder, and the heart) (fig. 2.2).

This paradigm of humoral medicine can be helpful in our understanding of art and literature as well as of medicine. A major example is the plays of William Shakespeare, whose central characters were sometimes based explicitly upon humoral concepts of temperament. Ophelia in *Hamlet,* for instance, is a textbook description of a melancholic personality suffering from an excess of cold and dry black bile. Similarly, phlegmatic Shylock in *The Merchant of Venice* is aging, cold, retentive, and thus unforgiving. By contrast, Kate in *The Taming of the Shrew* is a choleric personality. Her "taming" includes the Hippocratic principle of placing her on a diet that excludes meat, which is a hot and dry substance that "engenders choler [and] planteth anger," enflaming her fiery temper further.

Humoral theory is axiomatic and based on deductive reasoning from first principles. Central to all of its doctrines were the Mediterranean climate, which encouraged thinking in terms of four seasons, and its particular patient population, among whom afflictions that would today be labeled malaria and pneumonia were salient. This system of four humors and their qualities dominates the Hippocratic texts, and it was the view that was absorbed into the teaching of Galen and his followers. But there was some variety in the understanding of the humors within the Hippocratic corpus, and there were practitioners in ancient Greece and Rome who, for example, espoused numbers other than four or viewed humors as gases rather than liquids.

For all Hippocratic doctors, however, the foundation of health was "eucrasia"—a stable equilibrium of the humors and their corresponding

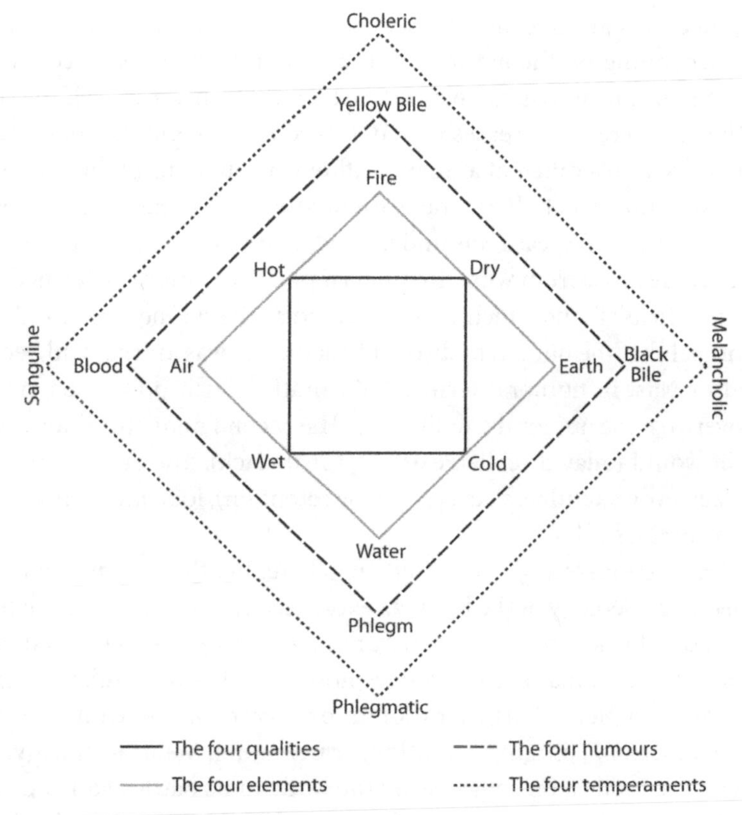

Figure 2.2. The four humors, elements, qualities, and temperaments according to the paradigm of humoral medicine. (Adapted by Bill Nelson.)

qualities. Up to a certain threshold, variations in the balance occurred both from person to person and within a single individual according to season, age, lifestyle, and sex. But beyond a certain threshold, a single humor could become either excessive or deficient, thereby unsettling the overall balance. This humoral derangement was termed "dyscrasia," and it constituted disease. For Hippocrates, therefore, disease was an imbalance of the humors through an excess or deficiency of one or more of them. In addition, a humor could become corrupted and then putrefy and poison the body.

In this humoral framework, there is no concept of individual diseases as discrete entities that modern medicine would classify as, for instance, typhoid, cancer, pneumonia, and influenza. The ancient view regarded disease as a holistic phenomenon of bodily equilibrium. For Hippocrates, in a sense, there was just one disease that could appear in various guises and degrees, depending on the nature and the extent of the imbalance. Furthermore, the humors were unstable and each could transform itself into one of the others. As a result, diseases were not fixed entities. On the contrary, each could evolve into a different ailment. Influenza could morph into dysentery.

If we know what Hippocrates understood by disease, what about the cause of disease? If disease was understood as singular, the causes were multiple. Disease arose from what, in modern parlance, might be termed "environmental insults" and which Galen codified as the six "nonnaturals." A first nonnatural that the human body could encounter was air that had been corrupted or was, in humoral terms, "miasmatic." Such poisonous air could then overturn the humoral equilibrium. The second nonnatural was motion, or what would today be called exercise (or its lack). The other nonnaturals were sleep (or wakefulness), excretion (or retention), food and drink, and the passions of the soul.

Therapeutic strategy was based on nature. In other words, Hippocrates believed in a teleology of the body, as expressed in the phrase *vis medicatrix naturae*, or "the healing power of nature." According to this understanding, the body strives to maintain or restore homeostasis—by regulating temperature, for example, or getting rid of an excessive, or "peccant," humor by sweating, sneezing, purging, vomiting, or urinating. Medical strategy, therefore, was the modest one of assisting the body in its battle against disease.

To that end, the physician's first tasks were to decipher the signs or symptoms that indicated the underlying disease process and to take a case history. In order to specify the nature of the imbalance, the Hippocratic healer closely observed the patient's body by taking the pulse, palpating the chest and listening to its sounds, examining the tongue, assessing the temperature of the skin, and testing the urine. The urine merited special attention, and Hippocrates carefully monitored its color and density, its smell and taste, and the presence of blood or froth. All of these clues could indicate the patient's underlying humoral condition. The goal was never to treat a single symptom, but rather holistically to assess the overall state of the sufferer. Furthermore, the particularity of the patient determined treatment. Therapy respected the unique condition of each person. Disease was not an entity in itself but a process that unfolded in a variable and individual manner.

As a result, humoral medicine was relatively indifferent to diagnosis. Instead, it stressed providing an answer to the patient's eternal question, "Am

I going to be all right, doctor?" Prognosis was uppermost in the mind of Hippocratic physicians.

Having determined the nature and severity of the sufferer's condition, the physician then followed the therapeutic principle that opposites best treat opposites. Thus, if the patient had an excess of black bile, a cold and dry humor, then the treatment was to ingest food or herbs that were hot and wet. In this case the quality "hot" did not necessarily mean warm to the touch, just as one speaks of spicy food as being "hot" regardless of its temperature.

Diet, then, was one of the principal tools available to the physician because all foodstuffs were deemed to possess qualities—hot, cold, wet, or dry—that could balance the opposite defect or excess in the human body. But other therapeutic interventions were also available. One was the prescription of exercise or rest, just as a modern doctor might recommend a health spa or a sanatorium. Other remedies could include a change of environment, moderation in sexual activity, and calm for the emotions. A doctor could also turn to herbal medications because Hippocratic physicians were avid practitioners of internal medicine. It was standard practice, for example, to promote the evacuation of an excessive humor by the administration of emetics, purgatives, sudorifics, or diuretics. Alternatively, since the blood was a humor and one that bore other humors in its flow, a powerful strategy was venesection, or bloodletting, which became a hallmark of the Hippocratic physician and a mainstay of medical practice for more than two millennia—at least through the end of the nineteenth century. Humoral therapeutics were above all a process of addition and subtraction. Physicians sought to add what was deficient and to subtract what was in excess.

A modern reader is likely to be skeptical of venesection, but it is important to remember that the procedure possessed significant advantages in the view of Hippocratic physicians. It was systemic, like disease itself; it was immediate in its effects; and the practitioner could control those effects. Furthermore, its limits were readily apparent to experienced doctors. If the pulse weakened, the color of the blood suddenly altered, or the patient fainted, the doctor could immediately stop drawing the blood. Finally, there were standard contraindications that led physicians to avoid the procedure entirely, such as old age, a deficiency of blood, summer heat, or the preexistence of abundant evacuation by other routes. Thus carefully surrounded by advantages and well-understood limits, venesection and its instrument—the lancet—were central to the orthodox medical profession (fig. 2.3).

Figure 2.3. The lancet owned by Edward Jenner (1720–1800). Such a blade was used for venesection and early vaccinations. (Science Museum, London. CC BY 4.0.)

Galen and the Primacy of the Text

If the Hippocratic corpus provided the first embodiment of humoralism as "scientific medicine," a profound alteration occurred at the hands of the "second father of medicine." This was Galen of Pergamum (129 to ca. 210 CE), a Greek physician who practiced primarily in Rome.

To understand the emergence and persistence of Galenism as a doctrine, one needs to consider Galen's personal qualities, which were very different from those of Hippocrates. Since so little is known about the historical Hippocrates, it is important to remember that Galen's Hippocrates is a stylized synthesis of the multiply authored Hippocratic corpus. The composite peripatetic Hippocrates was a great observer and empiricist. Galen, by contrast, based his authority on his mastery of Hippocratic texts and on deductions from philosophical principles. He virtually deified his master and reified his every word. In doing so, he appointed himself the high priest or official interpreter of the Hippocratic teachings, which he transformed into dogma. A Renaissance poem satirically attributed this hubristic outburst to Galen:

> Remove Hippocrates, first then shall I be.
> My debts to him are many, but so are his to me.
> Things left undone, obscure, by him, I leave complete
> A thousand volumes, crystal clear and neat.
>
> A tiny Island bore him, me the mighty land
> Of Asia; he a few things, I a myriad penned.
> He gave us building blocks, from which a citadel
> I built for medicine; Apollo keeps it well.[7]

At the same time, however, Galen performed the important function of Romanizing Hippocrates in the sense of making his work known to a Latin-speaking audience throughout the Roman Empire.

Galen came from a prominent family and enjoyed the benefits of an excellent education and a private fortune. He began his ascent to prominence when he worked as a physician to gladiators in Pergamum, and he continued his meteoric rise when he moved to Rome in 162 CE. He gained appointment as personal physician to the emperor—a position that is important in explaining the extent of his influence. A man of overweening confidence, Galen proclaimed himself a polymath—an ideal physician, philosopher, linguist, and scientist—and the sole worthy successor to Hippocrates. He had nothing but withering scorn for critics and colleagues, whom he regarded as amateurs unversed in Hippocratic wisdom. He turned his animosity particularly on two rival sects he called the Empiricists and Methodists.

In addition, Galen possessed immense knowledge. His influence is intelligible only if we remember that he gained an encyclopedic mastery of all the branches of science that existed at his time. Furthermore, he sustained his productivity throughout a long life, dictating volumes to his scribes well into his eighties. Only half of his written works survive, as the loss of his texts began even during his lifetime when a fire in 192 destroyed his personal library. The extant works alone, however, fill twelve hefty tomes of a thousand pages each. These two factors—his extraordinary productivity and his longevity—also are important to his intellectual stature. Vivian Nutton describes Galen's place in history well:

> To describe the fortunes of Galen over the centuries is almost to write the history of medicine since his death. Not only did his ideas constitute the basis of formal medicine in Europe at least until the seventeenth century, and arguably until the nineteenth, but . . . they constitute a major medical tradition in the modern Muslim world. . . . Galen's conception of Hippocrates and Hippocratic medicine not only dominated until recently historians' approaches to their medical past but, more subtly, continues to influence modern perceptions of what medicine is and how it should be practised.[8]

Nonetheless, Galen harbored a view of scientific knowledge and advancement that is foreign to current understandings. To Galen, Hippocrates was the permanent fount of medical science, and the main tenets of his doctrine were unalterable. There was no room in Galen's system for the possibility of the adoption of new paradigms. He considered the writings of Hippocrates, and his own, to be valid forever. They could only be refined and perfected, which was Galen's self-appointed life's work.

It is this frozen doctrine that constitutes the essence of Galenism. In Galen's hands, Hippocrates became a cult figure, an object of veneration and, eventually, of worship. Ironically, a man about whom so little was known apart from his writings was retrospectively endowed with all manner of apocryphal virtues. Galen idealized Hippocrates as a model of wisdom, courage, temperance, compassion, and honesty. Legends also circulated about his religious piety, heroism, and hard work. A myth even developed that he was descended on his father's side from the Greek god of medicine, Asclepius, and on his mother's side from Hercules. Hippocrates also became a great patriot who saved Athens from pestilence, a hero of such rectitude and wisdom that he scorned money and fame. Thus elevated, Hippocrates took his place beside those other intellectual icons of antiquity—Socrates, Plato, and Aristotle. In the process, however, the original foundation of the Hippocratic corpus—direct observation at the bedside—was replaced by mastery of the texts of both Hippocrates himself and of his authoritative interpreter, Galen. In this manner "beside medicine" was transformed into "library medicine." The source of medical knowledge was no longer the body of the patient, but the authority of the text.

Humoral Legacies

The question of the efficacy of humoral medicine inevitably arises. However intelligible the system, one wants to know, "Did it work?" How could a medical philosophy survive for millennia unless it was therapeutically successful?

A first point is that humoral physicians were not consulted only about illness. "Therapy" as Galen understood it also encompassed maintaining health. Indeed, one of his most important works is titled *On the Preservation of Health* and concerns hygiene. A humoral doctor following Galenic principles, therefore, spent a good part of his time advising patients on issues that one would today call matters of lifestyle. In ancient medicine, they were thought to have a major impact on preventing disease. These matters, which are discussed at length in the texts, include sleep, exercise, diet, sexual activity, bathing, voice exercises, and moral or psychological advice. All of these issues were thought to affect the emotions and therefore a person's overall balance or health. Greek or Roman physicians were not judged purely or even primarily on the success of their cures; in many of their activities they assumed roles that today would be regarded as those of a trainer, psychological counselor, or dietician.

A second point is that, when it formed the basis for a curative strategy, humoralism had a number of impressive strengths. A quantum leap separated it from magical thinking. It also appealed to contemporaries because its teachings were consonant with natural philosophy. Furthermore, Hippocratic and Galenic doctors exercised therapeutic modesty. They did not, for example, undertake surgery except for setting bones, lancing abscesses, and opening veins for phlebotomy. The internal cavities of the body, they understood, were off limits.

It is also important to remember that, even today in the clinic of a primary care physician, a large majority of patients have self-limiting ailments, frequently even psychosomatic ones. What they need above all is reassurance that all will be well. A major point about humoral medicine is that an experienced physician accustomed to seeing and treating the ill would become proficient at prognosis and skilled at reassurance. Hippocratic doctors also refused to treat cases they regarded as hopeless and had a referral system for the most serious cases beyond their skill (see below).

But humoralism, especially in its Galenic version, also had a number of major weaknesses. The first of these was that it constituted a closed system. Humoralism was based on deductive reasoning, and in the end, with the fading away of Hippocratic empiricism, it adopted a cult of personality—the cult of Galen and, through Galen, of Hippocrates. In this manner humoralism evolved into a cult of antiquity while knowledge ossified into a form of revealed truth. Galenism stressed authority and tradition, and in time it produced an elite medicine of university-trained physicians educated, above all, in the study of the classics. Its answer to the question of how to train physicians was that they should read Hippocrates and Galen in the original languages.

Temple Medicine

Despite stressing the importance of the breakthrough in ancient Greece to a secular and naturalistic medical philosophy, we can also see a possible tension between naturalism and religiosity. Hippocrates and Galen lived in a world of gods and their temples, and both men were devout believers. Gods had substantial importance in Greco-Roman society, and particularly for ancient healers, the Greek god Asclepius was significant.

Asclepius, the son of the god Apollo and a mortal woman, was a great healer, and a popular cult grew around his name and temples. In the ancient world physicians called themselves "Asclepiads," meaning the sons of Asclepius, and they regarded him as a "patron saint" of the profession. By the time of Alexander the Great, three hundred to four hundred healing temples (called *asclepieia*) were dedicated to the god throughout Greece. The largest and most famous of these were located at Athens, Cos, Epidaurus, Tricca (now Trikala), and Pergamum.

The key to understanding the relationship of Asclepius to ancient medicine is that he never practiced magic. He was merely a skilled physician, one who made use of the same principles that his followers, the Asclepiads, relied on as well. Some people note many similarities to the story of Christ, for the cult of Asclepius was in fact a major competitor with Christianity for several centuries.

For peripatetic healers like many Greek and Roman physicians, Asclepius was highly useful because he provided doctors with a badge of identity, a source of authority, and a collective presence. Hippocratic doctors were recognizable as members of the same "guild." Asclepius vouched for doctors' ethical conduct in a context in which physicians were often wandering practitioners whom patients invited into their homes. Such practitioners needed an authority to vouch for them. Asclepius could guarantee that they were qualified, that they were honest, and that they had special care for the poor who lacked means of payment.

In some respects the temples of Asclepius were precursors to health spas, sanatoria, and hospitals. They provided care for the poor and the seriously ill. Patients entered the temple precincts after a period of preparation in which they bathed, fasted, prayed, and offered sacrifice. They would then sleep in the temple, and Asclepius would appear to them in a dream and disclose the therapeutic regimen that they should pursue (this practice was called "incubation"). But this therapeutic strategy contained only what a skilled humoral physician would have prescribed; the incubation did not interfere with the naturalistic therapies that followed. Asclepius never prescribed treatment by magic, worked miracles, or engaged in practices that were not accessible to the ordinary doctor.

Conclusion

The various interpretations of disease that we have considered in this chapter—the divine, the demonic, and the humoral—were not temporally sequential. On the contrary, the establishment of a "scientific" philosophy of medicine did not replace the demonic and divine interpretations. All three persisted side by side for millennia, sometimes coexisting in the minds of individual thinkers. All three views survive today as ongoing parts of our cultural inheritance. Indeed, it is still possible in the Indian subcontinent to be treated according to humoral principles by the practitioners of Unani medicine.

This survey of the philosophies of medicine that prevailed in Europe to the nineteenth century prepares us in particular for exploring bubonic plague. The societies afflicted by plague understood the scourge in terms of the ideas of disease that they had inherited. It is impossible to comprehend the history of the plague without taking into account the intellectual framework within which it was experienced and that conveyed its meaning.

CHAPTER 3

Overview of the Three Plague Pandemics

541 to ca. 1950

Bubonic plague is the inescapable reference point in any discussion of infectious diseases and their impact on society. In many respects plague represented the worst imaginable catastrophe, thereby setting the standard by which other epidemics would be judged. In later centuries, when societies experienced new and unfamiliar outbreaks of disease, they waited anxiously to see whether they would equal plague in their devastation. Particularly feared diseases, such as cholera in the nineteenth century and both the Spanish influenza and AIDS in the twentieth, were said by some to be "the return of the plague." Similarly, tuberculosis, the leading killer of the nineteenth century, was widely referred to as "the white plague." Indeed, the term "plague" has become a metonym for societal calamities in general, even when the crisis is not infectious, as in such phrases as "a plague of accidents" or a "plague of bank robberies."

What features of bubonic plague, and of the responses of communities to it, marked it as so distinctive and so fearful? The most striking aspect of plague is its extraordinary virulence. "Virulence" is the capacity of a disease to cause harm and pathological symptoms. It is the measure of the ability of a pathogen to overcome the defenses of the human body to induce sickness, suffering, and death. Plague in that sense is extremely virulent. It strikes rapidly, causes excruciating and degrading symptoms, and, if untreated, invariably achieves a high case fatality rate (CFR), which simply means the kill rate of a pathogen—the ratio of deaths to illnesses. In the era before the discovery of antibiotics, plague normally killed more than half of the people it infected, for a CFR of at least 50 percent—a rate attained by few other diseases. Furthermore, its progress through the body was terrifyingly swift. As a rule, plague killed within days of the onset of symptoms, and sometimes more quickly.

Other fearful characteristics of this disease were the age and class profiles of its victims. Familiar endemic diseases primarily strike children and the elderly. This is the normal experience of a community with infectious diseases such as mumps, measles, smallpox, and polio. But the plague was different: it preferentially targeted men and women in the prime of life. This aspect made plague seem like an unnatural or supernatural event. It also magnified the economic, demographic, and social dislocations that it unleashed. In other words, plague left in its wake vast numbers of orphans, widows, and destitute families. Furthermore, unlike most epidemic diseases, the plague did not show a predilection for the poor. It attacked universally, again conveying a sense that its arrival marked the final day of reckoning—the day of divine wrath and judgment.

Another distinguishing feature of plague is the terror it generated. Communities afflicted with plague responded with mass hysteria, violence, and religious revivals as people sought to assuage an angry god. They also looked anxiously within their midst to find the guilty parties responsible for so terrible a disaster. For people who regarded the disease as divine retribution, those responsible were sinners. Plague thus repeatedly gave rise to scapegoating and witch-hunting. Alternatively, for those inclined to the demonic interpretation of disease, those responsible were the agents of a homicidal human conspiracy. Frequently, vigilantes hunted down foreigners and Jews and sought out witches and poisoners.

This chapter presents an introductory overview of plague as a disease, including the public health policies deployed against it, the general impact of plague, and the three major plague pandemics that span some fifteen hundred years of history (fig. 3.1).

Plague and Public Health

Plague was also significant because it gave rise to a critically important societal response: the development of public health. Bubonic plague inspired the first and most draconian form of public health policy designed to protect populations and contain the spread of a terrible disease, that is, various

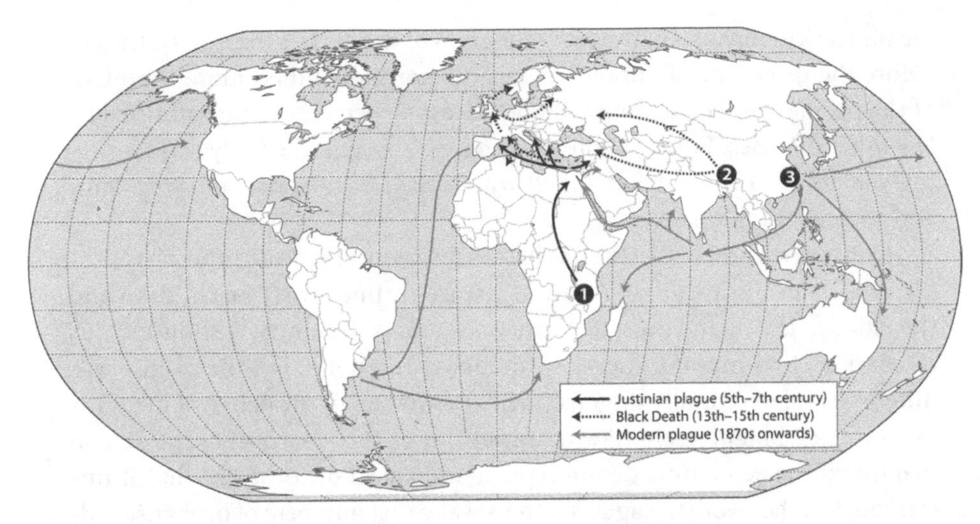

Figure 3.1. The three pandemics of bubonic plague. (Drawing by Bill Nelson.)

forms of enforced isolation of the sick (we omit Hansen's disease and the relegation of its victims to leper colonies on the grounds that leprosaria were noncurative institutions and did not contribute to the development of public health strategies). Antiplague measures of public health involved reliance on the military for their implementation. They mandated, first of all, sanitary cordons—military lines intended to isolate a population by preventing all movements of people and goods. In addition, defenses against the plague included pesthouses, known as lazarettos, and quarantine; health authorities variously known as health magistrates or boards of health were equipped with emergency powers to enforce the regulations. In some places stocks and scaffolds were erected to remind the population of the powers of these agencies.

Renaissance Italian city-states played a special role as the pioneers of these antiplague responses. This responsibility was thrust upon them by their vulnerable positions at the center of trade routes in the Mediterranean, where they received passengers and goods—and stowaway rats—from the Middle East and North Africa. Florence and the port cities of Venice, Genoa, and Naples were pioneers in the development of these policies, which were widely imitated.

In later centuries, another pattern emerged. When new and deadly diseases struck, such as cholera, yellow fever, and AIDS, one of the first responses of health authorities was to attempt to reactivate antiplague measures. It is said of military generals that they tend to fight the last war over again, thereby confronting new enemies with inappropriate strategies from the past. Much the same can be said of public health authorities over the centuries. This temptation is all the greater because the instruments of antiplague defense give the impression of a forceful and energetic response, thereby providing the population with some sense of protection.

The Impact of Plague

A leading feature of the plague as a disease is its enormous social impact. Certainly bubonic plague makes a strong case for not regarding infectious diseases as a narrow and specialized interest. Plague was part of the "big picture," as essential for understanding history during the periods in which it afflicted society as the study of war, religion, economics, and high culture. The point is not, of course, to make a case for disease determinism, or what one might call "microbial Marxism." The argument is more simply that certain diseases do have a transformative effect on society, and plague is one of them. Most other diseases, even massive killers such as influenza or polio, have not had a similar influence. A major task of our exploration of plague is to examine why such major differences exist in the realm of infectious diseases—why some leave a major cultural, political, and social imprint, and others do not.

Bubonic plague is one of the best examples of a disease that affects every aspect of society. It transformed the demography of early modern Europe. Its recurring cycles, with an epidemic every generation, constituted a major brake on population growth between the fourteenth and eighteenth centuries. It had devastating effects as well on economic life and development. And it substantially influenced religion and popular culture, giving rise to a new piety, to cults of plague saints, and to passion plays. Bubonic plague also deeply affected the relationship of people to their mortality, and indeed to God.

The plague in Europe led to an outpouring of sermons and religious pamphlets in which a central theme was theodicy—that is, the vindication of an omnipotent God's goodness in the face of evil and suffering. It was relatively easy to accept that God could be angry and would punish those who turned away from him and disobeyed his commandments. But how could one explain the gruesome suffering and mass death of innocents, especially children? It is true that plague led to an upsurge of piety, but it also generated a powerful undertow pulling in the opposite direction. For some, the experience of bubonic plague led to the terrifying conclusion that there might be no God. A loving and all-powerful being would not take the lives of half the population of a great city, indiscriminately slaying men, women, and children. The result was not so much atheism as a mute despair that was most often barely articulated—a psychological shock that, with historical hindsight and anachronism, one might call posttraumatic stress.

Plague also had major effects on the arts and culture. In literature, an entire genre of plague literature arose, including works by Giovanni Boccaccio, Daniel Defoe, Alessandro Manzoni, and Albert Camus. It also transformed the iconography of European painting and sculpture, and it deeply affected architecture, with the construction of major cathedrals and churches dedicated to the redeemer, the Virgin Mary, and the plague saints Sebastian and Roch. Plague columns, often built to celebrate the ending of plague in a city, appeared in Vienna and throughout central Europe to remind the population of God's mercy.

As late as the mid-twentieth century, the disease inspired Ingmar Bergman's 1957 film *The Seventh Seal*. At the height of the Cold War, Bergman was deeply concerned about the possibility of nuclear war. As a way of imagining the apocalypse, bubonic plague presented itself naturally as the ultimate experience of human calamity and therefore as a metaphor for atomic catastrophe.

Similarly inspired by the experience of plague in the seventeenth century, the city of Oberammergau in Bavaria initiated a German tradition of passion plays. The survivors of an epidemic there in 1630 took a vow: if the people were spared, the city council promised to perform a passion play involving the whole population of the town and to continue to do so at regular intervals in perpetuity. This vow gave rise to an ongoing, and controversial, series of plays that enacted the passion of Christ and sometimes incited viewers to anti-Semitic violence.

Plague also had a major intellectual impact on the medical paradigm of disease by profoundly testing the humoral explanatory framework. The doctrines of Hippocrates and Galen had difficulty in satisfactorily explaining the passage of bubonic plague. How was it possible that vast numbers of people experienced the same humoral imbalance at almost precisely the same time? Hippocrates and his followers invoked the possibility of what might be called, in present-day parlance, environmental insults. The orthodox explanation was that the atmosphere in a given locality had been "corrupted," creating an "epidemic constitution." Its cause was a deadly fermentation arising from decaying organic matter either in the soil or in nearby marshes and swamps. This poisonous effusion contaminated the air and sickened large numbers of susceptible people when they inhaled the poison or absorbed it through their pores.

A medieval variant of this view was propounded by astrologers, who suggested that the trigger to plague and other epidemic diseases was a dangerous alignment of the stars and planets. Disorder in the cosmos was then reflected in disorder in the microcosm of the body. Even those who did not regard the appearance of a comet or the conjunction of planets as the direct causes of an epidemic often believed that such celestial events could serve as portents. Similarly, unusual climatic events—earthquakes, floods, and fires—could presage a crisis in public health.

Girolamo Fracastoro, a sixteenth-century Italian physician, confronted the problem of explaining epidemics in a fundamentally different manner. He eliminated the mediation of the humors altogether, suggesting instead that epidemic disease was caused by a poisonous chemical transmitted—in ways that he did not understand—from one person to another. In the seventeenth century the German Jesuit Athanasius Kircher developed the idea further, suggesting that the plague was spread by what he called "animalcules" that somehow passed from an infected person to a healthy one. Fracastoro and Kircher were thus pioneers in developing the concept of contagion.

The idea of contagion appealed at first far more to the popular imagination than it did to the mind of elite, university-trained physicians who could find the suggestion nowhere in the classic texts. Only in the late nineteenth century did microbiology validate the heretical etiology of Fracastoro and Kircher through the work of Louis Pasteur (1822–1895) and Robert Koch (1843–1910), as discussed in Chapter 12.

A History of Three Plague Pandemics

It is important to distinguish among three closely related terms. Infectious diseases are normally placed along a continuum according to their severity in terms of the numbers of sufferers and the extent of their geographical reach. An "outbreak" is a local spike in infection, but with a limited number of sufferers. An "epidemic," by contrast, normally describes a contagious disease that affects a substantial area and a large number of victims. Finally, a "pandemic" is a transnational epidemic that affects entire continents and kills massive numbers of people. All three terms, however, are loose approximations, and the boundaries separating one from another are imprecise and sometimes subjective. Indeed, a contagious disease confined to a single locality is occasionally termed a pandemic if it is sufficiently virulent to afflict nearly everyone in the area.

In accordance with this terminology, humanity has experienced three pandemics of bubonic plague. Each consisted of a cycle of recurring epidemic waves or visitations, and the cycle lasted for generations or even centuries. So recurrent were these outbreaks of plague that they provided writers with a logical and convincing device to drive the plot of a story forward. A notable example is Shakespeare's tragedy *Romeo and Juliet,* which unfolds against the background of an outbreak of plague in the Italian city of Verona. The tale reaches its tragic denouement as a result of the disease, which interrupts communication between Verona and Mantua. When Friar John attempts to deliver Juliet's crucial letter to Romeo, who is exiled in Mantua, he is forcibly detained: "The searchers of the town, suspecting that we . . . were in a house where the infectious pestilence did reign, sealed up the doors and would not let us forth. So that my speed to Mantua there was stayed. . . . So fearful were they of infection" (Act 5, scene 2, lines 8–12, 17). Bubonic plague thus furnished an entirely plausible artifice for structuring the tale of the star-crossed lovers and their double suicide. As Shakespeare's audience well knew, plague in early modern Europe was an ever-present danger that could strike at any time and without warning.

The cyclical pattern of plague was marked also by a pronounced seasonality. Plague epidemics usually began in the spring or summer months and faded away with the coming of colder weather. Especially favorable were unusually warm springs followed by wet, hot summers. The modern explanations for these propensities are the need of fleas, which carried the disease, for warmth and humidity to enable their eggs to mature, and the inactivity of fleas in cold and dry conditions. Although this pattern predominated, the disease has also been known to erupt mysteriously in Moscow, Iceland, and Scandinavia in the depth of winter. These atypical eruptions of the disease posed serious epidemiological puzzles.

The First Pandemic, or the Plague of Justinian

The first appearance of bubonic plague in world history was the so-called Plague of Justinian, or Justinianic Plague, named after the Byzantine emperor Justinian I, under whose reign it first appeared. Some people held his alleged misdeeds, according to the historian Procopius, to be responsible for incurring divine wrath. The pandemic is thought by present-day geneticists, however, to have originated as a zoonosis, or epidemic disease that can be transmitted from animals to humans, in an endemic African "focus" (or local area of infection). In 541 CE it first erupted as a human affliction at Pelusium in the Nile Delta. Thereafter it lasted through eighteen successive waves over a period of two hundred years until 755, when it disappeared as suddenly and mysteriously as it had arrived.

This round of pestilence afflicted Asia, Africa, and Europe, and it left a dreadful but unquantifiable mortality in its wake. Few direct accounts of this disaster have survived, but the extant reports of such eyewitnesses as Gregory of Tours, John of Ephesus, Bede, and Procopius agree on the magnitude of the calamity. In Procopius's words, this was "a pestilence by which the whole human race was near to being annihilated."[1] Recent impressionistic assessments suggest a total fatality of 20–50 million.

This extensive mortality and the descriptions of the classic symptoms of bubonic plague—the hard bubo in the armpit, groin, or neck—are clear diagnostic indicators of the identity of the disease. In recent years, moreover, paleopathologists have been at work exhuming bodies from the cemeteries of late antiquity, extracting DNA from their dental pulp, and confirming the presence of the bacterium *Yersinia pestis,* which causes the disease. Scientists working in Bavaria in 2005, for example, identified the plague bacillus in skeletal remains from a sixth-century cemetery at Aschheim, strongly suggesting that the traditional diagnosis of bubonic plague is accurate (fig. 3.2)

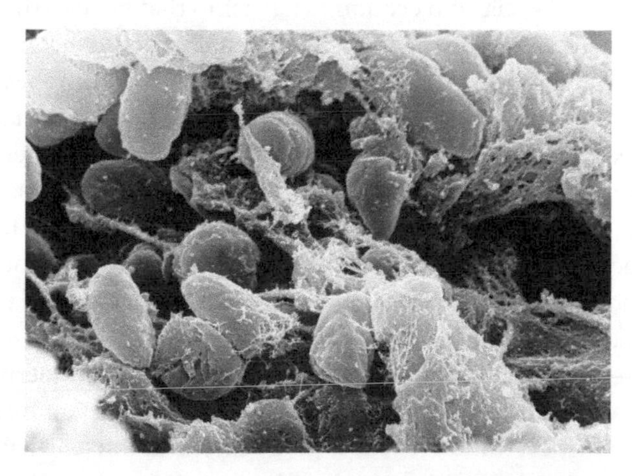

Figure 3.2. Scanning electron micrograph of a mass of *Yersinia pestis* bacteria, which cause bubonic plague, in the foregut of the flea vector. (Rocky Mountain Laboratories, NIAID, NIH.)

The Second Pandemic, or the Black Death

The second plague pandemic began in Central Asia in the 1330s, reached the West in 1347, and persisted for five hundred years until it disappeared in the 1830s. Its first wave in Europe, from 1347 until 1353, is often called the Black Death today, although this restricted terminology began only in the eighteenth century. Various fourteenth-century accounts instead refer to the disaster as the "great pestilence," the "plague of Florence," "the mortality," and the "plague." Partly for this reason, and partly because of the dark buboes and gangrene that are plague symptoms, many scholars still employ the original and more extensive meaning of Black Death—as a synonym for the whole of the second pandemic.

This pandemic is conventionally thought to have arrived aboard Genoese galleys that sailed from the Black Sea in the summer of 1347 and docked at Messina in Sicily. It then spread rapidly to the rest of the island, onward to Sardinia and Corsica, and then in a more leisurely manner to mainland Italy. There its work was assisted by ships bearing plague that docked at Genoa. Mainland Italy and the whole of mainland Europe were engulfed in the general epidemic carnage. That Italian cities were the first in Europe to be ravaged by plague was no coincidence: their early devastation reflected the geographical vulnerability of Italy's position at the center of Mediterranean trade.

At the time the Black Death arrived, Europe was in the throes of a lengthy period of social and economic hardship that favored the advance of the disease. The thirteenth century had been a period of economic expansion, demographic growth that doubled the numbers of Europeans between 1100 and 1300, and urbanization. Substantial towns with populations in excess of fifteen thousand inhabitants multiplied, and within them crowding and unsanitary housing became significant urban problems. Then, after about 1270, an economic recession set in as output stagnated, producing lower wages and contributing to poverty. Agricultural production fell sharply, leading to a classically Malthusian crisis—that is, population growth exceeded output, leading to famine.

Unprecedentedly sustained bad weather then fatally undermined a system that was already beginning to stall. Driven by persistent torrential rain occurring at critical times over successive years, and worsened by unseasonably cool temperatures that shortened the growing season, the production crisis culminated in a series of catastrophic crop failures. Widespread flooding, windstorms, and brutal winters compounded the impact, with "seedbeds sodden, crops and pastures under water, grain rotting, fish traps wrecked, dikes washed away, meadows too wet to be mown, turf too soggy to be cut, and quarries too swollen with the overflow to be worked for stone or lime."[2] Contemporaries feared the need for another Noah and his ark.

Rivaling the Egyptian famine predicted by Joseph in the book of Genesis, but without the Pharaoh's storage facilities or modern distribution networks, the late medieval "Great Famine" lasted—with ever-escalating impact—from 1315 to 1322. Causing the death of millions, it afflicted the whole continent north of the Alps, and it was followed by additional years of severe shortage and high prices between 1345 and 1348. Furthermore, in 1319 and 1320 a devastating disease thought to be rinderpest decimated cattle across northern Europe, seriously reducing access by most of the population to meat and milk and crippling production by destroying draft animals and their manure. This "Great Bovine Pestilence" therefore combined with recurring crop failures to undermine human nutrition, growth, and development.

The starkly unequal power relationships of late medieval society deepened the economic depression and enormously exacerbated poverty. Writing with specific reference to Sweden but in terms that applied to the whole of western Europe, paleoecologist Per Lagerås makes this point forcefully:

> The impoverishment of the population was also due to the inequality of medieval society. Even under normal conditions, heavy burdens of taxes, rents, tithes, and labour duties left little surplus for ordinary people. The number-one priority for the upper classes and the central power was to keep up their luxury consumption and life style while little resources were invested back into the agricultural system. When they experienced declining incomes due to poor yields their immediate reaction was to compensate by further raising taxes and rents. This counterproductive reaction is partly to blame for the stagnating economy and the non-sustainable agriculture. For the above-mentioned reasons people were on the edge of starvation.[3]

The results seriously compromised resistance to disease among people who were born after 1315. They suffered malnourishment in their developmental years and had become immunocompromised adults by the time plague-carrying Genoese ships docked at Messina.

During its relentless progress from Sicily across the whole of Europe, the Black Death outstripped even the Great Famine in its impact. During its first wave, between 1347 and 1353, it is estimated to have killed as much as half the population of the continent, creating what Lagerås calls "the worst disaster that has ever hit Europe."[14] One of the most famous local tragedies was the epidemic that decimated Florence in 1348 and is vividly portrayed by Boccaccio's *Decameron*. Other famous epidemics include Milan in 1630, which resulted in two major works of plague literature by Alessandro Manzoni—*The Column of Infamy* and *The Betrothed;* Naples in 1656; and the "Great Plague" of London during 1665–1666, which is the subject of Daniel Defoe's influential account *A Journal of the Plague Year.*

Then, for reasons that are much debated and that we discuss in Chapter 4, bubonic plague receded from western Europe between the end of the seventeenth century and the middle of the eighteenth. The last epidemics to strike Scotland occurred in 1640, England in 1665–1666, the Netherlands in 1710, France in 1720–1722, and Italy in 1743. Interestingly, outbreaks at Messina form convenient bookends for the second pandemic: it was the first place in the West to suffer in 1347, and the last in 1743.

Although, understandably, the demographic disaster so graphically described by contemporary writers and historians at the outset of the second pandemic has captured the imagination, it is clarifying to recognize that the virulence of the plague did not decline over the centuries. Some of the final epidemics of the second pandemic were among the most devastating and dramatic, such as the great disasters in London (1665–1666) and Marseille (1720–1722). What had changed was that these final attacks were localized events that failed to achieve the continental reach of the first invasions.

The Third Pandemic, or Modern Plague

The third and final pandemic of bubonic plague originated, like the second, from a Central Asian focus, erupting in the wake of social unrest and warfare in China in 1855. It attracted global attention when it attacked Canton and then Hong Kong in 1894 and later moved on to cities that were nodal centers in international trade such as Buenos Aires, Honolulu, Sydney, Cape Town, Naples, Oporto, and San Francisco. Unlike its predecessors, which devastated nearly every place they touched, the third pandemic was radically uneven in its impact as it followed instead the international fault lines of inequality, poverty, and neglect.

Bubonic plague during the third pandemic overwhelmingly scourged Third World countries while largely sparing the industrial nations of Europe and North America. Above all, this third round of plague ravaged India, where it caused as many as 13 to 15 million deaths between 1898 and 1910.

Before finally receding, it ultimately killed approximately 20 million people, according to conservative estimates, and touched five continents, though largely sparing the industrial West. Furthermore, in India and China plague was not a universal affliction as it had been during the second pandemic. In India, it had a predilection for the infamous tenements (*chawls*) of Bombay (now Mumbai) and the hovels (*bastis*) of Calcutta while barely affecting Europeans or the wealthy.

Europe experienced brief flare-ups in 1899 at Naples, Oporto, and Glasgow but lost a total of seven thousand people to plague for the entire half-century after 1899. Central and South America suffered thirty thousand plague deaths during the third pandemic. The United States recorded a mortality of approximately five hundred victims in minor outbreaks at San Francisco, New Orleans, and Los Angeles.

In the Americas the third pandemic had a restricted impact on humans, but it did leave a major environmental imprint by establishing stable reservoirs of infection among sylvatic rodents in the Southwest of the United States, the northeast of Brazil, and the south of Argentina. In those locations it persists to the present amidst massive cyclical die-offs of rodents and an ongoing trickle of cases of bubonic plague among humans who venture into the wrong place or whose pets exchange fleas with ground squirrels and gerbils. The Centers for Disease Control and Prevention (CDC) reports that between 1900 and 2016, there were just over one thousand cases of plague in the United States, concentrated in New Mexico, Arizona, Colorado, and California, mostly among hunters and campers.

The reservoirs in the Americas thus formed additions to the preexisting plague reservoirs on every other continent. As a result, the World Health Organization (WHO) reports that 3,248 cases of plague and 584 deaths occurred between 2010 and 2015, spread over four continents, but with concentrations in the Democratic Republic of the Congo, Madagascar, and Peru. The official statistics are, however, almost certainly significant underestimates because of misdiagnosis, concealment by communities and governments, and the absence of laboratory facilities in many settings.

Most influentially, the third pandemic marked the moment when the complex etiology of the disease with its interaction between rodents, fleas, and humans was unraveled. As a result, beginning in the early twentieth century new public health policies based on this knowledge were implemented. Authorities targeted fleas and rats with insecticides, traps, and poison, thereby replacing the draconian antiplague measures that characterized the second pandemic and the early years of the third.

CHAPTER 4
Plague as a Disease

The Etiology of Plague

The "etiology" of a disease refers to its origins—the path it follows to afflict human beings. Plague has a complicated etiology normally involving four protagonists. The first is the pathogen itself—the oval bacterium originally named *Pasteurella pestis,* but now universally termed *Yersinia pestis.* It was discovered simultaneously in 1894 in Hong Kong by Alexandre Yersin, a Swiss student of Louis Pasteur, and by the Japanese physician Shibasaburo Kitasato, a protégé of Pasteur's rival Robert Koch.

By 1898 Paul-Louis Simond discovered that, in addition to the bacterium, two vectors were normally responsible for conveying the disease to humans. These were commensal rodents, especially rats, and their parasitic cargo of fleas. Simond's insights were largely ignored until they were confirmed a decade later, as we will see in Chapter 16, by the exhaustive investigation conducted by the Indian Plague Commission into the epidemiology of the disease. The findings were conclusive for the third pandemic on the subcontinent, but both Simond and the commission made the mistake of assuming that the mode of transmission that prevailed in the third pandemic was the same in all epidemics of the disease. Since this view became the orthodox understanding of plague, the result was a misunderstanding of the Black Death. So much of the epidemiology of the second pandemic was incomprehensible in terms of the rat-flea nexus that, as we will see, it even led to skepticism that the disaster that swept Europe for four centuries after 1347 was plague at all.

There is, however, a universal consensus that plague epidemics begin as "epizootics"—unnoticed epidemics in permanent animal reservoirs of the disease. Especially important are wild rodents—marmots, prairie dogs, chipmunks, and squirrels in their burrows, where underground disasters unknown to humans take place. Plague, therefore, is best understood as a disease of animals by which humans are afflicted by accident and by way of exception. Human involvement can begin if hunters invade the reservoir of infection, contracting the disease directly while skinning their prey and providing a portal of entry for the bacterium to their bloodstream through a cut or other lesion. Warfare, ecological disaster, and famine can multiply the numbers of displaced persons entering the rodents' habitats. Alternatively, environmental changes, such as flood or drought, can send the animals scurrying over long distances closer to human settlements and above all bringing them into closer contact with the peridomestic rat. Pivotal in the second pandemic was the black rat or "ship rat," *Rattus rattus,* which lived in close proximity to humans and with whom it shared the same dietary preferences.

The bacterial exchange from wild rodent to domestic rat, from rat to rat, and from rat to human is mediated by the final plague protagonist—the flea. Two species are deemed to play key roles. The first is the so-called Oriental rat flea, *Xenopsylla cheopis,* which is naturally parasitic on warm-blooded animals and is a highly efficient vector for bubonic plague, enabling it to cross the species barrier from rodent to *Homo sapiens.* The other is the widespread "human flea," *Pulex irritans,* which does not colonize rodents but only people. It provides transmission from person to person.

In a single blood meal a flea of either species sucks up a quantity of blood equal to its own weight—an amount that contains millions of bacteria. Once engorged with blood and *Y. pestis,* the infected flea does not survive. The bacilli obstruct a valve regulating the movement of food into the insect's stomach, so the flea slowly dies of starvation and dehydration. This foregut blockage is consequential not only in the life of the flea, but also in disease transmission to humans. By causing the flea to regurgitate ingested blood mixed with plague bacteria, the obstruction ensures that each bite is infective. It also causes the flea to feed ravenously and repeatedly in a frenzied bid to survive. Before dying, an infected *X. cheopis* is a lethally efficient vector.

Furthermore, when a rat carrying infected fleas in its fur sickens and dies, the fleas leap to the warm body of another mammal—whether rodent or human being. The fleas' sensors—which are highly susceptible to warmth, vibration, and carbon dioxide—enable fleas to locate new hosts; their famous jumping prowess enables them to make the transfer successfully; and their ability, when they are uninfected, to wait for six weeks between blood meals explains the erratic appearance of cases during the course of an epidemic.

Xenopsylla cheopis demonstrates a pronounced preference for rats, and it moves to humans only in the absence of surviving rodents. For this reason, the mass die-off of a rat colony leads to the sudden availability of swarms of famished fleas to parasitize humans in the absence of their preferred alternative. This behavior explains the explosiveness of plague epidemics. A graph of plague mortality and morbidity exhibits a steep spike at the outset rather than the bell-shaped curve characteristic of most epidemic diseases.

After the crossing of the species barrier between rats and humans, foci are established when infected men and women share their fleas and their disease with family members and neighbors. Plague then begins, not as a disease of isolated individuals, but of households, neighborhoods in cities, and entire villages in rural areas. Living conditions, and especially population density and hygiene, are decisive. Overcrowding, with numerous individuals pressed close together in a single room and perhaps whole families sharing a bed, greatly facilitates the exchange of fleas. As the plague spreads, particular moments are especially dangerous, such as the laying out of and final attentions to the dead. As a corpse cools, the fleas infesting it grow desperate to escape to the next warm body.

To give rise to an epidemic, the early foci in Africa and Central Asia needed links to wider networks of contact, trade, religious faith, and commerce. One such link involved the garments of the sufferers. An item of clothing in the early modern world was precious, so the clothes of the dead, as well as bed linens, were reused by others or packed in crates and sold at markets and fairs, often with fleas still alive among the folds. Certain professions also came into regular and direct contact with the sick, the dying, and the dead, as well as with their ectoparasites. Street vendors, physicians, priests, gravediggers, and washerwomen were at serious risk in time of plague, and they conveyed disease from place to place as they moved about their duties. Other key figures in the transmission of plague were millers and bakers, because grain attracted rats.

Monasteries also played a significant part in the first pandemic and the late medieval outbreaks of the second, explaining the ability of the disease to decimate the thinly settled countryside as well as urban centers. Monasteries acted as nodes in the grain trade, linking settlement to settlement and village to village; they were substantial communities of people living in close quarters; and during times of disaster monasteries often served as places of refuge for people fleeing plague-ridden localities. On monastic grounds the healthy, the ill, and the bearers of infected fleas commingled. In such a context the flea readily established circuits of transmission for plague.

Fleas are severely restricted in their range, however, as opposed to rats, which are wonderful travelers. They hide in shipments of grain and are transported overland in carts or down waterways on barges and boats. But the rat also went much farther afield—by sea. Infected rats climbed aboard ships via ropes and gangplanks and were lifted aboard in crates of wheat and rice. In this way, shipping was essential to the spread of plague over long distances; this helps to explain the epidemiology of the disease—that is, its tendency to arrive in a country by ship and then to move inland by road and river traffic. For the black rat, the Mediterranean was not a barrier but a highway.

Istanbul (known as Constantinople between 330 and 1453) was a vital hub of trade and disease. It linked the whole of the Mediterranean—overland via the Balkans and by sea to Venice, Naples, Corfu, Genoa, Marseille, and Valencia. Sometimes there was havoc at sea when plague destroyed entire crews, and ghost ships drifted on the waves. More frequently, however, a ship would dock, and its cargo of rats would disembark via the same hoists, ropes, and gangplanks that had first brought them aboard. At the same time infected passengers and crew would go ashore, together with their fleas. Procopius already noted in the sixth century that the plague "always took its start from the coast, and from there invaded the interior."[1]

Not surprisingly, then, the first indication of plague was frequently a dramatic die-off of rats in the streets. This onset of pestilence is dramatically apparent in various works of plague art, such as Albert Camus's novel *The Plague*. In this book the emergence of innumerable sick and dying rats in the streets of the Algerian city of Oran is the prologue to the epidemic disaster, which serves as a metaphor for evil as embodied in the rise of Nazism and fascism.

Similarly, the neoclassical painter Nicolas Poussin (1594–1665) included rats in his painting *The Plague at Ashdod* (1630) (fig. 4.1). The biblical book 1 Samuel relates how the Philistines triumphantly displayed the stolen ark of the covenant in their temple to the pagan god Dagan, thereby declaring the superiority of Dagan over the God of the Israelites. God punished the Philistines by scourging them with plague and destroying the city. To heighten the terror of the scene, Poussin prominently depicted rats on the streets of doomed Ashdod. The seventeenth-century painter understood that for the viewer, the mass appearance of rats aboveground was a familiar premonitory sign of plague and impending disaster.

Figure 4.1. In his painting of 1630 *The Plague at Ashdod* Nicolas Poussin included rats, which were known to be a premonitory sign of impending disaster. Musée du Louvre, Paris.

Twentieth-century scientific investigations have carefully confirmed the link between plague and rats. Archeologists have recovered the bones of rats at plague burial sites dating from the Black Death, and as already mentioned, the Indian Plague Commission extensively documented the rat-flea nexus and its role in the modern plague.

Until the turn of the twentieth century, however, the link between the rat and bubonic plague was not deemed to be causal. It was thought that the rat was infected before humans because of its low physical stature. With its nose close to the soil or the floor, the rat was more quickly susceptible to poisonous effluvia rising from the earth and to plague-infested dust underfoot. Suddenly, before people were stricken, rats began to appear in the streets and in the middle of rooms indoors. In a dazed condition and with uncertain balance, these rodents demonstrated no concern for their natural predators and enemies. Instead, in a frenzy of thirst, they desperately sought water until they lost their strength and collapsed. They died where they fell with telltale buboes on their necks, their limbs splayed, and rigor mortis upon them even before death.

Miasmatic theory provided an explanation, since this doctrine held that plague originated in the soil, rising slowly from it and progressively killing different animals as it reached the air they inhaled. Thus it was logical that rats, with their snouts to the ground, should fall victim first, and that humans, with their superior height, should be stricken only later. Such an etiology indicated that human plague followed the disease in rats but was not caused by it.

Symptomatology and Pathology

Attention to the effect of a disease on the individual human body is not a matter of ghoulish curiosity. Epidemic diseases are not simply transposable causes of suffering and death. On the contrary, the history of each high-impact infectious disease is distinct, and one of the major variables is the specific manner in which it affected its victims. Indeed, a feature of bubonic plague was that its symptoms seemed almost purposefully designed to maximize terror; they were excruciating, visible, dehumanizing, and overwhelming.

After being bitten by an infected flea, a person experiences an incubation period varying from one day to a week, and then the classic symptoms of the disease appear, launching the first stage of bubonic plague. At the site of the flea bite, a black blister, or carbuncle, arises, surrounded by red pock marks. Along with the carbuncle, the infected person experiences high fever, shivering, violent headache, nausea, vomiting, and tormenting thirst, after which the second stage of the disease begins. Unlike malaria-transmitting mosquitoes, for example, fleas do not inoculate bacteria directly into the bloodstream but instead deposit them into the skin. As few as ten bacilli are now thought to be sufficient to produce an infection. The reason is a stealth mechanism, or "virulence factor"—the production of an enzyme that enables the bacteria to elude the body's defense mechanisms. Rapidly multiplying, *Y. pestis* invades the lymph system, drains into the regional lymph nodes nearest the infective bite, and causes the appearance of a bubo.

A bubo is an inflammation and swelling of a lymph node of the armpit, neck, or groin to form a hard mass sometimes as large as an orange beneath the skin. Its location varies according to the site of the infective flea bite, and it is not uncommon for more than one to erupt. The bubo is familiar as the classical symptom that appears almost infallibly and gives the disease its name as "bubonic" plague.

Buboes are a source of agony for patients. The sixteenth-century French surgeon Ambroise Paré explained that the bubo generates great heat and causes a "pricking pain, as it were with needles, burning and intolerable."[2] Defoe reported in his *Journal of the Plague Year* that the pain caused by the bubo was so violent that some victims in London dove into the Thames to escape it—an observation that Paré corroborated, writing from Paris that patients there hurled themselves naked from their windows. In the more measured parlance of the modern physician, the inflamed and suppurating bubo is described as "exquisitely tender."[3] There is also a consensus that the body and all of its excretions—pus, urine, sweat, the breath—have an overpowering stench, as if putrefaction precedes the individual's demise. Surviving accounts by attendants in pesthouses during epidemics describe the intolerable smell of the bodies of the sufferers as the worst aspect of their employment. The historian Jane Stevens Crawshaw, for instance, thus summarizes Father Antero Maria's account of his experience at Genoa, where he served in the plague hospital in 1575:

> the sick within the *lazaretti* stank horribly—so much so that a single patient could render a room uninhabitable. He wrote that individuals in the *lazaretti* fled the company of others because of the smell and confessed himself to having hesitated many times before entering rooms—not, he says, for fear of infection but because the smell was so foul. This was made worse by the vomiting brought on by the illness. This, he said, was so disgusting that it turned the stomach. He recorded it as the most difficult aspect of the conditions in the *lazaretto*, too abominable to describe in words.[4]

Meanwhile *Y. pestis* bacilli—still "the most feared pathogen of the bacterial world"—continue to proliferate exponentially, doubling their numbers every two hours.[5] In evolutionary terms, such expansion is positively selected because a bacteremia level of 10 to 100 million bacilli per cubic milliliter of blood is needed to ensure that a flea biting a human is infected. With the flea as a vector, extreme virulence was necessary to the transmission and survival of plague bacilli. An immediate consequence is that the ever-replicating bacteria rapidly overwhelm the body's defenses by preferentially targeting and destroying the cells—dendritic cells, macrophages, and neutrophils—that provide the body's immune response. In the words of a 2012 US Geological Survey report:

> *Y. pestis* uses a needlelike appendage to target a host's white blood cells [and] injects proteins . . . directly into the host white

> blood cells. These proteins act to destroy immune functions of the host and prevent it from developing an inflammatory response that would inhibit or prevent further growth of the bacteria. . . . *Y. pestis* can also inject into the host a different protein . . . that prevents the host from producing two of its own proteins that would be used to stimulate the formation of a mass of immune cells to surround the bacteria and prevent its growth. . . . In the case of plague, the host cells get a false message that tissue damage is under control, when, in fact, *Y. pestis* bacteria are rapidly taking over visceral organs, particularly the liver and spleen, leading to loss of function.[6]

Escaping from the lymphatic system into the bloodstream, the multiplying bacteria initiate the third stage of the disease: septicemia. Having gained access to the blood, the bacteria release a powerful toxin that is normally the cause of death. It attacks tissues, causing blood vessels to hemorrhage, giving rise to purple, subcutaneous spots—the so-called tokens of plague. They earned this name because many people thought them to be signs, or "tokens," of God's anger.

By producing degeneration of the tissues of the heart, liver, spleen, kidneys, lungs, and central nervous system, the systemic infection initiates multiple organ failure. At this point patients have wild bloodshot eyes, black tongues, and pale, wasted faces with poor coordination of the facial muscles. They experience general prostration, teeth-shattering chills, respiratory distress, and a high fever that normally hovers in the range of 103° to 105°F, but in some patients reaches 108°F. In addition there is progressive neurological damage manifested by slurred speech, tremors in the limbs, a staggering gait, seizures, and psychic disturbances ending in delirium, coma, and death. Pregnant women, who are especially vulnerable, invariably miscarry and hemorrhage to death. Sometimes there is also gangrene of the extremities. This necrosis of the nose, fingers, and toes is one probable source of the terms "Black Death" and "Black Plague."

During the first European epidemic of Black Death at Messina in 1347, the Franciscan chronicler Michael of Piazza wrote a vivid description of the sufferings of the victim:

> The "burn blisters" appeared, and boils developed in different parts of the body: on the sexual organs, in others on the thighs, or on the arms, and in others on the neck. At first these were of

> the size of a hazelnut and the patient was seized by violent shivering fits, which soon rendered him so weak that he could no longer stand upright, but was forced to lie on his bed, consumed by a violent fever and overcome by great tribulation. Soon the boils grew to the size of a walnut, then to that of a hen's egg or a goose's egg, and they were exceedingly painful, and irritated the body, causing it to vomit blood by vitiating the juices. The blood rose from the affected lungs to the throat, producing a putrefying and ultimately decomposing effect on the whole body. The sickness lasted three days, and on the fourth, at the latest, the patient succumbed.[7]

As soon as anyone was seized with headache and shivering during a visitation of the plague, he or she anticipated a fatal outcome. The minority of patients who recovered from their ordeal faced a lengthy convalescence and an array of lasting or permanent sequelae. These included deafness, impaired vision, paralysis of the muscles of one or more limbs, inability to speak as a result of laryngeal paralysis, and loss of memory. Psychological trauma also persisted after so arduous an ordeal. The experience did not even confer an acquired immunity, as a survivor from an epidemic in one year could die from plague the next. Against this background of sudden onset, a fulminant course, and a cascade of excruciating symptoms that usually ended in death, it is small wonder that the very word "plague" came to be synonymous with calamity and the worst imaginable disaster. In the Islamic world, it was even widely known as the "great annihilation."

Types of Plague

Bubonic Plague

Bubonic plague arises as an infection of the lymphatic system, it is transmitted primarily by flea bites, and it has the symptomatology described above. It is also the most common form of plague and has had the largest historical impact in all three of the plague pandemics. But the disease can also occur in two other forms, known as septicemic and pneumonic plague. It is vital to stress that these are not three distinct diseases, but simply three different manifestations of the single disease of plague, and all three are caused by *Y. pestis*.

Septicemic Plague

Primary septicemic plague is the fulminant and rarest of the three forms of plague. Like bubonic plague, it is transmitted by flea bite; but unlike bubonic plague, primary septicemic plague begins with inoculation of *Y. pestis* into the bloodstream, without prior inflammation of the lymph nodes and the formation of buboes. The bacteria immediately metastasize, with rapidly fatal consequences. In some cases, the progress of the disease is so fast that the patient dies within hours even before the onset of symptoms. More commonly, however, the sufferer endures organ failure, severe nausea, fever, and abdominal pain followed within hours by death through multiple causes. In septicemic plague, the CFR approaches 100 percent.

More common is secondary septicemic plague, which simply denotes a stage in the normal progression of bubonic plague untreated with antibiotics. In this stage the bacteria, having already caused the classic symptoms of the disease, escape the lymph system and enter the bloodstream. There they initiate the multiplication, diffusion, and toxicity that lead inexorably to death.

Pneumonic Plague

Pneumonic plague is a severe infection, not of the lymphatic system or the blood, but of the lungs, so historically it has sometimes been called "pestilent pneumonia." The etiology may be the spread of the plague bacteria from the lymphatic to the respiratory system—a condition known as "secondary pneumonic plague." Alternatively and more important historically, pneumonic plague can be transmitted directly from person to person by inhaling droplets from the coughs and sneezes of a plague victim whose respiratory system is the original site of infection, resulting in "primary pneumonic plague."

Since the site of infection is the lungs, the symptomatology is significantly different from that of bubonic or septicemic plague. The portal of entry of *Y. pestis* via the lungs significantly affects its symptoms and, above all, its lethality and time course through the body. One reason is the different temperature of the flea gut as compared with that of the human body immediately before infection. In bubonic flea-to-human transmission, the temperature environment in which *Y. pestis* develops is that of the flea gut, 26°C, whereas in pneumonic human-to-human transmission the plague pathogen develops at 37°C. Recent research suggests that at the molecular level, the development of the bacterium at the higher temperature activates genes that express virulence. These genes cause the production of antigens that destroy phagocytes (white cells) and chemical reactions that enable the pathogen to elude detection by the receptors of the large white blood cells known as macrophages. The result is "an immunosuppressive environment in the lungs" that leads to "rapid bacterial proliferation" in the lung alveoli—the tiny sacs where the vital exchange of oxygen and carbon dioxide in the bloodstream takes place.[8]

Thus attacked, the pneumonic plague sufferer experiences symptoms similar to those of acute pulmonary pneumonia. With destruction of the lung alveoli, edema, and hemorrhage, the patient manifests severe respiratory distress, fever, chest pain, cough, nausea, headache, and frothy, bloody sputum. Furthermore, the condition is universally fatal, often within less than seventy-two hours.

A significant historical corollary of the mode of transmission of primary pneumonic plague is that it does not depend on the rat and the flea. Here is the probable solution to the epidemiological puzzles that led to a current of "plague deniers," such as Graham Twigg and Samuel K. Cohn. They postulate that the disease responsible for the second pandemic was not plague at all but anthrax or a combination of anthrax and an unspecified comorbidity. If plague was responsible, they ask, why did massive die-offs of rats not figure more frequently in plague literature, paintings, and chronicles of the Black Death? How could a disease dependent on the gradual movement of rats sweep so quickly across the European continent? How could bubonic plague burst out in Moscow or Iceland during frigid winters, when fleas are inactive? Why were crucial features of the epidemiology and virulence of the third pandemic so different from descriptions by contemporary observers of the second?

In this context, the example of Iceland appears especially confounding. With a slight delay due to its distance and isolation, Iceland suffered the first wave of the Black Death in 1402–1404, when perhaps 50 percent of the population perished. But there is a conundrum that adds significantly to the puzzlement already engendered by the onset of plague despite the island's climate. A feature of Icelandic fauna in the late medieval period was that it lacked a population of rats. How could a disease transmitted by rats and Oriental rat fleas be widely and rapidly diffused in the absence of both?

Osteoarchaeology—the systematic excavation and scientific examination of human bones from archeological sites—has provided conclusive findings with regard to the presence of plague across northern Europe, and genetic research has provided at least partial answers to most of the outstanding questions. Examination of bones and dental pulp extracted from bodies exhumed from plague burial sites has irrefutably established the presence of *Y. pestis*. In the terse comment of a recent researcher, "Finally, plague is plague."[9] These findings do not exclude the possibility of a second epidemic pathogen as well, but they provide robust proof that bubonic plague was a factor. Furthermore, to date, the DNA of anthrax or other epidemic pathogens has not been detected in plague burial grounds.

In addition, genomic research has identified mechanisms that resolve many of the difficulties confronting those who consider plague as almost certainly responsible for the medical events of the second pandemic. It is now known that there are different strains of *Y. pestis;* that these strains vary in their proclivity to produce bubonic or pneumonic plague; and that the strain implicated in the Black Death was highly virulent, in part because of its tendency to produce the pneumonic form of the disease.

These data shed new light on the Black Death. Person-to-person dissemination by droplets allows a far more rapid propagation of the disease than the comparatively slow long-distance movements of rats over land and sea. Pneumonic plague also spreads more easily in winter than bubonic plague because it does not depend on the activity of fleas but rather on the human behavior of congregating indoors during cold weather within close range of coughs and sneezes. It would therefore thrive in the winter environment of northern and eastern Europe.

Furthermore, droplets were not the only vehicle for direct human-to-human transmission without the mediation of the rat. The human flea *Pulex irritans* can also play a significant role apart from the rat flea *Xenopsylla cheopis*. Investigation of the small but famous sample of the Derbyshire village of Eynsham in its experience of plague in 1665–1666 revealed that human-to-human propagation by the human flea was far more prevalent than by the rat flea. Such direct human-to-human diffusion occurring during the second pandemic as a whole would help to explain the rapid spread of the disease, which far outran the onward progress of rats and the fleas they concealed in their fur.

Since the Black Death appeared in Europe as a sudden and unfamiliar invader from abroad, the population possessed no immunity against it. Genetic findings demonstrate that although the Plague of Justinian was also caused by *Y. pestis,* it was a strain that was genetically distant from the strains responsible for the second and third pandemics. Crossover immunity from one to the other is therefore unlikely. Consequently, the Black Death probably spread as a "virgin soil epidemic" analogous to smallpox in the Americas—a factor that also helps to explain its extraordinary virulence and rapid spread. Furthermore, a pneumonic disease spreading through air droplets would account for the paucity of contemporary references—in art, literature, and chronicles—to rodent die-offs that preceded the affliction of humans.

Meanwhile, the contrasting predominance of bubonic rather than pneumonic plague in the third pandemic clarifies some of the features that distinguish it from the second: the constant references to swarms of dying rats by all observers, in clear contrast to the Black Death; the slow and erratic movement of the disease; its tendency to persist even for years in the localities it invaded; and its exclusive preference for warm climates and mild seasons. Pending further research and greater clarity, the preponderance of current evidence suggests that the traditional attribution of all three pandemics to *Y. pestis* is correct—provided that allowance is made for different strains and due account is taken of the balance between bubonic and pneumonic plague.

The specific features of pneumonic plague also explain the interest of bioterrorists and germ warfare laboratories in this disease. It spreads rapidly, is readily aerosolized, and is nearly 100 percent lethal. Furthermore, it begins with mild flulike symptoms that delay recourse to diagnosis and treatment, and it frequently runs its course within the human body in less than seventy-two hours. The opportunity to deploy curative strategies is therefore exceptionally brief. This situation is rendered even more critical by the recent appearance of antibiotic-resistant strains of *Y. pestis*. By virtue of these characteristics, the CDC has classified *Y. pestis* as a "Tier 1 select agent"—a pathogen with the highest appeal for use as a weapon of biological warfare or bioterror.

Conclusion

In practice, plague sufferers often received no medical attention. In the early waves of the Black Death in particular, societies were caught unprepared by an unknown "emerging disease." No facilities—administrative, religious, or medical—were in place to cope with the tidal wave of disease and death. Physicians and attendants recognized that they were powerless to understand or treat the invading new affliction and that they were far too few in number to cope with the catastrophe that engulfed them. Their profession also exposed them disproportionately to risk, and they perished in great numbers during outbreaks.

Furthermore, overcome with terror like the population at large, many doctors joined patients' relatives and friends in the general flight from plague-threatened cities. Indeed, one of the horrors of an epidemic of plague was that it broke the common bonds of humanity. As a result, sufferers were often abandoned to face agony and death alone. The most famous and perhaps most harrowing plague testimony is that of Giovanni Boccaccio in the *Decameron,* which was based on the experience of Florence in 1348:

> Tedious were it to recount, how citizen avoided citizen, how among neighbors was scarce found any that shewed fellow-feeling for another, how kinsfolk held aloof, and never met, or but rarely; enough that this sore affliction entered so deep into the minds of men and women, that in the horror thereof brother was forsaken by brother, nephew by uncle, brother by sister, and oftentimes husband by wife: nay, what is more, and scarcely to be-lieved, fathers and mothers were found to abandon their own children, untended, unvisited, to their fate, as if they had been strangers. Wherefore the sick of both sexes, whose number could not be estimated, were left without resource but in the charity of friends (and few such there were), or the interest of servants, who were hardly to be had at high rates and on unseemly terms.[10]

An aspect of the dread felt by those threatened with plague was that it broke the framework developed in the Middle Ages for dealing with death. The historian Philippe Ariès has explained that populations across Europe had devised a series of beliefs, practices, and rituals to provide support in dealing with mortality. These strategies helped people to cope with loss, to heal the tear caused in a family or community by the death of one of its members, to express grief, and to pay their last respects to the dead. Taken together, the practices constituted an "art of dying" (*ars moriendi*) that was codified in instructions elaborated in paintings, engravings, sermons, and books that all explained how to die properly in accord with Christian doctrine. A work of the kind—called a *memento mori* (reminder of death)—explained who should be present at the last hour, detailed the last sacraments to be administered by the clergy, and dictated the proper funeral rites—the laying out of the body, the wake, the procession, the funeral service and burial ceremony, the interment in consecrated ground, and the funeral meal for surviving friends and relatives. In all of these rituals the objective was to allow communities to express the values of solidarity and human dignity.

The most famous of all writers on the theme of *ars moriendi* was the ascetic seventeenth-century Anglican Bishop Jeremy Taylor. His major works were *The Rule and Exercises of Holy Living* (1650) and *The Rule and Exercises of Holy Dying* (1651). The aim of both books, which were widely read in both Britain and America, was to remind the faithful that life on earth is unsafe and ultimately unimportant; therefore, believers should primarily spend their time preparing for eternal life. It was vitally important to die with one's worldly affairs in good order and with one's soul in a state of grace ready to confront judgment day. Taylor's books were instructional manuals on how to achieve both ends—the material and the spiritual. Taken together, they indicated the means to "tame death" (using Ariès's term) so that believers faced mortality confident in the knowledge of being properly prepared to confront it.

What made bubonic plague especially fearful was that it presented communities with the antithesis of the "art of dying" and robbed individuals of the opportunity to achieve a "tame death." It caused believers to face sudden death, or *mors repentina*, in which victims could be caught with wills unwritten in this world and unconfessed souls in a state of sin that would lead to damnation in the next. Death from plague was sudden; people died alone and without the attention of the clergy; and often they were denied funeral rites and proper burial.

Fear of sudden death from plague is therefore analogous to the terror described by Drew Gilpin Faust in her book *This Republic of Suffering: Death and the American Civil War* (2008). Faust places the terror of sudden death at the center of her account because it was widespread among soldiers on both sides of the conflict. The dread appears repeatedly in their letters home to loved ones and friends. In this respect, bubonic plague resembles total war, as both present unlimited possibilities for the sudden death that can come to us "as a thief" (Rev 3:3).

The New Testament book of Revelation provides a vivid account of the end times—with a day of wrath, a great apocalypse, and plagues and suffering. During the plague centuries, the visual arts made the depiction of "death unleashed" as described in Revelation central to their iconography. Many artists portrayed the "triumph of death," which takes the form of a universal plague complete with the Four Horsemen of the Apocalypse. Perhaps no better example of this terrifying genre can be found than the painting *The Triumph of Death* (1562–1563) by the Flemish master Pieter Bruegel the Elder. In the foreground and center of the picture Death himself drives a great ox-cart while wielding a scythe to reap his grim harvest. Before him the Angel of Death sets forth blowing a trumpet while people are dying all around and graves are opening to yield up their skeletons.

A second major aspect of plague iconography was "vanitas," that is, a symbolic expression of the idea that earthly life is fleeting and insignificant (fig. 4.2). The vanitas theme gained widespread currency in the wake of the first wave of the Black Death and then faded in the eighteenth century with the coming of the Enlightenment and the end of the second pandemic. The traditional Christian view of the transitoriness of life was expressed in the book of Ecclesiastes (1:2–4): "vanity of vanities, all is vanity. What profit hath a man of his labour which he taketh under the sun? One generation passeth away, and another generation cometh." Such paintings often display temporal goods embodying the hubris of human aspirations—gold, musical instruments, scholarly tomes, globes, and elegant garments. These were juxtaposed to striking symbols of the underlying truth that human

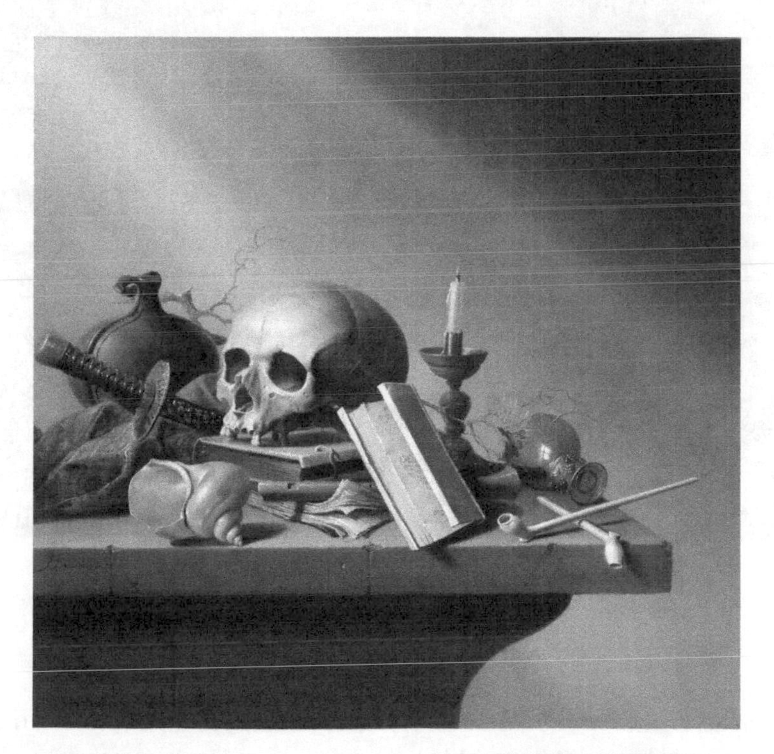

Figure 4.2. Harmen Steenwijck, *Vanitas stilleven* (ca. 1640), which symbolically depicts the transience of life and certainty of death. The *vanitas* was a popular art form during the Black Death. Museum De Lakenhal, Leiden.

Figure 4.3. German painter Lukas Furtenagel painted this *vanitas, The Painter Hans Burgkmair and His Wife Anna*, in 1529. Kunsthistorisches Museum, Vienna.

accomplishment is minimal and life is short—skulls, candles whose flame has just gone out, hourglasses marking the passage of time, crossbones, skeletons, and shovels. One example by the German painter Lukas Furtenagel shows the faces of a middle-aged couple reflected in a hand mirror as skulls (fig. 4.3).

Another artistic motif that coincided with the era of the plague was the *danse macabre*. Such works of art portray Death as a skeleton summoning people of all ages, ranks, and sexes to join in a merry dance. Sometimes Death is armed with a scythe, arrow, or dart, and he leads the dance while playing a musical instrument. Churches often enacted such a performance, transposing the representation of the fragility of life to the medium of theater. More recently, Bergman's plague film *The Seventh Seal* reaches its denouement when Death summons the protagonists to join in a jolly dance.

We have considered the cultural and physical impacts of plague on the populations it afflicted, but how did the authorities of church and state seek to contain the disaster? What administrative strategies and medical therapies did they deploy? We now turn to the collective responses of societies to the emergency.

CHAPTER 5

Responses to Plague

The first collective responses to the ravages of the bacterium *Yersinia pestis* were spontaneous and unorganized. Eventually, however, the first public health antiplague strategy was implemented and was followed by the first victory over a human disease when bubonic plague retreated from the West after the final flare-up of the second pandemic at Messina in 1743. The question to explore then becomes not only the impact of plague itself, but also the legacy of antiplague measures. To what extent was the victory over plague due to authoritarian policies of containment?

It is critical to remember that the conquest of plague, so dramatic from the standpoint of western Europe, was partial and geographically confined. It was partial because plague has never been eradicated. Animal reservoirs of plague bacteria remain on every continent except Antarctica, and the danger of reemergence persists. Furthermore, the species transfer of *Y. pestis* from rodent to humans continues to occur worldwide, resulting in an annual trickle of cases with occasional larger flare-ups. Finally, there is the ever-present danger of a man-made epidemic of plague through an act of bioterror. This occurred in China at the hands of the invading Japanese Army, and the capability to weaponize plague was available to both superpowers during the Cold War. The threat of bubonic plague remains.

Spontaneous Responses

Flight and Cleansing

In terms of community responses to the outbreak of plague, the first and most universal of all reactions was to flee. During the Great Plague of London in 1665–1666, wave after wave of people panicked and fled the city for their lives. Daniel Defoe (1660–1731) clearly depicts the terror that swept the city in *A Journal of the Plague Year* (1722):

> This was a time when every one's private Safety lay so near them that they had no Room to pity the Distresses of others; for every one had Death, as it were, at his Door, and many even in their Families, and knew not what to do, or whither to fly.
>
> This, I say, took away all Compassion; self Preservation, indeed, appear'd here to be the first Law. For the Children ran away from their Parents, as they languished in the utmost Distress: And in some Places . . . Parents did the like to their Children. . . .
>
> It is not, indeed, to be wondered at: for the Danger of immediate Death to ourselves, took away all Bonds of Love, all Concern for one another.[1]

Naples, which experienced one of Europe's major plague catastrophes in 1656, illustrates the reasons that propelled people away from cities. One of the largest and most densely crowded cities of the seventeenth century, Naples was especially vulnerable to plague because of its prominent position in Mediterranean trade and its densely crowded and unhygienic slums. In 1656, at the time of the plague's most devastating visitation, nearly half of its population of five hundred thousand perished. Every activity of normal life ceased amidst shuttered shops, unemployment, and hunger. As in the well-known plague adage, too few of the living were left to bury the dead. Corpses were abandoned, both indoors and in public spaces. In the end, reports suggest that tens of thousands of dead bodies were burned, and thousands more were dumped unceremoniously into the sea.

Against that backdrop, Italy's largest seaport was overwhelmed by the stench of decomposition while dogs, vultures, and crows picked at the dead. Along with disease, the city suffered a breakdown of law and order and the collapse of every public service. Thieves plundered the homes of those who died while death carts, with their fearful cargo, made their lugubrious way through the streets. Astrologers peddled advice and prognostications, charlatans hawked their nostrums, and healers of every description charged astonishing sums for practicing their arts. People inevitably believed that the end of the world was at hand.

The decision to flee was mediated by medical understandings of epidemic disease. Indeed, the humoralist doctrines of Hippocrates and Galen, reinforced by the advice of contemporary physicians, sanctioned flight. Classical medical doctrine explained that epidemics were caused by a mass imbalance of humors triggered by corrupted air. Rotting organic matter emitted dangerous effluvia that rose from the earth and degraded the atmosphere in a locality. Since disease was so closely tied to a given place, it made sense to flee, thereby escaping both the poison and the pestilence it caused.

The responses of populations reflected how they interpreted their encounter with disease, or—in a more recent phrase—how they "socially constructed" their experiences. If disease was a defilement of the atmosphere, then there were other options as well as flight. One was to search for the corrupting element. For those who had a miasmatic medical philosophy, suspicion fell first on foul odors, which were plentiful in an early modern city. People hurled night soil from their windows and doors; butchers swept offal into the streets; leatherworking and other manufacturing of the time resulted in noxious products. A logical response was an urban cleanup, and authorities everywhere frequently undertook sanitary measures for the purpose of combating odors. They collected refuse, closed certain workshops and trades, swept the streets, halted work in abattoirs, and ordered the prompt burial of corpses.

Furthermore, in Christian Europe water and its cleansing action had symbolic as well as literal meaning. Water was a purifying element because of its role in baptism, where it cleansed the soul. Therefore, throughout Europe cities ordered their streets cleaned with water during times of plague for religious reasons rather than strictly sanitary ones. Fire, smoke, and certain aromatic agents were part of other purifying strategies, and those charged with containing disease tried to make use of them to cleanse the air directly. To that end, they lit bonfires with aromatic pinewood or burned sulfur. Frequently, antiplague strategies also included the firing of cannon, with the idea that gunpowder would purify the air.

Self-Defense

Individual citizens followed this lead. Even if divine anger was the final cause of disaster, the proximate cause was a poisoned atmosphere. It was therefore

Figure 5.1. A doctor's protective plague costume, Marseille, France, 1720. (Wellcome Collection, London. CC BY 4.0.)

wise to carry a vial of aromatic spices and herbs around one's neck or a small bottle of vinegar to sniff periodically. Tobacco enjoyed a certain popularity for similar reasons, as people sought to smoke their way to health. It was also advisable when at home to shut windows and doors and hang heavy draping over them as physical barriers to prevent miasmatic poison from wafting indoors. The garments of people who had been infected also fell under suspicion because fatal effluvia were thought to cling to them, just as the scent of perfume could adhere to clothes.

For this reason, people also tried to protect themselves with plague costumes, especially physicians, priests, and attendants whose duties brought them into contact with plague victims (fig. 5.1). It was thought that dangerous atoms would not adhere to leather trousers and gowns made of waxed fabric. A wide-brimmed hat could defend the head, and a mask with a protruding beak extending from the nose could carry aromatic herbs that would protect the wearer from the fatal miasmatic smells. A doctor donning a plague suit was likely also to carry a rod—a secular counterpart of a verger's staff. This stick had a twofold purpose: to prod the people one encountered, reminding them to keep a safe distance, preferably downwind; and to search patients at a distance for buboes and other plague tokens in order to determine whether they should be taken to a pesthouse. To complete the outfit, some wearers carried braziers of burning coal to purify the air immediately surrounding them.

Although it was important to defend oneself from a menacing external environment where lethal miasma swirled, it was also vital to strengthen the body's inner defenses. Classical medical teaching, reinforced over the centuries by popular culture, taught that a body became susceptible to disease if its organic disposition, which determined the balance of the humors, was disturbed. In such perilous circumstances, it was important to avoid depleting emotions such as fear, distress, and melancholia; to eat and drink sparingly; to avoid excesses of exercise or sexual exertion; and to guard against sudden chills and drafts.

Apart from such means of self-protection sanctioned by medical teaching, superstitions also flourished during the plague years. There was a widespread popular belief, derived from astrology, that some metals and precious stones—rubies and diamonds, for example—possessed talismanic properties. Certain numbers also offered reassurance. The number four was particularly in vogue as it alluded to all of the main determinants of health that appeared in groups of four, as we have seen—the humors, the temperaments, the evangelists, the winds, the elements, and the seasons.

Ritual Cleansing and Violence

In early modern Europe the idea of cleansing also suggested ritualistic and sinister possibilities, especially when it was associated with notions of sin and divine punishment. The defilement of a city, in other words, could be moral as well as physical, and survival could depend on placating divine wrath rather than seeking naturalistic remedies. Anxious and vigilant communities often sought to identify and cast out those who were morally responsible for so overwhelming a calamity. The death-inducing sin could be the abuse of food and drink; excessive sleep or idleness; immoderate, unnatural, or sinful sex; or blasphemous religious practices and beliefs. Those guilty of offending God should be found and punished.

Figure 5.2. Plague was often understood as divine punishment. Jules-Élie Delaunay, in *The Plague in Rome* (1869), depicts an angry messenger of God directing plague incarnate into an offender's home. Minneapolis Institute of Arts (gift of Mr. and Mrs. Atherton Bean).

The painter Jules-Élie Delaunay illustrated such moral possibilities in his terrifying picture of 1869 titled *The Plague in Rome.* It depicts the messenger of an irate God indicating the door of a hardened sinner to an avenging angel, who is the plague incarnate. Thus directed, the plague is about to invade the offending home in order to destroy the sinners dwelling within (fig. 5.2).

With such an understanding of their moral as well as their physical jeopardy, people felt that it was prudent to assist in God's work of purging communities. Who, then, were the sinners? Suspicion frequently fell on prostitutes. In many places angry crowds rounded them up, expelled them forcibly from the city, and closed brothels. Jews were also repeatedly targeted amidst waves of anti-Semitic violence. Religious dissenters, foreigners, and witches were also attacked. All of them were guilty of offending God and bringing disaster on the faithful. In addition, lepers and beggars had already been marked—the former by disfigurement and the latter by poverty—as grievously sinful.

Thus towns across Europe closed themselves to outsiders during plague years while within their walls undesirables were hunted down, beaten, and cast out. In many places, people were stoned, lynched, and burned at the stake, and full-scale pogroms were also launched—what one might today call ethnic cleansing. Manichean thinking, which emphasized a dualistic struggle between good and evil, reinforced such tendencies because a prevailing hysteria held that wicked poisoners and anointers were abroad, acting not as the messengers of God but as the agents of a demonic plot. For those who held such a view, the plague could be halted only by finding and punishing culprits.

Two notorious examples convey the violent possibilities that the fear of plague could set in motion. The first occurred at Strasbourg in Alsace, on Valentine's Day 1349. There the municipal authorities held the two thousand Jews living in the city responsible for spreading pestilence by poisoning the wells where Christian citizens drew their water. Faced with the alternative of conversion or death, half abjured their religion; the remaining thousand were rounded up, taken to the Jewish cemetery, and burned alive. The city then passed an ordinance banning Jews from entering the city.

A second example occurred at Milan in 1630. These events are carefully reconstructed in two famous nineteenth-century works by Alessandro Manzoni (1785–1873): the epic plague novel *The Betrothed* (1827) and the historical work *The Column of Infamy*, first published as an appendix to the novel (1843). In 1630 Milan was at war with Spain when plague broke out in the city. The search was on for "plague-spreaders," and four hapless Spaniards were discovered and accused of mass murder. They were charged with smearing poisonous ointment on the doors of Milanese houses. Under torture, they confessed to the crime and were convicted. The sentence was to have their hands cut off, to be broken on the wheel, and then to be burned at the stake. At the site of the execution a column—the "column of infamy" of Manzoni's title—was erected to deter anyone else from attempting such an outrage. A plaque was also posted with an inscription in Latin stating what the crime had been and the severity of the punishment that had been meted out. It also proclaimed that no building should ever be constructed on the site.

Piety and Plague Cults

A less murderous response to the plague and its unbearable tensions was the attempt to propitiate the angry divinity by penance and self-abasement.

Scripture provided reassuring guidance. In the book of Jonah, the Assyrian city of Nineveh, notorious for its wickedness and dissolution, was foretold by the prophet to be condemned to destruction. When the inhabitants turned to contrition and amended their conduct, however, God was appeased and spared the city. Surely, then, if even Nineveh could escape annihilation, there would be hope for lesser sinners.

One method of penitence was the outdoor procession to a holy shrine amidst rogations and confessions. Among the earliest and most flamboyant of the processions were those of the Flagellants. They crisscrossed Europe during the early phases of the second pandemic until—condemned by both political and church authorities—the tradition faded into oblivion late in the fifteenth century. Pope Clement VI officially condemned Flagellantism in October 1349, and the University of Paris, the king of France, and—in time—the Inquisition followed his action.

As a collective movement, rather than individual practice, this paroxysm of asceticism first emerged in Italy during the thirteenth century. It then flourished with the Black Death, when it reached central Europe, France, the Iberian Peninsula, and the British Isles. To appease God and save Christendom, the Flagellants took a vow that for the duration of their pilgrimage they would not bathe, change clothes, or communicate with anyone of the opposite sex. Thus pledged, they set off two-by-two on treks lasting either forty days (to commemorate Christ's passion) or thirty-three (one for each year of Jesus's life). As they tramped, they whipped their backs with knotted leather thongs tipped with iron until the blood flowed, all the while chanting penitential verses. Some marchers bore heavy wooden crosses in memory of Christ; others beat their fellows as well as themselves; and many knelt periodically in public humiliation. Townspeople often welcomed the Flagellants' intervention as a means of bringing the plague epidemic to a halt.

On occasion Flagellants externalized their expiation by diverting their violence from their own bodies onto those of Jews they encountered or sought out. Jews were guilty in the minds of many not only of killing Jesus, but also of a conspiracy to annihilate Christendom with the plague.

More placid outbursts of piety were marked by the cult of those saints deemed to be most willing to intercede on behalf of suffering humankind—Saint Sebastian, Saint Roch, and the Virgin Mary. Most notable during the plague centuries was the new veneration of Saint Sebastian. He was a third-century soldier and Christian martyr who had been persecuted and put to death under Diocletian because of his religion. From the time of the early church he had been revered, but principally in Rome—the site of his death.

The plague initiated an intense pan-European veneration of Sebastian. Most meaningful for the plague centuries was the symbolism of his martyrdom, since tradition held that he was tied to a stake for his faith while archers pierced his body with arrows, which were a conventional symbol for plague. The subliminal meaning of the proliferating images of Saint Sebastian's riddled body was that, like Christ, he so loved humanity that he offered himself as a sacrificial atonement for sin. A human shield, Sebastian absorbed God's shafts, diverting the plague to himself. Emboldened by this act of love, the faithful beseeched the martyr through the likes of this widely recited prayer:

> O St Sebastian, guard and defend me, morning and evening, every minute of every hour, while I am still of sound mind; and, Martyr, diminish the strength of that vile illness called an epidemic which is threatening me. Protect and keep me and all my friends from this plague. We put our trust in God and St Mary, and in you, O holy Martyr. You . . . could, through God's power, halt this pestilence if you chose.[2]

As a result of such devotion, the scene of Sebastian tied to a post with arrows lodged in his flesh became a leading theme of Renaissance and Baroque painting and sculpture. Such images proliferated across the European continent, as nearly every artist of note depicted his agony (fig. 5.3). In addition, the devout wore medals and amulets bearing Sebastian's image. The symbolic meaning was clear. As bonds of community broke down during the pestilence, the example of a heroic martyr who had confronted death without flinching was comforting. Furthermore, since propitiation of God required a sacrifice that was perfect and without blemish, Sebastian was often depicted as a supremely handsome, athletic, and naked young man whose beauty enhanced his appeal.

The second of the great plague saints to become the object of a new devotion initiated by the Black Death was Saint Roch, who was known as the "pilgrim saint." Little about Roch's life is known. According to tradition, he was a noble from Montpellier in France who grew up as a devout and ascetic Christian. On reaching his maturity, he gave away all of his possessions and went on a pilgrimage to Rome as a mendicant. Shortly after his arrival in Italy, the Black Death erupted, and Roch devoted himself to attending sufferers. He contracted the disease at Piacenza but survived, and after convalescing he resumed his attentions to the sick and dying.

Figure 5.3. Gerrit van Honthorst, *Saint Sebastian* (ca. 1623). Saint Sebastian was venerated as the protector of plague victims. National Gallery, London.

Three of his leading attributes made his intercession especially desirable: his demonstrated love for his fellows, the fact that he had survived the plague, and his celebrated piety. Furthermore, the church attested to his willingness to perform miracles on behalf of those who invoked his aid. When an outbreak of plague menaced the Council of Constance in 1414, the prelates prayed to Saint Roch, and the disease ended. An outpouring of hagiographies, in both Latin and the vernacular, further enhanced his reputation and elaborated the legends of his life.

Like Saint Sebastian's, Roch's image rapidly appeared everywhere during the plague centuries—in paintings and sculptures and on medals, votive objects, and amulets. Churches were built in his name, and confraternities were dedicated to him. A Venetian confraternity performed a near-miracle by suddenly producing his body, which it deposited in the church bearing his name. Thus sheltering his relics, and commemorating his life in a series of paintings by Tintoretto, the Church of San Rocco (as he was known in Italian) became a central focus of pilgrimage and a powerful means of promoting the cult through the reports borne by visitors to the city.

Roch's image was immediately recognizable. He always appeared amidst unmistakable clues to his identity—a staff in one hand, a dog at his side, a pilgrim's hat on his head, and his free hand pointing to a bubo on his inner thigh. By his own example, he provided assurance that it was possible to recover from plague and that the righteous would tend to the afflicted. Roch also gave believers hope that he would intercede with God to bring an epidemic to a end and spare a community.

The third major plague cult was dedicated to the Virgin Mary. Unlike the cults of Sebastian and Roch, worship of Mary was not new. Her function for Christians had long been that of interceding with God on judgment day to temper his indignation with mercy. But the plague centuries gave a sudden urgency to her mission of intervening on behalf of a sinful and suffering humanity. Often, she is depicted beseeching the Lord on humanity's behalf, accompanied by Sebastian and Roch.

The Marian cult was especially important in Venice during the famous plague epidemic of 1629–1630, when as many as 46,000 Venetians perished in a population of 140,000. The disease arrived in the spring of 1629, and by the fall it gave no indication of abating despite the prayers of the faithful and processions bearing icons of Saint Roch and Saint Mark, the patron saint of the city. Equally unavailing was the invocation of all three plague saints by Bishop Giovanni Tiepolo, who ordered the exposure and benediction of the host in all the Venetian churches dedicated to their memory.

When these measures failed, the Venetian Doge and Senate turned to the Virgin Mary, who had always held a special place in the prayers of the republic. The commercially minded Venetian authorities offered a deal. If Mary spared the city by her grace, the Senate vowed to build a great church in her honor and to make an annual procession to the shrine in perpetuity. The result, as the pestilence at last receded, was the commission awarded in 1631 to the architect Baldassare Longhena to construct the monumental church of Santa Maria della Salute (Saint Mary of Health) on a spectacular site at the entrance to the Grand Canal. There the church dominates the urban landscape, reminds the people of their providential survival, and proclaims the Virgin's act of mercy in restoring Venice to health. Four hundred years later, the church of Santa Maria, whose dome alludes to Mary's heavenly crown, still receives the procession once a year.

Public Health Responses

One of the most important results of the second pandemic was the series of measures authorities adopted to prevent its return. These antiplague measures constituted the first form of institutionalized public health, and they were pioneered by the northern Italian city-states of Venice, Genoa, Milan, and Florence, which were exceptionally exposed by their geographical positions and had already experienced devastating epidemics. The Italian initiative was then imitated in France, Spain, and northern Europe. The measures that resulted demonstrated the influence of the heterodox idea of contagionism buttressed by the orthodoxy of the miasma theory; they also contributed to a first major advance in the conquest of disease—the exclusion of bubonic plague from western Europe.

The outline of the antiplague system was established during the early epidemic cycles of the second pandemic, and it then became increasingly sophisticated and comprehensive through the fifteenth and sixteenth centuries. Initially, the weakness of the system was that it was local in scope. The quantum leap that made the system effective and led to success was taken in the seventeenth and eighteenth centuries by the emerging early modern state, which backed the effort with bureaucratic and military power and extended coverage over a larger geographical area than one city alone.

Interestingly, authorities took action although there was no medical understanding of the mechanisms governing the disease they were facing. They acted in the dark, and in the process they took steps that were sometimes extreme, often wasted resources, and were frequently counterproductive. By the end of the eighteenth century, however, the path that they followed led to a first major victory in the war against epidemic disease.

Boards of Health

The first antiplague measure was the establishment of an institutional framework of officials capable of taking extraordinary action to defend the community throughout an emergency. Created under specially drafted "plague regulations," the new authorities were termed "health magistrates." They exercised full legislative, judicial, and executive powers in all matters relating to public health under the ancient precept *salus populi suprema lex esto* (the health of the people is the highest law). Originally, the health magistracies were temporary agencies, but by the end of the sixteenth century the cities in the vanguard of the war against the plague instituted permanent agencies, plague commissioners, and—as they were called ever more frequently—boards of health.

Lazarettos and Maritime Quarantine

Thus armed with nearly unlimited legal authority, the magistrates were charged with, as their primary goal, preserving the community from invasion by the pestilence, or, if it had already broken out, containing its further spread. One of the earliest to take action, the Office of Health in Venice, relied on three major institutions: quarantine, lazarettos, and sanitary cordons. The Venetian idea was that the republic could be insulated from invasion by sea. To that end, the Office of Health during the fifteenth century constructed two large-scale institutions—the Lazzaretto Vecchio and the Lazzaretto Nuovo—on outlying islands in the lagoon to which ships arriving from the eastern Mediterranean were directed. There, vessels from suspect areas were impounded to be scrubbed and fumigated. At the same time crew and passengers were taken ashore under guard and isolated. The cargo and the passengers' personal effects were unloaded, turned out in the sun, fumigated, and aired. Only at the end of forty days were the goods and passengers released to enter the city.

The period of confinement, termed "quarantine" after the Italian word *quaranta* (forty), constituted the core of the public health strategy. Its duration was based on Christian Scripture, as both the Old and New Testaments make multiple references to the number forty in the context of purification: the forty days and forty nights of the flood in Genesis, the forty years of the Israelites wandering in the wilderness, the forty days Moses spent on Mount Sinai before receiving the Ten Commandments, the forty days of Christ's temptation, the forty days Christ stayed with his disciples after his resurrection, and the forty days of Lent. With such religious sanction, the conviction held that forty days were sufficient to cleanse the hull of a ship, the bodies of its passengers and crew, and the cargo it carried. All pestilential vapors would be harmlessly dispersed, and the city would be spared. Meanwhile, the biblical resonance of quarantine would fortify compliance with the administrative rigor involved and would provide spiritual comfort for a terrified city.

Simple in principle, the enforcement of maritime quarantine presupposed effective state power. A lazaretto was a place such as Jarre Island at Marseille or Nisida Island at Naples where hundreds of unwilling passengers and crew had to be confined, provisioned, and isolated from all outside contacts. It also required a strong naval presence in order to compel recalcitrant and possibly terrified sea captains to anchor within its waters and to prevent attempts at escape or evasion. Furthermore, within the lazaretto complex and detailed protocols were needed to ensure that passengers at various stages of their confinement would be separated from each other and that all items unloaded from the ship would be aired and fumigated. Quarantine thus presupposed the economic, administrative, and military resources of the state.

Of course today, we know that the medical theories underlying the Venetian system were flawed: there were no pestilential miasmas to be dissipated, and many of the rituals of purification had no effect. But the idea of prolonged and militarily enforced isolation imposed on all vessels arriving from the East worked well in practice. Forty days exceeded the incubation period of bubonic plague and therefore provided sufficient time to guarantee that any person in good health released into the city was free of contagion. At the same time forty days were long enough to guarantee the death of infected fleas and of the plague bacteria, especially after exposure to the sun and the air. Thus an inaccurate theory combined with scriptural belief to produce effective public health procedures. The Venetian lazarettos, backed by the Venetian fleet, seemed to demonstrate in practice that they could protect the city, and its economy, from disaster.

Plague did breach Venice's sanitary ramparts twice after the construction of the lazarettos—in 1575 and again in 1630, when major epidemics struck. But the republic did enjoy prolonged periods of apparent immunity, and these—combined with the determination of other powers to do something in self-defense—promoted the Venetian containment strategy as the standard for public health in combatting plague. Other European ports—such as Marseille, Corfu, Valencia, Genoa, Naples, Amsterdam, and Rotterdam—imitated Venice and built their own lazarettos.

Often, lazarettos were temporary wooden structures built for the emergency or existing facilities that were requisitioned and repurposed. But others were permanent fortresses in continuous use. By the mid-sixteenth century ships from the Levant were regularly compelled to call at these facilities, as western Europe rose to the challenge of plague imported by sea. The disease continued to arrive, but it was largely contained with ever fewer major disasters. By the end of the seventeenth century, the Black Death had nearly ended in western Europe. Just two failures occurred after 1700. These demonstrated that, even in the eighteenth century, the defensive structures were occasionally porous.

In 1720 plague reached Marseille, and suspicion has long fallen on the merchant vessel *Grand Saint Antoine,* which carried precious fabrics from Smyrna and Tripoli, where the disease prevailed. Having lost eight sailors, a passenger, and the ship's doctor to plague at sea, the *fluyt* dropped anchor off Marseille on May 25 to undergo quarantine. Under pressure from local merchants, however, the health authorities released its cargo and crew on June 4 after a much abbreviated *petite quarantaine.* According to the traditional but recently contested account, pestilence spread by this means from the lazaretto to the city, where it killed sixty thousand of the one hundred thousand inhabitants of Marseille. A further fifty thousand people subsequently perished of plague in the hinterland of Provence and Languedoc.

Coming full circle, the second pandemic erupted for the final time in the West in 1743 in Sicily, where the Black Death had first broken out in 1347 at Messina. As at Marseille in 1720, a ship involved in trade with the Levant has long been deemed responsible for the disaster that ensued. Messina had no lazaretto, and the ship was allowed to dock in the city's unprotected harbor.

Land-Based Quarantine and Sanitary Cordons

If the sea presented significantly less danger after the introduction of antiplague measures, there was still the overland menace as people and goods were set in motion by trade, pilgrimage, and labor migration. Already from the days of the Black Death, communities motivated by fear more than by reasoned medical philosophy had resorted to self-help by forming vigilante parties that patrolled city walls during times of plague, turning away outsiders with the threat of violence. In later years this practice was regularized and made official with the deployment of troops around the perimeters of cities and towns. Their orders were to use their bayonets and rifle butts to repel anyone who approached, and, if necessary, to shoot.

These lines of troops, with sentry posts at regular intervals, were known as sanitary cordons (*cordons sanitaires*), and their use proliferated at municipal boundaries and national borders. Sanitary cordons were military barriers intended to protect a territory by halting all overland movement of goods, people, and therefore diseases until quarantine could demonstrate that they were medically safe. On occasion, as at Marseille in 1720, the church reinforced the physical barrier of the cordons with a spiritual edict by decreeing the excommunication of anyone crossing the lines by stealth.

The decisive and most visible cordon, however, was the one set in place by the Hapsburg Empire to confront the peril of the overland trade from

Turkey through the Balkans. The Austrian cordon, which operated continuously between 1710 and 1871, is perhaps the most impressive public health effort of the early modern period. It was a great, permanent line of troops stretching across the Balkan peninsula to implement a novel imperial instrument known as the Military Border. Reinforced in time of plague, the Austrian cordon stretched for a thousand miles from the Adriatic to the mountains of Transylvania. Punctuated with forts, lookout stations, sentry posts, and officially designated crossing points equipped with quarantine facilities, the Military Border was ten to twenty miles wide. Patrols moved between stations looking for fugitives.

Within the territory of the Military Border, all male peasants were conscripted for special duty manning the border so that the empire could field a force of up to 150,000 men without bearing the expense of a permanent deployment of regular soldiers. The peasant reserves did not need to be trained or equipped for combat but rather for police duty in an area they knew well. A graduated series of three levels of vigilance determined mobilization and preparedness. The prevailing level was set by an imperial intelligence service operated by diplomats and sanitary intelligence officers stationed in the Ottoman Empire, where they made direct observations, questioned travelers, and received informers. An emergency high alert signaled that troop numbers were to be increased and that the quarantine time should be nearly doubled, from twenty-eight to forty-eight days. While the emergency lasted, offenders such as smugglers and escapees from quarantine were tried and sentenced on the spot by military judges, with conviction entailing immediate death by firing squad. The cordon was finally dismantled amidst the complaints of liberals against its oppressive nature; of economists and farmers concerned with the impact of conscription on the agriculture of the border area; and of medical personnel, who pointed out that by the 1870s the plague had receded from the Turkish domains on the far side of the cordon. For a century and a half, however, one of the great powers of Europe assumed the onerous task of permanently isolating western and central Europe from the plague foci of the Ottoman Empire by land, just as the maritime lazaretto isolated them by sea.

Confronting the Danger Within

Troops, naval vessels, and the threat of excommunication protected a city from without, but the problem remained of what to do if, despite protective efforts, the plague burst out within it. Here the "plague regulations" across

Europe authorized sanitary authorities to undertake draconian measures of repression to confront the danger. The first task was to locate all victims of the disease. Given the enormous mortality of an epidemic of plague, the city was threatened with the danger of countless unburied corpses abandoned in houses and on the streets. Under the prevailing miasma theory, these decomposing bodies emitted a poisonous effluvium that caused the medical catastrophe, so their prompt removal and proper disposal were essential to public health. As part of their defensive measures, boards of health recruited a workforce of searchers, watchers, carters, and gravediggers who were identified by badges and a distinguishing sash. These municipal officials were charged with locating plague victims, whom they identified by their telltale tokens, or purple spots, and then transported to a lazaretto, which functioned as a plague hospital as well as a place of isolation and observation of suspect travelers. If the sufferers had already perished, carters were summoned to convey the corpses to plague burial grounds aboard death carts that lumbered through the streets with a bell clanging to warn passersby to make way.

All aspects of the lazarettos surrounded these institutions with an evil reputation. Pesthouses by definition were places to which many people were taken but from which few returned. Recent studies suggest that more than two-thirds of the patients confined at the Lazzaretto Vecchio and Lazzaretto Nuovo at Venice died on the premises. Confinement was therefore widely regarded as a death sentence to be served alone and forcibly cut off from friends and family.

As the death toll mounted within cities, the pesthouses were forced to resort to desperate expedients to cope with the ever-growing numbers of the dead. Often they dumped corpses unceremoniously into hastily dug pits where gravediggers stacked the bodies in layers or burned them on mass funeral pyres. A nighttime glow, thick smoke by day, and an overpowering odor made the pesthouse and its environs places of dread and horror. The rigid discipline within the establishment and the severe penalties inflicted on anyone who tried to escape reinforced the fear. For those who survived detention, the period of confinement in many places also brought financial ruin in its wake because patients were often charged for the cost of their maintenance during a prolonged residence. Alternatively, survivors faced increased taxation and special assessments as authorities sought to recoup the expenses incurred in combating the plague. Some pesthouses also carried a dishonoring stigma because they were used as places of punishment where the authorities relegated those whom they regarded as noncompliant with their regulations.

Visits by searchers and body-clearers to residents' homes magnified the terror. These minor functionaries practiced a notoriously dangerous occupation surrounded by popular hostility and the risk of contagion. Sometimes they fortified themselves with liquor and arrived at work foul-mouthed and abusive. Not a few thought of their positions as an investment and collected a return—by threatening healthy people with confinement, taking bribes from sufferers to be left with their families, plundering vacant homes, and relieving wealthy patients of their personal possessions.

In such a context, it is hardly surprising that numerous contemporary commentators wrote of lazarettos primarily as instruments of social control, set up along with prisons and workhouses to discipline and punish. Following their lead, historians have often taken the same approach. More recently, however, detailed studies of particular institutions have revealed that they were far more complex in their approach to plague, responding as religious and philanthropic institutions dedicated to providing care and promoting recovery. At Venice, for example, the city spared no expense in furnishing the lazarettos with medications and in hiring an array of healers—from surgeons and physicians to apothecaries, barbers, and attendants—under the direction of a prior.

Taking a holistic approach, the medical personnel tended to the spiritual and emotional needs of their patients, because such strong passions as fear and anger counted as "nonnaturals" that would affect a patient's humoral balance and undermine the possibility of recovery. To foster calm and confidence, the prior hired members of the clergy to minister to the sufferers and took steps to ensure, as far as possible in the midst of mass death, that an atmosphere of order prevailed within the walls of the institution. Some of the clergy acquired universal recognition, such as Carlo Borromeo, the archbishop of Milan. During a time of famine in 1578 that was followed by pestilence, Borromeo organized the archdiocese to provide grain for thousands of people and to tend to patients afflicted by the plague. Partly for these efforts, the Catholic Church canonized him in 1610.

The prior promoted a sense of order through other means as well. One was diligent recordkeeping so that patients and their possessions were carefully tracked and fraud did not occur in provisioning. At Venice a married prior was appointed and assisted by his wife, who had the title of prioress. The prior oversaw the overall care of adults, while the prioress supervised the attentions given to children. This helped to create a reassuring sense that the lazaretto was a great family.

Therapeutic strategy throughout the second pandemic was consistent with the dominant humoral understanding of the disease. In humoral terms, plague was caused by an excess of blood, the hot and wet humor. The leading therapy therefore was bloodletting to help the body expel the poison responsible for the disease. Since the patient's body was already attempting to expel the poison through vomit, diarrhea, and sweat, surgeons and barbers could assist nature through phlebotomy, although there were intense debates about timing, the best veins to open, and the amount of blood to draw.

In addition to venesection, the depletive strategy involved the administration of powerful emetics and purgatives so that the poison could pour forth more copiously. Plague doctors also promoted perspiration by piling covers on their patients' already fevered bodies and placing swine bladders filled with hot water under their armpits and against the soles of their feet. For the same reason a common practice was to lance, cauterize, or dry-cup buboes and to apply hot compresses to them in the hope that they would burst and discharge the excessive humor that caused disease.

Apart from depletion, plague doctors had recourse to internal medicines. The most famous was theriac, the nearest equivalent to a panacea of the period. Theriac was a complex concoction of various ingredients, including opium, cinnamon, gum arabic, agarics, iris, lavender, rapeseed, fennel, and juniper—all pulverized and mixed with honey and, ideally, the flesh of a viper. The preparation was then allowed to ferment and age. Since it was held to be an antidote to poison of every kind, theriac was indicated in the treatment of any malady triggered by a corrupted humor. It was therefore the remedy of choice for plague, but its complicated and lengthy preparation time made it expensive, scarce, and available only to the wealthy. Its use was most widespread at Venice, which was the center for its production and trade.

Other curative regimens based on orthodox humoral principles were more widely deployed. Sometimes the strategy of treating through opposites was used, such that corrupted blood, which was hot and wet, would be treated with medications that were cold and dry in an effort to restore the disturbed balance. An array of remedies and wide variety of approaches and ingredients were used, including endive, figwort, burdock, rose, chamomile, daffodil, rhubarb, ground pearl, flax, and vinegar. Ointments and poultices containing active ingredients such as these were frequently applied externally to draw off the poison from buboes and carbuncles.

Unfortunately, the therapeutic strategies of early modern plague doctors did little to prolong life, relieve suffering, or effect a cure. Indeed, the CFR in plague hospitals across Europe was normally in the range of 60 to 70 percent, which is comparable to the rate for untreated cases. On the other hand, mortality in lazarettos was not simply a reflection of their therapeutic strategies. Plague hospitals suffered the disadvantage of admitting large numbers of patients at an advanced stage of illness, and frequently the sheer numbers of victims overwhelmed the staff. On many occasions patients died or recovered without receiving medical attention or treatment of any kind.

Nevertheless, lazarettos were not similar to medieval leper colonies, which, as we have mentioned, functioned as death houses with no curative purpose. Plague hospitals, founded on religious and charitable intentions, attempted to provide the limited available therapies in the most difficult circumstances. The care administered in plague hospitals may have provided critically ill patients with a measure of hope and soothed their anxiety—with the proviso that they varied in their quality, staffing, and organizational efficiency. Inevitably, permanent lazarettos functioned more efficiently than hastily improvised facilities opened temporarily in the midst of an epidemic.

Everywhere, major epidemics caught authorities unprepared, leading to confusion, chaos, and improvisation. Even the best-run lazarettos lacked the capacity to cope with the sudden caseload of a plague emergency. In the face of overwhelming numbers of sick and dying patients, boards of health resorted frequently to the expedient of attempting prophylaxis alone and abandoning the hope of providing treatment. One widely adopted approach was to isolate patients, suspected patients, and their families in their homes. Searchers then marked the houses of plague victims with a cross daubed in red, sealed the premises, and posted a guard outside to thwart any attempt to enter or exit the sick house. This draconian measure condemned families and tenants to forced confinement with the ill, the dying, and the dead. In these circumstances there was no possibility of receiving further provisions or medical care.

Stringent restrictions applied also to dead bodies, which were thought to emit lethal miasmas, so it was critical to dispose of them as rapidly as possible while minimizing further risk to the living. Antiplague regulations therefore banned funerals, processions, and the final attentions of laying out and viewing the body. Instead, just as at plague hospitals, corpses found in the city were piled in common pits dug in unconsecrated ground. Before being completely buried, the top layer of bodies was covered with a thin layer of earth and caustic lye to promote disintegration and to prevent foul-smelling vapors from contaminating the air.

These regulations destroyed community bonds, since people lost all possibility of grieving, paying final respects, and assembling to fill the emotional hole in their midst. Furthermore, since the only certain indication of death in the early modern world was putrefaction, the ban on the laying out of the body along with hasty interment raised the fear of premature burial. Also, the rigor of public health rules prevented the administration of the last rites of the Catholic Church; this, combined with the fact that plague pits were located in unconsecrated ground, gave rise to deep anxieties about the future of the soul in the afterlife.

Finally, plague regulations imposed a series of miscellaneous prohibitions and obligations that varied from city to city. In Barcelona, for instance, authorities decreed that pet owners destroy their cats and dogs, that citizens sweep and wash the streets in front of their homes, and that believers fully confess their sins. At the same time citizens were forbidden to throw rags into the streets, to sell clothing of any kind, or to gather in public assemblies. Trades that produced noxious odors that could poison the atmosphere were carefully controlled. Tanning, for instance, was banned outright in many cities, and butchers faced many strictures. For instance, they were prohibited from hanging meat, keeping meat from more than one animal on a counter, allowing mud or dung in their shops, slaughtering animals that had not been vetted and approved, keeping flayed carcasses in a stable, and selling the flesh of animals after the day of their slaughtering. Those who failed to follow such rules faced severe penalties.

Thus a city besieged by a major plague epidemic became a perfect dystopia. Bonds of community and family ties were severed. Religious congregations found their churches bolted, sacraments unavailable, and bells silent. Meanwhile, economic activity halted, shops closed, and employment ceased, increasing the threat of hunger and economic ruin. The political and administrative practices of normal life did not survive as authorities fell seriously ill, died, or fled. Worst of all, above every other concern towered the menace of sudden and painful death made vivid by the stench in the streets and the dying sufferers who often lived out their final anguish in public.

An Assessment

To what extent were the antiplague measures responsible for the victory over plague in western and central Europe after the disease receded in the eighteenth century and never returned? A definitive answer is impossible, and considerable debate surrounds the topic. It is important to take into account the fact that the plague defenses created some negative results. Because they were so stringent and created so much fear, they frequently provoked evasion, resistance, and riot. By causing people to conceal cases, to evade the authorities, and to resist, the measures—or even the mere rumor that they would be imposed—at times even had the effect of spreading the disease farther afield. By encouraging populations to hide the sick in their midst, they also deprived the authorities of accurate and timely information regarding the emergency they faced.

Bombay (now Mumbai) provides a perfect illustration of the negative possibilities that could result from invoking traditional plague regulations. The disease struck the western capital of India and its more than eight hundred thousand inhabitants in September 1896. By December, more than half the population had deserted the city. Municipal authorities carefully noted that what most powerfully drove the evacuation was not fear of the plague itself but fear of the martial means deployed to repel it. The final tally of gains and losses achieved by the antiplague measures was therefore contradictory. They were, in effect, a blunt sledgehammer rather than a surgical instrument of precision. The plague regulations, the city council insisted, helped Bombay to survive the plague, but the fleeing residents carried the disease with them, thereby spreading it and turning the port into a danger not only to the rest of the subcontinent, but also, in the era of rapid travel by steamship, to seaports everywhere.

Apart from the effects of antiplague measures, other factors were undoubtedly at work that played significant roles in the recession of the plague. One influence was what might be termed "species sanitation." In the early eighteenth century the brown rat, or Norway rat (*Rattus norvegicus*), arrived in Europe as an outside invader from the East. Large, fierce, and prodigiously prolific, it rapidly drove out the native black rat (*Rattus rattus*) from its ecological niche and extirpated it. Finding ample food supplies and no natural enemies, the brown rat overran every country it reached. Readily finding accommodation aboard ships, it also progressively extended its range around the world. For the history of plague, the diffusion of the brown rat is significant because, unlike the black rat, it has a furtive temperament and scurries rapidly away from humans. Thus distancing itself from people, the brown rat is considerably less efficient as a vector of disease. Indeed, the difference in the human response to the two species played a role as well, as reports on the third pandemic in India made clear. There, the more furry, more familiar, and more friendly black rat was regarded by many as a domestic pet. It was common for people to tame, feed, and play with the rodent—with tragic consequences as the unfolding of recurring public health disasters demonstrated. By contrast, the aggressive and antisocial brown rat attracted no such affection or intimacy. Thus, as colonies of *Rattus norvegicus* violently drove out *Rattus rattus,* the species barrier between rodent and humans became more difficult for plague bacteria to cross. It is almost certainly no coincidence that the arrival of the brown rat in central and western Europe coincided chronologically with the recession of plague, and that one of the factors determining the geography of the major outbreaks during the third pandemic was the demarcation of the zones where *Rattus rattus* still prevailed, including the global epicenter of the pestilence, the Bombay Presidency in India.

A second influence is climate. The time when plague began to recede in the seventeenth century was also the era of the "Little Ice Age" when winter temperatures across Europe plunged. Dutch painters of the period, beginning with Hendrick Avercamp and continuing with Pieter Bruegel the Elder and Pieter Bruegel the Younger, even created a genre of winter paintings with depictions of deep snow and skaters enjoying the frozen canals of Amsterdam and Rotterdam. In England, the Thames regularly froze during the seventeenth century and supported not only skaters, but also ice festivals and frost fairs. Even the Baltic Sea froze so that it was possible to travel by sled from Poland to Sweden. The Little Ice Age lasted in northern Europe from 1350 to 1850, but it had three troughs of cold temperatures, the second and deepest of which began in 1650 and therefore coincided with the end of the plague in northern Europe. The new cold reduced fleas and *Y. pestis* to extended periods of inactivity. The fact that the third pandemic began after the end of the Little Ice Age and in synchrony with renewed warming after 1850 is suggestive.

A third possibility is changes in hygiene. Improved housing helped prevent rat infestation. One example is the declining use of the thatched roof and its replacement by tiles, which were less welcoming to nesting rodents. Replacing mud floors with concrete distanced humans from the rodent warrens beneath, and locating grain storage facilities at sites remote from human habitations also kept rats and humans farther apart. The thinning out of population density in urban dwellings also diminished the sharing of vermin. Lower occupation densities made homes, beds, and the flesh of their occupants less suitable as sites for flea infestations. Improved personal hygiene also contributed. The development and use of the soap bar along with the practice of bathing in the eighteenth century radically reduced the prevalence of fleas and lice as human ectoparasites.

There was even scope for hygienic myths to generate real-life effects. For instance, it was widely believed that the Great Fire of London in September 1666 had purified the British capital, thereby serendipitously extinguishing the plague of the previous year and permitting the rebuilding of urban rookeries with more spacious constructions that presented marked advantages in terms of solidity, ventilation, light, and sanitation. Two centuries later in British India, this way of remembering the Great Fire served as an inspiration for the antiplague program of burning pestilence out of its sanctuaries in the filthy tenements known as *chawls* in Bombay and the unsanitary huts called *bastis* in Calcutta. The inhabitants of these dwellings were evicted, and the buildings torched to follow the London precedent and preserve the cities from plague by means of filth-devouring flames and reconstruction.

Finally, there are hypothetical possibilities. Did mutations to *Y. pestis* occur that enhanced the resistance of rodents or fleas? Were ecological factors at work in sylvatic plague reservoirs that affected the movements of burrowing rodents? Did improvements in human nutrition provide enhanced resistance? What changes occurred in the distribution of the dangerous rat flea *Xenopsylla cheopis* relative to other flea species that are inefficient vectors of the plague bacterium?

Whatever the role of these various influences, there seems little reason to doubt that the power of the early modern state to impose rigorous measures of segregation through quarantine played a significant and perhaps decisive role in the passing of the second pandemic. The antiplague measures were also enormously influential because they seemed to be effective and to provide solid defensive bulwarks against the disease. The reaction of political and public health authorities in later centuries is therefore intelligible. When new, virulent, and poorly understood epidemic diseases emerged, such as cholera and HIV/AIDS, the first reaction was to turn to the same defenses that appeared to have worked so effectively against plague. It was unfortunate that antiplague measures, however successfully deployed against bubonic plague, proved to be useless or even counterproductive when used against infections with profoundly different modes of transmission. In this manner the plague regulations established a style of public health that remained a permanent temptation, partly because they were thought to have worked in the past and because, in a time of uncertainty and fear, they provided the reassuring sense of being able to do something. In addition, they conferred upon authorities the legitimating appearance of acting resolutely, knowledgably, and in accord with precedent. We shall see the results of this temptation in later chapters.

Plague regulations also cast a long shadow over political history. They marked a vast extension of state power into spheres of human life that had never before been subject to political authority. One reason for the temptation in later periods to resort to plague regulations was precisely that they provided justification for the extension of power, whether invoked against plague or, later, against cholera and other diseases. They justified control over the economy and the movement of people; they authorized surveillance and forcible detention; and they sanctioned the invasion of homes and the extinction of civil liberties. With the unanswerable argument of a public health emergency, this extension of power was welcomed by the church and by powerful political and medical voices. The campaign against plague marked a moment in the emergence of absolutism, and more generally, it promoted an accretion of the power and legitimation of the modern state.

CHAPTER 6
Smallpox before Edward Jenner

Comparing Epidemics

It is now time to look for comparative purposes at a second high-impact epidemic disease—smallpox. Why smallpox? And why cover it at this stage in our study? The answer to this second question is that smallpox followed plague as the most dreaded disease of eighteenth-century Europe. The answer to the first is that smallpox is a very different disease from plague, and one goal of this book is to examine the influence of different types of infectious diseases. Each is a distinctive and special case with its own mechanisms and its own historical impact; therefore, it is important to think comparatively.

It is crucial, however, to approach the issue of comparison in a systematic and explicit manner. For readers for whom the study of epidemic diseases is new, the aim is to specify the chief variables that govern the nature and extent of the effect that a particular infectious disease has on society. To that end, it is useful to pose a set of questions to guide the student of history.

Let us be clear from the outset that this list of questions is not canonical, and I do not suggest that these are the only questions that need to be asked. It is simply a starting point for orientation as we encounter each new disease, and it is hoped that a cascade of further issues will arise.

1. *What is the causative pathogen of the disease?*
   Bubonic plague, as we know, was caused by the bacterium *Yersinia pestis*. As we move forward we will encounter three

different categories of microbial pathogens: bacteria, viruses, and plasmodia. In a medical course on infectious diseases we would also have to consider prions, which cause such diseases as "mad cow disease" and kuru; but here, the diseases we consider are all bacterial, viral, or plasmodial.

2. *What is the total mortality and morbidity of the epidemic?*
"Mortality" indicates the total number of deaths, and "morbidity" the total number of cases, both fatal and nonfatal. Clearly, the numbers of deaths and cases are one of the measures of the impact of an epidemic. These statistics, for example, provide some support for the argument that Spanish influenza, which caused perhaps 50 million deaths in 1918–1919, was a more significant event than the 1995 outbreak of Ebola at Kikwit in the Democratic Republic of the Congo. For all of its international high drama at the time, the Ebola epidemic caused "only" 250 deaths and 315 cases, and it has left a slender legacy.

   On the other hand, mortality and morbidity figures are no more than a first and coarse assessment of historical significance, which can be established only by detailed case-by-case analysis that is as much qualitative as quantitative. It would be morally desensitizing and historically wrong-headed, for instance, to conclude a priori that by the measure of total mortality only major catastrophes such as those created by bubonic plague and Spanish influenza are important events. Indeed, one can argue forcefully that epidemics on a far smaller scale, such as outbreaks of Asiatic cholera in which "only" a few thousand perished, were decisive occurrences that cast a long historical shadow. There is no simple "quick fix" to the problem of weighing historical influence. But morbidity and mortality count in the scale and need to be considered.

3. *What is the case fatality rate (CFR) of the disease?*
This question addresses the virulence of the pathogen. CFR is determined by mortality as a proportion of morbidity—or, to put it another way, the percentage of cases that end in death. CFR is therefore the "kill rate" of a disease. One of the reasons that bubonic plague caused such terror and disruption was its exceptionally high CFR, which varied from 50 to 80 percent. At the other extreme, the great Spanish influenza that accompanied the end of the First World War gave rise

to an unparalleled morbidity, but it had a low CFR. This difference is important in assessing the variation in the popular responses to the two diseases.

4. *What is the nature of the symptoms of the disease?*
Symptoms that are particularly painful or degrading in terms of the norms of the society that experiences the affliction—such as those associated with plague, smallpox, and cholera, for instance—can contribute to how the disease is experienced and assessed. Smallpox, for example, when it did not kill, maimed, disfigured, and frequently blinded its sufferers. Tuberculosis, by contrast, was agonizing, but it was held by society to make its victims more intelligent, more romantic, and sexually more alluring. This is an important factor in helping to explain the paradox that pulmonary tuberculosis, the chief killer in nineteenth-century Europe, caused so little terror, while cholera, which had a limited demographic effect, was the most feared disease of the century.

5. *Is the disease new, or is it familiar to a population?*
Familiar diseases tend to be less terrifying than the sudden arrival of an unknown invader. Furthermore, a population is likely to have some degree of immunity against a recurring affliction, and the disease may have adapted itself to its human host by mutating and becoming less deadly.

   As examples, one can think of the so-called childhood diseases such as mumps and measles. Relatively mild afflictions in European society, where they were endemic, these diseases gave rise to devastating catastrophes when they were first introduced to new populations. This is the phenomenon of "virgin soil epidemics" such as the terrifying die-offs of native peoples in the Americas and Maoris in New Zealand when they first encountered measles and smallpox. For these reasons "emerging diseases" such as cholera and HIV/AIDS have tended to unleash more terror than the familiar afflictions of malaria and influenza.

6. *What is the age profile of the victims—does it affect the young and the elderly or people in the prime of life?*
This variable helps determine whether an epidemic is regarded as "natural" in its passage through a population or whether it seems particularly menacing because it magnifies

its effect by multiplying the numbers of orphans and widows it leaves in its wake. Cholera was especially terrifying because it seemed to target the people who were the economic mainstays of families and communities.

7. *What is the class profile of the sufferers, that is, is it a disease of the impoverished and the marginal, or is it a "democratic" affliction?*
   We have seen that a feature of the second pandemic of bubonic plague was that it was a universal affliction—a factor that helped to make its social history very distinct from that of cholera, which was clearly marked as a disease of poverty. Such associations generate class and social tensions that a more universal ailment such as influenza fails to unleash. Similarly, the early history of AIDS in the United States reflected its apparent predilection for male homosexuals rather than the general population.
8. *What is the mode of transmission of the disease—through person-to-person contact? contaminated food and water? vectors? sexual contact? droplets in the air?*
   The salience of this question is clearly demonstrated by the examples of syphilis and AIDS. As everyone knows, their impact and societal ramifications are intelligible only on the basis of the fact that they are sexually transmitted. Scapegoating and stigmatizing are much less likely to occur with regard to an epidemic transmitted through the air, such as influenza.
9. *How rapidly does the disease normally progress in its course through the human body—that is, is its course slow and wasting or fulminant (coming on quickly and strongly)?*
   Tuberculosis, syphilis, and AIDS are normally afflictions that endure, whereas plague, cholera, and influenza lead to rapid resolutions, whether by death or through recovery. Alternatively, malaria can assume either course. Again, the simple measurement of time is no magic key to unlocking the secret of historical significance, but it has a place and its weight on the scales.
10. *How is the disease understood or "socially constructed" by the population through which it passes?*
    As we have seen with plague, it makes a major difference whether a disease is viewed by contemporaries as divine

punishment, as the work of malevolent poisoners, or as a purely biological, naturalistic event. The ideas in the minds of a population, of public health authorities, and of medical professionals have a powerful influence on the unfolding of an epidemic emergency.

11. *What is the typical duration of an epidemic caused by the disease in question?*
Epidemic diseases exhibit enormous diversity in this regard. An influenza epidemic typically lasts for a number of weeks in a given locality; cholera and plague epidemics persist for months; and tuberculosis is an epidemic in slow motion, such that a single visitation endures for generations or even centuries and raises the issue of the distinction between epidemic and endemic diseases.

A Viral Disease

With our list of questions in the background, we are ready to deal with smallpox, the "speckled monster." This disease is caused by a virus that belongs to a group known as *Orthopoxviruses,* which include the variola virus—variola major and variola minor—that causes smallpox. Variola major, which was first seen under the microscope in the early twentieth century, causes "classical smallpox," the disease that is of most concern to historians. (Variola minor causes a far less virulent form of smallpox, and its societal impact has been limited, so it will have no further part in our discussion.)

Another *Orthopoxvirus,* cowpox, is distinct from smallpox and is primarily a minor, self-limiting disease of cattle. In humans it produces mild, flulike symptoms, but it is important historically because, as Edward Jenner discovered at the end of the eighteenth century, it provides its sufferers with a robust and lasting crossover immunity to smallpox. Cowpox, therefore, was key in the development of the public health policy of vaccination that Jenner pioneered (see Chapter 7).

Since smallpox is our first viral disease, we need to clarify our terminology and to note a biological distinction. "Microbe" is a general term for microscopic organisms, and it includes both bacteria such as *Y. pestis* and viruses such as variola major. Bacteria are unicellular organisms that are definitely and unequivocally forms of life. They contain DNA, plus all of the cellular machinery necessary to read the DNA code and to produce the proteins required for life and reproduction.

Viruses are completely different, but there is a possible source of confusion lurking for the student of medical history. The word "virus" is ancient. In the humoral system, diseases, as we know, were seen to arise as a result of assaults on the body from outside. One of the major environmental factors leading to disease was believed to be corrupted air, or miasma. But another important cause was a poison that was no more clearly identified than "miasma" and was termed a "virus." So when bacteria were first discovered in the late nineteenth century, they were commonly considered to be a form of virus. When a distinct category of microbes was discovered—the one that modern-day usage terms "viruses"—they were at first designated as "filterable viruses," meaning that they could pass through filters that were fine enough to retain bacteria.

For the remainder of this book, the term "virus" will be reserved for these tiny filterable microbes that are parasitic particles some five hundred times smaller than bacteria. Elegant scientific experiments established their existence by 1903, but they were not actually seen until the invention of the electron microscope in the 1930s, and their biological functioning was not understood until the DNA revolution of the 1950s.

Viruses consist of some of the elements of life stripped to the most basic. A virus is nothing more than a piece of genetic material wrapped in a protein case or, in the definition of the Nobel Prize–winning biologist Peter Medawar, "a piece of nucleic acid surrounded by bad news."[1] Viruses are not living cells but cell particles that are inert on their own. They contain a small number of genes. A smallpox virus, for instance, may contain two hundred to four hundred genes, as opposed to the twenty thousand to twenty-five thousand genes found in a human being. Thus streamlined, viruses lack the machinery to read DNA, to make proteins, or to carry out metabolic processes. They can do nothing in isolation, and they cannot reproduce.

Viruses survive as microparasites that invade living cells. Once inside a cell, they shed their protein envelope and release their nucleic acids into the cell. The genetic code of the virus (and the virus is almost nothing more) hijacks the machinery of the cell, giving it the message to produce more viral progeny. In this way the virus transforms living cells into virus-producing factories. In the process the viruses destroy the host cells they have invaded. The newly formed mature viral particles then exit the host cells and move on to attack and invade new cells. As viruses produce ever more viruses and destroy more and more cells, the effect on the body can be severe or even catastrophic, depending on the capacity of the immune system to contain or destroy the invasion. Here we have, in a sense, the opposite of the Hippocratic or Galenic idea of disease, with the body assaulted not from outside, but from a parasitic pathogen deep within. Some of the most virulent and historically significant diseases—such as smallpox, measles, rabies, yellow fever, polio, influenza, and HIV/AIDS—are viral.

The functioning of viruses has given rise to an exotic debate over the issue of whether they are living organisms. Those who argue that viruses are alive note that they are capable of transmitting genetic material, which is one of the definitional indications of life. But those who claim that they are not living note that, on their own, viruses are incapable of carrying out any metabolic processes. In this sense, they are the ultimate parasites. Perhaps the decision about whether viruses are alive is a matter of disciplinary perspective or personal preference. The reassuring point is that, except perhaps theologically, the answer hardly matters.

With regard to variola major, a critical point is that it affects only human beings. This factor is vitally important to the vaccination campaign that ultimately led to smallpox being declared eradicated in 1980—the first and still the only example of the intentional eradication of a human disease. One reason that made eradication possible was that the disease had no animal reservoirs that could reignite the disease whenever variola major found the means to cross the species barrier between humans and animals.

Transmission

Smallpox is highly contagious. People who have it shed millions of infective viruses into their immediate surroundings from the smallpox rash or from open sores in their throats. A patient is infective from just before the onset of the rash until the very last scab falls off several weeks later. Of course not everyone who is exposed is infected. Leaving aside people with immunity, one can estimate that the probability is about fifty-fifty that a susceptible member of a household will contract the disease from a sufferer in the home.

The disease is transmitted in three ways. The first is by means of droplets exhaled, coughed, or sneezed by a sick person and then inhaled by someone in close contact to the sick person. Smallpox is normally spread in a context of intense and close contact over a period of time—as in a family setting, a hospital ward, an enclosed workplace, a school classroom, an army barracks, or a refugee camp—especially during cool, dry winter months. As CDC director William Foege carefully observed during the smallpox eradication effort in the 1970s, a major vulnerability of the virus is that it is normally transmitted only within what one might call a three-foot "radius of infection" encircling a patient. Within this perimeter, moreover, the virus is both fragile and fastidious. It is fragile in the sense that it cannot survive for an extended period in the environment, and it is fastidious in that it infects only human beings and has no other animal reservoirs. Within the transmission perimeter, however, smallpox is capable of "vertical" transmission: an infected woman can transmit it transplacentally to her unborn child, who is then born with "congenital smallpox."

The only possibility of wider diffusion is by means of the physical movement of contaminated inanimate objects known as "fomites" that carry the virus across the patient's perimeter. Typically, these objects include bed linens, eating utensils, and clothing. If these materials contain scabs from the patient's lesions, variola virus can survive on them for two to four months, depending on temperature and humidity.

For a disease spread in these ways, a variety of societal factors can create favorable conditions for its propagation. These include whatever circumstances cause large numbers of people to assemble in crowded spaces, such as urbanization, congested housing, crowded workplaces, and warfare. Western European cities during the eighteenth and nineteenth centuries provided such conditions in abundance under the pressures of industrial development, mass migration to cities, laissez-faire capitalism, warfare, and colonization. In the leading European cities smallpox was above all a disease of childhood, such that about a third of all childhood deaths in the seventeenth century were due to this single cause.

Symptoms

A significant aspect of the dread of smallpox was the gruesomeness of its course in the body and the lifelong scarring and disfigurement that resulted. These features were as important as the fear of death itself in spreading terror. The very word "smallpox" has a strong and disturbing resonance in popular imagination even today.

The question may arise of which of the diseases that we encounter in this book is the most painful for the patient. Fortunately, this question does not allow for empirical verification because no one has ever suffered all of the afflictions that we consider. On the other hand, it is meaningful to register the impression of those who lived through the times when smallpox claimed its legions of victims. The physicians who treated smallpox patients were convinced that this was "the worst of human maladies," and one doctor wrote in 1983 that "in the suddenness and unpredictability of its attack, the grotesque torture of its victims, the brutality of its lethal or disfiguring outcome, and the terror that it inspired, smallpox is unique among human diseases."[2]

Here is one of the reasons that smallpox appeals as a weapon of bioterror. It is well known that a major outbreak of the disease would spread death, maximize suffering, and lead to widespread fear, flight, and social disruption. Thus the symptoms of the disease are an integral part of its history and social impact. From a historical standpoint, too, a familiarity with its symptoms is important in order to appreciate its distinctiveness and its lasting imprint.

Preeruptive Stage

The incubation period for smallpox normally lasted for approximately twelve days, during which there were no symptoms. This in itself was an important fact of epidemiology because it facilitated the spread of infection. Infected people had ample time to travel and to contact others before falling ill themselves. Symptoms began with a "viral shower" as the pathogen was released into the bloodstream and spread systemically, localizing eventually in the blood vessels of the dermis just below the superficial layers of the skin. The viral load released and the efficiency of the body's immune response determined the severity and course of the illness.

At the onset of symptoms, the fever rose to 101–102°F. The patient experienced a sudden malaise, and with it began a month of exquisite suffering and the danger of contagion. Early symptoms also included nausea, severe backache, and a splitting frontal headache. Children frequently experienced convulsions.

In some cases, the infection was so overwhelming that death occurred within thirty-six hours with no outward manifestations, although postmortem examination revealed hemorrhages in the respiratory tract, the alimentary tract, or the heart muscles. One description of such a case of "fulminating smallpox" notes that

> after three to four days the patient has the general aspect of one who has passed through a long and exhausting struggle. His face has lost its expression and is mask-like, and there is a want of tone in all the muscles. When he speaks, this condition becomes more apparent. He speaks with evident effort and his voice is low and monotonous. He is listless and indifferent to his surroundings.

> The mental attitude is similar. There is loss of tension, showing itself in a lengthening of the reaction time and a defective control. In the most fulminant cases the aspect of the patient resembles that of one suffering from severe shock and loss of blood. The face is drawn and pallid. The respiration is sighing or even gasping. The patient tosses about continually and cries out at frequent intervals. His attention is fixed with difficulty and he complains only of agonizing pain, now in his chest, then in his back, his head, or his abdomen.[3]

Eruptive Stage

Although an early fulminant death was possible, the patient normally survived to enter the "eruptive stage," which exhibited the classic symptoms of smallpox as a disease (fig. 6.1). On the third day after onset, patients usually felt considerably better, and in mild cases they frequently returned to normal activities—with unfortunate epidemiological effects.

Concurrently with the apparent recovery, a rash appeared with small round or oval lesions known as "macules" that were rose-colored and up to one-quarter inch in diameter. These appeared first on the tongue and palate, but within twenty-four hours the rash spread to cover the entire body—even the palms of the hands and the soles of the feet. On the cheeks and forehead the appearance was that of severe sunburn, and indeed the sensation patients felt was one of scalding or intense burning.

On the second day of the rash the lesions began to alter in their appearance. The centers of the macules became hard, and they gradually rose into structures called "papules" with a flattened or sometimes indented apex. The papules were said to feel to the touch like buckshot embedded in the skin.

On the third day the lesions, now called "vesicles," grew in size. They also changed in color from red to purple, and they were no longer solid but filled with fluid, like a blister. This process of "vesiculation" took approximately three days and lasted a further three. At this stage, with the disease presenting its characteristic and distinctive appearance, the physical diagnosis of smallpox became certain. Patients experienced ever-increasing difficulty in swallowing and talking because of extensive and painful vesicles in the mucous membranes of the palate and the throat.

By the sixth day of the rash the vesicles began to fill with yellow pus, which caused the indentation to disappear as the pockmarks became globular

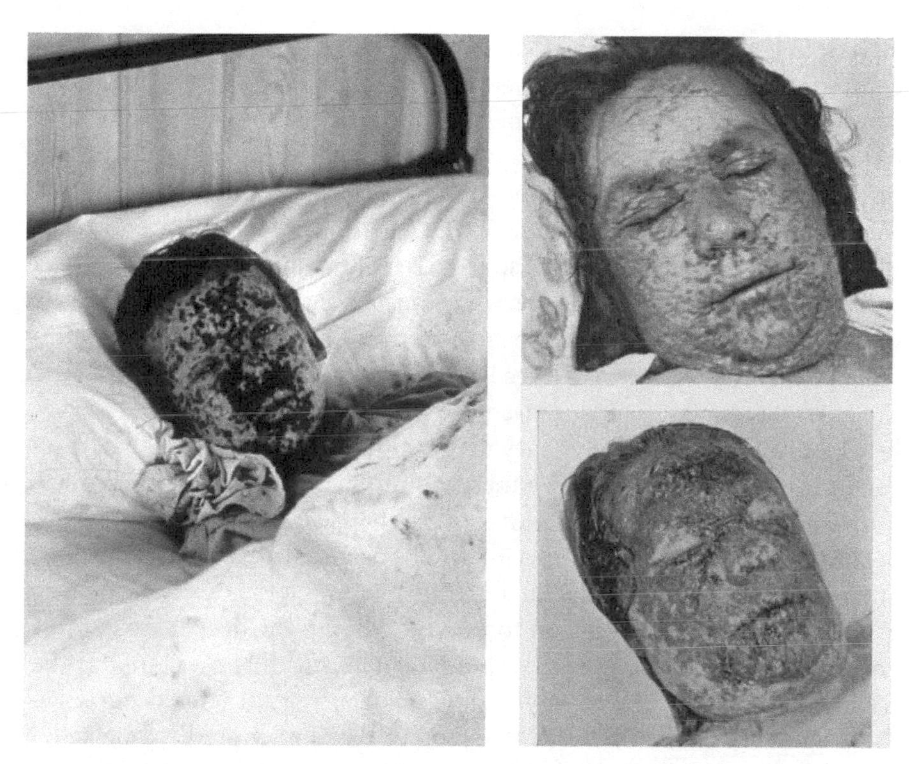

Figure 6.1. (left) Gloucester smallpox epidemic, 1896—J. R. Evans, age ten years, a smallpox patient (Photograph by H.C.F., 1896. Wellcome Collection, London. CC BY 4.0.); (right) mature smallpox pustular eruption on a woman's face. From *The Diagnosis of Smallpox*, T. F. Ricketts, Cassell and Co., 1908. (Wellcome Collection, London. CC BY 4.0.)

in shape. This process took forty-eight hours to reach full maturation. At this point the patients felt markedly worse. Their fever spiked, and the eyelids, lips, nose, and tongue became tremendously swollen. Patients were unable to swallow and slowly deteriorated as they became drowsy most of the time and extremely restless at night. Often they lapsed into delirium, thrashed about, and even tried to escape. These psychological effects were not simply the side effects of high fever but resulted from the involvement of the central nervous system in the infection. For this reason, even if patients survived, neurological sequelae were likely to be lasting and to result in permanent impairment.

On the ninth day the pustules became firm and deeply embedded; for this reason they were likely to leave permanent scars and deep pits wherever they appeared on the body. The lesions were described as soft, flattened, velvety, hot, and tender to the touch. In women, uterine hemorrhages were common, and pregnant women were certain to miscarry.

Another unpleasant aspect of smallpox was that a terrible sickly odor developed—one that physicians claimed was inexpressibly overpowering and offensive. At this point it was impossible for patients to drink, as even milk caused intense burning sensations in the throat. They experienced fearful weight loss and may actually have suffered from starvation. In addition, there was a complete loss of muscle tone, while the face of the still living body took on the appearance of a corpse that was nearly unrecognizable. The entire scalp sometimes became one large lesion entangled with hair. Pocks under the nails of the fingers and toes were a source of particular agony. The eyes became acutely sensitive; often they were themselves pocked, such that a person who survived could be permanently blind.

From ten to fourteen days after the first appearance of the rash, scabs began to appear. These contained live smallpox virus, were highly infective, and played a major role in the further transmission of the disease. Large areas of the skin may have begun to peel off, leaving the deeper tissues raw and exposed. These areas were exceptionally painful and contributed to the frightful appearance and misery of sufferers. Death most often occurred at this stage of the disease, as this stripping of the epithelium became widespread and could rapidly lead to a general toxemia and secondary streptococcus and staphylococcus infections. Attentive nursing, good hygiene, and sound nutrition reduced the likelihood of such complications, and as a result, prosperous and well-nursed patients were the most likely to recover.

The appearance of the patient was described as "mortification." In other words, the still living sufferer had the appearance of having been mummified, the face fixed in a death mask with the mouth constantly open. In terms of prognosis, the appearance of the lesions crusting over was favorable, but it led to a final torment: intolerable itching. Indeed, a portion of the scarring that resulted from smallpox was undoubtedly due to patients scratching and tearing at the scabs. Finally, however, among the survivors, the pustules dried out, the scabs fell off, and a protracted phase of recovery began.

The severity of the disease varied in conformity with the appearance, extent, and distribution of the pustules. Physicians described four characteristic patterns. The first was "discrete smallpox," in which the pocks were separated from one another by areas of intact skin. The CFR in such cases was less than 10 percent. The second pattern was "semiconfluent smallpox," in which the lesions began to merge into one another, especially on the face and forearms. For these patients the CFR rose to 40 percent. Third was "confluent smallpox," in which the eruptions intersected to form extensive lesions surrounding small islands of intact skin. Such a pattern indicated an unfavorable prognosis and a CFR of higher than 60 percent. Finally, and most disastrous of all was "hemorrhagic smallpox," the rarest form. It was so named because the natural clotting mechanism of the blood failed in such cases, and victims died of massive internal hemorrhaging into the intestines, the uterus, or the lungs. Mortality in hemorrhagic smallpox reached nearly 100 percent. As an overall average of all forms, the CFR of smallpox has been estimated at 30–40 percent.

Smallpox, then, most obviously attacked the skin and the throat, and the lesions there were indeed fundamental both personally in terms of the outcome for a single patient and epidemiologically with regard to a whole society. The outer lesions on the skin, in the throat, and on the eyes created the danger of death from the complications of secondary infection, caused inexpressible agony, generated dehydration, and left in their wake disfigurement and often blindness. The lesions of the throat were the source of the viruses that transmitted the disease via droplets in the air, while the scabs on the skin spread microbes as they shed. At the same time, however, the virus also attacked the lungs, intestines, heart, and central nervous system. There it could lead to severe or even fatal hemorrhages, bronchial pneumonia, mental derangement, and permanent neurological sequelae.

Treatment

How did physicians at the height of the prevalence of smallpox in the eighteenth and nineteenth centuries deal with so severe an infection? Since perhaps the tenth century, doctors thought that smallpox could be treated by surrounding patients with the color red (known as the "red treatment"). It was deemed important to hang red curtains around the sickbed, to bring red furniture into the sickroom, and to wrap sufferers in red blankets. Medical journals suggested that red light was soothing to the eyes and that it reduced scarring of the skin.

A variety of heroic measures were also attempted on the basis of humoral reasoning. One was to drain the pustules with a golden needle or to cauterize them. Whatever their efficacy, these measures were severely painful.

A further humoral idea was the "hot regimen," which consisted of piling sufferers high with blankets to induce profuse sweating that would rid the body of the overabundant humor. Alternatively, patients were immersed in hot water. According to this therapeutic fashion, light and fresh air were harmful, so patients were kept in the dark without ventilation. Meanwhile, internal medication was administered, especially sudorifics, to heighten the evacuation of the poison. Frequently, purging and bloodletting were also attempted.

In accord with the logic of the hot regimen but on the basis of the opposite conclusion, some physicians recommended a cold regimen, in which the sickroom was kept as cool as possible. Patients' bodies were repeatedly sponged with cold water, and ice bags were applied to their faces and extremities.

Since the possibility of disfigurement was one of the chief preoccupations of patients, considerable ingenuity was deployed to reduce the likelihood of pockmarking and scarring. One theory was that scarring on the face could be reduced by causing more intense irritation of the skin elsewhere on the body. To that end, mustard plasters, mercury, and corrosive liquids were applied to the back and the extremities. There were also fashions in every kind of local application—nitrate of silver, mercury, iodine, and mild acids. Ointments and compresses of virtually every known substance were used. Some inventive doctors applied glycerin to the face and then covered it with a mask, leaving holes for the eyes, nose, and throat; or they wrapped the face and hands in oiled silk. Others recommended splints as restraints in the late stages of smallpox to prevent patients from scratching their faces. For the same reason it was common to tie delirious patients to their beds.

Perhaps after learning of these various treatments for smallpox, one can appreciate the thinking of the English physician Thomas Sydenham (1624–1689). He postulated that the wealthy and the noble, who received extensive medical attention for smallpox, perished of the disease more frequently than the poor, who had no access to official attention. He therefore advised that the best physician was the one who did the least, and he himself made few interventions except to give his patients fresh air and light bed coverings.

CHAPTER 7
The Historical Impact of Smallpox

We now turn to a discussion of the historical impact of smallpox and the importance of the "speckled monster" as viewed from three vantage points. The first is to consider the role of smallpox in Europe; the second is to deal with the influence of the disease in the Americas; and the third is to discuss the part played by smallpox in the development of a new strategy of public health—prophylaxis by mass vaccination.

Smallpox in Europe

Unknown Origins

According the Centers for Disease Control and Prevention (CDC), the origins of smallpox are not known, though there are descriptions that resemble the disease from parts of the world other than Europe (China, India, and Asia Minor) from the fourth to the tenth centuries. It probably became established in Europe during the Crusades of the eleventh and twelfth centuries, with the dispersal of significant numbers of soldiers returning from the Levant and bringing far-flung new diseases with them. As we have said, warfare contributes to the spread of disease.

For our purposes, the exact origin of modern smallpox is of limited importance. What is more important is to note the rise of smallpox to ever-greater prevalence during the seventeenth and eighteenth centuries, when it replaced bubonic plague as the most feared of all killers. Once established, smallpox thrived in conditions of rapid social dislocation, urban development, congestion, and demographic growth.

Patient Population

The facts of immunology go a long way toward explaining the epidemiology of smallpox in Europe during the early modern period. A crucial factor was that smallpox sufferers who survived the ordeal thereafter possessed a robust lifelong immunity to the disease. No one was infected twice. In this context, the typical pattern that emerged was that in cities and towns across the continent smallpox became so ubiquitous that most people who survived childhood had been exposed to it and a large proportion had contracted it. The adult population of urban Europe therefore possessed a substantial "herd immunity" to the disease. In cities, only children, immigrants from the countryside who were immunologically naive, and those adult residents who had escaped contracting smallpox during their childhoods were susceptible.

In such a context, smallpox was largely an endemic disease of childhood. Perhaps once a generation, however, it would erupt as a substantial epidemic among the general population. This again reflected the dynamics of immunology. Since not every child contracted the disease, over time the number of nonimmune adolescents and adults would slowly rise. Furthermore, early modern European cities were so unhealthy that they sustained or expanded their populations only by the large-scale influx of people from the outside—peasants driven off the land in search of work, refugees escaping failed harvests or warfare, and migrant laborers—who would then be susceptible to the disease. Smallpox, then, was endemic year in and year out, thriving as it did in overcrowded cities with poorly ventilated houses and workshops; it became an epidemic when "combustible" circumstances accumulated to ignite when a spark was struck.

Figures of the number of people smallpox killed across Europe during the eighteenth century are unreliable. This single disease, however, is believed to have caused a tenth of all deaths during the century across the whole of Europe and a third of all deaths among children younger than ten. Furthermore, half the population of adults is estimated to have been scarred or disfigured by it, and it was the leading cause of blindness. In total, approximately half a million Europeans died annually from this one cause. This was the equivalent of witnessing the largest city in Europe vanish every year from the disease through the whole of the eighteenth century.

It was for this reason that the nineteenth-century English politician and historian Thomas Macaulay famously described smallpox as "the most terrible of all the ministers of death" and wrote:

> The havoc of the plague has been far more rapid; but the plague has visited our shores only once or twice within living memory; and the smallpox was always present, filling the churchyards with corpses, tormenting with constant fears all whom it had not yet stricken, leaving on those whose lives it spared the hideous traces of its power, turning the babe into a changeling at which the mother shuddered, and making the eyes and cheeks of the betrothed maiden objects of horror to her lover.[1]

Because of its mode of transmission through the air, smallpox, like influenza, was a disease with no particular predilection for the poor or any subset of the population. Even wealthy aristocrats and royal families suffered from it, including, for instance, Louis XIV (1647) and Louis XV (1774) of France, William II of Orange (1650), Peter II of Russia (1730), and the Holy Roman Emperor Joseph I (1711).

Indeed, smallpox was directly responsible for a dynastic change in England when it extinguished the House of Stuart. The last Stuart heir, eleven-year-old Prince William, died of smallpox in 1700. The result was a constitutional crisis. This crisis was resolved through the Act of Settlement of 1701, which prevented the crowning of another Catholic and brought in the House of Hanover.

Despite such a ubiquitous presence and the dread it inspired, smallpox elicited a social response that was profoundly different from that of bubonic plague. The disease did not give rise to mass hysteria, riots, scapegoating, or religious frenzy. The reasons why are evident. Unlike plague, smallpox was not a sudden outside invader that caught a population unaware, nor did it maximize its fury by targeting young adults and the middle-aged, who were the chief economic supports of families and communities. Since smallpox was ever-present, it was considered almost "normal," especially since it was like other familiar diseases in claiming its victims primarily among infants and children. Everyone had some experience of smallpox either firsthand or through the experience of family members, and half of the people on city streets were pockmarked—a reminder of its passage. So familiar was smallpox that it even bred a kind of fatalism—the belief that contracting it was an inevitable rite of passage. This pervasive attitude led some parents of healthy children to expose them intentionally to mild cases of the disease in the hope that they would then be protected from more serious danger later in life.

Popular attitudes to smallpox are embedded in English literature. When an eighteenth- or even nineteenth-century author wished to introduce a sudden plot change, smallpox was available as an instrument that no one would question as a clumsy or artificial contrivance. Henry Fielding deployed the disease as just such an artifice in *The History of Tom Jones, a Foundling* (1749), while in *The History of the Adventures of Joseph Andrews* (1742) it seemed natural that the heroine should be pockmarked. Similarly, William Thackeray, who set *The History of Henry Esmond* (1852) in the eighteenth century, chose smallpox as a device to propel the plot.

Charles Dickens also placed smallpox at the center of the narrative in *Bleak House* (1852–1853). Dickens's protagonist Esther Summerson contracted the disease while caring for the urchin boy Jo, and the author carefully noted many of the key features of her illness—chills and fever, a hoarse and painful voice, prostration, involvement of the eyes and temporary blindness, delirium and mental confusion, weeks of suffering, a close approach to death, and a prolonged convalescence. Most worrying to Esther, however, was her disfigurement, which she thought would distance friends and make her unlovable. As she explained in her narration, her attendants had removed all mirrors from the sickroom in order to avoid alarming her. She was thus very afraid the first time she gathered the courage to face a looking-glass:

> Then I put my hair aside, and looked at the reflection in the mirror . . . I was very much changed—O, very, very much. . . . I had never been a beauty, and had never thought myself one; but I had been very different from this. It was all gone now. Heaven was so good to me, that I could let it go with a few not bitter tears. I knew the worst now, and I was composed to it.[2]

Esther was both transformed and permanently scarred by the illness. What is most interesting, however, is that neither she nor Dickens felt the need to specify the disease from which she suffered. Smallpox was ubiquitous, and there was no reason to name it. *Bleak House* treated smallpox, which afflicted all social classes, as symbolizing the illness of Victorian society as a whole in its avarice and want of human sympathy. Smallpox, in other words, needed no explanation; it was always and unremarkably just there—an everyday aspect of the human condition.

In literature as in life at the time, private destinies rather than public cataclysms were determined by smallpox. There was no scope for the eighteenth-century equivalent of Daniel Defoe's account of public calamity during a plague year. Smallpox did not cause great European cities like London to empty of their populations, and there was no urge among people to seek scapegoats for an event that was all too daily and natural an occurrence.

But smallpox was still inseparable from private anxiety and fear. In Thackeray's novel *Henry Esmond,* for instance, this dread informs the narrative. The heroine, Lady Castlewood, contracts the disease as an adult. Her husband had been a brave soldier in combat, but he was unable to face a malady that he could not fight and that threatened him not only with death, but also with disfigurement. Unwilling to put his pink complexion and his fair hair at risk, Lord Castlewood took to his heels and deserted his household for the duration of the outbreak. But he was not part of a mass exodus, even though Henry Esmond declares that smallpox was "that dreadful scourge of the world" and a "withering blight" and "pestilence" that "would enter a village and destroy half its inhabitants."[3]

We also learn that Lady Castlewood's beauty, like Esther's, "was very much injured by the smallpox," such that, on her gallant husband's return, he no longer loved her as he once had. The effects of smallpox thus had a major impact on the marriage market. By disfiguring people, it made them far less likely to succeed in the nuptial sweepstakes. Thackeray continues that "when the marks of the disease cleared away . . . , the delicacy of her rosy colour and complexion was gone: her eyes had lost their brilliancy, her hair fell, and her face looked older. It was as if a coarse hand had rubbed off the delicate tints of that sweet picture, and brought it . . . to the dead colour. Also, it must be owned, her Ladyship's nose was swollen and redder."[4] To have such discoloration, scars, pockmarks, and bald patches was a source of considerable torment and unhappiness. Thackeray's narrative unfolds against this background.

Smallpox in the Americas

The European history of smallpox is marked by suffering, death, and private anguish, but there is also a more dramatic facet of the history of the malady. This is the story of what occurred when the disease was introduced to parts of the world where it was an outside invader among populations that possessed no immunity to it and no means to contain its ravages. In such cases, smallpox unleashed a "virgin soil epidemic." This catastrophe accompanied European expansion to the Americas, Australia, and New Zealand. In these settings, smallpox had a transformative importance in clearing the land of its inhabitants and thereby promoting settlement by Europeans who possessed a robust immunity. In some accounts, these biological events were said to have had a greater impact on European expansion than gunpowder.

The Columbian Exchange and Hispaniola

The "Columbian exchange" is the term used to describe the process by which the European encounter with the Americas brought about the large-scale transfer of flora, fauna, culture, and people from each side of the Atlantic to the other. That movement involved plants such as the potato, maize, and the tree bark from which quinine is made moving from the Americas to Europe, but microbes also moved in the reverse direction. Europeans, in other words, introduced smallpox and measles to the Americas.

The momentous impact of the Columbian exchange in microbial guise is well illustrated by the experience of Hispaniola, where it first occurred. This mountainous Caribbean island, now divided between Haiti and the Dominican Republic, was the famous site of the landing by Christopher Columbus in 1492.

The aboriginal inhabitants of the island were an Indian tribe known as the Arawak peoples, who are estimated to have numbered a million when Columbus's ships arrived. Columbus described the island as an earthly paradise of great natural beauty, and he reported that the inhabitants were a welcoming and peaceful people who greeted the Spaniards with generosity and warmth.

The kindness, however, was not reciprocal. The Spanish were interested in profit and international power politics. Hispaniola was strategically located, and it possessed fertile soil, a favorable climate, and land that the Spanish crown coveted for cultivation. The Spaniards militarily dispossessed the Arawaks of their territory and then enslaved them. In this process the Europeans were decisively assisted by two massive assets: gunpowder and the epidemic diseases of smallpox and measles, to which native peoples lacked all immunity.

What occurred biologically was spontaneous and unintended. There is no indication that the Spaniards engaged in a genocidal plot or deployed disease in accord with a plan to depopulate the land and resettle it. The results were nevertheless momentous as the native population of Hispaniola experienced a terrifying and unprecedented die-off. Between 1492 and 1520 the native population was reduced from 1 million to fifteen thousand. Agriculture, defense, and society itself disintegrated; survivors succumbed to terror and the fear that the Europeans were either gods themselves or worshipped divinities more powerful than their own. The diseases brought to the Americas by the Columbian exchange therefore cleared Hispaniola for European colonization and Christian conversion almost before a shot was fired.

Paradoxically, although smallpox and measles thus established the Europeans' power over the island, the diseases also thwarted the invaders' original intention of enslaving the indigenous peoples to work their plantations and mines. The near-disappearance of the native population drove the Europeans to find an alternative source of labor. The resulting turn to Africa was facilitated by the fact that Africans and Europeans shared some immunities, such that Africans were resistant to some of the same epidemic diseases that had destroyed the Native Americans. The immunological factors underlying these social and economic developments were of course unknown at the time, but empirical experience made the results and the risks all too apparent. In this way disease was an important contributing factor in the development of slavery in the Americas and the establishment of the infamous Middle Passage.

The process began almost without delay. The year 1517 marked the first importation of African slaves to Hispaniola and the beginning of the gradual rise of the island to wealth and international prominence over the following two centuries. Over time, after the plantation economy had been securely established, Saint-Domingue, as the western half was renamed when the island was partitioned with the French in 1659, established itself as the most profitable possession in the French empire.

Wider Developments

The story of the depopulation, resettlement, and development of Saint-Domingue is an important illustration of the impact of smallpox on the "big picture" of historical change, but in a way that was fundamentally different from its impact on European society. That illustration is itself a reason for dwelling on the Columbian exchange that followed 1492. At the same time, the story of Saint-Domingue is just a small and particular example of a much larger process that involved the whole of the Americas. Just as the Arawaks were decimated by virgin soil epidemics, the civilizations of the Aztecs in Mexico and of the Incas in Peru were destroyed by epidemics of smallpox and measles following the arrival of Hernán Cortés and Francisco Pizarro, and all of North America was cleared for conquest and resettlement.

This much larger story has been well told by others. Interested readers can find several such treatments in the bibliography, as well as items that deal with related developments affecting the aboriginal peoples as part of the British colonization of Australia and New Zealand. But the biological underpinnings of the European conquest of the Americas and the Dominions are clearly visible through the lens of developments on Hispaniola.

We must add here the fact that spontaneous die-offs of Native Americans were sometimes reinforced by intentional genocide. The precedent was established by the British Army officer Sir Jeffery Amherst, who introduced genocide to North America when he deliberately gave blankets infected with smallpox to Native Americans in order to "reduce" them. This precedent needs to be remembered because, at a time when biological terror is an ever-present threat, its successful use in the eighteenth century helps to explain the reason that smallpox is placed high on the list of biological agents most feared by public health authorities.

Smallpox and Public Health

Inoculation

In addition to the impact of smallpox on culture and society, an important reason for studying its history is that it gave rise to a new and distinctive style of public health—the prevention of disease initially by inoculation and then by vaccination. Of the two practices, inoculation was a much older folk art that emerged in various parts of the world as a result of two simple observations. The first was that, contrary to orthodox medical philosophy, smallpox clearly seemed to be contagious. Second, people everywhere noticed that those who recovered from the disease never contracted it again. This observation was relatively easy to make because the smallpox victims were so numerous and most of them were physically marked by their experience. Thus the idea emerged of inducing a mild case of smallpox artificially in order to protect a person from contracting a severe or life-threatening case later. This practice was known variously as "inoculation" or "variolation," or—to use a gardening metaphor—"engrafting."

The technique of inoculation varied from culture to culture and from practitioner to practitioner. But the dominant strategy was to draw liquid from the pustule of a person afflicted with a mild case of smallpox and no congruent or semicongruent lesions. The practitioner inserted a thread into a pustule and allowed it to become saturated with the yellow material. Using a lancet, he or she then made a superficial cut in the arm of the person to be inoculated, inserted the thread into the incision, fastened it there, and allowed it to soak for twenty-four hours. In successful circumstances, the person fell ill twelve days later from a similarly mild case, suffered the successive stages of the illness for a month, convalesced for a further month, and then enjoyed a lifelong immunity. In the eighteenth century the practice was common in the Middle East and Asia, but not in Europe.

A decisive figure in bringing inoculation to England and from England to western Europe more broadly was Lady Mary Wortley Montagu (1689–1762), the wife of the British ambassador to Turkey. Her beauty had been severely compromised by a severe attack of the disease, and on learning of the practice of inoculation at Istanbul, she resolved to have her own children protected by it. Exhilarated by the idea of providing immunity to the great eighteenth-century killer, Lady Montagu returned to England in 1721 determined to educate cultivated English society in the exotic procedure she had learned in Turkey.

Lady Montagu's social standing, intelligence, and obvious devotion to the cause gained an audience for her message. Her advocacy convinced the Princess of Wales, who had her own daughters inoculated—an event that was decisive in making inoculation acceptable. Since it rapidly became evident that the practice provided a real measure of protection against the most ruthless killer of the century, inoculation gained an ever-widening appeal as the first apparently successful strategy to be adopted against any epidemic disease since the antiplague measures of the Renaissance. Since England stood at the epicenter of the smallpox epidemic, it was logical that inoculation gained its first measure of enthusiasm in the British Isles. But it was soon taken up by progressive thinkers in France, Italy, Sweden, and the Netherlands. The procedure gained a place in the teachings of the French philosophes, who saw it as a victory for reason and progress. Voltaire and Charles-Marie de la Condamine were ardent advocates on the Continent, as were Benjamin Franklin and Thomas Jefferson in the United States. After considerable hesitation, George Washington even took the risk of ordering the inoculation of his army and perhaps thereby made a decisive contribution to the victory of the American Revolution. In Russia, Catherine the Great imported an English physician from London to inoculate her, and the nobility rapidly imitated her example. Thus, both in Europe and in America, the cresting wave of the smallpox epidemic in the eighteenth century was met headlong by the first practical measure of public health against it.

As a prophylactic procedure against smallpox, however, inoculation was highly controversial. On the positive side, it could provide a robust immunity against a fearful disease. Its successes were based on several carefully observed aspects of the procedure. The first was the rigorous selection of mild cases as the source of the infected pustular material because a mistake in this regard could lead to a fatal or disfiguring illness in the person inoculated. A second important guideline was that those subject to the procedure were chosen only if they were clearly in robust health. There was also a preparatory period of several weeks after selection and before inoculation during which the candidate was placed on a strict regimen of rest, exercise, and diet to enhance resistance. Finally, a serendipitous contributing factor to the mildness of the person's infection was the portal of entry of the virus into the body when the inoculation took place. The use of lancet and thread caused the variola major virus to enter the body through the skin in a way that does not occur in nature and that is now believed to attenuate the virulence of the microbe. For a large proportion of inoculated people the result was an unpleasant but mild case of smallpox that led to lifelong immunity, little or no disfigurement, and freedom from further anxiety.

On the other hand, inoculation was a flawed procedure that involved serious risk for both the person being inoculated and the community. It was expensive and required three months to be completed—one devoted to preparation, one to the illness, and one to convalescence, with an obligation of watchful care at each stage. Such a complicated measure was available only to people with financial means and leisure. In addition, there was always a danger to the person being inoculated. The calculation of expected virulence was always imperfect, and the procedure could result in a severe case or even in death or maiming. Variolation is estimated to have killed 1–2 percent of those who adopted it, as compared with 25–30 percent of those who contracted the disease naturally.

Furthermore, since inoculation resulted in actual smallpox, there was always the possibility that it would set off a wider outbreak or even unleash an epidemic. For precisely that reason a Smallpox and Inoculation Hospital was opened in London with the twofold purpose of caring for people who had been inoculated and isolating them until they were no longer infectious and a potential hazard to the community. Thus inoculation was always surrounded by a spirited debate as to whether, on balance, it saved more lives than it destroyed.

Vaccination

It was in this context of the massive prevalence of smallpox in eighteenth-century England and of the disappointments and anxieties surrounding inoculation that one of the decisive discoveries in the history of medicine occurred: the development of vaccination by the doctor Edward Jenner (1749–1823). To understand this breakthrough, we need to recall that the smallpox virus (variola major) is a member of the *Orthopoxvirus* genus, which also includes cowpox. Smallpox exclusively infects humans, but the cowpox virus primarily attacks cattle. Under the right conditions, however, the cowpox virus can be transmitted across the species barrier from cows to human beings. In people, it induces a mild illness while also providing a lasting crossover immunity to smallpox.

In Britain in the eighteenth century the people most likely to contract cowpox were milkmaids. Jenner, who practiced medicine in the Gloucestershire town of Berkeley, made a simple observation that could have been made only by a doctor with a rural practice in dairy country at a time when smallpox was prevalent. He observed that the milkmaids who got cowpox never seemed to contract smallpox. He was not the first physician to make this observation, but he was the first to confirm it by taking case histories of milkmaids and then to take the next step of experimentation. He made his experiment in 1796, when he persuaded his gardener to allow his eight-year-old son to be vaccinated first with cowpox obtained from a milkmaid, and then with live smallpox virus. He called the procedure "vaccination" after the Latin word *vaccinus*, meaning "from cows."

Under present-day ethical standards, an experiment of such a kind with a child would not be permissible. Fortunately, however, the boy contracted a mild case of cowpox and then remained robustly healthy after the challenge vaccination with actual smallpox. It was simple prudence that led Jenner to carry out his initial experiment on a single subject and then to wait two years before he repeated the procedure on fifteen additional volunteers. On the basis of these successful efforts, Jenner assessed the potential of vaccination in his slender but monumental work of 1798, *An Inquiry into the Causes and Effects of the Variolae Vaccinae, a Disease Discovered in Some of the Western Counties of England, Particularly Gloucestershire, and Known by the Name of the Cow Pox.*

Jenner's genius was that he fully recognized the significance of the experiment—that it opened up the possibility of eradicating smallpox from the earth. As he presciently declared in 1801, "The annihilation of the smallpox, the most dreadful scourge of the human species, must be the final result of this practice."[5] It was for this reason that the British Parliament soon afterwards declared vaccination one of the great discoveries in the history of medicine. It was the foundation of a new style of public health that has proved its effectiveness in dealing not only with smallpox, but with a host of other diseases, including polio, tetanus, rabies, influenza, diphtheria, and shingles. And research is under way in the hope that it will be possible to develop vaccines against other infectious diseases, such as malaria and HIV/AIDS.

After 1798, Jenner devoted the rest of his life to the smallpox vaccination crusade. In the process he rapidly made influential converts who established vaccination as a major instrument of public health—Pope Pius VII at Rome, Luigi Sacco in Italy, Napoleon in France, and Thomas Jefferson in the United States.

Unlike inoculation, which used the smallpox virus, vaccination had a low risk of serious complications for the individual and posed no risk to the community because it involved cowpox rather than smallpox. It did, however, pose problems that made the campaign against smallpox a difficult and protracted one. It was feared that Jenner's initial technique of arm-to-arm vaccination of live virus could increase the danger of spreading other diseases, especially syphilis. In addition, Jenner made it a dogmatic article of faith that the immunity derived from vaccination was lifelong, steadfastly refusing to consider conflicting data. In fact, the immunity provided by vaccination proved to be of limited duration. The irrefutable evidence of disease among patients who had previously been vaccinated went a long way toward discrediting the procedure and promoting skepticism. (It has subsequently been demonstrated that smallpox immunity following vaccination lasts up to twenty years and that revaccination is required in order to guarantee lifelong protection.)

Such overreaching claims made by Jenner himself, as well as some failures and concerns about safety, helped to galvanize political opposition to vaccination, further slowing the advance of the campaign. The antivaccination movement in both Europe and the United States became one of the largest popular movements of the nineteenth century, and vaccination generated some of the most hotly contested debates of the era. Opposition crested on waves of libertarian resistance to the apparently encroaching powers of the state, of the religious sense that it was an unnatural violation of God's order intentionally to insert material from a cow into the human body, and of latent fears of science and its potential for harm. Cartoons were widely circulated giving expression to such anxieties by portraying vaccine recipients sprouting horns from their skulls as they undergo a bestial metamorphosis before the eyes of horrified spectators and the hapless vaccinator. One doctor of the time, Benjamin Moseley, took antivaccination hysteria to its moral nadir by imagining that Jenner's procedure would lead ladies who received it to "wander in the fields to receive the embraces of the bull."[6] There were even sly insinuations that Jenner was a covert admirer of the French Revolution who had a secret goal to subvert the social order of his country. Some inoculators even joined the opposition out of anxiety that vaccination would put them out of business.

In part it was alarm at what Charles Dickens regarded as the unacceptably slow progress of vaccination due to such irrational opposition that led him to write *Bleak House*. The near-death experience and disfigurement of Esther Summerson conveyed the warning of the consequences of ignoring Jenner's simple and readily available prophylaxis. Without it, Esther's misfortune could happen to anyone.

Ultimately, however, Jenner's vision was vindicated. Two hundred years after Jenner's first vaccination, technological advancements such as refrigeration, freeze-dried vaccines, and the air-powered jet injector simplified techniques and virtually eliminated the risk of cross-infection. Such advancements also made vaccine available even in tropical and resource-poor settings. Combined with the further unprecedented administrative step of a globally organized vaccination campaign launched in 1959, mass vaccination eradicated naturally occurring smallpox from the earth. The last case occurred in 1977 in Somalia, and the World Health Organization (WHO) proclaimed victory in 1980.

Reporting to the US Congress in 1998, Senator Dale Bumpers, chairman of the Senate Subcommittee on Labor, Health, and Human Services, reviewed the economic costs and benefits of the campaign. In his assessment, the total cost of the international eradication program was $300 million, of which the United States contributed $32 million. "That investment," he stated, "has repaid itself many times over. Beyond the humanitarian benefits of eliminating this vicious killer, we have enjoyed tremendous economic benefits. The United States alone has recouped the equivalent of its entire investment every 26 days since the disease was eradicated."[7]

Similarly, the US Government Accounting Office (GAO) calculated that the United States had saved a net total of $17 billion in other direct and indirect costs related to vaccination, medical care, and quarantine related to smallpox. According to GAO figures, the economic benefit between 1971, when the US achieved eradication and eliminated routine vaccination, and 1988 was an average annual return of 46 percent on its global campaign investment. As a consequence of this international effort, smallpox became the first, and thus far the only, infectious human disease to have been eradicated by intentional action.

CHAPTER 8

War and Disease

Napoleon, Yellow Fever, and the Haitian Revolution

In late 1804, after one of the largest of all slave rebellions and thirteen years of insurgency, Jean-Jacques Dessalines, a leader of the insurrection, declared Haiti an independent nation. It became the world's first free black republic and the earliest example of decolonization. In the "Haitian Declaration of Independence," Dessalines announced to the citizens of the new nation:

> It is not enough to have expelled the barbarians who have bloodied our land for two centuries; it is not enough to have restrained those ever-evolving factions that one after another mocked the specter of liberty that France dangled before you. We must, with one last act of national authority forever assure the empire of liberty in the country of our birth; we must take any hope of re-enslaving us away from the inhuman government that for so long kept us in the must humiliating torpor. In the end we must live independent or die.[1]

An island the size of Massachusetts with the population of present-day Louisville, Kentucky, defeated France, a world power with 20 million people (fig. 8.1).

With the loss of Haiti, formerly Saint-Domingue, Napoleon abandoned his vision of a powerful French empire in the Americas. Deprived of the

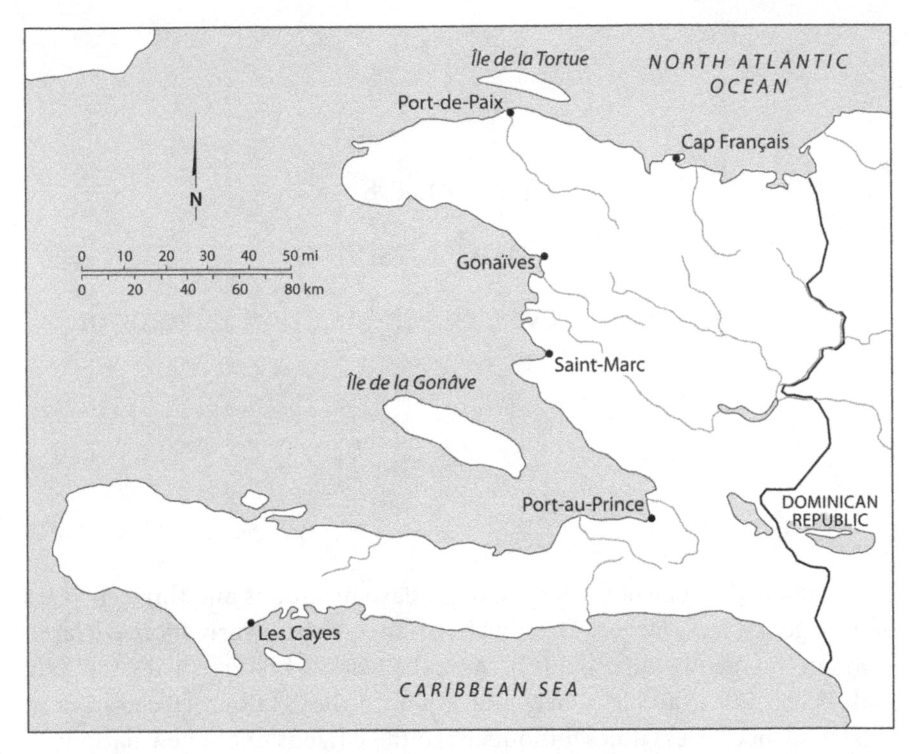

Figure 8.1. Haiti (at independence, 1804), where Napoleon's army was ravaged by the yellow fever epidemic of 1802–1803. (Map drawn by Bill Nelson.)

island as the forward base from which to project French power onto North America, and disillusioned with the complications of warfare in a tropical environment, Napoleon decided that the Louisiana territories were an indefensible liability. He therefore concluded the agreement known as the Louisiana Purchase with the United States. Through this 1803 transaction the United States doubled its size, acquiring 828,000 square miles of territory that in time became fifteen new states.

In 1803, however, so-called Louisiana was a fiction—a name on maps, not an area France controlled in reality. As Napoleon understood, enormous investment would be required to build the infrastructure needed to establish actual control, to promote settlement, and, if necessary, to vanquish a multitude of possible enemies—Britain, the United States, and a variety of Native American opponents. Since Haitian independence made the venture exponentially more risky and expensive, Napoleon concluded that the prudent course was to cut his losses, accept the money offered by Thomas

Jefferson, and move on to alternative projects elsewhere. Thus the successful slave rebellion directly influenced the rise of the United States to global power and determined the fate of large numbers of Native Americans.

This chapter examines the role in these developments of the fearful yellow fever epidemic of 1802–1803. In Haiti, yellow fever clearly demonstrated that infectious disease played a major role in the history of slavery, empire, warfare, and nation-building.

Saint-Domingue

By 1789 the Caribbean colony of Saint-Domingue on the island of Hispaniola had become a vital foundation of French wealth and economic growth. Saint-Domingue formed the western third of the island when it was partitioned by the Treaty of Ryswick in 1697. The treaty awarded eastern Hispaniola to Spain, forming what is now the Dominican Republic, while western Hispaniola went to France as Saint-Domingue. Some eight thousand Saint-Domingue plantations relied on a workforce of African slaves to produce sugar, coffee, cotton, tobacco, indigo, and cacao. These commodities were shipped from Haitian ports, especially Le Cap Français and Port-au-Prince, to Marseille, Nantes, and Bordeaux. Thus they supplied Normandy textile mills, drove the production of French shipyards, and satisfied the voracious appetite of half of European consumers for sweeteners and caffeine.

During the half century preceding the French Revolution, Saint-Domingue became the world's wealthiest colony, and Le Cap earned a reputation as the "Paris of the Antilles." Between 1780 and 1789 alone, raw material exports doubled, and sixteen hundred ships a year sailed from Saint-Domingue for the metropole. At the same time the importation of slaves rose rapidly—from ten thousand a year in 1764 to fifteen thousand in 1771, twenty-seven thousand in 1786, and forty thousand in 1787. Especially important was the eighteenth-century sugar and coffee boom, which fueled a heady upward spiral in land values. At the heart of this dizzying economic growth were the large-scale sugar plantations worked by as many as two hundred slaves each on the plain east of Le Cap and the smaller coffee estates rising on the hills above.

An insoluble problem was that the island embodied explosive contradictions. It was said that an acre of land on a Saint-Domingue plantation yielded more wealth than an acre anywhere else on earth. At the same time, the same area enclosed what many regarded as the highest concentration of human misery. Saint-Domingue plantation slaves endured punitive conditions that included whipping, shackles, imprisonment, rape, and branding. It was more economical for a plantation owner to replace a slave than to provide humane living standards. Such an inhumane policy also reflected the fact that most planters were absentee owners. They preferred the amenities of Paris to a tropical climate and the risk of contracting yellow fever, the most lethal of West Indian microbial threats. As a result, estate management was left to middlemen whose ambitions stressed short-term profit over long-term stability. As a result, on the eve of the French Revolution, the majority of slaves imported from Africa died within five years of disembarking at Le Cap, and their relationships with the plantocracy were structured by fear and mutual hostility.

On the plantations, deaths regularly exceeded births so that the growth of the eighteenth-century slave population depended on the continual arrival of ships from Africa. This high death rate among black slaves resulted from overwork, industrial accidents, deficient diet, crowding, filth, and disease—particularly dysentery, typhoid, and tetanus. Rarely, however, did the death of a black person result from the disease most feared by Europeans—yellow fever, which underpinned the daunting reputation of the Antilles as a white man's grave.

"Bitter Sugar"

Yellow fever was variously known as "bronze john," on account of the jaundice it induced; "black vomit," in recognition of its most feared symptom; "malignant fever," acknowledging its severity; "Siam sickness" (*mal de Siam*), nodding to its origins in the colonial world; and "yellow jack," referring to the yellow flag raised to signal quarantine when it was diagnosed. Its most lasting name—yellow fever—references both jaundice and quarantine. The disease first arrived from West and Central Africa aboard slave ships, together with the mosquito species whose adult females transmit it—*Aedes aegypti*. Both the slaves and the mosquitoes were reservoirs of yellow fever virus. Slave ships in effect served as super vectors bearing a fever that changed the history of the West Indies.

Sugar cultivation transformed the ecology of Saint-Domingue. The results were hardly edenic for slaves, but they did create an earthly paradise for microbes and *Aedes aegypti*. Sugarcane entailed a variety of operations that favored the diffusion of the fearful yellow fever insect vector. First was the process of forest clearing, which destroyed the natural habitat of the insect-devouring birds that kept mosquito numbers in check. Land clearance thus enabled *Aedes aegypti* arriving as stowaways to establish the critical mass necessary to survive in the Caribbean. Deforestation led in turn to soil erosion, siltage, flooding, and the formation of marshes along the coasts that were the delight of flying insects.

After the forest was cleared, the planting and cultivation of the sugar fields created further opportunities for *Aedes aegypti.* The mosquito does not require large expanses of water for breeding, preferring the sides of containers for laying its eggs at or slightly above the waterline. Thus, cisterns, water barrels, pots, and broken crockery were ideal. The innumerable clay pots that plantations used for the first stages of refining sugar and extracting molasses were also perfect, and the sweet liquid was an excellent nutrient for newly hatched larvae. Buckets were widely available alternatives. They transported water to meet the needs of slaves and work animals and to irrigate vegetable gardens.

Equally important were the living conditions of insect vectors on the estates. All female mosquitoes require blood meals in order for their eggs to mature. *Aedes aegypti,* moreover, will not willingly feast on mammals other than humans. The large concentrations of slaves on a plantation promoted breeding, while the constant importation of yellow fever virus in slave bloodstreams ensured that the mosquitoes were infective and able to transmit disease.

Although slaves of African birth were often immune to yellow fever, the vertiginous growth of the economy furnished *Aedes aegypti* with a constant supply of immunologically naive Europeans. Ship crews, merchants, officials, soldiers, artisans, peddlers, and shopkeepers arrived continuously from Europe to service the requirements of a bourgeoning sugar-and-slave economy. Le Cap, the economic capital, was unique with its population of twenty thousand, but smaller ports sprouted along the coast within mosquito flying range. Additionally, *Aedes aegypti* can adapt readily to an urban environment. Thus, in both town and countryside, nonimmune whites stoked seasonal outbreaks during the hot, rainy summers. Normally limited, these outbreaks burst into conflagrations of disease whenever large numbers of European sailors and soldiers arrived en masse, supplying fuel.

The plantation economy provided most of the requirements needed to sustain transmission, but the tropical climate of the West Indies was also favorable. Its hot, humid rainy season from May through October suited the life cycle of both vector and virus. Furthermore, some scholars point to the long-term warming trend that marked the end of the cooling known as the Little Ice Age. Its onset and end were gradual and not precisely delineated, but the cooling phase is understood to have begun in the early seventeenth century and ended in the early nineteenth. The warming tendency that followed brought increased heat and rainfall. Such climate change facilitated the introduction of yellow fever and the mosquitoes that transmitted it.

Slaves who came from regions of Africa where yellow fever was endemic normally possessed a combination of acquired, crossover, and genetic immunity. Acquired immunity occurred when people contracted the disease as children and thereby enjoyed lifelong immunity. Alternatively, people who had previously suffered from dengue fever, which was also endemic in much of Africa, enjoyed crossover resistance. This is because dengue fever virus is a member of the same *Flavivirus* genus as yellow fever virus, and it confers a significant crossover immunity to yellow fever. The mechanisms are analogous to those by which cowpox, as Edward Jenner so famously discovered, provides crossover immunity to smallpox. Finally, many observers believe that selective evolutionary pressure on African populations produced genetic immunity, just as the sickle cell trait and thalassemia arose among African and Mediterranean people subject to intense malaria transmission.

The Africans' immunity and resistance contrasted dramatically with the vulnerability of whites around them. For Europeans, Caribbean islands such as Hispaniola were notoriously unhealthy, and yellow fever was the most feared of the dangers they presented. Thus a starkly contrasting medical history divided Europeans and African slaves. The contrast was dramatic and universally noticed, as the English writer Robert Southey confirmed in 1850 when he observed, "Diseases, like vegetables, choose their own soil; as some plants like clay, others sand, others chalk, 'so the yellow fever will not take root in a negro, nor the yaws in a white man.'"[2]

This differential immunity between those of European and African descent powerfully affected the history of Saint-Domingue. It dictated the ongoing need to import plantation laborers from Africa rather than relying on impoverished Europeans or Native Americans; it inflamed the social tensions that led to revolt; and it drove the violent yellow fever epidemics that periodically swept the European populations of the Caribbean. As we will see, the arrival of an armada of European troops also played an outsized role in unleashing the most deadly epidemic of yellow fever in the history of the Americas in 1802. This epidemic produced a suite of momentous consequences, including the defeat of Napoleon's aspirations in the Americas.

Social Tensions

Given the brutal conditions prevailing on Saint-Domingue's sugar plantations, the colony experienced a history of slave resistance. Outsiders such as the French Enlightenment writer Guillaume Raynal predicted a major uprising under a "new Spartacus," while the French abolitionist Comte de Mirabeau compared Saint-Domingue to the volcano Vesuvius. Slave resistance took a variety of forms, including individual insubordination, counterviolence, the circulation of subversive ideas at Sunday masses, and the formation of bands of "maroons"—the term applied to escaped slaves who preserved the idea of freedom through a life of hiding in the hills and forests. More threateningly but less frequently, there were also collective actions when slaves opposed egregiously violent overseers and drivers by withdrawing as a group from a plantation and negotiating the terms of their return.

Most feared of all by planters were those occasional seismic shocks of all-out rebellion that punctuated the life of Saint-Domingue's sugar plantations. The most famous of such revolts was the Mackandal Conspiracy of 1758, when enraged slaves poisoned their masters and settled scores with slave drivers. Although François Mackandal, a Vodou practitioner, was burned at the stake, a belief persisted that he had miraculously survived the flames and would return to lead his people to freedom. His memory inspired the insurgents of 1791 to 1803, including François Dominique Toussaint Louverture.

The impulse to rebel was reinforced by additional features of the colony's demography. Unlike slaves in the American South, the plantation slaves of Saint-Domingue formed the overwhelming majority of the population. Numbering half a million, they constituted the largest slave population in the Caribbean, dwarfing the two hundred thousand slaves of Jamaica, the nearest rival. The remaining Saint-Dominguans consisted of thirty thousand mulattoes, or *gens de couleur,* and forty thousand whites. Furthermore, Saint-Dominguans of African descent formed the most densely concentrated slave population in the world. A contrast with the United States illustrates the point. By the fall of the Bastille in 1789, seven hundred thousand black slaves lived in the whole of the United States, while the five hundred thousand slaves of Saint-Domingue occupied an area that was only the size of Massachusetts.

This concentration of black Saint-Dominguans, moreover, had another dimension. Sugar plantations were highly labor intensive and deployed a workforce that could number hundreds of men and women. Subversive political and religious ideas could spread rapidly across a plantation in such conditions, as well as from a single plantation to other dense concentrations of slaves nearby. In these ways demography and work patterns provided a potential for successful rebellion unrivaled in other slave societies.

In addition, at critical moments the colony's blacks often found allies among the mulattoes. Under the prevailing racial definitions, mulattoes, or "coloreds," occupied a complex intermediate position. They were legally free, and whites relied on their collaboration as slave drivers, overseers, and militiamen. At the same time, however, people of mixed race were subjected to the humiliating indignities of segregation; they were forbidden by law to own their own weapons or hold public office; and they were normally confined to the lower rungs of the economic ladder. Their simmering discontent was a perennial weakness of the system.

Unhappily for those who sought to perpetuate hierarchies based on race, whites were so vastly outnumbered that even poorly armed black men and women could plausibly imagine turning the whites' world upside down. This vision was all the more alluring since the whites did not constitute a united force—being deeply divided along lines of class, education, and inherited wealth. The numerous *petits blancs*—as the slaves disparagingly called the amorphous mass of humble white artisans, sailors, shopkeepers, and street vendors—had interests that diverged sharply from those of the wealthy planters and merchants at the apex of the island's social pyramid. Even dissimilar interests within the plantocracy mattered, setting lowland sugar planters against the owners of much smaller coffee estates on the hills.

A further factor that destabilized Saint-Domingue at the end of the Ancien Régime was the fact that the majority of slaves had not been "broken" and trained to accept their plight. Because of the mortality on the plantations, the majority of slaves on the island in 1789 had been born in Africa and therefore remembered life before servitude. Men and women who had recently arrived in chains did not regard slavery as natural or permanent. On the contrary, they preserved memories of freedom and links to others from their homeland. They maintained their native languages, their customs, and elements of their religion. Many even had a background of military experience. For such reasons these "veterans of Africa" were well known to be prone to flight, and they were kept under the watchful surveillance of the *maréchaussée*—a rural police force commanded by whites and manned by mulattoes. Its responsibilities were to enforce labor discipline, punish malcontents, and capture maroons.

Paradoxically, a sinister light was cast on social conditions in the mid-1780s by the Colonial Ministry in Paris. Alarmed by reports of rebellion in

Saint-Domingue, the ministry issued decrees intended to buttress social stability by curbing abuses and improving the lives of slaves on the sugar plantations. It is sobering that the crown deemed it necessary to introduce punishment for planters who murdered slaves, to establish a weekly day of rest, to limit work hours, and to rule that slaves should receive a diet sufficient to preserve life. It is also revealing that planters vociferously protested the new regulations and that the court at Le Cap refused to recognize them. In the words of one recent scholar, Saint-Domingue had achieved an unhappy primacy as "the most extreme and concentrated slave system that had ever existed."[3]

Especially menacing to the old regime on Saint-Domingue was the phenomenon of marronage, that is, slaves who left the plantations and formed communities of fugitives beyond the reach of authority. Periodically, bands of maroons emerged from their retreats to conduct raids against the plantations, where they pillaged, burned, and killed. But economic growth, which entailed a remorselessly increasing acreage of land planted with sugarcane, posed a growing threat to the safety valve of flight into the hills. The extension of cane and coffee cultivation progressively reduced the availability of wasteland and forest, making marronage more difficult. Consequently, slaves increasingly confronted the stark choice between submission and rebellion.

Slave Rebellion and the Black Spartacus

Revolution in France ignited the combustible social order of Saint-Domingue, which was the antithesis of "liberty, equality, and fraternity." The colony followed events in the metropole with rapt attention. The storming of the Bastille in 1789 produced celebrations in Le Cap, Port-au-Prince, and Saint-Marc, especially by the *petits blancs,* who followed these festivities by opening political clubs on the Parisian model. Ideas, news from abroad, and local murmurings passed from ear to ear through networks of sociability in the towns. The segregated taverns, churches, and markets where black slaves and coloreds spent their weekly day off were forums for the exchange of ideas. There was also a colonial newspaper directed at a white readership, but coloreds and even some black slaves were literate and served as conduits for disseminating information to a broader public that extended to the plantations.

Meanwhile, the setting up of the National Assembly in France spawned the establishment of a Colonial Assembly at Saint-Marc. There, debates about racial privilege and the conflicting interests dividing France and its colony sent ripples traveling outward across the Caribbean. Indeed, the National Assembly's abolition of feudalism and aristocracy raised derivative questions that were passionately discussed in Saint-Domingue. Was slavery as obsolete as feudalism? Should the "aristocracy of the skin" in Saint-Domingue go the way of the "aristocracy of the robe" and the "aristocracy of the sword" in France?

Especially incendiary was the "Declaration of the Rights of Man," which was incompatible with conditions in the colony. Vincent Ogé, a spokesman for Haitian mulattoes, explicitly made the connection when he rose in the National Assembly to stress the universalism of human rights, which were just as valid in Le Cap as in Paris. Already in 1788 Jacques Pierre Brissot, the leader of the radical Parisian club known as the Girondins, had founded the abolitionist group Société des Amis des Noirs (Society of Friends of the Blacks), which espoused the cause of equal rights for mulattoes. Ironically, in view of their name, the "Friends of the Blacks" evaded the question of emancipation for black slaves. Questioning the idea of defining human beings as property and facing pressure from the Girondins and the society, the National Assembly in 1792 granted political rights to people of mixed race in its colony. Inevitably, the idea arose of the rights of people with darker skin. More radical than the Friends, the rival political club known as the Jacobins specifically committed itself to outlawing slavery in the colonies.

Between 1792 and 1794 the French Republic moved steadily in the direction of abolition. As a consequence, subversive ideas from the metropole swept Saint-Domingue at the same time that plantation slavery lost the automatic protection of the French state. In the past the state had always provided the means to enforce the social hierarchies of its colony—the governor, the courts, the military garrison, and the police. Now the National Assembly no longer prioritized the colonists' power to dominate half a million slaves. Ogé returned to Saint-Domingue from his Parisian sojourn to lead a revolt in 1790 and to demand an end to discrimination by race. The revolt failed, and Ogé was captured and executed. In the process he became, like François Mackandal, a symbol of the cause of freedom and a model of opposition to the social order.

In this increasingly politicized world, a spark touched off revolt on the sugar-producing plains. At the end of August 1791, black slaves and maroons held a famous meeting in the woods known as Bois Caïman to plot insurrection. After the meeting, the conspirators implemented their plan, marching from plantation to plantation to plunder supplies, torch cane fields, and destroy cane-processing machinery. Such acts of defiance were familiar as limited, local events in the tormented life of Saint-Domingue. The movement of 1791, however, dwarfed all prior rebellions both in numbers of insurrectionists and in the extent of the area affected. News of the revolt spread and triggered an uprising across North Province and then engulfed all of the colony. Field hands, urban slaves, domestic slaves, and even free mulattoes made common cause. Religion, both Vodou and Christianity, played a critical role by creating a sense of community among the slaves, furnishing a vocabulary of shared aspirations and promoting a recognized leadership of elders.

Dutty Boukman, a coachman and Vodou figure with great influence among his fellow slaves, presided over the Bois Caïman meeting and initially led the insurrection. The rebellion is most closely associated, however, with the "black Spartacus," Louverture, who joined the uprising from its outset. Under Louverture's leadership, the insurgents had a larger vision encompassing general emancipation and independence from France. Tellingly, the insurgent fighters sang revolutionary songs, raised the tricolor flag, and responded to Louverture's call to defend freedom—which he called "the most precious asset a man may possess."[4]

Once ignited, rebellion spread with remarkable speed. Nothing of its like had ever threatened slavery in the Americas. As the historian Laurent Dubois writes,

> The insurgents were enormously diverse—women and men, African-born and creole, overseer and fieldworkers, slaves on mountain coffee plantations and sugar plantations. . . . Using violence against a violent system, they shattered the economy of one of the richest regions of the world. During the first eight days of the insurrection they destroyed 184 plantations; by late September over 200 had been attacked, and all of the plantations within fifty miles of either side of Le Cap had been reduced to ashes and smoke. In addition, almost 1,200 coffee plantations in the mountains above the plain had been attacked. Estimates of the numbers of insurgents varied widely, but by the end of September there were . . . up to 80,000 in the insurgent camps.[5]

Violence and destruction were the hallmarks of the insurgency, as the rebels sought to destroy the plantations. They set fire to buildings and fields, ambushed militia patrols, and massacred landowners and overseers. Some rebel leaders urged moderation, but others boasted of how many whites they could hang in a day.

Saint-Dominguan whites were distracted by the news from Paris and divided among themselves. While the fields blazed across the north, their representatives at Le Cap carried on furious but oddly timed debates over strategy. Thus encumbered, they were slow to act, thereby allowing the rebels time to organize, recruit new adherents, and arm themselves. This delay is an important factor in the success of the revolt.

When the slaveowners and their supporters did take action, however, they responded with equal violence and intransigence. Refusing to negotiate with "rebel Negroes," the Colonial Assembly demanded unconditional surrender, an ultimatum that transformed revolt into revolution.

Napoleon's War to Restore Slavery

A critical question was the attitude of the authorities in France. Having ruled out compromise, the "aristocrats of the skin" had too narrow a following to defeat the mass of slaves, who were increasingly armed, united, and organized under the leadership of Louverture. The only real hope of the planters was that the National Assembly would send a large military force to Le Cap to crush the rebels and restore slavery.

Their misfortune was that between 1789 and 1792 the French Revolution was progressively radicalized, until power fell to the two factions represented in the assembly—the Girondins and the Jacobins. Both believed that abolishing the aristocracy in France implied abolishing the "aristocracy of the skin" in Saint-Domingue. In 1792 the assembly clarified its views on plantation slavery by sending two commissioners to Le Cap—Léger-Félicité Sonthonax, who supported abolition, and Étienne Polverel, who advocated reform. Together, they transformed politics in the colony.

In addition to their moral abhorrence of slavery, Sonthonax and Polverel were motivated by reasons of expediency. They sought desperately to repel the counterrevolutionary intervention in Saint-Domingue of Spain and Britain, which both feared that the rebellion could spread to their own possessions. France's major colonial rivals were eager to crush the dangerous examples of emancipation and black self-governance. The considerations that Saint-Domingue was a colony of incomparable wealth and that a blow struck there would send shockwaves across revolutionary France fortified the resolve of the two great enemies of revolution.

Against this threat posed by Spain and Britain, the commissioners took daring and decisive action to rally the slaves to the defense of the French Republic. On their own authority they issued decrees in 1793 that legally emancipated the slaves and granted them rights of full French citizenship. Louverture and the troops under his command then swore allegiance to the republic that had set them free. In exchange, the commissioners provided Louverture and his men with arms.

To secure approval in Paris for the measures they had propounded in the colony, Sonthonax and Polverel sent a three-man delegation to the National Convention (as the National Assembly had been renamed) that consisted of a white man, a mulatto, and the ex-slave Jean-Baptiste Belley. Dominated by the Jacobin left at the time, the Convention met Belley and his announcement of racial equality in Saint-Domingue with thunderous applause. Immediately afterwards, on February 4, 1794, the Convention issued one of the decisive documents of the French Revolution, its famous Emancipation Declaration, which stated simply, "The National Convention declares the abolition of Negro slavery in all the colonies; in consequence it decrees that all men, without distinction of color, residing in the colonies are French citizens and will enjoy all the rights assured by the Constitution."[6] There was no provision to compensate slaveowners for their loss.

The long-term survival of emancipation, however, hung in the balance for a decade because of a rightward swing of the revolutionary pendulum during the summer of 1794. In France, on July 27, a revolt against the Reign of Terror brought about the downfall of Robespierre and the Jacobin Club. Robespierre and close associates were executed, and the abolitionists were ousted from power. The three regimes that followed did not attempt to reverse the work of the revolution. In that sense, they were not counterrevolutionary, but they did intend to bring the revolution to a halt, while prioritizing stability and order. Furthermore, since revolutionary France was constantly at war, it became increasingly dependent on the army—and its dominant figure, the Corsican officer Napoleon Bonaparte (fig. 8.2).

As First Consul, Napoleon viewed events in Saint-Domingue with dismay. Liberation from below and miscegenation were both abhorrent to him, and he held that blacks were incapable of self-government. Like Spain and Britain, Napoleon also agonized over the possibility that the example of rebellious slaves in Saint-Domingue could threaten slavery in France's other American possessions—Guadeloupe, Martinique, Réunion, and Guiana. Committed to preserving racial boundaries, he began to entertain the idea of crushing the rebellion by force and restoring slavery. He gave a hint of his intentions when he ordered that whites who fraternized with Louverture be exiled, together with white women found guilty of having liaisons with coloreds or blacks. Circumstances had made Louverture governor of

Figure 8.2. Jacques-Louis David, *The Emperor Napoleon in His Study at the Tuileries* (1812). National Gallery of Art, Washington, D.C.

Saint-Domingue, but Napoleon regarded him as an impudent black who had risen above his station and needed to be put in his place.

Furthermore, Napoleon detested the political style of his nemesis—because it was far too similar to his own. Within his smaller domain, Louverture was a military leader of brilliance and an authoritarian dictator who, besides being known as the "black Spartacus," some called the "black Napoleon." Although officially still a French colony even after a decade of insurgency, Saint-Domingue under Louverture acted like a sovereign nation. It drafted a constitution, pursued a foreign policy that took little account of French interests, and declared Louverture governor for life. Although the constitution asserted that Saint-Domingue remained French, it also announced that the island was subject to "special laws" of its own creation.

Such defiance bordered on treason. It was especially unacceptable since the island played so powerful a role in the French economy. Napoleon calculated that the rich plantations of the island could be revived. His intention was to distribute the land as a reward to loyal supporters with the financial means to restore the damage to the estates and resume production. Napoleon therefore filled the ministries of the navy and the colonies with staunch defenders of slavery and began to plan a massive expedition. He was happy, he wrote, to yield to "the forceful advice of the gentlemen who live in Saint-Domingue."[7]

The First Consul also harbored a grander geopolitical aspiration. Ever restless to expand his dominion and to achieve lasting glory, Napoleon boldly imagined a revived colonial role for France in North America. If he could crush the hated Negro revolt in Saint-Domingue, he could then use the island to extend French power into Louisiana and up the Mississippi. In the process, France could enrich itself while winning the contest for power with Great Britain. Since Napoleon preferred to trust in his personal genius and intuition rather than detailed plans, he left the specifics to the future. But the first step toward making France a North American power was clear: to annihilate the blacks who stood between him and his ambitions. War with Louverture loomed.

It would be misleading to define the clash of Napoleon and Louverture as a simple contest between slavery and freedom, or between white supremacy and racial equality. Louverture was a pragmatist who envisaged the establishment of a hybrid, semifree plantation system. In his view of the future, wealthy whites, coloreds with capital, and elite blacks would be encouraged to cooperate, invest in sugarcane, and manage a docile workforce. Louverture insisted that under the new system, farmworkers would be free to change employers and to live without fear of corporal punishment. On the other hand, blacks and coloreds, though legally free, would be compelled, by force if necessary, to toil for low wages under harsh discipline, and they would not be permitted to migrate to the towns. Louverture feared that field hands had learned to regard farm labor as enslavement and that once free they would lapse into sloth.

His priority was to revive the island's productivity by all means necessary short of actual reenslavement, racial segregation, and French rule. He viewed plantations from a complex perspective. He was not only a former slave, but also a prosperous planter who resorted to forced labor and had once owned slaves himself. Furthermore, Louverture included former slaveowners among his closest advisors. His vision was not one of social equality and political freedom but of a stern autocracy that could impose the discipline needed to restore a strong economy after a generation of warfare and destruction. As governor, his power extended to life and death over his subjects. Indeed, Louverture ruled with so iron a hand that Napoleon's move against him was marked by an ironic political calculation. Napoleon reasoned that the authoritarian black Spartacus had so alienated his followers that many of them could be won over to France, provided that the eventual restoration of slavery was concealed.

Louverture's fear was that the land of Saint-Domingue would be parceled out to freedmen in tiny plots devoted to subsistence crops rather than to sugar. Such a transformation, he believed, would doom the island's economy. Therefore, the black Napoleon did not endorse the heady idea of freedom envisaged by the black rebels at their emancipation in 1794. His goal was a halfway regime of semifree wage labor, legal equality without regard to skin color, and national independence.

Buoyed by British and US endorsement, Napoleon decided in the spring of 1801 to strike the decisive blow. He informed his minister of foreign affairs that his purpose was "to crush the government of the blacks," and he wrote to his brother-in-law and general, Charles Victoire Emmanuel Leclerc, "Rid us of these gilded Africans, and there will be nothing left for us to wish."[8] Sparing no expense, he assembled an armada of sixty-five ships that would sail from seven ports and gathered scores of planters who had fled the rule of Louverture, some thirty thousand soldiers, and a full complement of sailors. This armada, headed by Leclerc, set sail in December 1801 as a first wave to be followed by twenty thousand reinforcements in the spring.

A wily opportunist, Napoleon gave his commanders, Leclerc and Rochambeau, authority to determine when and how to revoke emancipation in the colony. The first priority was to reimpose direct French rule over the unruly colony. Leclerc could then move in a more leisurely manner to restore slavery, depending on his assessment of how much opposition restoration would arouse. The goal was fixed, but the pace would be determined pragmatically.

As the troops embarked, their mood was ebullient. Victory was a foregone conclusion, and the soldiers believed its attainment would open the way to preferment, business opportunities, and great wealth. Nicolas Pierre Gilbert, medical officer-in-chief of the expedition, recalled that when plans for the invasion were known, the Ministry of War was besieged by men with an overpowering desire to sail for the new El Dorado. But things did not go their way, and something besides revolutionaries stood in their way of their anticipated victory and success.

The French Army Destroyed

What followed was a stunning and unexpected defeat. Napoleon's Haitian intervention failed for a variety of reasons. One was the novelty of the ex-slaves' style of war. France deployed a regular army drilled and equipped to fight conventional battles in Europe, but suddenly it confronted a people's war in the tropics. Napoleon and his generals opposed a people aroused by political ideals, religious beliefs, and the expectation that defeat meant total annihilation. The conflict in Saint-Domingue was fought with unmitigated ferocity because both sides understood it as a war of extermination.

A second factor in the failure of Leclerc's expedition was the fact that the forces he confronted were no longer the rebellious slaves armed with only machetes who had overpowered the planters in 1791. By February 1802, when Leclerc landed at Le Cap, Louverture's troops were disciplined, well-armed veterans. Furthermore, the generals who commanded them, including Louverture himself, Jean-Jacques Dessalines, and Henri Christophe, were brilliant tacticians. They knew the rugged terrain of the island intimately and how to exploit it for surprise attacks after which they vanished back into the woods.

Louverture's crucial insight was that victory depended on avoiding a pitched battle with Leclerc's army in the two to three months between its landing and the onset of the summer rainy season. The rebel forces lacked the fire power and training to defeat Napoleon's troops in a set-piece confrontation, so it was far better to harass the French by ambushes until the Haitian environment itself confronted the European army in the summer. In devising his military strategy, Louverture took account of his deep empirical knowledge of what might be called the medical climatology of Saint-Domingue. He had little formal education, but he had acquired some knowledge of African healing practices and had developed an interest in matters of health while working in a Jesuit hospital.

Louverture therefore knew full well that newly arrived Europeans perished of yellow fever every summer while black men and women remained stubbornly healthy. He also believed that the disease was the product of miasma—an environmental poison. So the appropriate strategy was to allow time for the Haitian summer and its airborne poisons to destroy the French. In this sense he was faithful to the legacy of Mackandal. Louverture intended to use poison—a classic example of what the political scientist James Scott calls the "weapons of the weak" in his book of the same name—to defeat the French and liberate Haiti. As a recent study argues, Louverture had "an awareness of when and where the fevers would strike his European enemies. He . . . knew that by maneuvering the whites into the ports and lowlands during the rainy season, they would die in droves. In a letter to Dessalines, he wrote, 'Do not forget that while waiting for the rainy season, which will rid us of our enemies, we have only destruction and fire as our weapons.'"[9]

To confound the French, Louverture took full advantage of two critical sources of information. The first was his detailed knowledge of Haitian medical topography and climate. The second was the extensive intelligence network he had built to track the movements of the invaders. Although the French also understood that the hills and mountains were far healthier in the summer than the lowlands and the towns, they underestimated the danger of yellow fever and never anticipated an epidemic as violent as that of 1802–1803. Furthermore, as Louverture understood, Leclerc needed to prioritize defending the ports, especially Le Cap, as they were key to French control, supply lines, and reinforcement. They could not be lost without losing the colony as a whole. Inevitably, therefore, Leclerc's strategy would combine holding the well-fortified towns and carrying out search-and-destroy expeditions into the interior with the goal of rapidly confronting his enemies in an all-resolving climactic battle.

Against that background, Louverture's counterstrategy was to avoid a pitched battle entirely, correctly calculating that time and disease were his best allies. If he could conduct hit-and-run engagements and allow Leclerc to experience the tropical summer on the lowlands and in Le Cap, the French could be destroyed by disease more surely than with musket fire.

Louverture himself was undone by agreeing to meet the French to negotiate a possible settlement. Betrayed by his supposed interlocutors, he was captured and deported to France. There, fearing unrest if Louverture became a martyr, Napoleon chose to let him die slowly while imprisoned in the Jura Mountains. Louverture perished in the spring of 1803 after months on a regimen of minimal food, heat, and light. Fortunately for the Haitian revolution, Dessalines, Louverture's successor as commander, fully understood the concept of warfare by miasma. As he prepared his troops for their first confrontation with the French in March 1802, Dessalines told them, "Take courage, I tell you, take courage. The whites from France cannot hold out against us here in Saint Domingue. They will fight well at first, but soon they will fall sick and die like flies."[10]

On the French side, Leclerc was a source of weakness for the expeditionary force. He was an inexperienced but cocksure officer who acquired his command by marriage to Napoleon's sister rather than his own abilities. Politically, he was unequal to the task of maneuvering successfully in the crosscurrents of race and patriotism that buffeted the island. More importantly, he led his army into a military cul-de-sac. With no colonial background, Leclerc expected an easy triumph over the *nègres,* as he derisively called them, whom he fatally underestimated. Louverture and his men, he reasoned, were an ignoble rabble who would drop their weapons and fly from a regular army. Leclerc therefore ignored the prescient wisdom of a subordinate who warned that one hundred thousand men rather than the thirty thousand available would be needed to subdue Saint-Domingue.

From the outset of his nine months of service, Leclerc met unwelcome surprises at every step. The landing near Le Cap, which the insurgents held until the French put ashore, provided the first. Before withdrawing inland, Louverture's forces set fire to the port city, leaving it a smoking heap of ashes with only its stone buildings and fortifications still standing. Rebuilding would prove a painful drain on French resources, and the sudden necessity of reorganization delayed French military action at a critical moment when time was of the essence. It was also a powerful gesture of psychological warfare.

Disastrously delayed while spring approached, Leclerc moved to implement his strategy of enveloping and destroying the rebel forces. His plan was to deploy five divisions that would march inland separately and converge on the center of the island. To that end, French forces marched from their positions on February 17 and headed for the unknown world of the Haitian interior. Curiously nonchalant, they had no maps for guidance.

What the French discovered in yet another painful surprise was the ingenuity of their opponents in exploiting the unwelcoming topography of the island. The landscape of the interior was a disorienting maze of hills pierced by steep ravines populated with all manner of stinging and biting insects. As they advanced, Leclerc's troops were tormented by heavy rain, a shortage of boots, and improper woolen uniforms. Bathed in sweat in the heat of the day, they bivouacked in the mud by night, still wet and suddenly cold. Worse still from Leclerc's perspective, the rebels ignored the conventions of European warfare and its duels on open battlefields. Most unchivalrously of all, ex-slave women joined the fray, ready to settle scores of their own and prepared to die rather than return to the humiliation of the plantations.

Carefully tracking the invaders' movements, the insurgents ambushed them with unparalleled ferocity. The term "guerrilla warfare" was coined only a decade later in Spain. After years of insurgency, however, the Saint-Dominguan rebels were already experts in the art.

Napoleon of course contributed to Leclerc's poor preparation for war in the tropics. As a Corsican, the First Consul knew of the danger of insect-borne diseases in warm climates. Like neighboring Sardinia and Sicily, nineteenth-century Corsica was devastated by mosquitoes and the parasites they transmitted. His advisors also warned Napoleon of the specific peril presented by yellow fever in the West Indies. Taking that information partially into account, Napoleon arranged the expedition's departure so that it coincided with the winter months—on the optimistic assumption that victory would be rapid. Just like Leclerc, Napoleon never entertained the possibility that the conflict would carry over into the summer fever season, and neither of them made any preparations for the medical consequences of such a delay. This hubristic miscalculation by both the First Consul and his brother-in-law played a large role in the disaster that followed.

Slow-moving columns of infantry—weighed down by heavy kit and artillery that they dragged themselves—proved to be ideal targets for lightning assaults. After seventy-six days of bloody but inconclusive clashes, the spring offensive lost all momentum and purpose. As the temperature rose, the rains unleashed their summer fury, and the first French soldiers fell ill. Leclerc admitted failure and retired to Le Cap.

The ineffectiveness of traditional military methods in the tropics against an opponent who refused to be drawn into battle drove Leclerc to adopt an alternative strategy. Since his prey had eluded pursuit, Leclerc devised a policy that would now be called counterinsurgency. His new goal was to drive the civilian population into submission. He therefore resorted to systematic reprisals, ordering his soldiers to burn the crops wherever rebel forces were most active. The hope was to starve them into submission. In addition, thwarted and frightened soldiers began to vent their rage by committing atrocities against the unarmed black population as well as the active rebels. Rape as an instrument of war came into its own. As so often in Saint-Domingue's history, sexual violence played its part in the radically unequal power relationships between European men and women of African descent.

By June 1802 the French commander acknowledged that both of the original military strategies—the blitzkrieg and the reprisals—were failing to reduce the rebellion. Adapting to necessity, he instituted a third approach—a program of disarmament that lasted until August. Under its terms the

French instituted summary execution for anyone who resisted authority or was caught with weapons in his or her possession. Leclerc also turned his attention to mulattoes, whom he found disappointingly unreliable, announcing that the overseers of any estate where guns were found would be shot. Furthermore, to underline French determination, the army changed its method of execution. Until the summer of 1802, the standard of capital punishment was death by firing squad. Now Leclerc favored public hanging. The intention was to terrorize blacks into surrendering their arms.

What further radicalized French policy in Saint-Domingue was an inopportune combination of extreme political decisions emanating from Paris and the outbreak of epidemic disease. In a series of momentous policies announced between May and August 1802, Napoleon reversed the ringing universalism of human rights proclaimed in 1794. Revealing his disdain for that "false philosophy," as he termed abolitionism, Napoleon took action. In May he confirmed the legality of slavery in those colonies, Martinique and Réunion, where it had never been abolished. He also imposed restrictions on people of color in metropolitan France. Then, in quick succession, he authorized the resumption of the slave trade in the colonies and the restoration of slavery in Guadeloupe and Guiana.

Napoleon prevaricated with respect to slavery in Saint-Domingue by stating that the labor regime to prevail there was still undecided and would be determined within ten years. He dissembled on the practical grounds that an immediate attempt at reenslavement would only inflame the rebellion.

Simultaneously, epidemic disease transformed French strategy. In one sense, Leclerc's spring campaign had been successful. He was able to traverse the colony, proclaiming metropolitan control and formally placing Saint-Domingue under martial law. In practice, however, the crucial questions were how to make the control real and how to maintain it. Both issues were soon moot as yellow fever rapidly overtook politics. The first cases came to the attention of military physicians in late March. Beginning in April, the number of cases increased daily as the outbreak gathered momentum. By early summer, when Leclerc had just introduced his disarmament program, the epidemic erupted across the whole colony. Although yellow fever had an impact on both sides engaged in the conflict, it disproportionately ravaged the French because they lacked all immunity. Napoleon's army began to melt away as his microbially defenseless troops began to die in terrifying numbers. Even the approximate statistics available tell a story of impending catastrophe.

In his book *Mosquito Empires: Ecology and War in the Greater Caribbean, 1620–1914,* John R. McNeill estimates that Napoleon dispatched sixty-five thousand troops in successive waves to suppress the revolt in Saint-Domingue. Of these, fifty thousand to fifty-five thousand died, with thirty-five thousand to forty-five thousand of those deaths caused by yellow fever. Thus in the late summer of 1802 Leclerc reported that he had under his command only ten thousand men, of whom eight thousand were convalescing in hospital, leaving only two thousand fit for active duty. Two-thirds of the staff officers had also succumbed. Nor could the convalescent troops be expected to return to duty in the near future because recovery from yellow fever is lengthy, complicated, and uncertain.

Yellow fever's exorbitant kill rate maximized its effect on Napoleon's army. Precise statistics on morbidity and mortality were not kept in the emergency conditions prevailing in Saint-Domingue, but the virulence of the disease was extraordinary. Few of the French soldiers who fell ill of the disease recovered. Almost as if it were a weapon intentionally deployed by the insurgents, fever targeted Europeans and killed a substantial majority. Indeed, one of the distinctive features of 1802–1803 that was carefully recorded in horror at the time was the fulminant course of yellow fever in soldiers' bodies.

During previous outbreaks in both the Caribbean and the United States, the disease normally progressed in two stages. The first was the onset, which was usually sudden with no premonitory signs. Patients would be seized by chills and high fever, an agonizing frontal headache, nausea, and general fatigue. Then, after approximately three days, they would seem to recover, even chatting with friends or attendants as the symptoms abated. In mild cases, the recovery lasted and a lengthy convalescence began.

In severe cases, however, the remission ended after less than twenty-four hours, and the virus then attacked the body with full intensity. In this second stage the clinical manifestations presented the classic tableau of yellow fever: high temperature with paroxysmic chills that rack the patient for hours at a time; vomit blackened by congealed blood that resembles coffee grains; profuse diarrhea; tormenting headache; skin yellowed by jaundice; hemorrhages from the nose, mouth, and anus; violent and prolonged fits of hiccough; a terrifying general prostration; delirium; and skin eruptions all over the body. Far too frequently the patient lapsed into coma and death, usually after twelve days of misery. For those who survived, convalescence lasted for weeks, during which sufferers were subject to extreme fatigue and to sequelae that followed from the involvement of the central nervous system—depression, memory loss, and disorientation. Even during convalescence, patients were still vulnerable to relapses, dehydration, and potentially lethal complications. In Haitian conditions, the most common of the complications were pneumonia and malaria. In aggregate, early-nineteenth-century physicians confronting a yellow fever epidemic anticipated a CFR ranging from 15 to 50 percent.

A marked feature of the epidemic at Saint-Domingue was the absence of mild cases. Directing the treatment of Leclerc's soldiers, Gilbert and his medical colleagues expected patients to fall into one of three "degrees" of yellow fever—mild, moderate, or severe. But to their horror, patients passed directly from the moment of onset into the most fulminant form of the disease. They then died with a rapidity that confounded Gilbert and left him in despair at his inability even to alleviate suffering. He reported that the disease ran its full course in the bodies of the French soldiers in no more than a few days. Since Gilbert and his medical team were overwhelmed, they paid little attention to record-keeping; however, informed retrospective guesses suggest a CFR in excess of 70 percent. Gilbert, understandably stunned by the soldiers' mortality, wrote that nearly all of his patients died.

We can only form conjectures to explain the exceptional violence of the yellow fever epidemic that afflicted Napoleon's expedition. It is possible that a mutation of the virus enhanced its virulence. Viruses are famously unstable, as their almost instantaneous replication multiplies mutations; but whether this contributed to the violence of the Saint-Domingue epidemic is speculative. Other contributing causes are more certain. One, of course, was the general circumstance that the epidemic ran its course as virtually a "virgin soil epidemic" in a population entirely lacking herd immunity. Some sixty-five thousand newly arrived French soldiers were "unseasoned" and highly vulnerable.

Factors specific to the military environment played their part in the medical catastrophe as well. Two are biological—age and gender. Yellow fever is unusual in the age profile of its victims. Instead of demonstrating a preference for children and the elderly in the "normal" pattern of most diseases, this *Flavivirus* preferentially afflicts robust young adults. Apart from senior officers, some of whom were middle-aged, the French expeditionary force precisely matched this criterion.

The virus also displays a marked gender preference for males. A partial explanation for this gender gap is the role of sensors located in the antennae of *Aedes aegypti*. The adult female mosquito locates her prey in part by detecting the enticing scent of human sweat. Since soldiers do heavy manual labor, a large population of newly arrived soldiers and sailors was at severe biological risk. At the same time, the expeditionary force was not completely male. Significant numbers of women accompanied the army—wives and servants of officers, cooks and purveyors of supplies, and prostitutes. But in comparison with other major outbreaks of yellow fever, the Saint-Domingue epidemic was an outlier because the population at risk consisted overwhelmingly of young European males engaged in exhausting physical toil during a tropical summer. Exceptional morbidity and mortality therefore followed.

The contemporary understanding of yellow fever was that it resulted from the effects of environmental poisoning on bodies with susceptible constitutions. Gilbert, for example, explained the disaster at Le Cap by reference to the noxious miasmas that contaminated the air. And indeed, the city reeked with pestilential odors, as did all nineteenth-century cities. A variety of local factors augmented this inevitable stench. One was that no sanitary provision had been made for the sudden influx of troops. As a result, the soldiers relieved themselves in abandoned buildings throughout the city, which soon stank of excrement and urine. In addition, there was the problem of the municipal cemetery, where sanitary regulations prescribing burial depths were ignored because of the overwhelming numbers of bodies. The dead were hastily laid in shallow graves, permitting noxious fumes to waft across the town. Finally, the prevailing onshore winds blew exhalations rising from decomposing matter in neighboring marshes into the city. In contemporary understanding, therefore, the air of Le Cap was itself lethal.

During yellow fever epidemics, the most common response of city dwellers was to flee. Moved by contemporary miasmatic doctrine and prodded by their doctors, people escaped the poisoned air by taking refuge in the countryside and on high ground. Philadelphia during the severe epidemic of 1793 is an example, and one that was followed carefully by observers in Saint-Domingue. When fever attacked Philadelphia, half of its population departed in a sudden exodus. The lesson of that city was not lost on the French army crowded in its quarters at Le Cap. Indeed, Leclerc followed traditional medical advice together with his wife and small son: they took up residence in a farmhouse set in the pure air of the hills overlooking the port. There the frightened and demoralized commander repeatedly penned letters requesting to be relieved of his duties and transferred to Paris.

Rank-and-file troops, however, were not free to follow their cautious general. Le Cap was the nerve center of the French military expedition. The city contained Leclerc's headquarters, the army and navy barracks, two large military hospitals, the docks where reinforcements arrived, and fortifications. Economic and political power were also centered there. As a result of these considerations, military units were concentrated at Le Cap and other ports, making only intermittent forays into the uplands. The result was enhanced vulnerability to yellow fever.

An additional condition affecting mortality at Le Cap was the standard of care the troops received. Historically, in many epidemics, this variable was of limited significance because so many patients had no access to medical care and because health facilities had no surge capacity to cope with the dizzying spike in the numbers of sufferers. As a population subject to a violent outbreak of contagious disease, the soldiers and sailors serving in Saint-Domingue were an exception. Acknowledging both the hazards of warfare and the evil reputation of the Caribbean for the health of Europeans, the French Ministry of War had established two large hospitals at Le Cap. But as the soldiers were decimated by fever as the summer progressed, these establishments were stretched beyond their intended capacities and filled to overflowing. They coped by allocating two patients per bed and consigning any remaining patients to mats on the floor.

Unfortunately, then as now, there was no effective treatment for yellow fever, and the medical profession had not reached any consensus regarding the best standard of care. According to learned medical theory of the time, epidemic fevers were produced in susceptible bodies by atmospheric poison and heat. These factors corrupted the bodily humors and gave rise to a plethora of blood. This result was crystal clear at the bedside. Blood is the humor that combines the elements of heat and humidity, and the yellow fever patient was invariably hot to the touch and drenched in perspiration. The proverbial "healing power of nature" provided guidance because the body seemed spontaneously to make every effort to expel the excess humor of corrupted blood by evacuating bloody vomitus and stools and by oozing blood from the mouth and nose. The indication for treatment was therefore to second the work of nature, enhancing the flow through venesection, purgatives, and emetics. Reinforcing the appeal of this therapy was the usual presumption that so violent a disorder requires an equally forceful remedy.

In discussing his own strategy during the Philadelphia epidemic, the influential American physician Benjamin Rush had adopted what he termed a "grand purge" by means of a powerful emetic composed of mercury and jalap. He administered this medication several times daily in so large a dose that his colleagues shuddered, remarking that it was more than enough to kill a horse. Undaunted, Rush supplemented the purge by opening veins and administering emetics, with results that he pronounced as the sovereign therapy for "bilious remitting yellow fever," as he called the disease in the title of his 1794 book.

Judged by modern standards, the strategy of bleeding and purging sufferers who are prostrate and sinking into a coma is dubious. Gilbert, however, had prepared himself for the expedition to Saint-Domingue, where he knew yellow fever to be endemic, by reading Rush and consulting medical doctors who had also experienced the disease in Philadelphia. Under his direction, therefore, doctors at the military hospitals at Le Cap followed Rush's guidance. Thus it is plausible that the therapeutic strategy of the French military physicians was a factor, initially, in the unprecedented mortality of the outbreak. Indeed, Gilbert writes that he was depressed by the results of his efforts and his inability to save his patients. "My care, and my repeated visits every day," he noted, "were useless. I was despairing."[11]

These early attentions of the French doctors may have hastened the demise of their patients. Over the course of the outbreak as a whole, however, inappropriate medical treatment is unlikely to have played a major role. The military physicians in Saint-Domingue rapidly changed course when they observed that their methods were counterproductive. They were conscientious and held regular meetings to compare findings and agree collectively on new approaches in a desperate situation. They even experimented with Vodou remedies known to the hospitals' colored attendants. By the end, however, they abandoned the hope of cure entirely, resorting instead to a strategy of alleviating the most painful symptoms by modest means—cool flannels, tepid baths, lemon-flavored water, small doses of quinine, and mild laxatives. As the epidemic reached its acme, the fever even negated medical and nursing care entirely by killing off the hospital staff. At Fort Liberté, for instance, all caregivers perished, leaving patients to their own devices.

Whatever the cause of the virulence of the disease afflicting his soldiers, Leclerc despondently acknowledged that the combination of reenslavement and disease placed the expedition in a position that was daily growing more untenable and wrote to Napoleon:

> The malady is making such frightful progress that I cannot calculate where it will end. The hospitals at Le Cap alone have lost this month 100 men a day. My position is no better; the insurrection spreads, the malady continues.
>
> I entreated you, Citizen Consul, to do nothing which might make the rebels anxious about their liberty until I was

> ready. . . . [But] suddenly the law arrived here which authorizes the slave-trade in the colonies. . . . In this state of affairs, Citizen Consul, . . . I can do nothing by persuasion. I can depend only on force, and I have no troops.
>
> My letter will surprise you, Citizen Consul. But what general could calculate on a mortality of four-fifths of his army and the uselessness of the remainder?

In a revealing imputation of recklessness, Leclerc complained to Napoleon, "My moral ascendancy has collapsed now that your plans for the colonies are well known."[12] What Leclerc concealed from Napoleon was that he had highhandedly rejected the prophylactic advice of his medical advisors. Accepting that there was no effective therapy, his medical council had suggested that, since the disease was miasmatic, it might be prevented by a multipronged program. This program would remove filth from the city, station troops on hills above the mephitic breezes below, move the cemetery, and cleanse streets and public spaces. Why Leclerc refused to act is unclear, but the measures he cast aside probably would have made some difference. Philippe Girard, a leading historian of the rebellion, argues that folding his hands with regard to hygiene was "Leclerc's greatest error as commander-in-chief."[13]

Unable to implement the disarmament policy he had announced in June, Leclerc launched an even more violent phase of the expedition. Realizing that he could no longer hope for conventional military victory, he gambled on a strategy of extreme terror to be carried out by surviving French troops and a complex assemblage of auxiliaries. The latter consisted of planters and merchants who formed their own militia, sailors from the fleet whom he pressed into infantry service, and those Haitian mulattoes and blacks whom he could induce to collaborate. Although he lacked able-bodied white soldiers, Leclerc had funds, and he used them to purchase allies.

Even in the racially charged conditions of Saint-Domingue, the divide between black and white was never absolute. At least until the French announced their intention of restoring Caribbean slavery, some blacks found reasons apart from pecuniary inducements to take up arms on behalf of Leclerc. Some were opportunists who calculated that France would emerge victorious and that a day of reckoning would then be at hand for those who had taken the other side. A number of former slaves persisted in their regard for the country that had first abolished slavery. There were also farm laborers who, as Napoleon had guessed, violently opposed Louverture's plans for labor, which they regarded as a restoration of all the essentials of slavery except the whip. Others had local scores to settle and ambitions that Louverture had thwarted. Some acted under duress when the French threatened their families. With such multifarious collaboration, the French commander enlisted units of black and colored men to join his counterrevolution.

The new strategy was one that Leclerc described as a "war of extermination." It bordered on genocide as he began to contemplate the importation of indigent white farmworkers from Europe to replace black "cultivators" on the land. "We must," he announced, "destroy all the Negroes in the mountains, men and women, keeping only children under twelve; we must also destroy those in the plains, and leave in the colony not a single man of color who has ever worn an epaulette."[14]

Before he could witness the outcome of the new course, Leclerc himself died of yellow fever on November 2, 1802, despite the supposed safety of his elevated farmhouse. The command passed to Rochambeau, who persevered with his predecessor's draconian strategy. He even earned enduring notoriety for his ingenuity in devising the means to inflict gratuitous pain. Leclerc, Rochambeau lamented, had been too restrained, and he accused his predecessor of a "negrophilism" that could ruin the colony. True to his promise to show no mercy, Rochambeau constructed a wooden arena at Le Cap where he revived memories of the Roman Coliseum by arranging gladiatorial combats between black prisoners and packs of fierce guard dogs imported from Cuba. A further sadistic twist was Rochambeau's revival of *noyades*, or drownings, to intimidate peasant rebels. During the Terror in France, envoys of the National Convention had drowned thousands of insurgents at Nantes by loading them on barges that were then sunk in the Loire River. Now in 1802–1803 Rochambeau ordered that blacks suspected of rebellion be transported offshore, shackled, and thrown overboard. Increasingly desperate and vindictive, Napoleon's general even introduced crucifixion to the West Indies and used the holds of ships as gas chambers where he murdered prisoners by asphyxiation with sulfur.

Despite Rochambeau's best efforts, the tactic of intimidation by atrocity collapsed early in 1803. Although twelve thousand reinforcements arrived in time to help launch the strategy of terror, they rapidly sickened, and the remaining white troops were too few and too demoralized to implement the French general's genocidal vision.

Meanwhile, all hope of further reinforcements disappeared when renewed war with Britain led the Royal Navy to blockade French ports. The British acted not to hasten Haitian independence but to weaken the French economy and limit Napoleon's military options. Napoleon recognized that the game was up in the Americas and turned his attention to other matters. He is alleged to have muttered, "Damn sugar! Damn coffee! Damn colonies!"[15]

Thus abandoned by France, Rochambeau found that few blacks and mulattoes were willing to assist in his race war. Far from crushing the insurgency, he withdrew to the fortified port towns, only to be besieged there by Dessalines and his implacable ally, *Aedes aegypti*. One by one towns fell to the insurgents, and on November 18, 1803, Rochambeau lost a final engagement with Dessalines's troops at the Battle of Vertrières near Le Cap. The next day Rochambeau negotiated surrender in exchange for clearance to set sail on November 19 with his surviving soldiers. They departed, however, as prisoners aboard British warships, and Rochambeau remained captive until 1809.

Conclusion

Events in Haiti clearly show the influence of epidemic disease on war. By its defeat in Saint-Domingue, France directly lost fifty thousand people—soldiers, sailors, merchants, and planters. In the process it also disappointed the hopes of the adventurers who expected an easy colonial victory to open their way to fortunes. Instead, almost no French profited from the war to reimpose French rule and slavery.

A cascade of consequences followed on the geopolitical level. A major purpose of Napoleon's effort to reestablish French hegemony in Haiti was for the island to become the springboard to renew French empire in North America. Having been humiliated in the Caribbean, however, Napoleon concluded that the territories ceded by Spain had become worthless. Without Saint-Domingue, France would be militarily vulnerable, and Napoleon had a new conception of the danger posed in the Caribbean by disease.

He therefore sought to fulfill his restless ambitions in other directions. To the east he had a vague but grand idea of overturning the British hegemony in India. To achieve that goal, however, he needed to destroy Russia, which stood in his path. He would find, however, that epidemic diseases—this time dysentery and typhus—would again play a role in thwarting his vision.

CHAPTER 9

War and Disease

Napoleon, Dysentery, and Typhus in Russia, 1812

After assembling the largest military force ever created, Napoleon Bonaparte launched the decisive campaign of his career—the invasion of Russia in the summer of 1812. In Russia as in Haiti, however, epidemic diseases destroyed his men and defeated his ambitions. Dysentery and typhus determined the course of war far more than strategic skill and force of arms. In order to understand that result, we need to examine the conditions to which the emperor subjected his troops throughout the fatal trek to Moscow and back (fig. 9.1). Those conditions created an ideal environment for microbes to flourish and men to die.

Having put the disastrous expedition to Saint-Domingue behind him, Napoleon had turned his attention elsewhere. For a time he was overwhelmingly successful in augmenting his domestic power in France and expanding the empire across Europe. Between 1805 and the Russian campaign in 1812 he reached the peak of his power through victory in a dizzying series of battles. Through the Treaties of Tilsit in 1807 he imposed humiliating terms on both Prussia and Russia. His domains extended beyond France from Italy to Holland and into dependent states and satellite kingdoms such as the Confederation of the Rhine and the Kingdom of Naples. As military commander Napoleon had created an aura of invincibility, and as emperor he had choreographed his own near-apotheosis in two weeks at Dresden. There, a multitude of German kings, queens, and dukes joined the Austrian emperor

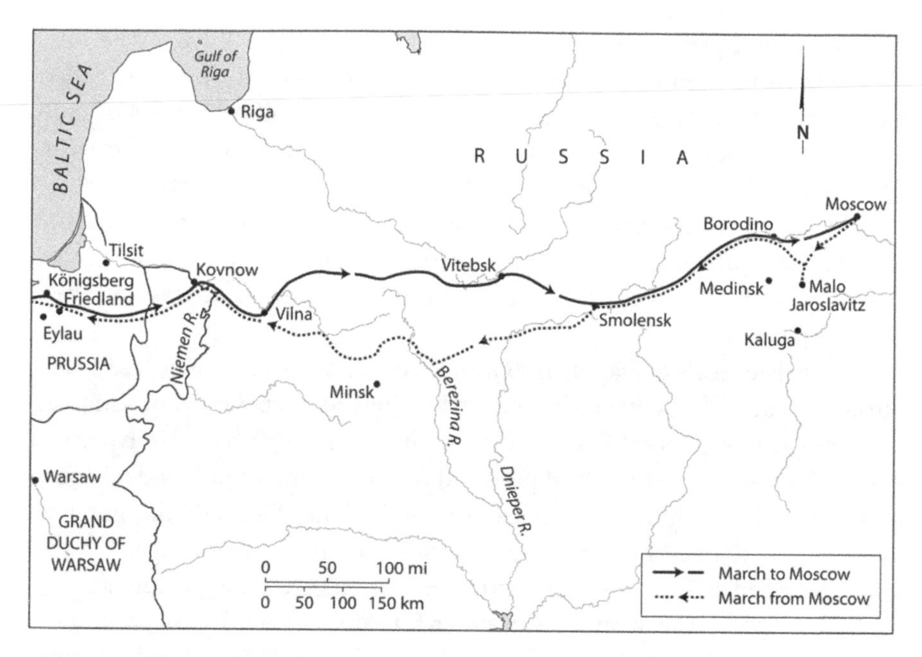

Figure 9.1. During Napoleon's campaign in Russia (1812), the Grande Armée was defeated by dysentery and typhus. (Map drawn by Bill Nelson.)

to pay him homage in May 1812. His only matters of pressing concern at that time were the ongoing military conflict in Spain, the irreducible enmity of Britain with its control of the seas, and the festering discontent of those dominions subject to French exactions, especially Germany.

What puzzled Europe at the time and scholars ever since is why, flush with such astonishing success, Napoleon made the momentous mistake to overreach by invading Russia in June 1812. This campaign was different from any other Napoleonic war. It was an act of naked aggression or, in Eugene Tarle's phrase, "more frankly imperialistic than any other of Napoleon's wars."[1] There were two pretexts for the onslaught. One was that France sought to liberate Poland, and for that reason Napoleon dubbed the invasion the "Second Polish War." As we know from his conduct in Haiti, however, the emperor was by no means committed to emancipating subject peoples. Indeed, beyond the empty announcement that he had come as a liberator, Napoleon gave no indication of the nature of Polish emancipation or of the boundaries of the new Polish state. His promise of freedom is best understood as a propaganda ploy to aid him in recruiting Poles to serve in his Grande Armée.

A second pretext was the claim that invading Russia was the only means for him to pursue economic war against England by the so-called Continental System. Proclaimed by the Berlin and Milan Decrees of 1806 and 1807, the Continental System imposed an embargo on trade between Britain and Europe. The goal was to cripple British industry because French goods would replace British ones in the marketplace. When the Russian tsar flouted the decrees, Napoleon held that his attack was defensive. It was simply a means to bring the Russians to their senses and then restore the peace.

The sheer scale of Napoleon's Grande Armée suggests that he had larger purposes than Polish emancipation, in which he did not believe, and enforcement of the Continental System, for which the assault was entirely excessive. In Napoleon's centralized personal dictatorship, all policy decisions, both political and military, emanated from him alone. Although the emperor never clearly defined his objectives, he indicated to his entourage that the unprecedented assemblage of men under arms reflected the grandeur of his goals—to destroy the army of Alexander I, to dismember Russia, and then to march from Moscow to India, where he would topple the British empire. In his words to the Count of Narbonne:

> Now we shall march on Moscow, and from Moscow why not turn to India. Let no one tell Napoleon that it is far from Moscow to India. Alexander of Macedon had to travel a long way from Greece to India, but did that stop him? Alexander reached the Ganges, having started from a point as distant as Moscow. Just suppose, Narbonne, that Moscow is taken, that Russia lies prostrate. . . . Then tell me, will it be so impossible for the French army to reach the Ganges? And once the French sword touches the Ganges, the edifice of England's mercantile greatness will tumble in ruins.[2]

It mattered nothing to Napoleon that his advisors, including Armand de Caulaincourt, his ambassador to Russia, unanimously opposed the adventure. Even the more lucid of Bonaparte's enemies saw further than he. Count Vorontsov, the Russian ambassador to England, for instance, summarized the probable military results of Napoleon's inexplicable gamble:

> I do not fear the military events at all. Even if the operations should prove unfavourable to us in the beginning, we can win

> by stubbornly waging a defensive war and continuing to fight as we retreat. If the enemy pursues us, he will perish because the farther he proceeds from his stores of provisions, his arms and munitions depots, the farther he penetrates a land without passable roads or provisions which he could seize, the more pitiful will be his condition. He will be surrounded by an army of Cossacks and in the end will be destroyed by our winter, which has always been our loyal ally.[3]

Defiantly, however, Napoleon even carried with him into Russia a copy of Voltaire's morality tale *History of Charles XII*. In it Voltaire recounted the megalomania of the Swedish tyrant and the manner in which it led to doom when he invaded Russia in 1708. Having ignored all advice, Charles lost his army to the rigors of the Russian winter and was decisively routed at the Battle of Poltava in 1709.

It was France's misfortune that by 1812 Napoleon, still only forty-three years old, was no longer the commander he had been. Physically, he had grown stout, and he suffered from a painful, chronic dysuria. This made riding difficult and distracted him, sometimes at vital junctures during a campaign. Since dysuria is a symptom frequently associated with sexually transmitted diseases, it is possible that he suffered from tertiary syphilis. This hypothesis gains plausibility from the irrational quality of his decision-making, both in launching the adventure and then in leading it to destruction. His marshals noted in dismay that the emperor no longer seemed in full command of his faculties. Members of Napoleon's general staff reported a new hesitation and inability to focus in the emperor that left him irresolute at moments of crisis. Whatever the diagnosis, Napoleon appeared to his staff to be mentally impaired.

Compounding his problems, Napoleon had fallen victim to sycophantic adulation and now believed that his every impulsive thought was the prompting of his genius. Accepting the myth of his own invincibility, he disdained advice, reduced his chief of staff—the hapless General Louis-Alexandre Berthier—to a bearer of orders from above, and rarely consulted his marshals. Crossing into Russian territory, therefore, the Grande Armée possessed no coherent strategy, and no one except Napoleon himself understood the goals of the expedition. Perversely, he welcomed the dangers ahead because overcoming them would only heighten his glory. In just such a spirit he had proclaimed in 1808, "God has given me the necessary strength and will to overcome all obstacles."[4]

Crossing the Niemen River

On June 24, 1812, about nine hundred miles from Paris, the Grande Armée began to stream over the Niemen River that demarcated the western boundary of the tsar's domains. Making use of three bridges, the crossing took three days and nights to complete. Estimates of the number of troops under Napoleon's command are conflicting, but the consensus is that they totaled more than half a million, together with one hundred thousand horses, one thousand cannon mounted on caissons, thousands of supply wagons and officers' carriages, and as many as fifty thousand camp followers—servants, wives, and mistresses of officers; attendants; cooks; and prostitutes. Marching such a force across the Niemen was the equivalent of moving the entire city of Paris. The Grande Armée was also multinational and multilingual, which reduced its cohesion. Its core consisted of Frenchmen, but there were large contingents from other countries, including ninety-five thousand Poles and forty-five thousand Italians. Each non-French contingent was commanded by French generals.

In numbers, the Grande Armée represented a decisive break from the traditions of eighteenth-century warfare, which it replaced with a new style of "total war." Eighteenth-century field armies seldom exceeded fifty thousand men. Ten times that size, the Grande Armée originated with the French Revolution's great military innovation—*levée en masse,* or mass conscription. This pivotal measure produced wars of entire peoples, and its purpose was not merely the defeat but the extermination of an opposing army.

In pursuit of this objective, the Grande Armée was too large to be commanded as a single unit, so it was broken into subunits—the corps, each of which was composed of four or five divisions numbering five thousand men apiece. But the Grande Armée differed from armies of the Ancien Régime not only in size and organization, but also in tactics. It conducted its battles by exploiting to the full its capacity to move with unprecedented speed. This mobility was due to the fact that Napoleon's army traveled light, unencumbered by the usual impedimenta required to meet the provisioning and sanitary needs of so numerous a force.

During the Italian campaign of 1796–1797 that first brought him fame, Napoleon introduced the practice of having his army live off the land as it moved, requisitioning animals, fodder, and crops from the surrounding countryside and its inhabitants. This practice worked successfully in western and central Europe. Indeed, the "system of plunder" provided a powerful morale boost for the troops, who learned to think of military campaigns as the means of enriching themselves. General Philippe de Ségur, who accompanied Napoleon to Russia, explained that

> Napoleon was . . . well aware . . . of the attraction which that mode of subsistence had for the soldier; that it made him love war, because it enriched him; that it gratified him, by the authority which it frequently gave him over classes superior to his own; that in his eyes it had all the charms of a war of the poor against the rich; finally, that the pleasure of feeling and proving himself the strongest was under such circumstances incessantly repeated and brought home to him.[5]

Economic factors also influenced the decision to live off the land. Since a nation-at-arms was unprecedentedly expensive to provision, it was a major saving to outsource the task to the soldiers themselves. For the same reason, the army opted to equip its men with mass-produced uniforms made of thin and flimsy material that provided scant protection from the elements and disintegrated during a long campaign. Worst of all was the impact of financial considerations on footwear. French soldiers set off for Russia with shoddy boots whose soles were glued together rather than stitched. As a result, the boots began falling to pieces by July, and by the Battle of Borodino in September, many of Napoleon's infantrymen marched barefoot into combat. When winter struck in November, they trudged through the snow with their feet wrapped in rags.

Napoleon's strategy as he crossed into Lithuania at Kowno was to take advantage of a massive gap in the Russian defensive alignment facing him. The Russian forces consisted of some three hundred thousand soldiers defending against Napoleon's assault by twice their number. But the Russian troops were divided in half between a northern army under General Mikhail Barclay de Tolly and a southern army under Pyotr Bagration. Napoleon's strategy was to employ his classic technique of finding the weakest point in the enemy's position and then striking it with overwhelming force. In this case he planned to move with dizzying speed between the two Russian forces, cutting them off from each other. Massively outnumbering each as a result, he would envelop and destroy first one and then the other. After a spectacular victory akin to the emperor's triumph at the Battle of Austerlitz in 1805, the Grande Armée would find the road open to both Russian capitals—St. Petersburg and Moscow. The war, Napoleon anticipated, would then end rapidly as Tsar Alexander, his forces annihilated, would sue for peace on any terms. Meanwhile, he gave no thought to mundane matters of sanitation, diet, and health.

During the first ten days after the Niemen crossing, the French did indeed achieve a first objective. By means of his "blitzkrieg" strategy, Napoleon succeeded in splitting the two Russian armies. But the expected result did not follow. Advancing in pursuit through Lithuania, the emperor expected a decisive clash. To his surprise, the Russian commanders declined to give battle. Instead, they retreated, allowing the French to occupy Russian territory but refusing to risk their main forces. The French warlord, they remembered, had crushed them at the Battle of Friedland in 1807, and now his army threatened to overwhelm them again with numbers, fire power, and tactical skill. Like Peter the Great against Charles XII, Tsar Alexander decided to avoid a direct trial of strength and to rely instead on his most highly prized assets: space and climate.

Writing from Napoleon's headquarters, General Raymond A. P. J. de Montesquiou-Fezensac noted in dismay that the emperor was unwilling to alter his plans to take account of the reality before him. He became obsessed instead with the mirage of a great decisive battle, and he followed the mirage willy-nilly deeper and deeper into Russia. The result was that he awakened each day expecting "that the Russians would stop their rearward movement and give battle . . . and without bestowing a thought on the fatigue which his troops underwent."[6] In a pattern repeated day after day, the Russians withdrew and the French pursued. By the end of a month, Napoleon had overrun the whole of Lithuania, but in an eerie manner. He encountered harassing firefights with the Russian rearguard but failed to engage the main forces of the Russian army for three months.

Into Russia

As he moved farther and farther east, the emperor did not consider the difficulties that would arise in lands where roads were few, the population was thinly settled and impoverished, and resources were scarce. Napoleon, in other words, ignored the natural and social environments surrounding his army, and the medical consequences that could follow.

Furthermore, assuming that he could anticipate his enemies' countermeasures, Napoleon launched the campaign without considering the possibility of complications and surprises. Of these the worst was that, as they retreated, the Russians systematically laid waste to the landscape. Their strategy was to force the Grande Armée to confront the insoluble problem of foraging in the midst of charred fields, empty villages, and towns reduced to smoldering piles of ashes. Count Ségur marveled at the determination of the Russians, who withdrew "as at the approach of a dire contagion. Property, habitations, all that could detain them, and be serviceable to us were sacrificed. They interposed famine, fire, and the desert between them and us. It was no longer a war of kings . . . but a war of class, a war of party, a war of religion, a national war, a combination of all sorts of war!"[7]

The Prussian military theorist Carl von Clausewitz, serving as advisor to the tsar's army, explained the Russian strategy and urged its uncompromising implementation:

> Bonaparte must totally fail, by virtue of the great dimensions of the Russian empire, if these should only be brought sufficiently into play, i.e. if its resources were husbanded to the last moment, and no peace accepted on any condition. This idea was especially put forward by [General Gerhard von] Scharnhorst . . . whose expression was that, "The first pistol shot must be fired at Smolensk"—an idea that could not fail to be beneficial if carried out to the extent of not shrinking from evacuating the whole country as far as Smolensk, and only beginning the war in earnest from that point.[8]

Ironically, Napoleon carried a warning with him in his carriage: Voltaire's account of Charles XII's sufferings a century before—*History of Charles XII, King of Sweden*. Tsar Peter the Great had also deployed a scorched-earth policy as the means to destroy the Swedish invader. This knowledge, according to General Fezensac, "created in Napoleon a disposition to ill-humour."[9] War, the general recalled, had formerly been "sport" to the French emperor, who loved to match his wits against fellow kings and commanders. But the conflict in Russia was unsporting and confounding. Napoleon, Ségur wrote, "was daunted, hesitated, and paused."[10] But always, after vacillating for a moment, he marched on.

During this unconventional campaign, the first result was a crisis of French morale. The troops found no opportunity for spoils to offset their travails, among which hunger and thirst were prominent. In eastern Europe the Grande Armée experienced an ever-increasing shortage of food and potable water, undermining the resistance of the soldiers and their horses and exposing the men to the microbes that lurked in the marshes and streams they crossed. In time, it led also to hunger and dehydration.

After six weeks of fruitless tramping, the Grande Armée reached Smolensk and Russia proper. Expecting rapid conquest, the troops began to grumble. Why had they been force-marched so relentlessly "to find nothing but muddy water, famine, and bivouacs on heaps of ashes; for such were all their conquests . . . as they marched on amid . . . vast and silent forests of dark pines."[11] Disillusioned and weary soldiers deserted and drifted away in a steady stream. Often, they deserted in armed groups that established bases in villages near the line of march. There they defended themselves and lived by marauding.

Measures designed to keep the army light proved unavailing in the vastness of the Russian steppe. General Jean-Baptiste de Gribeauval, the military engineer, earned lasting renown by devising a series of technical innovations to facilitate Napoleon's lightning tactics. As if by magic, Gribeauval made heavy artillery mobile by featuring in his designs light, narrow carriages for the cannon; thin, short barrels; a screw mechanism to raise or lower the breach; better aiming devices; and an improved propellant. These initiatives halved the tonnage of the French guns, which could now move rapidly during combat, without compromising accuracy and volume of fire. The results were speed, surprise, and a devastating barrage. This "Gribeauval system" was a fundamental component of Napoleon's tactics, as he relied heavily on big guns and howitzers, having been an artillery officer himself.

Unfortunately, even Gribeauval's stripped-down guns were too cumbersome for a trek from the Niemen River to Moscow. Napoleon's reliance on cannons presupposed tens of thousands of horses to pull them. The horses, however, were no better provisioned than the men, and already on the outward journey they began to sicken and die by the thousands. As a result, the French were progressively compelled to abandon their artillery and transform their cavalry into infantry. The Grande Armée thus lost its initial advantage in firepower and the lightning cavalry charges for which it was famous. In time, the French also advanced without the reconnaissance patrols and intelligence on which their tactics depended. Heavy casualties and bewilderment followed as the army trudged forward—light, but also blind, outgunned, and poorly nourished.

Dysentery

Casualties and the ever-growing shortage of horses were not, however, the only consequences of traveling light. As it prepared to invade the tsar's domains, the Grande Armée had made the stark choice of sacrificing health to the overriding priority of rapid movement. Just as they jettisoned provisions, so too the French commanders opted not to transport medical, sanitary, or surgical supplies into Russia. This decision was ironic given that the French surgeon-in-chief, Baron Dominique Jean Larrey, was famous for having devised measures to save lives by evacuating the wounded from the battlefield speedily and placing forward hospitals near combat zones. With these innovations, the surgeons under his command were able to amputate mangled limbs with the haste needed to minimize blood loss and prevent gangrene.

In Russia, however, the French medical teams lacked supplies of every kind. Larrey had no splints, bandages, or bedding for his patients, and the only food available was a thin soup of cabbage stalks and horse flesh. Since he also had no medicines to administer, he and his colleagues were reduced to foraging for medicinal herbs in the forests through which the army passed. French field hospitals soon earned a frightening reputation as places of filth, overcrowding, overwhelming stench, and death. They also became hotbeds of disease.

Most seriously, the troops did without sanitation and without tents, bivouacking instead—first in the mud moving east, and then in the snow during the retreat. During the advance, the troops marched as many as fifteen to twenty miles a day in the summer heat, carrying packs weighing sixty pounds, a flintlock musket as tall as a man, a cartridge belt around their necks, a bayonet, and a sword. Sweating profusely, exhausted by endless marching, ill fed, and thirsty, soldiers began to suffer dehydration and malnourishment. Already in Lithuania, the army began to see its manpower melt away even without combat. Stragglers dropped from prostration, fell behind, or were taken to field hospitals.

A variety of circumstances prevailing during the trek into Russia created perfect conditions for microbial diseases to flourish—especially dysentery, which was a much feared affliction of nineteenth-century armies. Also known as Shigellosis, dysentery is a bacterial disease caused by four species of the *Shigella* genus. Like typhoid and cholera, dysentery is transmitted by the oral-fecal route through the ingestion of food and water contaminated by excrement. The Grande Armée, which stretched out for a distance of twenty miles from end to end, lived day and night in an environment that it systematically befouled. The soil of the road on which it plodded and the ground on either side were churned by the passage of innumerable troops, horses, carts, and cannons. This earth became a tangled, stinking mixture of mud and human and equine excrement where infinite numbers of flies feasted and bred amidst the dung. In these conditions the troops marched, ate, slept without changing their clothes, and occasionally fought. They were constantly outdoors, but in a microenvironment of their own creation that rivaled the appalling sanitary conditions of a dense urban slum.

At the end of a day's footslog, units found a space beside the road and claimed it for their own. There, crowded together, the men lit a fire and hastily prepared whatever food they possessed. They also relieved themselves haphazardly nearby and greedily drank the water of any source available, regardless of odor, viscosity, and color. Survivors recorded in their memoirs that some men were driven by thirst even to drink the urine of the horses that trekked at their side and shared their living space.

For an army so constituted, every moment was beset with medical hazards. One was the water they drank. Often, marshes and streams were already saturated with microbes when the troops arrived. But then half a million men and a hundred thousand horses randomly fouled that water as they passed. Many soldiers defecated directly into the flow of streams or, if suffering from diarrhea, rinsed themselves and their garments as best they could in the communal water source.

Eating was equally problematic. Any food that troops shared with one another by means of long-unwashed hands conveyed bacteria. Furthermore, since soldiers on the march were exhausted and sometimes frightened, they were unlikely to pay attention to the omnipresent flies swarming about them. The men were accustomed to the insects, were unaware of the existence of microbes, and knew nothing of the role that flies played in disease transmission—carrying bacteria on their hairy appendages and in their guts. Soldiers took little care to shield their provisions, allowing the flies to shed microbes on the food they sampled and trampled.

Beasts fared no better than men, and multitudes of overworked, poorly fed, and ill horses were left where they fell—to decay, impede the passage of the ranks trudging behind, and nourish abundant maggots. Growing ever more emaciated as the campaign progressed, men also began to stagger and die.

Having once gained access to the Grande Armée, *Shigella* bacteria were richly endowed to overcome the body's defenses. A notable feature of dysentery is the small number of bacteria needed to infect a human host. In addition, Shigellosis provides patients who recover with no acquired immunity as a defense against further episodes. Nor is there crossover immunity from one species of the *Shigella* genus to another. In fact, nineteenth-century physicians believed that one attack of dysentery left a sufferer more likely to succumb in the future; therefore, the disease did not allow the population to develop the kind of herd immunity we have seen with smallpox. The army thus experienced Shigellosis as a virgin soil epidemic. Furthermore, dysentery produces asymptomatic carriers who are infective without displaying clinical manifestations. As a result of such "Typhoid Marys," dysentery circulated silently long before it attracted medical attention.

There is no way to know when the first cases of Shigellosis broke out, because Napoleon's troops were not under close medical surveillance. But the disease was well known to the physicians of the Grande Armée, who diagnosed it and reported its impact. They did so, however, only after it was already widely prevalent. In early August 1812, when the French paused at Vitebsk—just 250 miles beyond the Niemen River—Ségur estimated that there were three thousand cases of the disease. The infection was clearly gathering momentum and earning its name as "epidemic dysentery." Larrey penned laconically that there was "a large amount of sickness."[12]

In the human body, the progress of dysentery is rapid but variable, ranging from hours to perhaps a week. *Shigella* bacteria pass into the colon, where they invade the epithelial cells of the intestinal lining, causing tissue damage and releasing a powerful toxin. Initial symptoms are fever, a hard and rapid pulse, severe abdominal pain, and watery diarrhea "discharged with more or less explosive force." Often, the evacuations are stained with blood. Its appearance "exactly resembles water in which raw flesh has been macerated," and its stench is overwhelming.[13] Nausea, a furred tongue, sunken eyes, and skin covered in clammy sweat follow. The body also emits a cadaverous odor as if to warn of impending putrefaction. Such violent attack and copious evacuations are reminiscent of Asiatic cholera. Fluids gush from patients' bodies while they suffer tormenting but unquenchable thirst. Even if water is administered, it is not retained but expelled as vomit.

On the Russian front in 1812, military doctors possessed neither the means nor the time to provide patients with supportive care, and any understanding of rehydration therapy and antibiotics lay far in the future. Inevitably, the affected soldiers sank into shock. Soon they were prostrate in a condition physicians termed "adynamic" or "asthenic" inanition, or stupor, followed by delirium, coma, and death.

The Grande Armée possessed no means of keeping medical records and statistics, so it is impossible to determine the CFR of the dysentery that ravaged the troops. There was a consensus, however, that a large majority of those who contracted the disease perished within a week, although some men with resilient cases weathered the attack and began a lengthy convalescence.

Observers attributed the remarkably high mortality in part to the abuse of alcohol. A popular belief at the time was that alcoholic drinks cleansed the bowels, so the soldiers medicated themselves with vodka at the onset of the affliction—with lethal results. Cruelly, too, many convalescents either relapsed or, in their weakness, contracted the disease anew and were carried away by the second assault. Larrey himself contracted dysentery, but recovered.

Inevitably, the diagnostic category of "dysentery" that Larrey and his colleagues applied does not coincide precisely with the Shigellosis known to modern clinicians. Nineteenth-century diagnoses were shifting and uncertain; they were based on physical diagnosis alone; and the French medical men were too overwhelmed by the flood of disease to insist on diagnostic precision. The label "dysentery," therefore, is best regarded as a loose umbrella term that encompassed Shigellosis but probably included other severe gastrointestinal diseases.

By late August, Ségur reported, dysentery was "was progressively extending its ravages over the whole army."[14] Instead of the three thousand cases diagnosed at the start of the month, by then four thousand men died *daily* of this one cause. Dysentery, in other words, rapidly dissipated the massive numerical advantage the French had enjoyed when the Russian campaign began. When on September 14 the Grande Armée at last reached Moscow—fifteen hundred miles as the crow flies from Paris—it had lost a third of its men to desertion, combat, dehydration, and disease. But the worst was dysentery. This steady depletion was all the more problematic because Napoleon had no means to make good the losses. Furthermore, the tsar's forces did not suffer to the same extent. With their shorter supply lines, Russian commanders were able to provision troops and horses and to mobilize reinforcements.

Napoleon's staff officers assumed at every pause in the great eastward slog—at Vilna, Vitebsk, and Smolensk—that the emperor would call a halt until the spring. The army needed time to recover, rest, and replenish its ranks and supplies. Caulaincourt, who had spent four years at St. Petersburg as French ambassador, was especially fervent. He pleaded the case that the French troops—feverish, poorly clad, and badly provisioned—would find themselves in dire peril if the Russian winter overtook them in the field. He especially hoped that Smolensk, the first city in Russia proper, would mark the outermost bounds of Napoleon's ambitions for the 1812 campaigning season. With each stop, however, the emperor grew more impatient. "It was then, especially," wrote Ségur, "that the image of captive Moscow besieged him: it was the boundary of his fears, the object of his hopes; possessed of that, he would possess everything."[15] Indeed, Ségur began to suspect that the very factors that should have caused Napoleon to halt—the distance, the climate, the unknown—were in fact what attracted him most. Napoleon's pleasure grew in proportion to the danger he faced.

Borodino

Before reaching Moscow, the French and Russian armies met at last and fought the only major engagement of the Russian campaign—the Battle of Borodino—on September 7. The most ferocious battle of the Napoleonic era, Borodino was the all-out clash that Napoleon had sought between the two armies. Ironically, however, the engagement took place at a time and in a manner he had not envisaged. It did not occur because the emperor had willed it, but because the Russian commander, General Mikhail Kutuzov (1745–1813), had decided that the opportune moment had come. The Russian armies of the north and south, which the French had intended to keep apart, had finally reunited just west of Moscow.

At this juncture Tsar Alexander conferred overall command of his now unified forces upon the experienced and wily veteran Kutuzov. As Leo Tolstoy describes Kutuzov in *War and Peace,* his epic novel of the 1812 campaign, the Russian general was by character the polar opposite of his French opponent. Kutuzov was modest in temperament and had no inclination to compete with Napoleon in displays of tactical genius. Having been routed by the French emperor at Friedland in 1807, he feared Napoleon and subscribed to the strategy adopted first by Peter the Great and now by Tsar Alexander—to retreat into the depths of Russia, whose vastness and climate would destroy the enemy. In Clausewitz's words, he was content to allow the first pistol to be fired after Smolensk. In the parlance of military theory, he implemented a strategy of "force preservation."

With his back to Moscow, however, Kutuzov was finally ready to stand and fight. His intelligence networks had kept him informed of the Grande Armée's travails; his own forces were united and well supplied; and he hoped to block the road to the Russian capital, which lay just seventy miles farther east. Arriving four days before Napoleon, he entrenched his forces behind two redoubts on elevated positions commanding the plains below. It was the last militarily advantageous position before Moscow, and the Russians fortified it with trenches, wooden palisades, six hundred artillery pieces, and "wolf pits," as they termed the holes they dug to break the legs of charging horses and infantry. Thus positioned, the Russians waited for the French to attack. The Grande Armée, now barely superior in numbers, found itself outgunned and fighting uphill. Thus, on a battlefield three miles across, 134,000 onrushing French engaged with 121,000 Russian defenders.

A slaughter of grand proportions produced what one historian calls "the deadliest engagement in the annals of warfare to that date."[16] The two sides opened fire at first light on September 7, and the killing lasted until dark fourteen hours later. Men died by the thousands from cannon balls, shards from antipersonnel canisters, musket fire, bayonet thrusts, and saber slashes. As dusk brought the sanguinary day to a close, Kutuzov ordered his troops to withdraw, leaving Napoleon in command of the field and technically the victor. Feverish, tormented by dysuria, and indecisive, the emperor mysteriously failed to commit his elite troops—the Imperial Guard—to the fray. Both Clausewitz and Napoleon's marshals believed that, ordered into battle at the proper moment, the Guard would have carried the day.

A simple but ironic twist of fate may explain the emperor's failure to seize his opportunity. The Grande Armée had been purpose-built with the goal of concentrating overwhelming force and firepower at the enemy's weak point. Surveying the battlefield with his spyglass from above, Napoleon normally directed his men with deadly precision and unmatched tactical acumen. Borodino, however, constituted the limiting case in which the sheer press of numbers defeated generalship. Within the confines of a small field of battle, the clash of two armies of such unparalleled size created a dense screen of dust that concealed the contest from view. Smoke puffs rising from a thousand artillery pieces and a hundred thousand flint muskets, earth thrown up by ninety thousand cannon balls, dust raised by ten thousand charging horses and advancing infantry—all this made it impossible for Napoleon to follow the scenes below him. Total war created a literal fog of war that nullified Napoleon's tactical genius just when it was most needed.

In his study of the Russian campaign, Tolstoy stressed the irony of fate by raising the question of "how far Napoleon's will influenced the Battle of Borodino." His view was that the French emperor was reduced to playing the part of a "fictitious commander." "The progress of the battle was not directed by Napoleon," Tolstoy concluded, "for no part of his plan was carried out, and during the engagement he did not know what was going on before his eyes."[17]

Blinded, Napoleon not only allowed Kutuzov to retreat in good order, but he also failed to launch a pursuit. What the French gained by the end of the day was thus a pyrrhic victory. The field of battle belonged to the Grande Armée; the forty thousand Russian dead outnumbered the thirty thousand

French who fell during the fray; and Kutuzov had retreated. Only then did the French surgeons set to work, uninterruptedly performing amputations for the following twenty-four hours. Larrey himself amputated two hundred limbs in the immediate aftermath of Borodino. Victory at such a cost in dead and wounded irremediably depleted the Grande Armée while Kutuzov repaired his losses with reinforcements.

Furthermore, the Russians' morale soared after Borodino; they had absorbed the worst blow Napoleon could deliver and had survived. The French, by contrast, were despondent. Ségur provides an illuminating commentary:

> French soldiers are not easily deceived; they were astonished to find so many of the enemy killed, so great a number wounded, and so few prisoners, there being not 800 of the latter. By the number of these, the extent of a victory had been formerly calculated. The dead bodies were rather a proof of the courage of the vanquished, than the evidence of a victory. If the rest retreated in such good order, proud, and so little discouraged, what signified the gain of a field of battle? In such extensive countries, would there ever be any want of ground for the Russians to fight on?
>
> As for us, we already had too much, and a great deal more than we were able to retain. Could that be called conquering? The long and straight furrow which we had traced with so much difficulty from Kowno, across sands and ashes, would it not close behind us, like that of a vessel on an immense ocean? A few peasants, badly armed, might easily efface all trace of it.[18]

So many French field officers perished that General Fezensac was transferred from the general staff to regimental duty, where he assessed the soldiers' state of mind. There reigned, he found, an "atmosphere of despondency" as "the moral constitution of the army had never before been so materially shaken." The emperor, however, refused to acknowledge the impact of so savage a bloodletting. In Fezensac's terse assessment, Bonaparte "would see nothing—would hear nothing."[19]

The Russians, Tolstoy among them, ever after regarded Kutuzov as a national hero and celebrated Borodino as an essential component of their "Great Patriotic War of 1812." Kutuzov had preserved his army to fight another day; he had inflicted major losses on the French; and he now faced the Grande Armée in a position of parity. Furthermore, Napoleon had lost the psychological weapon of apparent invincibility. Two of his experienced marshals summed up the events of September 7 with dismay: "Joachim Murat said he had not recognized the Emperor that whole day. Michel Ney said the Emperor had forgotten his trade."[20]

In measuring the relative role of dysentery in destroying Napoleon's army, one should set Borodino in context. By the time the Grande Armée arrived in Moscow, it had lost 150,000 to 200,000 men to three factors: combat, desertion, and disease. Although deaths in battle and losses from desertion drained the army's strength, dysentery had the greatest effect: the Grande Armée lost 120,000 troops—4,000 every day—to disease during the final weeks before Moscow.

Moscow

After Borodino, Kutuzov withdrew to defensive positions east of Moscow, leaving the city undefended. But even as the Russians abandoned their capital, they presented the French with yet another unpleasant surprise. Napoleon, who had entered so many European capitals in triumph, expected that the capture of Moscow would follow the familiar pattern in which humble delegations of dignitaries came forward to offer submission and the keys to the city.

Instead, on reaching Moscow on September 14, Napoleon discovered that Kutuzov had defiantly evacuated its 250,000 inhabitants. Worse still, the next day incendiaries began to implement a fearful plan. After destroying all fire-fighting equipment, Russian arsonists set the city aflame by means of gunpowder explosions. With the fires fanned by gale-force winds, a conflagration engulfed the city, destroying 80 percent of the buildings and sparing only stone structures—the Kremlin, the city's churches, and its cellars. Tolstoy, who questioned the role of intentionality in human affairs, thought that the fire was inevitable even without a plot. "A great city built of wood, and deserted," he said, "is fatally certain to be burned."[21]

Napoleon's generals viewed occupying Moscow in such circumstances as a "hollow victory," and Larrey superstitiously considered the fire a portent of doom. The city reminded the French officer Césare de Laugier of the deserted ruins of ancient Pompeii. It was a sardonic irony that the band of the Imperial Guard paraded into Moscow playing the tune "Victory Is Ours!"

Napoleon, however, clung fast to illusion, mistaking territorial conquest for victory. Rebuffing his generals, he insisted that seizing Alexander's capital would compel the tsar to sue for peace. He therefore dispatched envoys to St. Petersburg to discuss terms and beguiled the time by reading fiction and reviewing his troops. Occasionally he fretted over the possible consequences of tarrying in a shattered city while winter approached, but he was unwilling to take the ignominious decision to retreat.

Napoleon's officers, on the other hand, regarded the occupation more as trap than triumph. By their reasoning, the war had not been won, and the Grande Armée had two viable options: either retreat forthwith before frigid weather set in, or overwinter in Moscow and resume campaigning in the spring. Paralyzed by wishful thinking, Napoleon made no decision at all. Instead, he idled recklessly for weeks and was even deceived by the fact that early October was unseasonably mild that year. From September 14 to October 19 he ruminated, cursed his painful urination, and impatiently awaited the tsar's surrender. Ségur observed that Bonaparte "prolonged his meals, which had hitherto been so simple and so short. He seemed desirous of stifling thought by repletion. He would then pass whole hours half reclined, and as if torpid, awaiting with a novel in his hand the catastrophe of his terrible history."[22]

Nor did the French troops benefit from their commander's delay. While Kutuzov's forces grew relatively stronger, disease continued to decimate the Grande Armée. The military hospitals that Larrey hastily set up in the city were soon filled to overflowing with troops stricken with diarrhea and fever. Propelled by infection and death, the military pendulum swung silently away from Napoleon. No longer conquerors, the French found themselves besieged. Beyond the perimeter of the city, Cossacks attacked foraging parties, killed cavalry on reconnaissance, and severed communications with Paris. By the time they vacated Moscow, the French set off not as the pursuers but the pursued.

Meanwhile, weeks of occupation undermined French military preparedness. The most obvious factor was the impact of epidemic disease, which was favored by warm weather and the unsanitary conditions of the crowded French encampments and burned-out houses that the troops commandeered. But a new factor—plunder itself—played a prominent part. Although the city lay in ruins, its cellars and their contents powerfully attracted marauding soldiers, who found ample stocks of vodka that they drank to the detriment of their already compromised health.

As an additional consequence of a month spent pillaging an abandoned capital, the bonds of discipline necessary to an army's cohesion began to dissolve. Officers and enlisted men alike developed a habit of thinking only of enrichment. They behaved, observers noted, like greedy merchants at the great "Moscow fair," where they loaded themselves down with all that glittered. Larrey bitterly lamented that the fine weather held, allowing the men to give no thought to the winter ahead. Instead of prudently seeking woolen garments, gloves, and fur coats, they hoarded silk, trinkets of gold and silver, precious stones, and objects of religious devotion. Officers filled their carriages with trophies, while foot soldiers discarded useful items from their knapsacks, preferring to weigh them down with baubles. Jean Baptiste François Bourgogne, a sergeant of the Imperial Guard, made an inventory of the items he collected in his own cowhide rucksack:

> I found several pounds of sugar, some rice, some biscuit, half a bottle of liqueur, a woman's Chinese silk dress embroidered in gold and silver, several gold and silver ornaments, amongst them a little bit of the cross of Ivan the Great. . . . Besides these, I had my uniform, a woman's large riding cloak . . . ; then two silver pictures in relief, a foot long and eight inches high, all in the finest workmanship. I had besides, several lockets and a Russian prince's spittoon set with brilliants. These things were intended for presents, and had been found in cellars where the houses were burnt down. No wonder the knapsack was so weighty![23]

No admirer of the French emperor, Tolstoy found the Grande Armée's conduct incomprehensible. The simplest course, he wrote, was

> not to allow the army to engage in pillage, to prepare clothing for winter (there was enough in Moscow for the whole army), and to get together the provisions which were of such quantity that they would have sufficed to supply the French troops for at least six months. . . .
>
> And yet Napoleon, this genius of geniuses, did nothing of the sort . . . ; but he used his power in favor of measures which were of all possible measures the most stupid and the most disastrous.[24]

Larrey anticipated Tolstoy's judgment. It would have been "ordinary forethought," he said, to have stockpiled furs and woolen stuffs wherever they could be found.[25]

By the time of departure, the Grande Armée found itself heavily laden with useless trinkets. In addition, it acquired a long tail of thousands of refugees. These were French citizens—businessmen, diplomats, actresses, artists—who lived in Moscow but now feared that they would be slaughtered when the Russians returned. They therefore sought safety by joining the camp followers trailing behind the fighting men.

Finally, on October 15 cold weather arrived and the first snow fell three inches deep. Stunned into sudden appreciation of his exposed position, Napoleon decided to depart. On October 18, the appointed day, the French army, reduced to one hundred thousand men, set off. Filing out of Moscow in the morning, the army evoked the chaos of biblical exodus, not a disciplined military retreat. Ahead lay seven weeks of Dantean torment.

Flight

Almost immediately the French were beset by misfortune. Ever vigilant, Kutuzov was well informed of his enemies' movements. He was also fully aware that the French faced a choice between two roads leading westward. To the south lay the road to Kaluga that traversed countryside which had been neither torched by the Russians nor pillaged by the French. It therefore beckoned with the possibility of provisions and sustenance. To the north lay the road to Smolensk that both armies had already devastated and therefore had no resources left to support the returning French.

Once fearful of Napoleon's wizardry, Kutuzov now knew that he held the upper hand and that victory was his if he denied the Grande Armée the supplies it needed. So the Russian general force-marched his troops to the small town of Malo-Jaroslavetz, where a stronghold barricaded the Kaluga road where it passed through a narrow ravine. No longer seeking a full-scale battle, however, Napoleon avoided a second trial of strength. He limited himself to small probing actions that the Russians repulsed. Then he turned resignedly north on the road to Smolensk that he already knew too well.

By forcing the Grande Armée onto the Smolensk road, the Battle of Malo-Jaroslavetz of October 24 constituted a vital Russian triumph. Although small in scale, it transformed the French retreat into a rout—a desperate race against time before winter arrived in full force. Ségur called Malo-Jaroslavetz "this sinister field, where the conquest of the world came to a halt, where twenty years of victories were undone."[26]

Napoleon and Kutuzov now held radically altered positions. Napoleon, the strategic genius, had no plan except to flee with all possible speed. By contrast, Kutuzov implemented a clear and consistent strategy, adopting the approach of a hunter toward a wounded but powerful wild beast. His goal was to avoid cornering the animal, which could turn suddenly and prove still dangerous. Instead, he intended to harry it relentlessly but at a careful distance until it fell at last from sheer exhaustion. Then the hunter could safely approach to strike the final, deadly thrust.

Staggering to Smolensk, however, the French encountered an enemy more implacable than Kutuzov. As cold weather set in, a second epidemic disease erupted with deadly fury. This was typhus. Dysentery may have destroyed a third of the Grande Armée en route to Moscow, but typhus killed most of the remainder during the retreat. Fewer than ten thousand of the one hundred thousand men who evacuated Moscow survived the trek. In terms of deaths per capita, the historian Stephan Talty calls this epidemic catastrophe "a die-off that had scarcely any parallel in the history of the world."[27] And these deaths occurred without a major battle.

Typhus

As with all gastrointestinal diseases, the force of dysentery wanes with cold weather, which renders transmission difficult. Typhus, however, is different. A marching army in winter provides perfect circumstances for any louse-borne affliction. A hundred thousand shivering troops huddled together in filthy bivouacs facilitate its transmission. Indeed, the affinity of the disease for military environments is suggested by its popular nineteenth-century names: "war pest," "camp fever," "war plague."

Typhus is not transmitted by direct contact, through the air, or via the oral-fecal route. It depends instead on a complex interrelationship among humans, body lice, and the bacterium known as *Rickettsia prowazekii*. Human beings are the most important reservoirs for the typhus pathogen, which is conveyed from person to person by the louse species *Pediculus humanus corpus,* which feeds exclusively on human blood. As soon as it hatches, this human body louse begins to feed voraciously.

If infected by Rickettsiae, lice do not inoculate bacteria directly into the bloodstream in the manner of mosquitoes. Instead, while feeding, they evacuate feces laden with microbes. They also inject an anticoagulant into the wound that irritates and provokes a tormenting itch. The itch normally causes the host to scratch the lesion, thereby contaminating it with excremental matter and initiating a new infection. Scratching also drives the lice to escape to the outer surfaces of the victim's clothing, thereby facilitating the transfer of the vermin to a neighboring body, where it establishes yet another breeding site. This sequence of events is important to the transmission of typhus because lice are inefficient vectors. Unable to fly, they have a tightly circumscribed range of movement, extending only as far as they can crawl. Body lice therefore need close contact between people in order to maintain the parallel chains of infestation and transmission.

The conditions prevailing on the road to Smolensk were perfect for body lice and microparasites. As temperatures plunged, the troops donned ever heavier and more numerous layers of clothing. Such attire created nesting places for lice, which demonstrate a preference for the seams of garments. There they cling fast to fabric by means of claws that seem almost custom-made to ensure their security. Happily for them, the troops seldom removed their clothes and never washed either their garments or their bodies for the whole of an eighty-three-day tramp. Furthermore, the soldiers huddled for warmth—at resting places, at night while they sat by the fire to eat, and finally when they hunkered down on the snow to sleep. Such close proximity gave insects infinite opportunities to migrate from body to body. The result was a universal infestation by as many as thirty thousand lice per soldier. Of the many torments of the retreat, the intolerable itching inflicted by these creatures figured most prominently in the memory of survivors. "In the evening," one recalled,

> when we huddled round the campfire, life would return to the insects which would then inflict intolerable torture on us . . . , made all the more powerful by the disgust which it inspired. . . . From the beginning of the retreat it had become a calamity . . . as we were obliged, in order to escape the deathly cold of the nights, not only not to take off any of our clothes, but also to cover ourselves with any rag that chance laid within our reach, since we took advantage of any free space by a bivouac which had become vacated by another. . . . These vermin had therefore multiplied in the most fearful manner. Shirts, waistcoats, coats, everything was infested with them. Horrible itching would keep us awake half of the night and drove us mad. It had become so intolerable as a result of scratching myself that I had torn the skin off part of my back, and the burning pain of this . . . seemed soothing by comparison. All my comrades were in the same condition.[28]

It was no consolation to the men that the vermin which tormented them soon perished as well.

By mid-November the temperature plunged to −9°F accompanied by a strong, biting wind, while snow "enveloped the troops like a winding-sheet."[29] Blinded by glare from the snow, half-frozen soldiers with icicles hanging from their beards stripped the bodies of those who had succumbed. Nothing could have better spread the epidemic. This intermingling of garments contravened contemporary medical advice that it was "prudent to touch as few of the clothes of any person who may be sick with this fever as possible."[30]

Having gained access to a patient's bloodstream, Rickettsiae are carried by the circulatory and lymphatic systems to the small capillaries of internal organs—the brain, lungs, kidneys, and heart. There they invade the epithelial cells lining blood vessels and replicate by fission. Thus bacteria accumulate within host cells in ever growing numbers until they destroy the cells by lysis, or rupture of the cell membrane. This releases the microbes into surrounding tissues, where they renew the process of replication and destruction. After ten to twelve days of incubation, the early clinical manifestations appear—a high temperature of 102–104°F, a telltale skin rash, severe headache, nausea, chills, and myalgia. Sharp pains shoot across the back and groin, and the body begins to emit an odor akin to ammonia.

Meanwhile, having dispersed throughout the major organs, the bacteria reach astonishing numbers, coagulate, and obstruct the circulation. The results are hemorrhage, vascular dysfunction, and disturbance to vital bodily functions. Patients have blue lips, a parched tongue, tormenting thirst, eyes that are suffused and glassy, a persistent dry cough, and black, insufferably fetid diarrhea. They also lose control of their muscles. The resulting state of ataxia gave the disease two of its diagnostic labels—"adynamic fever" and "nerve fever"—both referring to a patient's staggering gait and poor mobility.

In addition, the involvement of the pulmonary system provokes bronchial pneumonia as the air sacs fill with fluid, leading to dyspnea, or oxygen hunger. Blockage of blood vessels causes the fingers and toes to blacken with gangrene, while lesions to the central nervous system produce mental confusion, seizures, and delirium.

In the Grande Armée's field hospitals, typhus patients added a disturbing presence because of their fits of laughter, sudden shouts, and animated dialogues with imaginary interlocutors. This mental fog is the etymological origin of the diagnostic term *typhus*—the Greek word for haze or stupor.

Since multiple bodily systems are simultaneously affected, symptoms change in a kaleidoscopic array of agonies and potential causes of death.

In Russia, a Belgian surgeon noted that the end came "in a lightning-fast manner."[31] Death commonly ensued from swelling of the brain (encephalitis) or heart failure. As the army retreated, it also experienced a growing number of suicides, attributable both to the cognitive impact of the disease and to a more general despair that spread through the ranks.

Singularly infectious and highly virulent, typhus invariably produced a high rate of mortality before antibiotics, when CFR exceeded 50 percent. But the conditions of the winter retreat in 1812 greatly exacerbated the death rate by destroying all possibility of recovery and convalescence.

Furthermore, in its filth the Grande Armée acted as a giant petri dish in which various species of microbe competed for hegemony. At a disadvantage when cold weather set in, dysentery played an ever-declining role during the autumn. But venereal diseases, hepatitis, and diarrhea continued to ravage the French. In addition, recent evidence demonstrates that trench fever, a debilitating but rarely fatal disease borne by the same body lice that transmit typhus, also afflicted Napoleon's troops. Thus multiple comorbidities compounded the misery of the retreating troops and compromised their resistance.

Furthermore, typhus is well known to be especially lethal when it runs its course in populations that are undernourished. The nineteenth-century epidemiologist Rudolf Virchow reminds us that the disease fully earned yet another of its many nicknames—"famine fever." The classic case is Ireland, where famine and typhus accompanied one another in successive crises between the end of the eighteenth century and the potato famine of 1846–1848. In Virchow's words of 1868: "Since now almost two hundred years Ireland may be considered as the principal seat of the famine fever. It is not too much to say, that as Egypt was from the plague, so has Ireland ever since 1708 been desolated with every new visitation of this most malignant of epidemics—the typhus fever. . . . No other country in the world can be even distantly compared with it in this respect."[32] Typhus therefore followed in the wake of the potato blight in Ireland, but it also spread more broadly across the continent, ravaging Flanders and Upper Silesia in particular.

In an analogous manner, conditions prevailing during the Grande Armée's flight encouraged three of the Four Horsemen of the Apocalypse—famine, plague, and war—to ride forth together. The road that led west to Smolensk and then onward to the Niemen River offered none of the means to sustain human life. Its landscape, already systematically pillaged, was now frozen hard and covered in deep snow and ice. Since foraging had become impossible, the French faced starvation.

The day of reckoning approached earlier in the retreat than Napoleon's staff officers had foreseen because the march slowed to a painful hobble. The army made ponderous headway as soldiers trudged through snow and slipped on ice. The remaining horses also had poor traction because the army had neglected to rough-shoe them with iron studs for winter. They therefore made no better progress than the men but slid and toppled to the ground.

At the head of the struggling column of troops, the forward echelons made the plight of those who followed even worse than their own. Compacting the snow underfoot, they transformed it into a sheet of ice. In addition, they strewed the road with a multitude of obstacles that impeded the advance of those behind. Horses and men collapsed and died of multiple causes—disease, exhaustion, hypothermia, and dehydration—and their carcasses were left where they fell. As draft animals grew scarce, units abandoned wagons, caissons, carriages, and cannon. In a bid to lighten their burden, soldiers unloaded the loot from the "Moscow fair." Many also dropped muskets and cartridges that were useless to men with frozen fingers.

Dead bodies and discarded articles of every kind thus formed hazardous obstructions dangerously submerged in drifting snow. Meanwhile, the column lengthened, with more than sixty miles separating the vanguard from the tail. Conditions also worsened immeasurably on November 6, when the mercury plunged further and a blizzard blanketed the ground to the depth of a yard.

In this environment, the all-consuming task of the soldiers was survival. Above all else, they sought food, and they grew increasingly desperate. Foraging was no longer an option, both because there were no foodstuffs to be found and because leaving the column meant sudden death. Cossacks harried the flanks and rear of the retreating army, and they summarily killed all foragers and stragglers who fell into their hands. Famished soldiers therefore turned to horse flesh. Whenever a soldier had cut meat from a carcass, he halted to light a fire and roast a morsel. As the number of horses diminished, men even carved steaks from living animals and drank their blood mixed with snow. Such a meal excited envy, and it had to be gulped in haste before it was snatched away. Men with meat in their hands were lucky, Fezensac noted, "if their comrades did not carry off from them this last resource. Our soldiers, famishing, hesitated not to seize by force the provisions of every isolated man they came across, and the latter considered themselves lucky if their clothes were not also torn from their backs. After having ravaged the whole country, we were thus reduced to destroying ourselves." No longer a fighting force, the Grande Armée had become a dwindling and violent rabble. The road to Smolensk became a battleground where French soldiers fought one another. "The want of food," Fezensac continued, "was alone sufficient to destroy the army, without any of the other calamities which followed in its train."[33]

His description of the men he commanded effectively conveys the disintegration of the Grande Armée, which had become

> a confused mass of soldiers without arms, staggering at each step, and sinking beside the carcasses of horses and the lifeless bodies of their companions in misery. Their countenances bore the stamp of despair, their eyes were sunken, their features drawn, and black with dirt and smoke. The fragment of a sheep's skin or a shred of cloth was wrapped round their feet, and supplied the place of shoes. Their heads were enveloped with rags, their shoulders covered with horse-rugs, with women's petticoats, or with half-burnt hides of animals. As soon as a man dropped, his comrades stripped him of his rags to clothe themselves. Each night's bivouac resembled on the morrow a field of battle, and on awaking one found those dead around by whose side you had slept the night before.[34]

At length the final step in the Grande Armée's degradation occurred when troops resorted to eating each other. Atomization had proceeded so far that the men were engaged in a war of all against all. When the last inhibitions gave way, starving men tasted human flesh. Though denying having practiced cannibalism himself, Sergeant Bourgogne understood the dire necessity that drove men to the deed. Cloaking judgment in a soldier's rough humor, he wrote: "I am sure that if I had not found any horseflesh myself, I could have turned cannibal. To understand the situation, one must have felt the madness of hunger; failing a man to eat, one could have demolished the devil himself, if he were only cooked."[35]

Thriving in these conditions of filth, cold, and famine, typhus swept through the French ranks from October 18, when the Grande Armée left Moscow, until December 11, when the remnants of the army again reached the Niemen River. These weeks of flight form a homogeneous period dominated by the remorseless passage of the typhus bacterium through an army no longer capable of making war. By November 1, the army numbered seventy-five thousand of its original half million. On November 9, when it reached Smolensk, the figure had fallen to thirty-five thousand, then fifteen thousand at the Berezina River crossing on November 26, and finally ten thousand when the tattered survivors recrossed the Niemen.

Napoleon preferred not to experience the denouement of the tragedy. On December 5 he disguised himself as plain "Monsieur Reynal." Accompanied by bodyguards, he escaped by sleigh, heading for Paris and abandoning his soldiers to their fate.

Conclusion

Napoleon's 1812 campaign demonstrates the capacity of warfare to unleash epidemic diseases by creating precisely the sanitary and dietary circumstances in which they flourish. It also illustrates that the causal chain could operate in the opposite direction—in other words, that disease can determine the course of war. In Russia, dysentery and typhus combined to annihilate the largest military force ever created and to award victory to Tsar Alexander.

Just as yellow fever in Saint-Domingue stopped the western expansion of Napoleon's empire, so dysentery and typhus halted its advance to the east. Indeed, the two diseases played a major part in causing regime change in France. After the Russian fiasco, Napoleon was permanently weakened and never able again to construct an army of comparable power.

Furthermore, Alexander's victory multiplied Napoleon's enemies by destroying the myth of invincibility that had once so intimidated his opponents. Here the most important example was the German "national awakening" when intellectuals such as Johann Gottlieb Fichte and Karl Wilhelm Friedrich Schlegel successfully roused German national awareness. The historian Charles Esdaile aptly summarizes the impact of the Napoleonic wars, which, in his view,

> left in their wake both a very different Europe and a very different world. Prior to 1789 France had been unquestionably the strongest of the great powers. . . . By 1815, however, all this had been swept aside. France's domestic resources remained very great, but the establishment of a new German confederation . . . had ensured that the capacity to dominate the "third Germany" that had been central to the Napoleonic imperium . . . was no more. Across the seas, meanwhile, much of France's colonial empire had been swept away, together with Spanish control of the

> mainland of Central and South America. Ironically, then, the greatest hero in French history had presided over nothing less than a total collapse of France's international position, leading Britannia to rule the waves and the rest of Europe to contend with the emergence of what would ultimately become an even greater threat to its security than France had been.[36]

Thus the Russian campaign played a prominent part in undermining French power in Europe and around the world. And disease was a critical factor in this outcome.

CHAPTER 10
The Paris School of Medicine

The study of epidemics is far broader than the examination of one infectious disease after another. Several major themes arise directly from the experience of Western societies with epidemics. One of these has already emerged—the development of public health strategies as societies organize to defend themselves against epidemic invasions. In line with that theme, we have noted the first form of public health, which consisted of draconian antiplague measures of quarantine, lazarettos, and sanitary cordons enforced by military means. After bubonic plague, smallpox ushered in a second major style of public health, which is that of vaccination as pioneered by Edward Jenner. Public health is inseparable from the history of infectious disease, and we will be returning to it to examine the emergence of a variety of strategies in addition to quarantine and vaccination.

A further theme to consider is the intellectual history of medical ideas. Indeed, running parallel to the history of epidemics is the career of scientific medicine through several incarnations, from humoralism down to the modern biomedical paradigm. We considered humoralism in the work of Hippocrates and Galen in Chapter 2, and in later chapters we will examine the filth theory of disease, contagionism, and the germ theory of disease. But now we turn to another crucial moment in the development of medical science that occurred in Paris between the eruption of the French Revolution in 1789 and the middle of the nineteenth century: the establishment of what is universally known as the Paris School of Medicine.

The Paris School was so important that it is occasionally referred to—a shade too concisely—as the moment of transition in medicine from the Middle Ages to modernity. In order to understand what happened in Paris, we consider three aspects of the new development: (1) the intellectual crisis of the humoral system of Hippocrates and Galen, whose thought confronted serious challenges in the seventeenth and eighteenth centuries; (2) the intellectual and institutional preconditions that made the Paris School possible; and (3) the consequences, and the limitations, of the medical innovations that took place in Paris.

The Crisis of Humoralism: Paracelsus

In the first incarnation of scientific medicine in antiquity, Hippocrates and his followers established the importance of naturalistic causes of diseases while rejecting all magical explanations, both divine and demonic. This orientation had enormous importance for medical epistemology—in other words, for what medical science can know, and how it can know it. What are the sources of medical knowledge?

The Hippocratic corpus asserted that knowledge in matters of disease is derived from direct observation at the individual patient's bedside. For this reason humoralism is often referred to as "bedside medicine." It locates the source of knowledge at the bedside through empirical observation of the patient's body. This doctrine had as its corollary a clear program for medical education. Hippocratic medical students learned their craft by apprenticeship to doctors whom they accompanied on their peripatetic rounds and under whose guidance their central task was to observe.

A subsequent development in the course of scientific medicine, as we saw in Chapter 1, was that of Galenism. In Galen's writings humoralism grew less concerned with direct observation and relied more on the authority of ancient texts, which Galen considered almost infallible. The work of the ancients could be carefully explicated, and Galen regarded himself as the person best suited for that task, but it could never be overthrown. For Galen there was no scope for a paradigm shift or fundamental innovation. His view, which conceived of medical knowledge as learned and authoritative commentary on the Hippocratic texts, is frequently called "library medicine." Galen's understanding of medical epistemology also had an educational corollary. Under his influence, medical education consisted essentially of carefully reading the ancients in the original languages, supplemented by lectures interpreting the classics and delivered in Latin.

An early challenge to Galenism was one of the most radical. This was the work of the Swiss physician and alchemist Paracelsus (1493–1541), who earned the epithet "the Martin Luther of medicine" because he rejected the authority of the ancient texts, he did so at the height of the Reformation, and his medical theories had intensely religious implications. Appealing more to barber-surgeons and apothecaries than to learned physicians, Paracelsus regarded the elite, bookish arts of orthodox medicine as impiously naturalistic in seeking to restore health by purely materialist methods. In their place he urged a philosophy of medicine that included the first causes of disease located in the divine and spiritual realms of the macrocosm as well as the immediate causes situated in the microcosm of the body and its interactions with nature. The Paracelsian practitioner acted as the agent of a transcendent power. In therapeutics Paracelsus introduced chemical distillations as remedies to supplant the venesection and botanical simples that orthodox practitioners used to restore the equilibrium of the humors.

In stark apparent contrast to Galen, Paracelsus opposed theory and system in the name of a return to empiricism. In odd contradiction with this critique, however, he replaced humoral theory with an a priori system of his own. Paracelsus regarded the body as constituted of three chemical substances imbued with spiritual properties. The body became ill, he argued, not because of an internal imbalance of humors but because of environmental assault from without. The remedy was to administer chemicals and minerals when distillation had brought forth their inherent spiritual properties and when they were used in accord with the idea of treating "like with like" instead of the Galenic procedure of treating by means of opposites.

Such is the salience of Paracelsus's medico-religious critique of orthodox medicine in the sixteenth and seventeenth centuries that it informs the whole structure of Shakespeare's play *All's Well That Ends Well.* The plot hinges on the failure of orthodox medicine to cure the agonizing and apparently mortal fistula of the king of France. As the doomed and despairing king explains,

> . . . our most learned doctors leave us, and
> The congregated college have concluded
> That labouring art can never ransom nature
> From her inaidible estate. (Act II, Sc. 1, lines 115–118)

Hearing the king's lament, Helena, the protagonist and the daughter of a Paracelsian physician, announces that her father's remedy can, "with the help of heaven," effect a cure where Galenic physic had failed. In her words,

> I am not an imposter, that proclaim
> Myself against the level of mine aim,
> But know I think, and think I know most sure,
> My art is not past power, nor you past cure. (Act II, Sc. 1, lines 154–157)

It is the ability of her Paracelsian therapy to cure the king's fistula that enables the orphaned commoner Helena to upend established gender roles. As her reward from a grateful sovereign, she asks for the hand in marriage of the reluctant aristocrat Bertram. Shakespeare then allows Paracelsus to subvert social as well as gender hierarchies through her marriage as a penniless commoner to a noble courtier. Helena also displays a piety at odds with the materialism of the medical faculties. Shakespeare himself is clearly sympathetic to the notions of Galen's heretical critic.

Scientific Challenges to Orthodoxy

In the medical profession, however, the challenge of Paracelsus represented an assault on orthodoxy from the outside that had limited influence on developments in elite, academic medicine. Challenges with more lasting impact on orthodox medicine arose from other sources. One was the overall ethos of the scientific revolution. Among the intellectual elite, the empirical and experimental methodologies that emerged from the time of Francis Bacon generated a democratic ethos that was incompatible with the hierarchical faith in authority on which Galenism depended. This direction was reinforced by the "middling sort" of craftsmen whose practices, intellectual exchanges, and inventiveness alone made scientific discovery possible and whose contributions eroded social hierarchy.

Specific scientific developments also eroded the edifice of "library medicine" at its base. In human anatomy Andreas Vesalius (1514–1564), the Flemish doctor who taught at Padua, produced in 1543 the monumental work *De humani corporis fabrica* (*The Fabric of the Human Body*), which by symbolic coincidence appeared within a week of the revolutionary work of Nicolaus Copernicus. With magnificent anatomical drawings by one of the leading artists of the day and the author's own textual commentary, *De Fabrica* marked a major shift from traditional medical teaching. Although Vesalius expressed reverence for Galen, his direct observations from human dissection indicated that some two hundred corrections needed to be made in the work of the master, who had based his human anatomy on extrapolations from the dissection of animals.

The decisive factor, however, was not the anatomical amendments that Vesalius promoted, but rather his approach to the subject of medical science. Refusing to accept the authority of Galen's text a priori, he turned for knowledge to direct observation of what he called, in a famous phrase with subversive implications, "that true Bible of nature, the human body." This approach inspired other leading anatomists in Italy, where Vesalius taught, such as Gabriele Falloppio and Girolamo Fabrizio. Together with Vesalius, they set anatomy solidly on non-Galenic empirical foundations, despite their own disclaimers. In practice, if not in rhetoric, their work constituted a radical break with ancient authority.

Even more important than anatomy, developments in physiology provided a crucial impetus for new departures in medical science. Here the major influence was William Harvey's 1628 book *De motu cordis* (*Exercitatio anatomica de motu cordis et sanguinis in animalibus* [An Anatomical Exercise on the Motion of the Heart and Blood in Living Beings]), which established the modern understanding of the circulation of the blood. Orthodoxy held that the blood did not circulate at all but ebbed and flowed in two separate circuits, one in the veins and the other in the arteries, between which there was no exchange except minimally through pores in the septum of the heart. In this Galenic conception, the heart was not a pump but a secondary organ in the visceral hierarchy of brain, liver, and heart. For Galen, the movements of the heart were driven by the flow of the blood, as a millstone is driven by the currents of a river. Harvey revolutionized human physiology and cardiovascular anatomy. Through observation and experimentation he conclusively demonstrated that the heart was a pump that circulated the blood through two intersecting circuits—to the body from the left ventricle and to the lungs from the right. He further established that there was no Galenic seepage between the ventricles through the wall separating them.

So radical was this departure that Harvey waited twelve years between the conclusion of his experiments in 1616 and the decision to publish the results in 1628. His concerns proved to be well founded. The elite of the medical profession in Britain shunned Harvey's work, and English-language texts took no notice of his findings until after the English Civil War. At the same time, authorities in France, Spain, and Italy condemned Harvey and his work, with the orthodox Galenist Jean Riolan most forcefully articulating the wholesale rejection of his views by the profession. Only in radical and republican Holland did Harvey's conclusions find an early welcome in the scientific community.

The reason for such resolute opposition was that Harvey's physiology threatened to overturn Galen entirely and therefore to subvert the authority of the profession. As in the case of Vesalius, it was not only the conclusions that mattered, but also the methodology. Harvey's exclusive reliance on experimentation, mathematical measurement, and direct observation rather than texts marked a sea change in medical epistemology. The implications were profoundly troubling both for humoralism and for the acceptance of established authority in other arenas, including politics and religion. It mattered little that Harvey was not a political or religious radical and that he did not draw out the possible consequences of his teachings for disciplines beyond anatomy and physiology. His epistemology was inherently radical and antihierarchical.

Parallel with developments in anatomy and physiology, major discoveries in the natural sciences had profound medical implications. The chemical revolution associated with Antoine Lavoisier, Joseph Priestley, and Jöns Jakob Berzelius threw into question the Aristotelian notion of nature as composed of four elements (earth, air, water, and fire). Indeed, Lavoisier by 1789 had established thirty-three elements as the foundation of what was to develop during the following century into the periodic table. There was no possibility of reconciling this new chemistry with the Aristotelian worldview of four elements, humors, temperaments, and properties. With regard to the properties, even simple devices—the thermoscope (devised by Galileo) and its successor the thermometer (invented by Giuseppe Biancani in 1617)—pointed to the conclusion that the "property" of cold was not in fact an independent property at all but simply the absence of heat.

Finally, among the intellectual challenges to medical orthodoxy, epidemic diseases also played an important role. It was conceptually difficult for humoral medicine to provide a convincing explanatory framework for sudden mass die-offs such as were experienced during visitations of plague, smallpox, or cholera. If disease was an imbalance of the humors in an individual human body, what could account for the sudden and simultaneous imbalance in the humors of so many? How could humoralism, which was fundamentally individualistic, convincingly explain the course of a disease as it progressed through an entire community?

The medieval addition of astrology to humoralism provided greater scope for sublunar calamities through the influence of celestial alignments and influences, but even astrology did not provide a robust account of pandemic disease. It was partly for this reason that the unorthodox idea of contagion emerged. Contagionism provided a more plausible account of the progress of an epidemic through a community, and it accorded more readily with the everyday observation that people seemed to fall ill after contact with ill people. The passage of epidemic diseases, therefore, gave rise to doubts about humoralism and helped prepare the ground for alternative medical philosophies.

Background to the Intellectual Revolution in Paris

During its period of ascendancy between 1794 and 1848, the Paris School of Medicine was responsible for a conceptual revolution in the understanding of disease and in medical epistemology. It also transformed medical education, gave rise to medical specialties, restructured the profession, and changed the medical marketplace by providing regular physicians with new claims to authority in competition with rival schools and doctrines. Paris became the leading force in Western medicine and the model for imitators across Europe and North America. It was here that "hospital medicine" succeeded "library medicine" and that a new paradigm progressively supplanted humoralism. What, then, were the sources of the "new medicine"?

Institutional Foundations

The challenges to humoralism on a variety of fronts provided the intellectual doubts that dethroned antiquity, but there were also preconditions that positively prepared the way for the emergence of a new medicine. Fundamental was the part played by the extensive network of Parisian hospitals. Hospitals, of course, already existed in Paris. Indeed, on the site of the greatest, and largest, of them—the Hôtel-Dieu—care had been provided continuously since the seventh century. Originally, however, the hospitals did not function primarily as places of treatment. They were welfare safety nets for the aged, the incurable, and the orphaned in association with philanthropy and the church. The Industrial Revolution and urbanization, however, radically increased the numbers of patients and transformed their ailments. Paris, the intellectual center of western Europe and one of its chief urban centers, thus became the locus for some of the largest and most famous

Figure 10.1. The Hôtel-Dieu Hospital, in Paris, was one of the hospitals that provided an institutional foundation for the Paris School of Medicine. (Wellcome Collection, London. CC BY 4.0.)

hospitals in Europe—not only the Hôtel-Dieu, but also others, including the Charité and the Pitié. By the end of the Ancien Régime, the Hôtel-Dieu boasted four wards that accommodated as many as four thousand patients, often sleeping several to a bed (fig. 10.1).

It is impossible to appreciate the Paris School and the understanding of medical science that emerged in the French capital without taking into account these massive institutions that underpinned the new ideas. A single ward in a Parisian hospital provided an almost endless supply of diseased bodies, and it seemed natural to group sufferers with similar ailments together. The hospitals also became places of instruction under the control of a centralized secular state whose bureaucratic ethos found it convenient to practice triage, grouping like with like. Furthermore, the hospitals in Paris were dedicated—and this was crucial—to furthering medical and scientific knowledge. They were in fact more concerned with advancing knowledge than with treating patients.

The hospitals of the city provided an institutional basis for the Paris School, but the school also had philosophical origins. A very generic but significant background factor was the ethos of the Enlightenment, with its questioning of authority, its intellectual skepticism, and its empirical orientation. A particular and vitally important figure was John Locke (1632–1704). His work *An Essay Concerning Human Understanding* of 1690 was so influential that many consider it the foundational text of the Enlightenment. Locke famously postulated that the mind at birth is a blank slate, or *tabula rasa*. The corollary is the idea of philosophic "sensualism," which holds that knowledge is not innate but is derived entirely from sense impressions and reflections on those impressions.

Here was a radical epistemology. Sensualists such as Locke and the French philosophe Étienne Bonnot de Condillac (1714–1780) believed that the source of knowledge is the information received directly from nature by the five senses, which provide the data for the reflections of the brain. In this view of epistemology, Locke considered not only the origin of human knowledge, but also the limits to what can be known. God, for example, lies outside the realm of sense-based knowledge. In addition, he established rigorous steps for being certain of the things that can be known.

More immediately important for medicine was the work of the seventeenth-century English physician Thomas Sydenham, who was Locke's close friend and is variously termed the "English Hippocrates" and the "father of English medicine." Politically, Sydenham was a radical—a left-wing Puritan who rebelled against the crown during the English Civil War and served as an officer in Oliver Cromwell's army. Sydenham was also radical in his medical ideas. His prescription for a reformed medical practice was rooted in the idea of sensualism. He called for a rigorous return to observation of the patient and the abandonment of theory. In his opinion, the advancement of medical science could proceed only by a systematic, empirical, case-by-case comparison without reference to the classics, to system, or to theory.

Somewhat in contradiction with this logic, Sydenham did not wholly reject humoral medicine. His medical practice, if not his theory, was still largely humoral. But when he sought new knowledge, he turned from the classics to direct bedside observation, believing that physicians should trust their own experiences and their reasoning based on those experiences. In many respects, he advocated skipping over Galen, who prioritized system, in order to revive the original Hippocratic reliance on observation. Although he was an Oxford graduate, he distrusted bookish learning and university education, and in return the medical elite of his day scorned him.

Sydenham also gave special attention to epidemic diseases and studied smallpox, malaria, tuberculosis, and syphilis. Indeed, his thinking is a good illustration of the impact that infectious diseases had in undermining medical orthodoxy and promoting a new scientific paradigm. In his work on malaria, for example, he reached the radically new conclusion that "intermittent fever"—as the disease was called—was not a holistic humoral imbalance at all but rather a specific disease entity. Thus Sydenham promoted the idea that diseases are specific entities rather than a generalized dyscrasia, and he even suggested that the time might arise when they would be classified according to the principles of Linnaeus. Thus he argued, "All diseases ought to be reduced to certain determinate kinds, with the same exactness as we see it done by botanic writers in their treatises on plants."[1] Revealingly, his famous work of 1676 bore the title *Observationes Medicae,* clearly stressing the importance of direct empirical observation.

Further demonstrating the role of epidemics in subverting humoralism, Sydenham embraced the idea of contagionism. He wrote, for example, of the plague: "Besides the constitution of the air, there must be another previous circumstance to produce the plague; namely the effluvia or seminium from an infected person, either immediately by contact or immediately by some pestilential matter conveyed from some other place."[2] Here too was a radical notion.

Sydenham is also famous for introducing a number of practices and therapies into medicine. He popularized the use of quinine for malaria, and he made use of opium to disguise its bitter taste and make the "Jesuit bark" more acceptable in Protestant England. In addition, he used cooling drinks and fresh air to treat fevers rather than bloodletting, and he pioneered this kind of "cooling regime" in treating smallpox. Equally radical was his advocacy of what might be termed therapeutic minimalism. Frequently, he wrote, the best thing a physician could do was nothing at all.

A further influential figure in the development of a new philosophy of medicine was Pierre Cabanis (1757–1808), who was a physician, physiologist, and philosopher. He was also an administrator of the Paris hospitals and an early supporter of the French Revolution. But the relevant factor here is that Cabanis, too, was a pioneering advocate of sensualism. Like Locke and Condillac, he believed that all mental processes are derived from the five senses and that physicians should therefore consult their own observations rather than rely on ancient texts. Rejecting dualism—the view that the mind (or soul) is separate from the physical structures of the brain—Cabanis held that the brain works in a manner analogous to the functioning of the stomach. In the stomach, food is taken in, and the result is digestion. In the brain, sensual impressions are taken in, and the result is thought. Here Cabanis took a philosophical and clinical stance that would be shared by all the leading figures of the Paris School.

The French Revolution

Besides institutional and philosophical foundations for the Paris School, it is important to consider a major political factor—the French Revolution. A general feature of the revolution was that it provided an opportunity to wipe the slate clean of established authorities. In the field of medicine, this meant the demolition of medieval medical corporations and the reconstruction of the profession. Stressing French nationalism, the revolution promoted the idea that instruction should no longer take place in Latin but in the vernacular, further undermining the authority of ancient texts.

A number of specific, contingent circumstances also made the revolution an important moment in the development of a new medicine. Most prominently, the period from 1792 to 1815 constituted a generation of nearly uninterrupted warfare, creating an urgent, practical need for more numerous physicians and more adequate hospitals. These needs in turn facilitated the reform of medical education and hospital administration. The hospitals were now centralized and controlled by the state rather than the church. Furthermore, within hospitals the wards specialized in specific patient populations. Pressed into state service, these reformed institutions served the overriding aim of promoting the progress of science.

This vision of the purpose of hospitalization had a major impact on the way in which patients were viewed. Since patients had a duty to promote the advancement of knowledge, their bodies were made wholly accessible to physicians and medical students both in life and—even more crucially—after death, when they were subject to postmortem examination. While still alive, patients furthered knowledge as physicians conducted physical examinations to study the signs and symptoms of disease with rigor and precision. In death, their lesions were seen as the underlying causes of symptoms, which were regarded as surface epiphenomena revealing specific diseases hidden within the body. Symptoms and lesions were therefore correlated and understood as the two sides of a single disease process. The centrality of the postmortem examination inevitably also furthered the understanding of anatomy and physiology, allowed surgeons to develop their techniques, and permitted pathologists to track the course of disease in the body with precision.

In the new world of Parisian medicine, the locus of medical education shifted almost entirely to the hospital ward as it became overwhelmingly clinical and practical rather than bookish. Medical students undertook three years of training on the ward, followed by a year of internship. Their professors in the faculty of medicine became full-time instructors appointed by competitive examination (the *concours*) and employed by the state.

Just as important, the Paris hospitals were suffused with a new set of values that promoted ability and merit rather than privilege, birth, and cronyism as the French Revolution provided a new dynamism and sense of open competition. The mottos of the French medical profession were "progress," "reform," "observation," and "precision." These values also included secularism as the revolution ended the hold of the church over the hospital system. Altars were removed from the buildings and crucifixes from the wards. At the same time, nurses—who were nuns—were rigidly subordinated to the attending physicians. Buildings were also renovated or repurposed in order to accommodate large wards, dissecting rooms, and auditoria for grand rounds.

Now under state control, the hospitals came under the direction of a new institution—the Paris Hospital Council—that regulated all aspects of hospital life and administration. Especially important was the role of the board's Bureau of Admissions as a centralized triage office. The bureau carefully divided patients into categories corresponding to their symptoms and then sent the patients of each category to the same hospital ward. Through the same process, certain hospitals ceased to be general hospitals and specialized instead in certain conditions. Thus, almost by bureaucratic rather than scientific intention, the Hospital Council and its Bureau of Admissions became agents in promoting the concept that diseases were separate and distinct entities rather than the single humoral imbalance or dyscrasia that underlay Hippocratic and Galenic medical philosophy.

Paris was crucial, in the words of historian George Weisz, in part because it also created something radically new—a large and unified research community with a sense of innovation and strong institutional support:

> [Paris represented] the reconstitution of medical institutions in the decades following the Revolution as a huge, interlocked, and prestigious network. At its center was the Paris Faculty of Medicine, with its more than two dozen full professors and many junior personnel (by far the biggest medical school in the world), and the municipal hospital system of Paris, with its many linked institutions and several hundred physicians and surgeons (including the vast majority of faculty professors) working with them.

This complex was "different from anything that had existed until then."[3]

The Paris School in Action

Built on such foundations, the Paris School became the Mecca of the new medicine. Students and physicians came to the Latin Quarter from all over the world to observe and to be trained. Indeed, substantial numbers of Americans made a pilgrimage to study in Paris, taking ideas back to the United States and placing themselves in a position to charge higher fees because of the prestige that went with time spent in the French capital.

Radically empirical in orientation, the Paris School adopted the motto *Peu lire, beaucoup voir* (Read little, but see a lot). Thus M. Gubler, professor of therapeutics at the Paris School, stressed in his inaugural lecture in 1869 that with regard to the method of acquiring medical knowledge, "the only valuable one was the Baconian method, that of strict scientific observation and induction . . . ; in other words, that positivist philosophy which had rallied all the scientific minds of the day." Professor Gubler advocated a return to "ancient observation."[4] But observation in Paris was not the more passive bedside approach of Hippocrates. Paris introduced instead the modern physical examination in which the body was actively interrogated through the techniques of percussion and auscultation. René Laënnec invented the stethoscope in 1816, and it rapidly became the emblem of the Paris School.

Laënnec's original monaural stethoscope was a wooden tube a foot long that he placed against the chest of the patient as he listened to the internal sounds of the heart and lungs (fig. 10.2). This was "mediated auscultation," which provided vast new sources of diagnostic information as compared with "immediate auscultation," in which the physician pressed an ear against the patient's chest. Laënnec, who devoted his attention above all to sufferers of pulmonary tuberculosis, devised a lexicon to describe the sounds that he heard—rales, ronchi, crepitance, egophony. In 1819 he published a treatise on chest diseases in which he sought to standardize and codify the internal sounds that he heard with a precision not unlike that of a trained musician (see Chapter 14). He even employed musical notations to indicate some of the sounds.

Furthermore, such major figures in the Paris School as Laënnec, François Magendie, Pierre Louis, and Marie François Xavier Bichat systematically combined the knowledge that they gained through physical examination on the wards with a more intimate examination by autopsy on the dissecting table. After a patient's death, they related the symptoms they had recorded on the ward to the lesions they exposed with the knife. Diseases were, they stressed, more reliably classified by lesions than by symptoms.

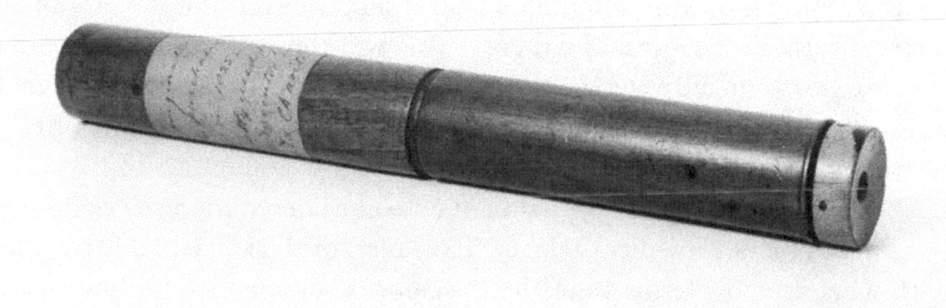

Figure 10.2. The Paris School introduced the modern physical examination, which was facilitated by instruments such as the monaural stethoscope devised by René Laënnec in 1816. (Wellcome Collection, London. CC BY 4.0.)

Instead of bodily fluids, which were the preoccupation of the humoralists, the Paris physicians based their work on the solid organs and tissues of the body. As a result, their medical philosophy was frequently known as "solidism" or "localism." It emerged in close relationship to the principal diseases then prevalent among the patient population in the French capital—phthisis (pulmonary tuberculosis), pneumonia, typhoid fever, heart disease, puerperal fever, and cholera.

The role of the Parisian wards in promoting the new medicine was clearly apparent. Laënnec's ability in systematizing the internal sounds of the chest reflected the fact that every year he examined some five thousand patients, most of whom suffered from pulmonary tuberculosis. With such a vast experience of tubercular patients both in the ward and on the dissecting table, Laënnec and his colleagues reached the settled conviction that they were dealing with a discrete disease entity. This was the revolutionary notion of disease specificity. Each disease, they argued, was distinct and immutable and could therefore be classified on Linnaean principles. This conclusion gave rise to the new discipline of nosology, that is, the study of the classification of diseases, and with it to new medical specialties, each dealing with a subset of diseases—for example, venereology, psychiatry, pediatrics, pathological anatomy, and internal medicine.

Paris also created a radically new conception of medical education. As in the past, lectures were still part of the curriculum, as were some texts; but the principal site of learning was the hospital ward, where famous professors like Pierre Louis conducted rounds with a large retinue of students.

The new medical education was practical, hands-on instruction. Students were taught to learn through their senses—sight, sound, and touch—and to distrust authority, dogma, and theory. This was the philosophical program of sensualism applied to medicine, and the discipline it spawned came to be known as hospital medicine, as opposed to the bedside medicine of Hippocrates and the library medicine of Galen and his followers.

The practitioners of hospital medicine considered themselves the embodiment of a new medical science. However, this science had little to do with what are now termed the "basic sciences" of chemistry, physics, and physiology. These were not central to the curriculum and were revealingly termed "accessory sciences." What was considered "science" in Paris was a combination of rigor, precision, unbiased direct observation, the establishment of numerical correlations among observed phenomena, and the confirmation of diagnoses by dissection on the autopsy table.

This new educational and research program proved capable of generating an enormous amount of new knowledge of diseases and their mechanisms. Indeed, it transformed the nature of medical science and practice. Major advances were made in fields such as diagnosis, pathology, classification, and surgery. It also transformed the profession, providing physicians with robust new claims to authority and to enhanced compensation in the medical marketplace. For these reasons the Paris School became the model for an international medical reformation. It was soon followed by the Vienna School, which adopted its methods; Guy's Hospital in London; the Harvard Medical School; the Massachusetts General Hospital in Boston; and the Johns Hopkins School of Medicine in Baltimore.

Unfortunately, however, the strengths of the Paris School did not result in significant benefits for patients treated on the wards. Therapeutics was the widely acknowledged weakness of the new medical science. Physicians amassed knowledge, but that knowledge did not lead to improved patient management. Visitors to France from Britain and the United States even expressed serious moral reservations about Paris and its priorities. They noted that all too often physicians were little concerned about alleviating suffering or preserving life; surgeons, it was said, regarded operations as primarily a means to achieve greater manual dexterity, and the curriculum did not stress that the primary mission was to heal. Knowledge and its advancement were all that counted. Patients were objects to be observed as if they were displays in a natural history museum or stage props in a theater, and their presence on a ward was principally a means to serve science. Dr. Griffon, a caricature of Pierre Louis in the 1842 novel *The Mysteries of Paris* (1842) by the French counterpart of Charles Dickens—Eugène Sue—tells students on his rounds to look forward to the lesions they can expect to see when the patient before them dies. Dr. Griffon, Sue wrote,

> considered the wards as a kind of school of experiments where he tried on the poor the remedies and applications which he afterwards used with his rich clients. These terrible experiments were, indeed, a human sacrifice made on the altar of science; but Dr. Griffon did not think of that. In the eyes of this *prince of science* . . . the hospital patients were only a matter of study and experiment; and as, after all, there resulted from his eyes occasionally a useful fact or a discovery acquired by science, the doctor shewed himself as ingenuously satisfied and triumphant as a general after a victory which has been costly in soldiers.[5]

This weakness in therapeutics meant that medical practice continued to rely on the traditional armamentarium and on practices such as venesection handed down from antiquity—despite vast new learning and the conclusion of eminent doctors like Louis that they served no positive clinical function. It was largely for this reason that by the middle of the nineteenth century, the Paris School began to lose its momentum. There was by then a widespread therapeutic skepticism and frustration at the combination of great power to diagnose and powerlessness to treat. There was also a sense that Paris was failing to give due recognition to some of the most notable developments in midcentury medical science, especially microscopy. International students began to desert Paris for other centers that regarded the laboratory as the dynamic site of new medical learning rather than the hospital ward.

CHAPTER 11

The Sanitary Movement

In the 1970s, the English medical historian and physician Thomas McKeown made a dramatic suggestion regarding the demographic explosion that occurred in the West after the onset of the Industrial Revolution. Britain, the first industrial nation, exemplified the phenomenon as the population of England and Wales doubled from 10,164,000 to 20,066,000 in the half century between 1811 and 1861, and nearly doubled again to 36,070,000 in the fifty years that followed. McKeown's interpretation of this population surge appeared in two much-debated but important works of 1976—*The Modern Rise of Population* and the more explicitly provocative *The Role of Medicine: Dream, Mirage, or Nemesis?* In both works he dealt with the problem of explaining the falling death rate and enhanced longevity characteristic of the West since the late eighteenth century.

McKeown accepted, like most demographers, that a central factor in this massive growth in population was a "demographic transition," meaning that infectious diseases ceased to be major causes of death, with the principal sources of mortality becoming instead chronic degenerative diseases—heart disease, cerebrovascular disease, cancer, dementia, and diabetes—that overwhelmingly afflict the elderly. He further agreed that even urban centers, those sites of massive mortality in the early modern world, have undergone a "mortality revolution." Instead of growing only by large-scale in-migration, industrial cities in the developed world have became salubrious places with low rates of death and high life expectancy.

In explaining these striking trends, McKeown made the controversial claim that medical science played only a minor role. Indeed, he argued that, until roughly the Second World War, physicians were powerless to prevent or to treat the major afflictions of their patients. By then, however, the demographic explosion in western Europe and North America had already occurred, as well as the great extension in life expectancy and the transformation of urban centers such as Paris, Naples, and London into healthy modern cities. Medical and scientific interventions could not be responsible. In their stead McKeown invoked social, economic, and infrastructural factors as the causes. Health and longevity resulted, he claimed, not from science but from more humble determinants—improvements in nutrition, wages, and sanitation.

In this way McKeown generalized from the turn-of-the-twentieth-century research of Arthur Ransome, who famously argued that the declining morbidity and mortality from tuberculosis were the result of "unconscious policy." Tuberculosis declined, in other words, as an indirect effect of social and economic uplift rather than as the product of medical science or public health measures. McKeown applied Ransome's analysis to the whole field of infectious diseases.

The debate begun by this "McKeown thesis" has largely run its course, and there is no need to assess his dual claims that (1) intentional, scientific intervention played a minor role in promoting human health and (2) diet was the most important of the key "spontaneous" improvements. But even though there is no consensus about the relative importance of the factors McKeown invoked to explain the demographic transition, there is agreement that he was correct in stressing the importance of the transition and in pointing to sanitation as one of its most significant contributing factors.

The "sanitary idea" emerged first at the turn of the nineteenth century in Paris and then, more systematically during the 1830s and 1840s, in Britain, where its most influential figure was Edwin Chadwick. Chadwick launched a public health movement that implemented the reforms implicit in the "idea" during the decades between 1850 and the First World War. This movement revolutionized morbidity and mortality rates in British towns and cities by means of a vast centrally inspired urban cleanup. The movement originated in France, flourished in Britain, and then moved onward—back to France and on to Belgium, Germany, the United States, Italy, and the rest of the industrial world.

Sanitary Science in Paris

An initial stimulus to the sanitary reform movement in England came from across the English Channel. In Paris, with the background of the Enlightenment and then the Paris School of Medicine, the late eighteenth century had produced the idea of urban reform. Especially important was the Paris School. The combination of disease specificity, the vast wards of carefully triaged patients suffering from the same illness, and the passion for keeping statistics led rapidly to the correlation of diseases with (1) the social backgrounds of the patients and (2) the localities where sufferers were overrepresented.

As Alain Corbin has explained in his study *The Foul and the Fragrant: Odor and the French Social Imagination* (1986), there was also a growing general awareness among the population of the intolerable stench of the French capital. Corbin conducts his readers on an olfactory tour of the innumerable fetid oozings of a huge city marked by cesspools; unpaved streets deep in mud, dunghills, and dead animals; walls of houses saturated with urine; slaughterhouses; butcher shops and their offal; night soil and refuse collecting in narrow and ill-lit lanes; ditches filled with stagnant and evil-smelling black water; crowded tenements with dark interiors that were never swept or cleansed; a multitude of crowded and unwashed occupants; and everywhere a shortage of water that prevented both individual ablutions and street cleaning.

Obsessed by the unrelenting assault on their senses and afflicted with what some historians have called "pollution anxiety," Parisian scientists sought to measure the smell with an instrument called an odiometer, or olfactometer, that they invented in order to analyze the chemistry of the effluvia. They also promoted a new science—osphresiology—devoted to odors and their impact on the sense of smell. Given the Enlightenment heritage of philosophical sensualism, it was logical that French philosophes heeded the unceasing olfactory bombardment of early nineteenth-century Paris. They also correlated the intensity of the odors with the incidence of disease, linking the two by means of the theory that disease (or "fever" in contemporary parlance) was the result of some still-unspecified airborne poison that wafted aloft along with the stench from decomposing matter. Alexandre Parent-Duchâtelet made an enduring reputation by studying the mephitic outflows of the Paris sewers and relating them to the health of the city.

The most significant French figure in gathering data and establishing correlations, however, was the physician Louis-René Villermé (1782–1863), who measured death rates across the twelve arrondissements, or districts, of Paris and correlated them with population density and income. Motivated by the hypothesis that miasma emanating from waste caused disease, Villermé urged action through a cleanup campaign that would remove refuse and clean public spaces in the most badly affected localities. He founded a public health journal to support the campaign—*Annales d'hygiène publique et de médecine légale* (Annals of Public Health and Forensic Medicine)—and he inspired the creation of a Paris Board of Health that was most active during the two decades after 1820.

As a municipal rather than a national phenomenon, the sanitary movement in Paris was limited in its reach. It was also never fully theorized, systematized, or institutionalized. In some respects its most lasting effect was that it helped to spawn a much broader and more influential movement across the English Channel. Chadwick, who possessed a command of the French language, steeped himself in the works of Villermé and the *Annales d'hygiène*. He admired their approach of assembling massive empirical data and of relating disease to salubrity. He understood their neo-Hippocratic methodology of stressing climate in the macroenvironment as the direct or "exciting" cause of epidemics, but he concerned himself instead with the microenvironment of specific localities. He viewed climatic variables such as temperature and humidity as affecting disease only indirectly by their effect on decomposition. For Chadwick, the real issue in the etiology of disease was filth, and the practical policy question was how best to remove it.

Edwin Chadwick and Poor Law Reform

Curiously, the founder of the sanitary movement was not a physician and had little interest in medicine. Edwin Chadwick (1800–1890) was instead a barrister from Manchester. He was also a disciple of the liberal political economist and social reformer Jeremy Bentham. By the time he turned his attention to health, Chadwick had already gained notoriety as the reformer of the "Poor Laws" that constituted England's provision for the health and welfare of the indigent. Relief during times of distress was the birthright of all Britons and was available in their native parishes. Chadwick's view was that public assistance through the Poor Laws created a vicious and counterproductive downward spiral in which welfare fostered demoralization, dependence, and indolence. The inevitable result was to breed further poverty. In the process, the system also imposed a heavy burden on local property owners, whose local real estate taxes, or "rates," funded public assistance. In a statement of one of his leading articles of faith, Chadwick proclaimed that "indigence cannot be made generally to disappear simply by grants of money."[1]

Chadwick's solution to the problem was embodied in the so-called New Poor Law of 1834, which was founded on laissez-faire principles and belief in the free market. The measure, drafted jointly by Chadwick and the economist Nassau William Senior, had two major objectives. The first was to centralize the administration of the system in order to achieve uniformity of policy. The second was to make welfare so repugnant that only the truly desperate—the "deserving poor"—would seek support. Recipients would be compelled to accept residence in workhouses, where they would receive the necessities of life but under conditions that would make them long for any available outside employment. Families in workhouses were separated—parents from children and husbands from wives. The food was purposefully cheap and unappealing, surveillance was constant, and the work was designed to be more unpleasant and boring than any to be found outside the walls of the institution. In his novel *Oliver Twist*, an incisive critique of the New Poor Law published serially during 1837–1839, Charles Dickens described workhouses as offering people the brutal choice of starving slowly within or rapidly outside. Chadwick and his followers termed the ethical basis of workhouses the principle of "least eligibility." The poor found the application of the principle cruel, but ratepayers approved.

Chadwickian Poor Law reform and the sanitary movement that followed were both applications of the Benthamite perspective of directing centralized state power to solve the problems of urban modernity. Rapid urbanization and industrialization led to a seemingly ever-growing train of social problems concentrated in Britain's towns and cities. Two were especially worrying—poverty and disease. Having dealt with poverty through Poor Law reform, Chadwick turned his attention to the other devastating affliction of the Victorian city—epidemic disease. In particular, he and his fellow reformers concerned themselves with the scourges of consumption (tuberculosis), cholera, smallpox, scarlet fever, and "typhus," which was a diagnosis that included the infection known today as typhoid fever. Poor Law reform and the sanitary movement were governmental initiatives to deal with the economic and the medical facets, respectively, of the Victorian urban problem so vividly portrayed in London by Dickens and Henry Mayhew and in Manchester by Friedrich Engels.

Sanitary and Poor Law reform were more, however, than just sequential efforts to confront the social pathologies of the mid-nineteenth-century English city. Poor Law reform in some measure actually led to the sanitary movement. The centralized state apparatus of relief after 1834 gathered extensive data concerning the ravages of disease in the overpopulated, poorly housed, and impoverished cities of the period. Indeed, the Poor Law bureaucratic machinery became the mechanism that was employed to document the catastrophic urban conditions that required urgent action. In Chadwick's case, the research he conducted in preparing the New Poor Law provided him with the awareness of health and disease in the towns. It was no coincidence that he was the leading figure first in one movement and then in the other. Poor Law reform helped to create a new consciousness of the extent of the sanitary problem.

The New Poor Law also powerfully affected the nature of the sanitary movement by determining the issues that public health policy did not address. In the 1830s when Chadwick turned his attention to disease, a leading current of medical opinion held that poverty was a major driver of ill health and that wage reform was therefore a necessary component of any campaign to ameliorate health. Most prominent among those who held this view was William Pulteney Alison, a philanthropist and professor of medicine at the University of Edinburgh. Alison argued that economic hardship was not simply one of many factors promoting disease but rather its leading determinant.

Chadwick and his movement, however, carried the day with the counterargument that the chain of causation ran in precisely the opposite direction—that disease generated poverty rather than the reverse. Individual irresponsibility was the source of both. Furthermore, Chadwick believed that the New Poor Law had already addressed the issue of indigence by providing the poor with incentives to help themselves while supporting those who were genuinely unable to work and therefore to earn. It was logical, therefore, that he and the other veterans of Poor Law reform resolutely excluded wage levels, working conditions, and economic exploitation from their purview as matters irrelevant to disease.

The Filth Theory of Disease: Thomas Southwood Smith

Given that Chadwick was not a man of science or medicine, the most important theorist in establishing the medical philosophy underpinning the sanitary movement was Thomas Southwood Smith (1788–1861). He was cofounder of the sanitary idea and Chadwick's closest associate and fellow Benthamite. Southwood Smith had received his medical education at Edinburgh, but the turning point in his understanding of the etiology of disease was his appointment to the London Fever Hospital in the impoverished East End. Practicing there for nearly the whole of his career, he had ample occasion to observe the abysmal living conditions and medical afflictions of the handloom weavers of Bethnal Green and Whitechapel. A Unitarian preacher as well as a doctor, he was appalled by the wretchedness of the workers, their ill health, and what he regarded as their moral and spiritual depravity. It made sense to him, as it did to Villermé and Chadwick, that the filth in which they led their lives destroyed their physical health just as it undermined their humanity, leading them to intemperance, indebtedness, and debauchery.

It was no novelty, of course, to link epidemic diseases to airborne miasmas. Until the Enlightenment, however, the dominant interpretation was that a miasma was a corruption of the air resulting from a major cosmic event—an inauspicious conjunction of the stars or a climatic change in temperature or humidity. What was new, and most fully developed by Southwood Smith, was the "filth theory of disease." This theory still held that disease was caused by miasma, but the origin of the miasma was decaying filth in the microcosm of a specific neighborhood, community, or village. In his most important work, *A Treatise on Fever* (1831), the Unitarian physician wrote:

> The immediate, or the exciting, cause of fever, is a poison formed by the corruption or the decomposition of organic matter. Vegetable and animal matter, during the process of putrefaction, give off a principle, or give origin to a new compound which, when applied to the human body, produces the phenomena constituting fever. . . .
>
> The more closely the localities are examined of every situation in which the plague prevails, the more abundant the sources of putrefying animal matter will appear, and the more manifest it will become, not only that such matter must be present, but that it must abound.[2]

In the filth etiology of disease, climatic factors had a presence, not as "exciting" causes but only as remote or "predisposing" causes that operated indirectly—as in the case of heat and moisture—by promoting more rapid and intense putrefaction and by weakening the resistance or "animal energy" of the population. As Southwood Smith explained, "Of the conditions which are ascertained to be essential to the putrefactive process of dead organic substance . . . , those of heat and moisture are the most certain, and as far as we yet know, the most powerful."[3] For Southwood Smith, more distant cosmological or astrological events had no place. As a rationalist both in his utilitarian philosophy and in his Unitarian religion, arbitrary astronomical events that caused suffering and sin among humans were unthinkable in a universe created by an almighty and loving God.

Given the filth theory, the means to prevent disease were immediately knowable. Like their colleagues in Paris, English sanitarians turned to the remedy of urban cleanups. As disciples of Bentham, however, they believed in the efficacy of using state power to tackle the problem centrally, forcefully, and systematically on a national level. It was no coincidence that the sanitarians in Britain brought their reforms to a climax through the passage by Parliament in 1848 of a Public Health Act and the creation of a General Board of Health (see more below). In the case of Southwood Smith and his fellow Unitarians, the imperative to act was all the more powerful because it was religiously grounded as well. Preventing disease and sin through cleanliness would demonstrate God's benevolent ways toward humanity by showing that suffering arose through human neglect and that it could be easily prevented in a society of individuals of goodwill. Sanitation, therefore, was a moral and humanitarian imperative.

The sanitary idea carried conviction for a series of reasons that were at once medical, religious, and epidemiological. But two additional factors enabled it to sweep the medical profession and informed opinion in Britain during the 1830s and 1840s. A strength of the filth theory was that it was simple and easy to comprehend. It also carried conviction because it was a localist modification of traditional miasmatic teaching rather than a radically new departure. Filth was noxious and ubiquitous in English towns and cities, and it was an easy step to believe that it was also poisonous, especially since the most unsanitary conditions were present wherever those fearsome killers of the era—consumption, typhoid, and cholera—claimed their victims. The jump from correlation to causation was small. A theory of disease was propounded just as a double wave of anxiety about filth and revulsion at stench had already swept the country.

The *Sanitary Report* (1842)

Having dealt by 1834 with one facet of Britain's great urban problem—poverty—Chadwick turned to confront its other major dimension—disease. As he did so, he was armed with three ideas: (1) a detailed knowledge of the sanitary idea that had first emerged in Paris, (2) the Parisian conviction that medical and health issues were best approached through statistics and the massive accumulation of data, and (3) Southwood Smith's filth theory, which carried complete, even dogmatic, conviction in Chadwick's mind.

Chadwick's approach to public health was to begin his labors by carrying out—at his own expense—an exhaustive study of the poverty and ordure that burdened the working classes throughout Britain. He believed that the sheer volume and shock of the information he collected would establish the case for the filth theory, silence potential opposition, and galvanize the country to act. These were the purposes of his monumental work of 1842, *Report on the Sanitary Condition of the Labouring Population of Great Britain*—known more briefly as his *Sanitary Report*. This report is one of the foundational texts of modern public health, and it instantly became a bestseller. Chadwick's procedure was to gather firsthand information through the bureaucratic apparatus of the New Poor Law system that he had helped to create. Together with his fellow Poor Law commissioners, Chadwick sent out questionnaires and solicited reports—thousands of them—from the Poor Law assistant commissioners and the physicians associated with them in localities throughout England. In Scotland, where the Poor Law system did not apply, he relied instead on an ad hoc network of medical officers, factory inspectors, and local doctors. In order to leave no informed and persuasive voice unheard, Chadwick also sought the views of prominent medical men outside the Poor Law administration and of notable men of letters. Having collected the reports from all of these sources over a period of three years, he sifted, edited, abridged, and arranged them. He also added his own commentary, based on an overview of the material and on personal tours of inspection that he conducted to observe and document conditions for himself. Before publication he also sent drafts to leading medical authorities for their approval and comments.

What emerged was a detailed and documented description of the horrifyingly sordid living conditions of workers throughout the nation—agricultural laborers, artisans, miners, factory workers, and handloom weavers. The impact of the tableau that emerged was especially powerful because in all cases the sources of the information were unimpeachable and, apart from Scotland, official. Everywhere the human consequences of the agricultural revolution, the Industrial Revolution, mass urbanization, and massive population growth were apparent. Overcrowding, poverty, filth, and stench were ubiquitous. Such results were not entirely surprising in huge cities such as London, Manchester, and Glasgow, but unexpectedly, similar conditions were found to prevail in small towns and villages as well. Chadwick's measured introductory words set the tone:

> The following extracts will serve to show, in the language chiefly of eyewitnesses, the varied forms in which disease attendant on removable circumstances appears from one end of the island to the other amidst the population of rural villages, and of the smaller towns, as well as amidst the population of the commercial cities and the most thronged of the manufacturing districts—in which last pestilence is frequently supposed to have its chief and almost exclusive residence.[4]

Furthermore, the social conditions that Chadwick laid bare mapped perfectly onto the geography of epidemic disease—"fever," "plague," or "pestilence" in the language of the *Sanitary Report*. So stark was the information that the Poor Law commissioners, who had originally agreed to issue the document jointly with Chadwick, backed down and insisted that it be published in his name alone.

Two examples from the West Country illustrate the pervasiveness of dirt and the link between such filth and the prevailing diseases of the locality. Reporting from Truro, for example, Chadwick's correspondent, Dr. Barham, noted:

> Passing into St. Mary's parish, the proportion of sickness and even of death is . . . as great, perhaps, as that of any part of Truro. . . . There is, however, no mystery in the causation. Ill-constructed houses, many of them old, with decomposing refuse close upon their doors and windows, open drains bringing the oozings of pigsties and other filth to stagnate at the foot of a wall, between which and the entrance to a row of small dwellings there is only a very narrow passage; such are a few sources of disease which the breeze of the hill cannot always dissipate.[5]

Similarly "morbific" effluvia led to corrupted air or "malaria" in Somerset, where the assistant commissioner enumerated the resulting "fevers" among the population—agues, typhoid, smallpox, and scarlet fever—all of which "are met with at all seasons of the year, but prevail as epidemics" at certain times of year.[6]

Such conditions, detailed at length for the whole of the country, served, as Chadwick had hoped, to prod Parliament and county authorities to action. Part of the reason was the fear of a return of Asiatic cholera, the most dreaded disease of the century, which had struck Britain with devastating ferocity in 1832. Interestingly, Chadwick edited cholera out of the report, which he viewed as a document intended to deal with diseases that were ever-present rather than with exotic visitors from abroad. Because of its capacity to cause massive suffering, political tension, and economic disruption, however, cholera played a substantial though hidden role in the *Sanitary Report:* its threat to return made the sanitary idea urgent.

Although the *Sanitary Report* dealt extensively not only with filth and disease, but also with poverty, Chadwick and his correspondents were nearly unanimous in stressing that filth caused poverty and not the reverse. In their view, dirt and indigence were mediated by drunkenness. Filthy living conditions and ill health demoralized workers, leading them to seek solace and escape at the alehouse. There they spent their wages, neglected their families, abandoned church, and descended into lives of recklessness, improvidence, and vice. The results were poverty and social tensions.

At the time that Chadwick was collecting his data, the poor and the working classes were held by the propertied and the state to be dangerous and subversive. The French Revolution had occurred within living memory; echoes of the revolutions of 1830 still reverberated; and the tensions that brought about the revolutions of 1848 were gathering. Riots, strikes, and socialist ideas were rife. Even England, which avoided revolution, was home to deep social tensions as various and repeated strikes, riots, and demonstrations amply revealed. Chadwick was therefore obsessed with the dangers of "the anarchical fallacies which appeared to sway those wild and really dangerous assemblages" that compromised the public peace by the "violences of strike after strike."[7]

Sanitation presented itself as a means of social control. A feature of the "wild assemblages" of workers, he noted carefully, was that they were led by the young. Older men with family responsibilities rarely took part. A vitally important role of cleansing, therefore, was that it would lower the death rate, extend longevity, and thereby alter the age structure of the general population. To explain Chadwick's idea by means of a modern analogy, one could say that older males played a role akin to that of boron rods in a nuclear reactor. Just as boron rods control fission to prevent nuclear meltdown, so middle-aged family men would calm passions to prevent social meltdown or revolution. Chadwick argued that the facts:

> show the importance of the moral and political considerations, *viz.*, that the noxious physical agencies depress the health and bodily condition of the population and act as obstacles to education and to moral culture; that in abridging the duration of the adult life of the working classes they check the growth of productive skill and abridge the amount of social experience and steady moral habits in the community: that they substitute, for a population that accumulates and preserves instruction and is steadily progressive, a population that is young, inexperienced, ignorant, credulous, irritable, passionate, and dangerous.[8]

Cleanliness, therefore, exercised a civilizing, even Christianizing, function. It would redeem the working classes, lifting them out of poverty and disease. It would also allow the wholesome influences of education, religion, and middle age to promote social stability and class harmony.

This political focus on the danger posed by workers explains a further aspect of the *Sanitary Report*—its strong age and gender inflections. The health that Chadwick promoted was not primarily designed as health for all. His preoccupation was with the longevity and productivity of young and middle-aged working males. Other possible meanings of public health were never broached. Women, children, and the elderly, together with their diseases, were of no major interest in the *Sanitary Report*. Even the middle classes were largely ignored. Chadwick intended that "middling sorts" should benefit from the sanitary revolution, but he said little about their burden of disease. The middle-class gains that he stressed were social stability, an improved urban environment, and an economy strengthened by a healthy workforce.

Indeed, economic factors were vitally important. Just as Chadwick's reform of the New Poor Law had the goal of lightening the burden borne by the ratepayer, so the sanitary movement would promote wealth and economic growth. Fever incapacitates and bears away men in the prime of life whose skills are thereby lost to their employers and whose dependents are then reliant on the rates. Simple prudence and self-interest on the part of the middle classes and not just humanitarian sentiment indicated reform.

Sanitary Reform

Strikingly, the *Sanitary Report* falls silent after presenting its exhaustive evidence regarding the interaction of dirt, poverty, and disease. At no point does it consider the sanitary measures needed to redress the crisis of filth in industrial Britain. The remedy that Chadwick envisaged is tacitly embedded throughout the document, but he does not explicitly elaborate a concrete plan of action. The task of devising a strategy was not to obscure the essential first step, which was to convince the nation that action was imperative, humanitarian, and prudent.

Public discussion of the *Sanitary Report* was followed rapidly by four initiatives intended to promote the cleanliness so clearly rendered necessary by industrial Britain's overwhelming filth. All involved Southwood Smith or Chadwick, or both men, in their implementation. The first was the establishment in 1843 of a Health of Towns Commission that included Southwood Smith; the second was the Nuisance Removal Act of 1846 giving towns wide powers to remove filth; the third was the passage of the landmark Public Health Act of 1848; and the last was the creation of a General Board of Health, also in 1848. The General Board of Health, which included both Chadwick and Southwood Smith in its membership, was created for the express purpose of carrying out the measures that followed from the logic of the *Sanitary Report.*

As in the case of the Poor Laws, Chadwick again invoked the power of the state to address the diseases of Britain's towns and cities. Only the national government had the means to finance his program and the authority to ensure uniformity and compliance throughout the land. Chadwick's vision was that of a vast and costly program of public works that would retrofit the nation with a carefully engineered hydraulic infrastructure underground.

His plan was directly influenced by William Harvey's discovery of the circulation of the blood, and indeed Chadwick referred to his famous reform as an "arteriovenous system." This circulatory analogy made the system easy to explain and emphasized the point that it was necessary for life itself. Newly laid water mains—the arteries of the system—would deliver an abundant supply of clean water—the first coefficient of health—to every town and city in Britain. Connector pipes as capillaries would then deliver water to every household, where it would achieve two things. The first purpose of household water was ready accessibility, which would enable inhabitants to cleanse their domestic spaces and their bodies, ending the neglect resulting from the old regime in which people fetched water in buckets from public pumps. Chadwick's gifts to all were running water and flushing toilets. The origins of the flush toilet are much debated, but it was patented by George Jennings in 1852, incorporated into Chadwick's system, and marketed by the superlatively named Thomas Crapper (fig. 11.1).

Figure 11.1. Model of the "Optimus" water closet, invented by Steven Hellyer in 1870. Early toilets contributed to the emphasis on sanitation as a way of combatting disease. Hellyer campaigned for better plumbing and wrote *The Plumber & Sanitary Houses* in 1877. (Science Museum, London. CC BY 4.0.)

The second goal was achieved as water flowed out of the home, eliminating cesspits, overflowing privies, and the practice of depositing refuse in the streets. An uninterrupted flow would carry waste safely away through the veins of Chadwick's anatomically construed system—the drainage and sewage collector pipes. As in the bloodstream, there would be no stagnation. Organic matter would have no opportunity to decompose in the open air and emit its poison. Water would move continuously into, through, and out of the home and the city. Modern technology, such as the design of egg-shaped sewers and drains, would maximize the flow, ensuring that the pipes were self-cleansing and free from clogs.

As with human physiology, the system also provided for the final disposal of waste. Raw human sewage would flow unimpeded to distant outfalls located in the countryside. There, farmers would purchase it as fertilizer, thereby improving their crop yields per acre. Meanwhile, dangerous emissions would rise into the air, but they would be wafted harmlessly away by the breezes that swept freely across the open fields. Because of these air currents, organic matter, in accord with the filth theory, could finally decompose without posing a threat to human health. In time, such sewage farming would also partially offset the enormous expense of installing and maintaining the network, and it would help to feed the ever-growing and increasingly urbanized population.

Simple in conception, the sanitary reform required a series of additional measures to accomplish its goals. One problem was leakage. City streets needed to be paved and drainage ditches filled in order to prevent water containing filth from seeping into the subsoil and giving off plumes of poisonous miasma. Similarly, the streets needed to be swept and flushed, and any discarded refuse had to be removed. After cleansing, the streets had to be safeguarded against discharges from noxious trades such as tanners' workshops, slaughterhouses, and butchers' shops. In the minds of the reformers, it was also wise to whitewash houses, to ensure that their ground levels were floored, and to make them impervious to evil winds from outside. At the same time the population needed instruction in sanitary principles in order to correct dangerous habits and to inculcate new ideals of personal and domestic cleanliness. Sewers and drains were thus a moral, civilizing force as well as an underground architecture of health.

The Impact of Sanitation on Health

The sanitary movement that Chadwick and Southwood Smith inspired lasted until the First World War, and it transformed life in Britain; however, it is not possible to calculate the deaths and illness that did not occur because of their water, drainage, and sewage provisions. This problem is compounded by the fact that the implementation of the sanitary regime did not happen at a stroke, but advanced by fits and starts over many decades. Similarly, it is impossible to quantify the relative contribution of sanitation to reducing morbidity and mortality from epidemic diseases in comparison with those other critical factors considered by McKeown and other demographic historians—diet, improved wages, the control of plague, and vaccination against smallpox. Even Chadwick himself, obsessed though he was with data and statistics, did not attempt such calculations. Casting a retrospective gaze at sanitation in 1877 near the end of his life, he confined himself to estimating the benefit in specific and limited settings such as orphanages, prisons, and ships. An aggregate figure for the nation as a whole was beyond reach then, as it is now.

Modern understandings of epidemiology, however, make it appear certain that the measures carried out by the sanitarians made a significant contribution to the battle against infectious disease. A secure water supply and sewer system are now known to be obvious and essential tools for combatting such major diseases transmitted by the oral-fecal route as typhoid fever, gastroenteritis, and cholera, all of which claimed their legions of victims before Chadwick's intervention. It is no coincidence, for instance, that the 1850s marked the last of the major invasions of Britain by cholera. Thereafter, the steady advance of the sanitary measures urged by Chadwick made Britain cholera-proof—in striking contrast to continental nations such as Spain and Italy. There, the sanitary gospel arrived late, and cholera epidemics continued to claim victims until the end of the nineteenth century and even into the twentieth.

Furthermore, the reduction in morbidity and mortality from diarrheal infections indirectly lowered the incidence of other conditions. Both directly and indirectly, therefore, sanitation made a major though incalculable contribution to the "mortality revolution" as well as to the establishment of public health as a field of policy. Indeed, providing an ample supply of clean water flowing constantly through mains, drains, and sewers has lasted since Chadwick's time as one of the most successful reforms in history, as the minimal requirement everywhere for public health, and as the gold standard for civilized living.

Chadwick and his associates also had a larger effect on health and disease. The visible, and above all "smellable," transformation of the urban environment was accompanied by an educational campaign conducted by physicians, pastors, teachers, and the advocates of cleanliness to produce a new sanitary consciousness among the general public. This popular awareness of the danger lurking in decomposition empowered individuals and families to take measures of self-defense against miasma that complemented the provisions made by municipalities and the state. Analyzing a later period when the germ theory of disease had replaced the theory of filth, Nancy Tomes in *The Gospel of Germs: Men, Women, and the Microbe in American Life* carefully charts the way that an awareness of the life-protecting importance of sanitation deeply marked everyday life. Ordinary people armed themselves against disease through multiple daily rituals of washing their bodies, food, utensils, clothing, and homes. Cleansing and vigilance became integral parts of daily routines.

In this regard the new availability of water revolutionized domestic practices. Until the time of Chadwick, families had relied almost exclusively on the broom to clean their homes, and it was deployed intermittently or sometimes not at all. The sudden abundance of water and the "zero tolerance" of dirt completely altered domestic life. This indeed was one of Chadwick's objectives. He had intended from the outset that the sanitizing of outdoor spaces in cities would work its effects in tandem with the cleansing of household interiors and with personal ablutions. The reformed habits of individuals contributed to the Victorian mortality revolution in precisely the manner the sanitarians expected.

What Chadwick had not foreseen was that the sanitary revolution disproportionately altered the lives of women, again setting a precedent for the developments that Tomes considers. Women had long taken the leading role in managing the family and the domestic sphere, but with the advent of sanitarianism, they acquired enhanced responsibilities and new tools to meet them. It became the charge of women to protect the entire family from disease by cleansing the home and by imparting to children the lessons of personal hygiene. For many, this change resulted in a new sense of worth both outside and inside the home. Eileen Cleere, in *The Sanitary Arts: Aesthetic Culture and the Victorian Cleanliness Campaigns,* documents this development in the United States in the wake of the sanitary movement's arrival. In the closing decades of the nineteenth century, urban and especially middle-class American women in centers like Baltimore, Philadelphia, and Boston came to regard what they called "municipal housekeeping" as a logical extension of domestic housework and their family responsibilities. They reported sanitary nuisances in the streets; they called for alleys and roadways to be cleansed; and they organized crusades for local sanitary improvements. In this way they developed a sense of mission, experience in organizing, and confidence. The sanitary movement thus unintentionally, but in ways that seem logical in hindsight, played its part in broadening women's social responsibilities.

Sanitation and the Arts

Beyond its direct and massive impact on health and disease, the sanitary idea touched more remote spheres of life as its indirect influences rippled outward. Three areas where its arrival was clearly felt were literature, painting, and interior decoration.

In literature Charles Dickens became the standard bearer for sanitary reform. Early in his career Dickens had bitterly opposed Chadwick and the New Poor Law, which he caricatured in *Oliver Twist.* From the early 1840s, however, he became a lifelong convert to sanitation as expounded by Southwood Smith and to its practical application in the reform program of Chadwick, his former *bête noire.* As he wrote to a mutual friend in 1842, after reading a copy of what he called the "excellent" *Sanitary Report,* "Pray tell Mr. Chadwick that . . . I heartily concur with him in the great importance and interest of the subject though I do differ with him, to the death, on his crack topic—the new Poor Law."[9]

Soon after meeting Chadwick in person, Dickens began to incorporate the miasmatic and filth theories into his novels, starting with *Martin Chuzzlewit* (1843–1844) and *Dombey and Son* (1848), where the concepts are clearly delineated. Then in 1850, when Chadwick, as a member of the General Board of Health, was at the height of his efforts to implement his cleansing program, the novelist took two steps forward as a sanitarian in his own right. First, Dickens founded the weekly periodical *Household Words* that gave pride of place to sanitation and transformed him into the foremost public spokesman for Chadwick's movement. In that guise he even addressed the Metropolitan Sanitary Association on the need for hygienic reform. Although Dickens was a socialist and Chadwick a staunch defender of the existing social order, they agreed that sanitation was the most effective means of reducing human suffering.

Dickens was not the only Victorian writer to make sanitary reform a significant element in his work. Some historians have identified what they call the "sanitary novel." In addition to works by Dickens, the list of candidates for the designation include George Eliot's *Middlemarch* (1871), Elizabeth Gaskell's *North and South* (1855), and Benjamin Ward Richardson's *Hygeia: A City of Health* (1876). All of these writings confront the evils of filth, and they call for cleansing as a means of promoting both health and beauty. In the case of Richardson, who was an eminent physician, a supporter of the National Temperance League, and a close associate of the physician John Snow (see Chapter 12), an activist commitment to the sanitarian movement accompanied his art. *Hygeia* portrayed life in a utopian city in which Chadwick's intentions had been implemented to perfection. There, both foul smells and alehouses were no more.

Sanitation also influenced painting. The most important figure is the art critic and social theorist John Ruskin. An ardent opponent of filth and admirer of the sanitary reformers, Ruskin was regarded even by Chadwick as a fellow sanitarian. In his *Modern Painters* (1843), which is laced with the sanitarian lexicon of filth, disease, and foul odor, Ruskin criticized the Old Masters in the name of health, which he raised to the level of an aesthetic ideal. He chastised Rembrandt because of such features as his dark palette, which was suggestive of dirt, and his interiors, which were lit by candles and filled with shadows rather than health-giving sunlight. Only sanitary art could reach the highest ideals of beauty, Ruskin theorized. Rembrandt's canvases, Ruskin specifically objected, were "unromantic and unhygienic." By contrast, he extolled J. M. W. Turner's landscapes for their bright colors extending even to white, and for their direct depiction of sunlight. Such work, according to Ruskin, was modern, hygienic, and romantic. Sanitation helped to further a change of style and sensibility in the arts, associating modernity with clarity of line, bright tones, and vivid colors. All that was dark was dirty, stinky, and abhorrent. It also provided a precondition for the emergence of the Pre-Raphaelites, who made use of recent chemical advances to create strikingly bright pigments.

Fashions in interior decoration also evolved rapidly under the force of the sanitary idea. Traditionally, the comfortable classes in mid-Victorian England filled their sitting rooms with all that was heavy, intricate, dark, and cluttered—drapes, sturdy furniture, rugs, paintings. Everywhere was a profusion of ornament, and every surface was covered with endless bric-a-brac. To sanitarian taste, such furnishings were dangerous receptacles for dust, traps for miasmatic atoms, and bearers of the silent imputation of filth. Postsanitarian decorators, in harmony with the priorities of Ruskin and the designer and writer William Morris, moved instead in the direction of simplicity, light tonalities, and clear lines. Modernity everywhere was bright, clean, and in harmony with the latest thinking in science and technology.

The Legacy of Sanitarianism for Public Health

Important though it was, the sanitary revolution established public health on a narrow basis. Chadwick and Southwood Smith were far removed from the contemporary alternative of social medicine as represented by the German physician Rudolf Virchow. Holding that medicine should not confine itself to individual diseases but should treat the collective pathologies of society, Virchow argued for the necessity of considering a broad spectrum of the social determinants of disease, including diet, wages, and working conditions.

By steadfastly insisting on a monocausal explanation for disease, the British sanitarians narrowed public health to the measures needed for cleansing. They gave no comfort, for example, to those who argued that the high incidence of disease among the poor resulted from their working conditions. In Chadwick's theory the workplace was a source of disease only if it stank. He therefore opposed regulating factories, controlling child labor, abolishing sweatshops, and limiting the length of the workday—just as he staunchly opposed labor organization and strikes as he saw improved wages as irrelevant to health. In this way, Chadwick was the polar opposite of his contemporary Karl Marx, for whom work and the environment in which it took place were all-important determinants of the intellectual, spiritual, and physical health of laborers. It was no surprise that mine owners and manufacturers embraced Chadwick's approach to public health, while workers' health, wages, and safety suffered.

Another result of the sanitary movement was a transformation in the power of the state and the state's relationship to citizens. In a fashion that philosopher and social critic Michel Foucault would have understood, the system required the constant "gaze" of the state. Public authorities needed to establish regulations for the construction of buildings, city planning to control the width of streets so that air and sunlight would penetrate, management of public spaces, and oversight to ensure routine maintenance. Interventions were also important to introduce sanitary standards into a wide variety of institutions—army camps and barracks, naval and merchant marine vessels, asylums, hospitals, cemeteries, and schools. Boarding houses also needed to be brought into conformity with the system. Chadwick's reforms were thus resolutely top-down and centralizing. They marked a step in the "Victorian governmental revolution" that significantly augmented the power of the state. Implementing the sanitary idea was not a one-time achievement. It necessitated a large and permanent bureaucracy to manage the reforms, ongoing taxes to pay for public works, and extensive regulations to control construction and private behavior.

CHAPTER 12

The Germ Theory of Disease

A revealing thought experiment in the history of medicine is to compare the profession at the start and at the end of the so-called long nineteenth century, contrasting 1789 with 1914. In 1789, on the eve of the medical revolution launched by the Paris School, the conceptual framework of the discipline was largely that established by Hippocrates and Galen. Humoralism was in retreat as doctors absorbed ideas about the circulatory and nervous systems; as the chemical revolution and the periodic table undermined Aristotelian notions of the four elements composing the cosmos; and as the idea of contagion gained traction on the basis of experience with epidemic diseases. Nevertheless, medical philosophy, therapeutics, and education were still cast in the classical mold, supplemented by the later development of astrology. Physicians and the educated public explained epidemics by the doctrine of miasma as a "corruption" or poisoning of the air.

By 1914 momentous developments had taken place in medical science. Indeed, more change had occurred in the decades since the French Revolution than in the all of the centuries between the birth of Socrates and the seizure of the Bastille combined. A scientific discipline whose basic principles were recognizably similar to those prevailing today had begun to emerge.

Furthermore, the speed of change in medical understandings gathered momentum as the century progressed. The closing decades of the nineteenth century—roughly between 1860 and 1900—witnessed a wholesale revolution in medical philosophy with the germ theory of disease as its central feature. With only slight hyperbole, one could argue that this theory constituted as important a revolution in medicine as the Copernican theory of the rotation of the earth around the sun was to astronomy, as gravitational theory was to physics, or as Darwin's theory of natural selection was to biology.

Conducting such a thought experiment is useful for an additional reason: it clarifies both the excitement and the resistance of contemporaries confronted with the ideas presented most forcefully by Louis Pasteur and Robert Koch. To Pasteur's son-in-law, René Vallery-Radot, for example, it appeared that the germ theory was the means to understand the origins of life itself and the meaning of death. In his 1900 biography *The Life of Pasteur,* Vallery-Radot explicitly invoked Galileo and Darwin to express his astonishment. He also comprehended the magnitude of the changes in understanding the world that the new doctrine required as key in explaining the resistance that it encountered.

In this chapter we define and explore this theory and the decisive events in its establishment as medical orthodoxy. These developments were associated above all with the famous trio of Louis Pasteur, Robert Koch, and Joseph Lister. As we move forward, however, we must avoid the pitfall of believing that the driver of scientific knowledge is ever a single genius working alone. Medical discoveries in the nineteenth and twentieth centuries were based on a complex collective process that required a long train of preconditions. Of these, the most important were conceptual, institutional, and technological.

Conceptual and Institutional Preconditions: From the Paris Hospital to the German Laboratory

The Paris School of Medicine, as we have seen, was devoted to the idea that diseases were discrete and stable entities that could be classified according to their symptoms observed in life and the lesions they produced that could be perceived on the dissecting table. Nosology, or the study of the classification of diseases, was a central feature of the Paris physicians, and it gave rise to the general principle of the specificity of disease. For the emergence of the germ theory, this concept was critical as it led to two corollaries that made it possible to regard microbes as pathogenic agents. The first was that, contrary to Hippocrates and Galen, disease was not a holistic phenomenon reflecting an overall imbalance or corruption of the four bodily fluids, but rather a local and specific ailment of the body's solid tissues. Second, the Paris School postulated that diseases could never morph one into another, in contrast to the widely held belief among practitioners before the establishment of the Paris School that cholera, for example, arose spontaneously in a locality as a heightened form of endemic summer diarrhea rather than as a fixed and specific disease in its own right.

The concept of species was familiar to zoologists and botanists, who understood that a willow, for example, never morphed into an oak or a salamander into a frog. But it took time for the idea to be applied to the world of microorganisms. Pierre-Fidèle Bretonneau was one of the early scientists to embrace the concept when he argued that each "morbid seed" causes a specific disease just as every seed in natural history gives rise to a specific species. He developed the idea in detail while studying diphtheria during the 1820s.

Not all of the leading theorists of disease specificity were French. William Wood Gerhard, for example, was a Philadelphian and a veteran of Paris who had studied for two years with Pierre Louis. Returning home during a typhus epidemic in 1833, he dissected the cadavers of hundreds of victims and reached the conclusion that the lesions of typhus bore no resemblance to those of typhoid, as contemporaries believed.

Gerhard's view was shared by William Budd in England, who gained lasting fame for his work *Typhoid Fever: Its Nature, Mode of Spreading, and Prevention* (1873). He stressed the specific and unchanging nature of the disease. In his view, this concept formed the essence of the idea of a species. Typhoid, therefore, was a fixed species that could not arise by spontaneous generation and could not transform itself into another disease of a different species.

Going further, the French physiologist Claude Bernard took the position that diseases were not only specific, but also dynamic. In other words, they exhibited a developmental process in the body. In that way he offered a novel critique of the Paris School itself. In *An Introduction to the Study of Experimental Medicine* (1865) he held that the cases of disease in hospitals were deceptive because they displayed diseases in their final stages and thus prevented physicians from observing their beginnings and their development. Furthermore, the hospital complicated scientific work by presenting an excess of variables. Prophetically, therefore, Bernard proposed an alternative: the laboratory. He saw the laboratory rather than the hospital as the proper locus for an "experimental medicine." In the laboratory alone it would be possible to test a single variable in a controlled setting. The lab rather than the library or the hospital world should therefore be the proper locus of medical epistemology—the source of medical knowledge.

Bernard's theory of a new medical epistemology made him a crucial transitional figure. He looked beyond the hospital ward, which had applied the idea of specificity to disease, to the laboratory, which alone would permit the exploration of the microbial world and its relationship to etiology, or disease causation. His idea also pointed to a geographical transition in European medical science. Germany was replacing France as its leading center, for it was Germany that was most fully committed to the scientific laboratory located in a university or research institute. Germany had also gone farthest in developing the figure of the full-time medical scientist.

Technological Foundations: Microscopy and "Animalcules"

Antonie van Leeuwenhoek

Another essential prerequisite for the germ theory of disease was the development of microscopy. Here the elusive Delft draper Antonie van Leeuwenhoek (d. 1723) is critical. In recent years historians of science have transformed the understanding of the scientific revolution and demonstrated that the discoveries of the canonical figures of eighteenth- and nineteenth-century science depended on indispensable foundations laid by humbler figures from the sixteenth and seventeenth centuries. These early figures were not elite, university-trained intellectuals; instead, they were artisans working in vernacular languages who possessed—and shared with one another—an inquisitive, hands-on interest in exploring the natural world. Leeuwenhoek was such a figure.

The contributions of these often forgotten men and women to the giants of science who followed—from René Descartes and Isaac Newton to Louis Pasteur and Robert Koch—were many. Deborah Harkness, a pioneering examiner of this scientific "history from below," has fruitfully suggested in her book *The Jewel House: Elizabethan London and the Scientific Revolution* (2007), that these people made three essential contributions without which the scientific revolution would have been impossible: (1) they forged communities interested in the exchange of ideas, practices, and the verification of hypotheses; (2) they established mathematical, instrumental, and print literacies—beyond the usual reading, writing, and arithmetic—that are essential preconditions for scientific advances; and (3) they developed the hands-on practices of experimentation and the investigation of nature.

Practical business concerns launched Leeuwenhoek on his scientific career. He wanted to examine the quality of the threads in the cloth he handled better than the magnifying glasses of his time would permit. With skills honed by an apprenticeship in lens grinding and metalworking, he devised a single-lens microscope that enabled him to achieve a magnification of 275×. Going beyond his immediate interest in fabrics, he used the new instrument to observe nature, becoming the first person to observe the world of single-celled organisms, which he christened "animalcules." He also reported regularly on his observations to the Royal Society in England. In that way he established both the technological foundations and the awareness of the microbial world necessary for the emergence of microbiology.

Although Leeuwenhoek's discoveries pointed in the direction of microbiology, further technological breakthroughs were required, especially that of the compound microscope. With two achromatic lenses, it achieved much higher magnification and corrected the visual distortion known as chromatic aberration that prevented clear observation.

Ignaz Philipp Semmelweis and John Snow

A figure who was much reviled in his own time but whose speculations and medical practice also paved the way for the germ theory was the Hungarian gynecologist Ignaz Philipp Semmelweis (1818–1865). In the 1840s, working at the Vienna General Hospital, Semmelweis was appalled by the rate of maternal mortality from puerperal fever, now known to be a severe bacterial blood infection and then the leading cause of death in lying-in hospitals. He also observed a curious fact. The obstetric service in the hospital was divided into two sections. In Section 1 deliveries were carried out by physicians and medical students, who also conducted autopsies as a central part of their medical and scientific work. In Section 2, however, deliveries were entrusted to midwives, who did not take part in dissections. In Section 1 the maternal death rate was 20 percent, but it was only 2 percent in Section 2.

Semmelweis observed that doctors and their students regularly reported to Section 1 immediately after conducting dissections—without washing their hands first. He began to suspect that they carried some kind of mysterious and invisible "cadaveric particles" on their hands that conveyed disease from the autopsy table to the women they examined or delivered. His suspicion hardened into near certainty when a colleague died of infection after being accidentally pricked by a scalpel during a dissection. Revealingly, his symptoms were identical to those of women who died of childbed fever on the obstetric ward. In 1847 Semmelweis therefore persuaded colleagues, midwives, and students to cleanse their hands with a chlorine solution before entering the maternity ward. The results were immediate and impressive as mortality plunged to 1.3 percent in both sections.

Semmelweis had made a clear and powerful demonstration of the contagiousness of puerperal fever and of the danger lurking in invisible "cadaveric particles." Unfortunately, however, the outcome was profoundly discouraging. Unable to identify the "putrid organic matter" that he claimed was the causative pathogen rather than the accepted miasma, he was reviled by the wider medical profession in Vienna as a charlatan and forced to resign his hospital appointment. He returned, therefore, to his native Budapest, where he cared for women in a local lying-in hospital. He continued to implement his life-saving antiseptic rituals, but in obscurity and with few imitators. In 1865 he suffered a nervous breakdown and was shut away in an insane asylum, where he was beaten by the attendants and died of the injuries. Recognition for his discovery has been entirely posthumous.

At almost exactly the same time that Semmelweis was speculating on the lethal effect of microbes in childbirth, John Snow (1813–1858) in London was theorizing that "animalcules" were responsible for transmitting Asiatic cholera. Snow was a general practitioner who made notable contributions in the fields of obstetrics and anesthesiology, but he also became one of the founding figures of epidemiology as a discipline through his work on cholera. His point of departure was the symptomatology of the disease.

Cholera, Snow reasoned, invariably begins with severe abdominal pain, diarrhea, and vomiting. Those early symptoms are consistent with a pathogen that, having been ingested, infects the gut. By interviewing cholera patients in the London epidemic of 1848, he noted that all of the people he examined recalled that the first symptoms of the disease that they experienced was a range of digestive problems. All of the later manifestations of the disease—faint pulse, dyspnea, the dark tarry appearance of the blood, heart failure, the livid appearance of the face, and "dishpan" hands—probably result from the loss of blood plasma in the profuse watery stools. The entire pathology of cholera, therefore, is consistent with the hypothesis of a living "germ" or "animalcule" that multiplied in the intestines after being swallowed in food and water. In Snow's words, "A consideration of the pathology of cholera is capable of indicating to us the manner in which it is communicated."[1]

To support this interpretation, Snow carried out tireless detective work into the epidemics of 1848–1849 and 1854 in London. In the earlier epidemic he meticulously compared the mortality among households supplied by two different water companies. The first, the Lambeth Waterworks Company, drew its water from the Thames River upstream from London before it was contaminated with sewage from the metropolis. The other company, the

Southwark and Vauxhall Company, pumped water from the river as it flowed through the capital at Battersea. Neither of these two businesses, unlike other suppliers in London, filtered the water. Since the two companies competed for customers on a house-by-house basis, their subscribers overlapped in terms of such variables as income, housing conditions, and sanitation. It was therefore highly suggestive that the households that drank downstream water suffered a mortality that was many times that of households that consumed the relatively pure water drawn upstream. Cholera mortality disproportionately followed the map of the contaminated Thames water supplied by the Southwark and Vauxhall Company.

With regard to 1854, Snow also focused on the neighborhood of Soho and the public pump on Broad Street from which residents drew their water. Tracing the diffusion of the disease within a radius of 250 yards from the pump, he reported what he termed "the most terrible outbreak of cholera which has ever occurred in this kingdom."[2] Snow discovered that more than five hundred deaths from cholera had occurred within a ten-day period, and the victims were people who had consumed water from the pump. Furthermore, when he persuaded the authorities to remove the handle so that the pump was no longer usable, the Soho outbreak suddenly ceased. Later investigation even established that, in modern terminology, "patient zero" was an infant who had fallen ill outside the neighborhood. The child, however, was taken to Soho by his mother, who then washed the contaminated diapers and disposed of the water in a cesspit located just a few feet from the pump.

Other considerations also affected Snow's thinking. One was his early experience of the epidemic of 1831–1832. At that time he was an apprentice physician in Newcastle, where he took an active role in treating coalminers, who suffered extensively from the disease. What he quickly wondered was whether the course of the disease among the mining population was compatible with miasmatic explanations. There were no swamps, sewers, or rotting organic matter in the mine shafts and no indications that there were alternative sources of poisonous effluvia. But the suffering of the miners, in both 1831–1832 and 1848–1849, was of enormous theoretical interest because they were the occupational group that, per capita, was most afflicted by the disease in England.

Similarly, Snow's interest in anesthesiology in London during the 1840s taught him to question medical orthodoxy in the matter of epidemic disease. He found it implausible that lethal vapors could exert so great an influence over the health of the population at substantial distances from their site of origin. Gases, in his view, did not behave in the manner required by the theory. Snow's observations during the cholera epidemic of 1848–1849 helped to turn his doubts about vapors into a certainty since the alternative explanation of contaminated water being the decisive factor resolved the difficulties and carried conviction as a far simpler and more consistent solution.

Snow published his findings and the maps that illustrated them in his major work of 1855 *On the Mode of Communication of Cholera*, which is now widely considered the foundational text of epidemiology as a discipline. During Snow's lifetime, although the book kept the emerging germ theory in public view, it failed to persuade the medical profession. Anticontagionism and miasmatism persisted as orthodoxy with regard to cholera. Indeed, William Farr (1807–1883), a prominent member of the General Board of Health, investigated the same London epidemic of 1848–1849 that Snow had examined but concluded that the disease was the product of poisonous effluvia and the conditions that made the population vulnerable to it.

Various factors explain the skepticism that greeted Snow's publication. A first problem for contemporaries was that his narrative strategy was not to refute orthodox opinion but to ignore it entirely while promoting his own theory. This strategy was especially problematic because Snow's interpretation seemed to reduce the etiology of cholera to a single cause—his unseen "animalcules"—whose existence and role in the epidemic he could not directly establish. In that respect the germ theory he espoused required more robust proof. Only the further development of microscopy, which rendered the *Vibrio cholerae* visible, and the experimental method, which demonstrated the role of microorganisms in inducing disease in animals, could establish the actual mechanisms of contagion. Snow painstakingly established correlation, but he was unable to prove causation. In the meantime, Farr's defense of orthodoxy, his reliance on multiple causation, and his extensive use of medial statistics carried the day. Snow's hypothesis may have intrigued many scientists and physicians, but like Semmelweis, he was ridiculed by numerous skeptics.

The Famous Trio

Louis Pasteur

If Leeuwenhoek first opened the world of microbes to exploration, and Semmelweis and Snow suggested the role of these animalcules in causing disease, Louis Pasteur (1822–1895) provided the experimental evidence necessary for a conceptual revolution in medical science. Pasteur was among the first to make systematic use of the microscope for medical purposes.

At the moment when Pasteur—a chemist rather than a biologist or physician—turned his attention to the problem of disease and its causation, the dominant theory was the "zymotic theory of disease." A close relative of miasmatism, zymotic theory held that epidemic diseases were caused by a chemical process. In other words, the fermentation of decaying organic materials emits miasmatic poison into the air under favorable conditions of soil, temperature, and moisture.

Closely related to zymotic theory was the popular idea of "spontaneous generation." This idea, which had been held since antiquity, when it was propounded by Aristotle, postulated that living organisms could arise from inanimate matter rather than from common descent from parent organisms of the same kind. Cases of epidemic diseases therefore were not necessarily linked to one another in a chain of transmission but could emerge from nonliving material. This idea had been tested by the Italian experimental philosopher Francesco Redi in a series of famous experiments that he described in his 1668 work *Esperienze intorno alla generazione degli insetti* (Experiments on the Generation of Insects). Redi placed meat and fish into flasks, one half of which he sealed while leaving the other half open to the air. The meat and fish in the open flasks soon pullulated with maggots while the flesh in the closed flasks did not. Redi's lesson was that the maggots appeared only if flies were able to deposit their eggs. Spontaneous generation *de novo*, Redi suggested, was not a valid concept. "From my many repeated observations," he wrote, "I am inclined to believe that the earth, ever since the first plants and animals that she produced in the first days on command of the supreme and omnipotent Maker, has never again produced from herself grass, trees, or animals perfect or imperfect."[3]

Nevertheless, two centuries later the idea that nonliving matter could give rise to life remained a widely accepted view, and it was vehemently defended—most notably by the eminent German chemist Justus von Liebig (1803–1873), one of Pasteur's bitterest opponents. The main difference from antiquity was that the field of spontaneous generation had been progressively diminished over the centuries. By the nineteenth century, large animals and even insects were known to be generated sexually, so the realm of spontaneous generation had become confined to the microscopic world. Part of the theory's tenacity reflected the difficulty of disproving the theory in a sphere where so little was known—and that seemed to lie in the liminal zone between the living and the nonliving where it seemed compelling to think that life would first be breathed into matter. In addition, spontaneous generation carried conviction because of its theological underpinnings. Creation—the origin of life itself—for some believers was simply the first act of spontaneous generation. To deny the phenomenon therefore threatened religious belief.

Pasteur himself was deeply religious, but he rejected the idea of spontaneous generation because the stakes were too high. Spontaneity introduced random chaos into the orderly world of nature. If spontaneous generation were true, the disease process would become arbitrary and unintelligible. Etiology, epidemiology, nosology, and preventive public health would have no basis in reality.

Pasteur began to develop an alternative theory during the 1850s. At that time he devoted his attention to two major and related problems of French agriculture: the spoilage of wine transformed by acetic acid fermentation into vinegar, and the spoilage of milk by lactic acid fermentation. This spoilage was universally considered to be a chemical process. Pasteur demonstrated instead that it was due to the action of living microorganisms—bacteria that he identified through the microscope and learned to cultivate in his laboratory. He also made a far-reaching connection, regarding the fermentation of wine and beer as analogous to putrefaction. Fermentation and putrefaction, in other words, were bacterial processes rather than chemical reactions triggered by a catalyst.

Furthermore, these bacteria were linked through descent from preexisting bacteria of the same type. By rigorous observation and cultivation Pasteur showed that bacteria differed in their morphology, nutrients, and vulnerabilities. This line of inquiry led to the discovery that, if heat was applied to destroy the bacteria, spoilage did not occur and the taste of the wine or milk was not affected. He experimented with heat after multiple attempts to find a suitable toxin had failed. This process involving heat came to be called "pasteurization."

Through this work Pasteur was fully transformed from chemist to founding father of microbiology. He published his results in three works that deeply influenced the emerging new discipline: an 1857 paper on lactic fermentation, *Mémoire sur la fermentation appelée lactique;* an 1866 study on wine (*Études sur le vin*); and an 1876 study on beer (*Études sur la bière*). The secrets of acetic acid, lactic acid, and alcoholic fermentation were uncovered.

After these works, which changed biology forever, Pasteur explored the implications for medicine and public health. With regard to epidemic diseases, advocates of spontaneous generation held that an outbreak of cholera, for example, could arise from local causes. In their view, as we have seen, the disease was simply an enhancement of preexisting "summer diarrhea." Combining the Paris School's idea of specificity with his observations of microbes, Pasteur postulated instead that a disease such as cholera that did not exist in a locality could arise only through the importation of the specific bacterium that causes it.

Pasteur demonstrated this principle in a simple experiment that elegantly illustrated Redi's concept. To destroy all microbial life, he boiled cultures in a sterile flask with a swan neck that would prevent air from gaining access (fig. 12.1). The question was whether microbes could appear spontaneously in the germ-free culture. What Pasteur discovered was that, if all previously existing microbes had been destroyed, germs would appear in the culture only if the neck of the flask was broken and air was admitted. Thus "seeded," the culture would teem with bacteria that reproduced luxuriantly. If carefully sealed, however, the flask and its culture would remain germ-free indefinitely. Pasteur immediately saw implications applicable to wound infections and disease. "There is no known circumstance," he said, "in which it can be affirmed that microscopic beings came into the world without germs, without parents similar to themselves. Those who affirm it have been

Figure 12.1. Louis Pasteur used the swan-necked flask to help disprove the doctrine of spontaneous generation. (Wellcome Collection, London. CC BY 4.0.)

duped by illusions, by ill-conducted experiments, by errors that they either did not perceive or did not know how to avoid."[4]

More decisive was Pasteur's work in the years 1865–1870 that he undertook despite suffering a major cerebral hemorrhage in 1868 that left him paralyzed on the left side. In this phase of his work, Pasteur turned to experiments involving disease in order to draw out the full implications of his earlier conclusions about fermentation. At the outset he turned to an unexpected experimental animal—the silkworm—and its pathology. Once again, he was drawn to research involving a disease that was decimating one of France's leading industries—silk. Through painstaking microscopy, Pasteur discovered, he believed, that the disease afflicting France's silkworms was in fact two separate diseases—*pébrine* and *flâcherie*—caused by bacteria whose role in producing these diseases he was able to demonstrate. Later discoveries making use of more developed microscopy and more refined methodologies revealed that *pébrine* was in fact caused by a fungus-like microparasite known as a microsporidian and *flâcherie* by a virus. The critical demonstration for contemporaries, however, was the role of microorganisms—seen and unseen—as the causes of disease.

Pasteur did not invent the concepts that underlay his investigations—the ideas of contagion; of living animalcules, or germs; and of disease specificity. There were other scientists, such as Casimir Davaine in France and John Burdon-Sanderson in England, who also advanced the idea that the microbes they saw with their microscopes were the agents of disease. Indeed, already in the early eighteenth century the Cambridge University botanist Richard Bradley suspected that microscopic living creatures he variously termed "animalcules" or "insects" were responsible for epidemics. Since he thought that all life originates from eggs or seeds, he even propounded his own theory for the disappearance of plague from London after 1665: the Great Fire of 1666 must have destroyed the eggs of the tiny animalcules that had caused pestilence.

As the very term "germ" suggests, botanists, agronomists, and horticulturalists provided elements for the germ theory. The first half of the nineteenth century witnessed extensive discussion about the role of fungi in plant diseases in ways that prefigured advances in medical science. Above all the potato blight of 1845 focused urgent attention on plant pathology and its causative organisms. Those discussions were known to Pasteur, and they provided the background to his pathbreaking consideration of the diseases of silkworms.

Particularly important was Pasteur's proof that the microbes he isolated and cultured were the causative pathogens of diseases that he could then reproduce in experimental animals through inoculation. Thus he confirmed the role of microbes in specific diseases and devised a methodology for further experimentation involving other diseases.

After his research on silkworm diseases, Pasteur turned his attention to chicken cholera and anthrax. Both are primarily diseases of animals, not major killers of humans; but the process of isolating microbes, cultivating them in culture, and then reproducing the disease was transformative rather than the importance of the diseases themselves in human health.

As a result of Pasteur's work, the germ theory of disease became the dominant paradigm by the late 1870s, although there was still considerable resistance among clinicians. They remained puzzled by the concept that invisible organisms could cause devastating epidemics. Even distinguished scientists such as Rudolf Virchow rejected the idea. The very profusion of terms used, almost interchangeably, to describe the pathogens was sometimes bewildering—corpuscles, germs, bacteria, infusoria, vibrios, viruses, animalcules, bacilli. Exactly what the germ theory meant to its adherents at a time when it was still controversial was expressed in 1880 by the American physician William H. Mays:

> I hold that every contagious disease is caused by the introduction into the system of a living organism or microzyme, capable of reproducing its kind and minute beyond all reach of sense. I hold that as all life on our planet is the result of antecedent life, so is all specific disease the result of antecedent specific disease. I hold that as no germ can originate *de novo*, neither can a scarlet fever come into existence spontaneously. I hold that as an oak comes from an oak, a grape from a grape, so does a typhoid fever come from a typhoid germ, a diphtheria from a diphtheria germ; and that a scarlatina could no more proceed from a typhoid germ than could a seagull from a pigeon's egg.[5]

In addition to demonstrating the truth of the germ theory, Pasteur developed the public health practice of vaccination, thereby helping to found the discipline of experimental immunology. As we saw in Chapter 7, Edward Jenner had pioneered the first vaccine nearly a century earlier through his work on smallpox. Pasteur developed a methodology that enabled him to produce an array of vaccines against diseases. He believed in the universality of what he called the principle of "nonrecurrence" or, in later terminology, acquired immunity. This principle, he thought optimistically, could be used as the basis for developing vaccinations against all infectious diseases.

Vaccination can be defined as the introduction into the body of either the whole or part of a disease-causing microorganism in order to "train" the immune system to attack that same organism should it reappear in the body through natural means. The vaccine primes the immune system to produce antibodies, or teaches immune cells to recognize and destroy the invading organism—a bacterium, a virus, or a parasite.

The problem, of course, is to stimulate immunity without causing disease. Jenner had discovered crossover immunity from cowpox to variola. Pasteur instead developed the concept of attenuation, whereby live pathogens are introduced into the body after being treated in such a way that their virulence is diminished. Heat provided one means of attenuating virulence in some pathogens; the prior passage of the microorganism through the body of a foreign host was another. Thus attenuated, the microorganisms stimulate an immune response without causing disease

In an era before the mechanisms of immunity were known, how did Pasteur conceptualize the process of nonrecurrence and the benefit conferred by attenuation? Once again, he had recourse to agricultural metaphors and in particular the analogy of seed and soil that was central to the germ theory. In farming, planting wheat in the same field in successive years exhausts the nutrients of the soil, which therefore becomes incapable of supporting further growth. Pasteur reasoned that, by analogy, an attenuated bacterium consumes the nutrients of the blood in the course of the mild infection that it causes. The bloodstream then resembles an arid field of exhausted soil no longer capable of supporting germination. If Pasteur then inoculated virulent seeds of the same disease in that experimental animal's bloodstream, the nutrients necessary for their growth and development would already have been consumed by the attenuated germs. The disease therefore would not recur. In today's terms, the animal had become immune.

Taken together, the two concepts of nonrecurrence and attenuation were two of the most important discoveries in the history of medicine and public health. Pasteur acknowledged his debt to Jenner, but he also recognized that he had generalized an approach that for Jenner had been limited to the single case of smallpox. Pasteur instead envisaged the development of additional vaccines that would produce immunity against a cascade of infectious diseases. And, indeed, successful vaccines have been developed against such various communicable diseases as measles, pertussis, tetanus, diphtheria, seasonal influenza, typhoid, rabies, and polio.

Inevitably, the question arose of whether the list could be extended to all transmissible diseases. Pasteur imagined that the limiting factor in each case would be whether a patient who had naturally recovered from the disease subsequently possessed a robust immunity against it, or, in his terms, whether the patient was subject to recurrence of the same disease. With regard to diseases to which a patient was subject to subsequent virulent attacks, as in the case of cholera and malaria, the development of an effective vaccine was problematic. Pasteur viewed vaccination as an effective strategy for the control or even eradication of many diseases, but he did not anticipate that it would be come a universal panacea.

Serendipity helped Pasteur to discover attenuation during his work with the bacterium that causes chicken cholera. But, as he repeatedly told his assistants, and informed gathered scientists at the University of Lille in 1854, "Where observation is concerned, chance favors only the prepared mind." After leaving a batch of the bacteria untouched for a week during a hot summer, he found—initially to his frustration—that the culture would not produce the disease when he attempted to infect healthy chickens. Starting again with fresh and highly virulent culture, he injected the same chickens with it while also injecting a number of random chickens. The result was the surprising discovery that the original chickens that had been injected previously with the old (in our terms, attenuated) culture remained healthy. In other words, they were immune. On the other hand, the random and previously untreated chickens all sickened and died.

Pasteur repeated the experiment, with the same outcome, and concluded that summer heat had altered, or attenuated, the virulence of the bacteria in the culture. This was one of the decisive discoveries of public health—that the virulence of microorganisms is not fixed but can be modified and controlled in ways that provide immunity against disease. In the broader context of the life sciences, Pasteur also contributed to transforming biology, then understood largely as natural history collections, into experimental biology conducted in the laboratory. He also drew out the practical implications of the Darwinian concepts of variation and mutation for pathology and medicine.

Later developments have expanded the repertoire of the means to produce immunity through vaccination. In 1886, also working with chicken cholera, the American scientist Theobald Smith discovered that vaccines consisting of heat-killed—rather than live-but-attenuated—chicken-cholera bacteria also induced immunity. Since that time medical scientists have used both killed and live-but-attenuated viruses and bacteria for prophylactic purposes. Vaccines have also been developed that do not use whole microbes but microbial subunits instead.

Having discovered the principle of attenuation through his work with chicken cholera, Pasteur proceeded to test the same methodology with other diseases. In one of the most famous experiments in medical science, he sought to produce an attenuated vaccine for anthrax, a zoonotic disease that primarily affects sheep, cattle, and goats. *Bacillus anthracis,* the bacterium responsible, had recently been isolated by Robert Koch. Pasteur reproduced the procedure Koch had employed with chicken cholera, heating batches of *B. anthracis* culture to 400°C (752°F). The result is vividly portrayed in William Dieterle's 1936 film *The Story of Louis Pasteur.* In May 1881 at the village of Pouilly-le-Fort, he first vaccinated twenty-four sheep with the attenuated bacterium. The next step was to challenge these vaccinated sheep with live, virulent bacilli as well as a control group of twenty-four other sheep that had not been vaccinated. All of the vaccinated sheep remained healthy; all of the unvaccinated sheep died.

Pasteur then extended the principle by devising other means of attenuation. This work began in the early 1880s and involved a series of experiments with rabies. Rabies, it is now known, is a viral disease rather than a bacterial one; but viruses were unknown at the time and were too tiny to be visible at the magnification available to Pasteur. As he himself claimed, however, serendipity favored his "prepared mind." Working with a class of microbes that was still unseen and undiscovered, Pasteur succeeded in attenuating the rabies virus. His procedure was to isolate it from rabid foxes and then to pass it through the bodies of a species that is not naturally susceptible to the disease—the rabbit. He inoculated a series of rabbits and then produced, at the end of the series, a variant culture. It would no longer cause infection in susceptible foxes but would successfully immunize them against cultures of wild rabies virus isolated directly from other rabid foxes. Rabies, of course, is not a high-impact disease among humans, but it was a disease of high drama and scientific interest because it was both excruciatingly painful and universally fatal.

The human trial of the rabies vaccine took place in July 1885, when nine-year-old Joseph Meister was bitten by a rabid dog. Pasteur's rabies vaccine was still in a highly experimental and untested developmental stage; even his assistants were divided over the ethics of vaccinating Meister. Emile Roux, his most gifted assistant, refused to take part in the experiment. But

Pasteur felt that Meister had been so badly mauled that the boy was certain to die an agonizing death. That settled the ethical question for Pasteur. Taking advantage of the long incubation period for rabies, Pasteur vaccinated the boy with his newly developed attenuated rabies virus. Meister survived and became a celebrity patient—the first person ever known to have lived after being severely bitten by a rabid animal.

The medical and public health implications of Pasteur's achievements with chicken cholera and rabies were immediately apparent. Cognizant of them, the French government founded the Pasteur Institute in 1887 with Louis Pasteur himself as its first director. By means of biomedical research both at its headquarters in Paris and at a series of satellite institutes elsewhere, the institute pursued the course of developing vaccines as a public health strategy, and even a possible means of disease eradication.

Edward Jenner had devised the first vaccine, and he had foreseen the possibility of employing it to eradicate smallpox. Nearly a century later, Pasteur developed the methodology of attenuation that unlocked the possibility of devising vaccines against an array of pathogens. In time, it proved possible to vaccinate against not only smallpox, anthrax, and rabies, but also, for example, polio, measles, diphtheria, tetanus, mumps, pertussis, and rubella in a seemingly ever-lengthening cascade. The strategy has eradicated smallpox, and the campaign against polio is presently at its critical moment. The question for public health arises of how far the strategy can be generalized. Is vaccination a viable strategy for all infectious diseases, or for certain classes only? What are the criteria, and why—more than two centuries after Jenner's discovery—has only one human disease been successfully eradicated? These issues will be pursued further in Chapter 18.

The founding of the Pasteur Institute provides the opportunity to notice a further aspect of modern medical science—its frequent tendency to become a focal point for competing nationalisms. This was clearly the case in the nineteenth century in the rivalry between Louis Pasteur and Robert Koch, who became the embodiments and icons of French and German medical science, respectively. The rivalry pitted not only the two scientists against one another, but the Pasteur Institute in Paris against the Robert Koch Institute in Berlin, and French science against German.

Robert Koch

Some twenty years younger than Pasteur and trained as a physician rather than a chemist, Robert Koch (1843–1910) was the second decisive figure in the establishment of the germ theory of disease. As a young scientist, his turn toward microbial pathogens was a logical choice in the context of the times. During the 1870s various "germ theories" were in the forefront of scientific debate, and his professor in the medical faculty at the University of Göttingen, Jacob Henle, was an early proponent of the idea of *contagium animatum*, the view that living organisms cause disease.

Koch's first scientific work, dating from the mid-1870s, involved the investigation of anthrax and the rod-shaped bacteria that cause it. Anthrax, which was widely known as "splenic fever" at the time, played a leading role in the establishment of the germ theory of disease, attracting the attention of both Pasteur and Koch, for a number of reasons. First was its economic impact on agriculture and animal husbandry. The disease was endemic in large areas of France, Germany, Italy, Russia, Spain, and America, where it decimated the populations of sheep and cattle and occasionally attacked those whose work was closely involved with them, such as shepherds, tanners, and cattle herders. For instance, during four years at Wöllstein, where Koch had his early practice, the disease had caused enormous damage, killing fifty-six thousand livestock.

Technical reasons also promoted the use of *B. anthracis* in the laboratory. The bacterium had the advantage of being unusually large and was clearly visible using the magnifying power available in the 1860s and 1870s. Even before Pasteur and Koch, the French physician Casimir Davaine (1812–1882) had pioneered the study of the disease with the microscope and discovered a suspicious microbe in the blood of infected animals. Both Pasteur and Koch built on Davaine's findings.

Koch first examined the blood of sheep that had died of anthrax, used the blood to inoculate healthy animals, observed that they contracted the disease, and noted that their blood and tissues teemed with *B. anthracis*. The next step in his research involved developing in vitro cultures of the microbe and passing it through successive generations both in culture and in the bodies of healthy experimental animals. After inoculation, all of these animals developed the symptoms of anthrax—fever, convulsions, and intestinal and respiratory disturbances. They also exhibited *B. anthracis* in their blood and suffered a high mortality rate. Here were clear indications that the bacterium, as Davaine had suspected before Koch, was in fact the cause of the disease. Complementing Pasteur's research, Koch's work helped to confirm the germ theory and to place it on solid, reproducible foundations.

Koch took a further step, however, in understanding the complex etiology of anthrax by discovering the formation of resistant spores that remained in the fields where diseased animals had grazed. He thereby solved the enigma that sheep and cattle contracted anthrax after grazing in fields where diseased animals had roamed. Indeed, the disease was transmitted primarily by grazing rather than directly from animal to animal. As a preventive measure, he recommended that the bodies of diseased animals be burned in order to prevent the formation of spores and further transmission. Koch published these results in his first paper, "The Etiology of Anthrax, Based on the Life Cycle of Bacillus anthracis," in 1876. This publication brought Koch international acclaim and is one of the foundational texts of medical bacteriology. It turned out, however, that sporulating bacteria are unusual among the major disease-causing human pathogens. Tetanus and botulism are others.

In order to move beyond anthrax in his investigation of microbes as possible causative agents of disease, Koch found himself hindered by the available technology. One factor was entirely personal—the still modest financial circumstances that compelled him to conduct his research on anthrax in the confines of a jerry-built lab in his backyard. More important were four objective limitations on the development of microscopy: inadequate magnification, poor lighting, the transparency of bacteria, and their motility in fluid.

Working with the Carl Zeiss optical company, Koch made use of new optical glass and oil immersion lenses that eliminated astigmatism from his microscopes. He also "fixed" bacteria to slides by drying them in a solution, resolving the problem of motility while he applied dyes such as safranin and methyl violet to circumvent the issue of microbial transparency. Since different species of bacteria absorbed different dyes, staining also provided a method of distinguishing species. With these methodological innovations Koch could observe the morphological details of microbes at a higher resolution, and he became the first scientist to publish photographs of bacteria.

To these improvements in microscopy, devised during the three years after his anthrax paper, Koch added the development of a solid culture medium. Initially, like Pasteur, Koch had used the practice of passing microbes through animal bodies as a means of obtaining pure cultures. While working on anthrax, however, he devised an alternative means of implementing the goal of laboratory research, which is to enhance the control of the scientist by simplifying the research environment and reducing the number of variables. To that end, he devised a solid medium that would enable him efficiently to grow microorganisms outside the bodies of animals. He first poured a liquid nutrient into a Petri dish and then solidified it with the gelatine agar-agar. Afterwards he could pour a fluid containing the bacteria he wished to study onto the now solid medium, with the result that he could more readily isolate microbes of one species from another, avoid the hazard of mixing them together, and observe the target organism under the microscope during the course of its development. This was a pivotal step in the history of microbiology as a discipline, and it provided a firm base for progress in the study of infectious disease. In the comment of a recent study, the result was that "the hitherto unruly and confusing world of bacteria could be easily subjected to the researcher's manual and visual control."[6]

At the same time, the limitations of Koch's personal circumstances were permanently resolved when he was invited to Berlin in 1879 to join the staff of the Imperial Health Office. From that time he possessed well-equipped laboratory facilities and the collaboration of three tireless and talented assistants—Georg Gaffky, R. J. Petri, and Friedrich Loeffler—who played a critical role in his later discoveries

Thus equipped, Koch turned his attention to tuberculosis, the greatest killer of the time, but one enveloped in a confounding mystery that particularly attracted his attention. The mystery was whether the various conditions involving tubercles, such as the pulmonary disease of phthisis and what is now called miliary, or "disseminated," tuberculosis, were separate diseases as Virchow taught or a single disease as René Laënnec argued. It was impossible, as the evidence stood, to resolve the debate because no causative agent had been discovered despite the diligent efforts of laboratories around the world. A major reason was that the microbe that came to be known as *Mycobacterium tuberculosis* was much more difficult to isolate than the large and readily visible *B. anthracis*.

The problem, Koch discovered, was not simply one of magnification, although the tubercle bacillus is much smaller than *B. anthracis*. A further difficulty was decisive—that it does not stain in the manner of other bacteria. As he wrote, "It seems likely that the tubercle bacillus is surrounded with a special wall of unusual properties, and that the penetration of a dye through this wall can only occur when alkali, aniline, or a similar substance is present."[7]

After devising an appropriate staining method, Koch became the first to discover the elusive *M. tuberculosis* and to observe that it was present in all infected tissues. Logically, however, he felt that he required more rigorous proof than correlation. He needed to demonstrate that it was not a coincidental presence in cases of tuberculosis but the causative agent. Even his own prior work on anthrax did not meet this new and more demanding standard.

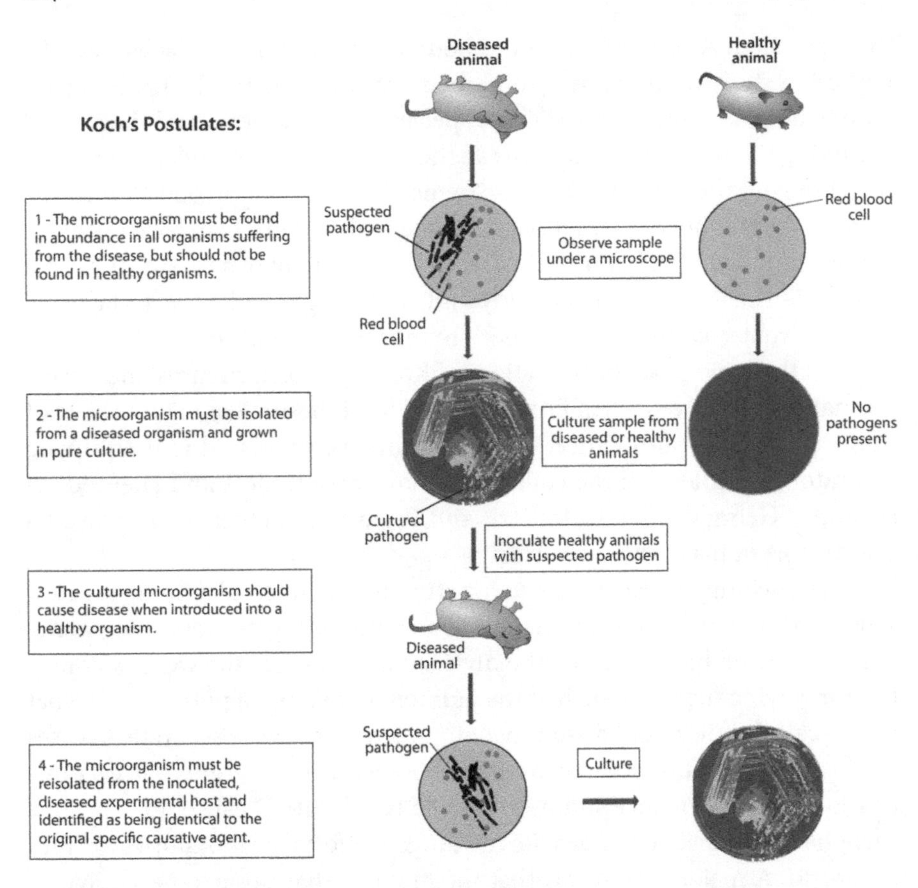

Figure 12.2. Koch's Postulates, designed to establish a causal relationship between a disease and the microbe that causes the disease. (Drawing by Mike Jones, CC-SA 3.0; adapted by Bill Nelson.)

Determined to demonstrate irrefutably the causative role of the bacterium he had isolated, Koch formulated rigorous protocols that he announced in his triumphant paper of 1882 "The Etiology of Tuberculosis," one of the most influential writings both in the establishment of the germ theory and in the history of medicine more generally. The four criteria he established to prove that a microbe discovered to be present with any disease is actually its causative pathogen are known as "Koch's Postulates" (fig. 12.2). These criteria were clear, and when applied successfully to tuberculosis in his lab, they convinced the scientific world that Koch and his team had in fact discovered in *M. tuberculosis* the cause of the most important disease of the nineteenth century. In doing so, he also imposed a uniformity of method on microbiology.

In 1883 the onset of an epidemic of Asiatic cholera in Egypt provided Koch with what he confidently believed would be the opportunity to use his newly developed methodology to isolate microbial pathogens and demonstrate their causative role in another major disease. Several national commissions—from Germany, France, and Belgium—rushed to Egypt in an international race to make the discovery first. At the outset, tragedy and poor timing struck the French commission when its leader contracted cholera himself and died, and then the epidemic ebbed away. Koch and his team moved on to India, where the epidemic still raged, and there they isolated *Vibrio cholerae* and provided epidemiological evidence that it was in fact the causative pathogen. Ironically, Koch announced that he had proved the role played by the newly discovered comma-shaped bacterium even though he could not meet the exacting rigor of his own postulates. The reason was that, since cholera is transmitted only among humans, it proved impossible to induce the disease by inoculating experimental animals. Koch's Postulates were decisive when they could be adopted, but cholera demonstrated that they were not universally applicable.

By 1883, then, the pathogens responsible for anthrax, tuberculosis, and Asiatic cholera had been isolated and their roles in disease causation had been demonstrated. Making use of the methodologies developed by Pasteur and Koch, scientists rapidly isolated a succession of microbes responsible for human disease—typhoid, plague, dysentery, diphtheria, scarlet fever, tetanus, and gonorrhea. The decades between 1880 and 1910 were therefore known as the "golden age of bacteriology," when the new techniques of microscopy unraveled many of the mysteries of disease etiology, definitively proved contagionism, and established the germ theory of disease.

Joseph Lister

Paradoxically, understanding the causes of epidemics provided few benefits for patients as medicine remained powerless to cure infectious diseases until the age of antibiotics began with penicillin and streptomycin after World War II. But if medical patients drew few benefits from the germ theory of disease at the outset, the same was by no means true for surgical patients, who were the immediate beneficiaries of a "surgical revolution" based on the new understanding. Here the decisive figure was the third major scientist responsible for establishing the germ theory—Joseph Lister (1827–1912).

Lister was professor of surgery at Edinburgh, where he was appalled by the numbers of patients who died of infection after otherwise successful operations. Pasteur's idea and the agricultural metaphor of germs and soil to which it gave rise immediately struck Lister as having lifesaving practical implications for surgical procedures. If the wounds of the surgical patient could be protected from contamination from dust in the air, then airborne germs would not appear in them—just as germs did not appear in the cultures protected inside Pasteur's swan-neck flasks.

Until Lister's time, surgery was severely limited by three major constraining factors—pain, the loss of blood, and the danger of septicemia. As a result, the major bodily cavities—abdominal, thoracic, and cranial—were considered surgically off-limits except in dire emergencies arising from trauma and war. Pain and blood loss were in principle subject to amelioration. Indeed, in the 1840s chemistry provided the first tools for anesthesia—ether and nitrous oxide—which allowed the first painless surgical operations in 1846 at Massachusetts General Hospital in Boston and at University College Hospital in London. Infection, however, was thought to be an inevitable and therefore irremediable part of normal healing. Contemporary understanding was that it arose by spontaneous generation through the release of toxins from dying tissues.

Already in the 1860s, following the logic of Pasteur's work on fermentation and spontaneous generation, Lister reasoned that Pasteur's thought had profound implications for surgery. Indeed, Pasteur himself drew out the relevance of his discoveries for surgery in his later work *Germ Theory and Its Application to Medicine and Surgery.* Infection did not arise from spontaneous generation within the body of the patient but through contamination from without by an airborne microorganism carried by dust. Remedies were therefore available. One solution, suggested by Pasteur and implemented by Lister, was antisepsis. This strategy was to prevent microorganisms from gaining access to the wound by destroying them. Lister proposed this solution in his work "On the Antiseptic Principle in the Practice of Surgery" of 1867. Patients, Lister noted, usually died after surgery not from their original ailment or the postoperative healing process, but rather from infections contracted as "collateral damage" during the surgery. This was iatrogenesis, or what Lister called "hospitalism."

Lister's revolutionary methodology was to scrub his hands before surgery, to sterilize his instruments, and to spray carbolic acid into the air around the patient and directly onto the wound to prevent suppuration (figs. 12.3 and 12.4). Lister worked tirelessly to promote this lifesaving methodology. He wrote articles, delivered countless lectures across Britain and the United States, and gave demonstrations to persuade the profession. At the outset

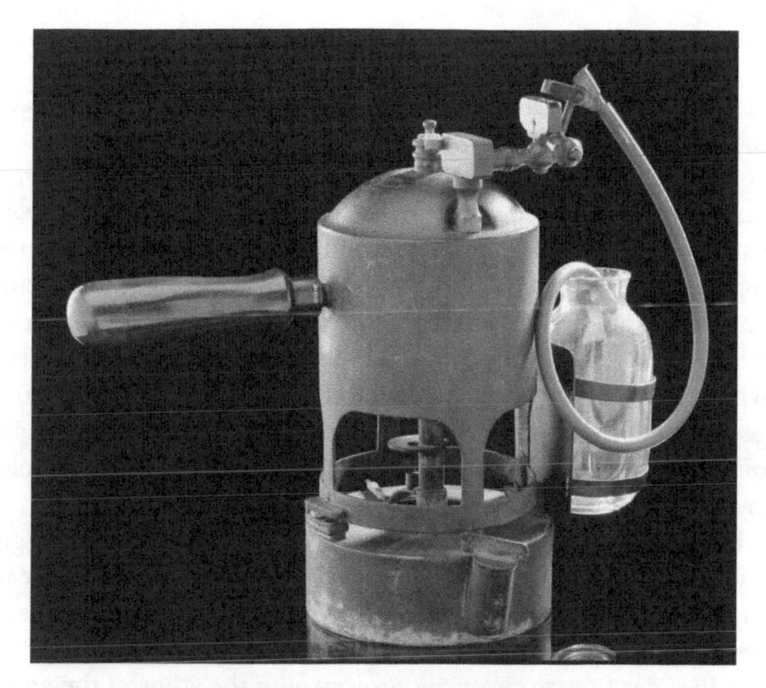

Figure 12.3. Joseph Lister was using this carbolic steam spray equipment for antiseptic surgery by 1871, when he used it to treat an abscess of Queen Victoria. The spray covered the operating room with a vapor of carbolic acid, creating an antiseptic environment. (Science Museum, London. CC BY 4.0.)

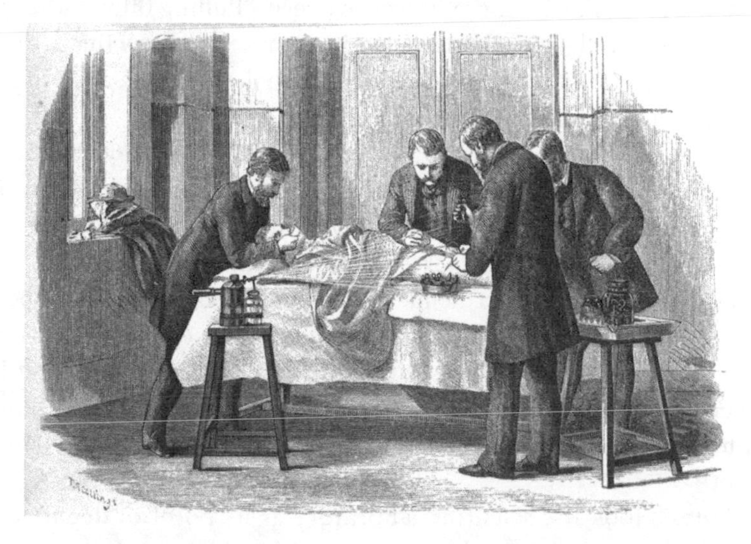

Figure 12.4. This 1882 illustration shows the use of the Lister carbolic spray equipment, nicknamed "Puffing Billy." (Wellcome Collection, London. CC BY 4.0.)

Lister's colleagues derisively rejected his ideas. Surgeons hated the fact that carbolic acid spray burned their hands and eyes, and they found the sterilization of instruments to be a time-consuming and unprofitable chore. Furthermore, the concept that tiny, invisible organisms caused the deaths of robust adults appeared implausible. But over time Lister's astonishing postoperative survival rates carried the day. Obstetricians followed suit and began a major roll-back of puerperal fever; and Queen Victoria herself legitimized his work—first by allowing him to lance a deep abscess in her armpit while following his methodology and then by knighting him in 1883. Surgery, which had been an emergency treatment of last resort, became a normal procedure, and the germ theory acquired highly visible and successful practical applications.

But "Listerism," as the set of practices pioneered by the Edinburgh surgeon was termed, soon confronted a scientific rival. Lister's antiseptic methodology incorporated insights gained from the experimental science of microbiology developed by Pasteur. It did not, however, match the rigor needed in order fully to reconcile surgery with the ethos of the emerging scientific laboratory. Lister, for a start, did not carry out experiments to test his results or keep statistics to validate the efficacy of his procedures. He registered progress by his impressionistic view that patients recovered far more often when carbolic acid was introduced, but he did not submit new tools to rigorous scrutiny. Furthermore, as we have seen, "Puffing Billy"—the sprayer that Lister deployed—was clumsy; it emitted an acidic spray that stung; and tests conducted by its critics revealed that carbolic acid was not a highly efficient antiseptic agent.

More fundamentally, by the 1880s surgeons in Germany devised an alternative to antisepsis. Their procedures were founded on the germ theory as well, but on the research of Robert Koch rather than of Louis Pasteur. Like Pasteur, Koch had taken an interest in wound infections that were considered at the time a form of "fermentation." Demonstrating that microbes were the cause of suppuration, Koch also paid attention to the implications for surgery. His logic, however, was different from that of Pasteur and Lister, and Koch's followers were polemical in terming his techniques "asepsis" in order to underline the difference from the "antisepsis" of the Frenchman and the Scot.

Asepsis took the scientific laboratory as its point of departure and sought to bring surgery into line with its underlying principles. Just as the laboratory placed the researcher in control by reducing the number of variables that could influence an experiment, so the proponents of asepsis sought to place the surgeon in control by transforming the operating room into a totally artificial environment to which germs were denied access. At the outset, antiseptic surgeons had often operated bare-handed, in street clothes, and in patients' homes or in theaters before an audience, relying on acidic spray to destroy germs during the operation. Taking advantage of lessons learned in practice since the early antiseptic operations of the 1860s, Koch's aseptic followers considered themselves more scientific because their technology of sterilization aimed to provide surgeons in the operating room with the control of a scientist in the laboratory.

In principle, antiseptic and aseptic surgical techniques were radically opposed strategies although both were based on the germ theory of disease. In practice, however, the two methodologies converged over time. By the 1890s surgeons who considered themselves Listerians donned gloves, masks, and gowns, and they sterilized their instruments. Now, a century and a half since Lister, aseptic surgeons administer perioperative antibiotics to their patients—a measure that is pure antisepsis, just without carbolic spray. Thus the approaches of both Pasteur and Lister on one hand and of Robert Koch and his followers on the other converged in establishing the methodologies of the modern operating room.

"Laboratory Medicine" and Medicine as a Profession

As Claude Bernard had prophetically suggested, the laboratory bench became the emblem of the newly emergent medical science and the locus of medical epistemology. After millennia of reverence for the classic texts of Hippocrates and Galen, and a century of hegemony for the hospital ward, authority in medicine moved to the laboratory, with its oil immersion lenses, dyes, petri dishes, and complement of full-time medical scientists and their assistants.

Germ theory had major consequences for the medical profession. First of all, it produced major reforms in medical education, which became increasingly based on the laboratory and the basic sciences following the German model. In the United States, medical schools such as those at Johns Hopkins, the University of Pennsylvania, Harvard, and Michigan were among the first to adopt the new perspective. At these institutions medicine also spawned an array of specialized subdisciplines, such as microbiology, parasitology, tropical medicine, pharmacology, and bacteriology.

Such rigorous training combined with the clout derived from the array of scientific discoveries furnished allopathic physicians with a new cultural authority in their competition with other medical sects. Since medicine in the form of the biomedical paradigm was also a powerful tool for knowledge and practical public policies, the standing of the profession was enhanced by the attention of the state, the pharmaceutical industry, and public health services.

Germ theory also slowly but radically transformed the doctor-patient relationship. With the rise of new technologies, no longer were diagnosis, therapeutic strategies, and case management based on the holistic narrative approach of traditional case histories. Medical records were reduced to charts with graphs and numbers derived from thermometers, microscopes, stethoscopes, and lab reports. The strategy of preference was often to treat a specific disease entity rather than the entire patient.

Effects of the Germ Theory on Domestic Life

As Nancy Tomes discusses in *The Gospel of Germs,* the discovery of the microbial world also transformed daily life in the home. As we have seen (Chapter 11), a first set of changes had already been instituted by the sanitary movement, with its call to remove filth and noxious odors as dangerous to health. The result was the appearance of the water closet, drains, wash basins, and mops as instruments of domestic hygiene.

The changes brought about by the germ theory were based on concepts different from those underlying the filth theory of disease and the sanitary revolution. In practice, however, the analogy of seed and soil on which the germ theory rested reinforced the sanitarian campaign to remove the filth that—according to the sanitary idea—led to decay, putrefaction, and disease-causing miasmas. Furthermore, like the sanitary idea, the "gospel of germs" was tirelessly explained to the general public by means of the press, magazines, pamphlets, circulars, and public lectures. It also produced campaigns against single diseases, such as the "war on tuberculosis." Under the press of these influences, the general public came to see the home as a place of danger where microbes lurked, ready to invade the body and cause disease. Domestic spaces and personal hygienic practices therefore stood in need of reform.

The discoveries of Pasteur and Koch led to the retrofitting of houses to facilitate the war on germs. Changes included watertight pipes, soil traps, porcelain toilets, bathroom tiles, washbasins, and linoleum kitchen flooring. These were innovations that had not been part of the earlier sanitary campaign against filth but that were mandated by the war on germs and the more exacting standards of cleanliness that it implied. Meanwhile, mindful of the tubercle bacillus, conscientious individuals covered their mouths when they coughed, avoided spitting, and washed their hands and bodies frequently. The general public was also afflicted with a new anxiety about microbes that found expression in "tuberculophobia," in the refusal to use common communion cups in church services, and in the work of some writers, such as Bram Stoker. Stoker's short story "The Giant" and his 1897 novel *Dracula* were Gothic expressions of Victorian fears about contagious disease.

Conclusions

The germ theory of disease was clearly a decisive advance in the history of medicine. It led to a new understanding of the nature of disease, the development of microscopy, the public health strategy of vaccination, and the transformation of daily life in accord with the demands of cleanliness. There were two respects, however, in which its impact had potentially negative consequences. The first is that the germ theory directed public health toward narrow-gauged "vertical" campaigns against single, specific microbial targets and away from "horizontal" programs that focused on the social determinants of disease—poverty, diet, education, wages, and housing. In such vertical approaches, it is possible to neglect the positive goal of promoting overall health and well-being in addition to the negative goal of attacking a specific pathogen and the single disease it causes. The sanitary movement had already led away from social medicine and the inclusion of the workplace and wages in the narrow war on filth. The advent of the germ theory provided an occasion for a still narrower focus that targeted microorganisms alone.

Another issue that arose prominently as a result of the germ theory is the ethical dilemma that it posed. For the first time in the history of medicine, laboratory research demanded the recourse to vast numbers of experimental subjects. Pasteur's research was dependent on the possibility of inoculating rabbits, mice, guinea pigs, sheep, dogs, cows, and chickens. Koch's Postulates specifically demanded the inoculation of healthy animals with virulent and potentially lethal microorganisms. In the absence of a code of ethics, the result in many cases was the unnecessary and unregulated suffering of research animals. Additionally, in some research projects the experimental animals employed were human beings. It was not until the scandals of Nazi medical "science" and the Tuskegee Syphilis Study that such research was subjected to scrutiny and strict regulation.

Looking farther ahead into the late twentieth century, the germ theory of disease led to new complexities in the understanding of disease itself. For a time it seemed that infectious diseases and chronic diseases belonged to two separate categories. More recent discoveries, however, have demonstrated that the distinction is far less clear because many "chronic" diseases are triggered by bacterial infections. This discovery began with peptic ulcers and led to a revolution in the way that ailment is understood and treated. Recent research has explored analogous mechanisms involving other chronic diseases, including various forms of cancer, type 1 diabetes, and Alzheimer's disease, that may prove to have microbial triggers. Thus the germ theory of disease has returned to illuminate pathology in ways its founders could not have anticipated.

CHAPTER 13

Cholera

Scholars debate whether cholera had long existed on the Indian subcontinent, which encompassed its endemic home—the delta of the Ganges and Brahmaputra Rivers. What matters for our purposes is that by the early nineteenth century cholera was endemic in India but unknown elsewhere until it burst into a major epidemic in 1817. Soon thereafter it moved beyond India, beginning a devastating international career that took it to Europe in 1830. Cholera, then, is the history of seven successive pandemics:

1. 1817–1823: Asia
2. 1830s: Asia, Europe, North America
3. 1846–1862: Asia, Europe, North America
4. 1865–1875: Asia, Europe, North America
5. 1881–1896: Asia, Europe
6. 1899–1923: Asia, Europe
7. 1961–the present: Asia, South America, Africa

At the start of the nineteenth century, the disease was imprisoned in India because the bacterium that causes it—*Vibrio cholerae*—is delicate and does not travel easily (fig. 13.1). During the succeeding decades a number of factors vastly increased the movement of people between India and the West, while revolutionizing the duration of the journey. Three developments were critical: British colonialism, as it set troops and trade in movement; religious pilgrimages and fairs, including the Haj, which took Indian Muslims to

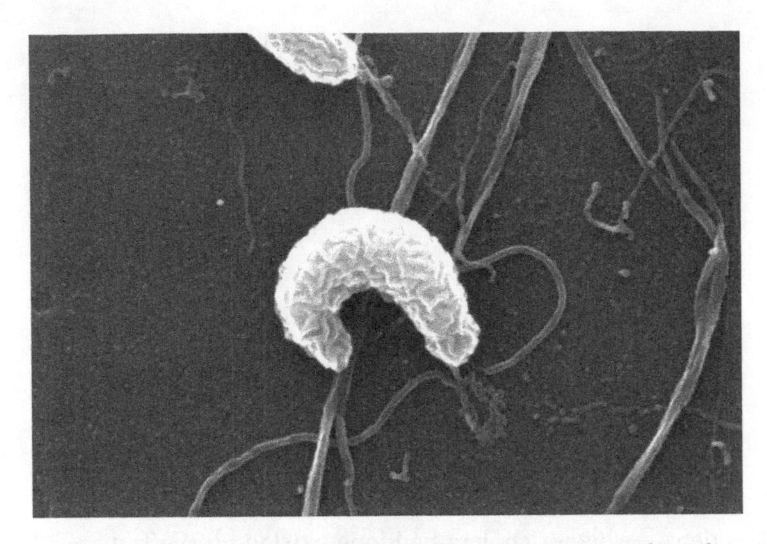

Figure 13.1. Electron microscope image of the comma-shaped *Vibrio cholerae* bacterium, which causes Asiatic cholera. (Photograph by Louisa Howard, Dartmouth College Electronic Microscope Facility.)

Mecca; and the transport revolution involving railroads, steamships, and the Suez Canal.

These developments enabled the vibrio to gain access to the West, but the bacterium also required welcoming conditions in which to thrive upon its arrival. The epidemic diseases that afflict a community are never purely random. They exploit features of a society that are social, economic, political, and environmental. In the case of cholera—a disease transmitted by the oral-fecal route—the Industrial Revolution and its pathologies created favoring conditions. Cholera thrived on such features of early industrial development as chaotic and unplanned urbanization, rapid demographic growth, crowded slums with inadequate and insecure water supplies, substandard housing, an inadequate diet, ubiquitous filth, and the absence of sewers. When the vibrio disembarked in the port cities of Marseille, Hamburg, Valencia, and Naples, it found ideal conditions awaiting it.

Plague was the most dreaded disease from the fourteenth century through the early eighteenth; smallpox succeeded it in that role during the eighteenth century; and cholera was the most feared disease of the nineteenth century. Indeed, much of the discussion surrounding the early encounters with cholera concerned the question of whether its arrival marked a "return of the plague." Some of the familiar names given to the disease capture the horror it inspired—cholera morbus, cholera asphyxia, "the gypsy," "the monster," "blue cholera," and "king cholera."

Cholera was feared for many reasons. One was its sudden appearance as an unknown invader from the East—indeed, it was called "Asiatic cholera." It also created alarm because of its gruesome symptoms, its high CFR, its sudden onset, and its predilection for attacking adults in the prime of life. So intense was the terror it inspired that cholera gave rise to a series of societal responses, some of which will be already familiar to readers, that magnified the disruption that it left in its wake—mass flight, riot, social hysteria, scapegoating, and economic disruption.

Since the nineteenth century was an era of acute societal tensions marked by a series of revolutionary outbreaks, it is often dubbed the "rebellious century." Its passage was punctuated by such drastic social upheavals as the revolutionary wave of 1830, the revolutions of 1848–1849, Italian and German unification, and the Paris Commune. Because of the fact that cholera accompanied these events and indisputably heightened political tensions, it was once common for historians to wonder whether cholera across Europe was a contributory factor in the coming of revolution. It is clear by now, however, that such a view is untenable. The causal chain worked the other way around; that is, the outbreak of revolution, war, and social disorder created ideal conditions for cholera to survive. Cholera moved in the wake of revolution rather than triggering it. Asiatic cholera accompanied the troops who were mobilized to crush revolutions rather than with the crowds that rioted to start them.

Although it is now clear that cholera did not cause the revolutions of nineteenth-century Europe, the pendulum can also swing too far in the opposite direction. Some argue that cholera not only failed to cause revolutions, but even failed to produce a lasting impact of any kind. This assessment holds that cholera was dramatic and led in the short term to considerable sound and fury; but in the long term, its legacy was small and nowhere near the order of magnitude of other epidemics such as plague and smallpox. In order to decide among these conflicting views, we first need to examine the etiology and symptoms of cholera, its treatment, and its epidemiology.

Etiology, Symptomatology, and Artistic Responses

Cholera is caused by the bacterium *V. cholerae,* which was discovered by Koch in 1883 and is famously gruesome in its pathology. In a majority of people who ingest the vibrio, the gastric juices of the digestive tract are sufficient to destroy the pathogen without ill effect. If, however, the number of vibrios swallowed is overwhelming, or if the digestive process is undermined by a preexisting gastrointestinal disorder, the consumption of overripe fruit, or overindulgence in alcohol, then the invading bacteria pass safely from the stomach into the small intestine. There they establish an infection as the vibrios reproduce and attach themselves to the mucous lining of the bowel.

In response, the body's immune system attacks the bacteria, but as they die they release an enterotoxin that is one of the most powerful poisons in nature. Its effect is to cause the wall of the intestine to function in reverse. Instead of allowing nutrients to pass from the lumen of the bowel into the bloodstream, it now permits the colorless liquid portion, or plasma, of the blood to drain into the digestive tract and to be expelled explosively through the rectum.

This loss of blood plasma is the source of the cholera patient's "rice-water stools"—so named because of their resemblance to the liquid present in a pot in which rice has been boiled. This fluid pours forth copiously—at a rate of up to a liter an hour—and it is normally supplemented by further fluid loss through violent retching during which, it is said, fluid pours forth from the mouth like water from a tap. The result constitutes the functional equivalent of hemorrhaging to death through hypovolemic shock.

Violent expulsion of liquid from the digestive tract marks the sudden onset of illness after a short incubation period that lasts from a few hours to a few days. In the extreme case, known as *cholera sicca*, the draining of fluid is so rapid and devastating that the result is instant death. In all patients, the onset occurs without warning so that people are often overtaken in public spaces—associating cholera particularly with fear because the gruesome and painful ordeal of the victim can be a public spectacle. A characteristic that marks the disease apart from others is the rapidity of its course in the human body. Apparently robust and healthy people can finish lunch and die in agony before dinner, or board a train and perish before reaching their destination.

Such sudden seizure also sets cholera apart as an affliction because it bears more resemblance to poisoning than to a "normal" endemic disease familiar to most people. The entire pathology of the disease even resembles the effects of ratsbane—the white, powdered arsenic used to exterminate rodents that was widely used in the nineteenth century. All the tormenting symptoms of the disease were therefore consistent with the idea that perhaps cholera was a crime rather than a natural event. This conclusion was all the more frightening because there was no cure and nearly half of nineteenth-century victims of cholera perished.

The appalling symptoms of cholera are a direct result of the loss of bodily fluids and the devastating systemic impact that ensues. Nineteenth-century physicians held that the disease passes through two phases after the incubation period ends: the "algid" (ice-cold) stage, and the "reaction" (countervailing) stage. The first is especially marked by high drama and lasts from eight to twenty-four hours, with a terrifying and sudden transformation of the patient. The longer the algid stage lasts, the more dire the prognosis. While fluid drains away, the pulse becomes progressively feebler, blood pressure plummets, and the body grows cold to the touch as the temperature falls to 95 or 96°F. The face rapidly takes on the livid, shrunken aspect of a death mask or a patient after a prolonged wasting illness. Often, it was said, the still living body of the patient is "cadaverized" and unrecognizable. The lifeless eyes are cavernous and surrounded by dark circles and are bloodshot beneath lids that are permanently half closed. The skin is wrinkled, the cheeks are hollow, the teeth recede from blue lips that no longer close, and the tongue, now dry and thick, resembles shoe leather. Unrelenting waves of dizziness, hiccups, and unquenchable thirst add to the torment.

As doctors discovered when attempting to practice venesection, the blood itself turns into a dark tar so viscous that it barely circulates. Deprived of oxygen, the muscles contract in powerful cramps that sometimes tear both muscles and tendons and cause searing abdominal pain. In the words of the Victorian-era cholera expert A. J. Wall: "In extreme cases, nearly the whole of the muscular system is affected—the calves, the thighs, the arms, the forearms, the muscles of the abdomen and back, the intercostal muscles, and those of the neck. The patient writhes in agony, and can scarcely be confined to his bed, his shrieks from this cause being very distressing to those around him."[1] Often the cramps themselves lead to death as they cause such severe contractions of the laryngeal muscles that they prevent both swallowing and breathing. Patients therefore experience a desperate sense of impending asphyxiation, and they thrash about wildly in frantic efforts to breathe. To both patient and onlooker the experience is all the more painful because the mind is unimpaired and every moment of pain is fully experienced as death threatens from cardiac arrest and suffocation.

Those who survive the algid stage pass into the reaction period, which is comparatively uneventful although the prognosis is no more favorable. The clinical signs of algid cholera subside or are even reversed. Body temperature rises or gives way to fever, cramps and evacuations diminish, pulse strengthens, and skin color returns. The overall appearance of sufferers suggests that they are on the path to recovery. Unfortunately, however, a severely weakened patient, now frequently delirious, is highly susceptible to a range of complications—pneumonia, meningitis, uremia, and gangrene of the extremities such as fingers, toes, nose, and penis. Most lethal of all is uremia, which accounted by itself for nearly a quarter of all nineteenth-century mortality from cholera. The condition results from the patient's dense blood, which mechanically obstructs circulation through the kidneys and leads to renal failure, suppression of urination, and toxemia.

Even after a patient dies, cholera continues to horrify. A ghoulish aspect of the disease is that, while the living patient resembles a corpse, the dead body of a victim seems alive. A haunting feature of the disease is that it produces vigorous postmortem muscular contractions that cause limbs to shake and twitch for a prolonged period. As a result, the death carts that collected the bodies of those who died of the disease seemed to be teeming with life, triggering fears of malevolent plots and of premature burial (fig. 13.2).

Figure 13.2. Gabriele Castagnola, *Cholera in Palermo* (1835). The dead were loaded onto carts, sometimes with their muscles still eerily twitching. Twenty-four thousand people are said to have died in Palermo during the second world cholera pandemic in the 1830s. (Wellcome Collection, London. CC BY 40.)

In a variety of ways, therefore, the nature of the pathology of cholera powerfully shaped the cultural response to it. This exotic invader from the East was too foul and degrading to give rise to extensive treatment in operas, novels, and paintings in the manner of some other infectious diseases. As we shall see in Chapter 14, tuberculosis generated an extensive literary and artistic production and was perceived as an appropriate starting point for reflections about the nature of beauty, genius, and spirituality. Whatever the reality of the misery tuberculosis caused, it was an affliction of the lungs that wasted the body away in "obedience" to Christian theology and the prevailing "norms" of bourgeois sensibility.

Syphilis also had its romantic devotees. This was paradoxical because the disease was disfiguring, morally troubling, and often lethal. But it struck all strata of society without distinction, and even its dissolute connotations enabled it to provoke levity. Libertines could regard "the clap" as an iconoclastic badge of honor, of free thinking in defiance of hypocritical conventions, and of sexual conquest. The nineteenth-century writers Gustave Flaubert and Charles Baudelaire, for instance, treated the disease with bravado.

Even plague, the most cataclysmic of all afflictions, had its redeeming artistic features. The great mortality it caused was universal in the sense that, unlike cholera, it afflicted all echelons of society from top to bottom, and its symptoms, however agonizing, were not directly scatological. As we have seen, it prompted outpourings of works in all media contemplating the relationship of God to Man, and the meaning of life.

But cholera was an irredeemably filthy, foreign, and lower-class disease. A cholera epidemic was degrading, vulgar, and stigmatizing, both for the victims and for the society that tolerated such squalor and poverty in its midst. In the later pandemics, when the mechanisms of cholera and its oral/fecal mode of transmission were largely understood, the necessary societal remedies were both readily apparent and far from lofty. Sewers, safe water, and flushing toilets were required—not repentance or divine intercession. Similarly, it is impossible to imagine a choleraic opera heroine reaching her denouement while pouring forth her insides onto the stage as a counterpart to the beautiful death of consumptive Mimì in Giacomo Puccini's *La Bohème,* which premiered in 1896.

Cholera did attract artistic treatment, but in ways that are distinctive and revealing. One narrative strategy was to concentrate on the social impact of the disease while omitting its medical aspects. Thus, in his naturalistic novel *Mastro-don Gesualdo* (1889), which traces the course of cholera across the island of Sicily, the Italian author Giovanni Verga chose not to enter the sickroom or to explore the suffering of individual patients. In the same spirit the Colombian writer Gabriel García Márquez opted, in his work *Love in the Time of Cholera* (1985), to keep the disease as a looming background presence rather than as a central feature requiring vile description.

Perhaps the clearest indication of the differences that set cholera apart from other epidemic diseases is to contrast two works by the same author—Thomas Mann, who dealt with tuberculosis in *The Magic Mountain* (1924) and cholera in *Death in Venice* (1912). In *The Magic Mountain,* the German author examined every nuance of life in an elegant sanatorium and carefully traced the medical career and intellectual enlightenment of the protagonist, Hans Castorp. When dealing with cholera in the outbreak of 1910–1911, however, Mann presented the disease as emblematic of the final "bestial degradation" of the sexually transgressive writer Gustav von Aschenbach. Revealingly, however, Mann spared von Aschenbach the final indignity of revealing his symptoms, allowing him instead to make medical history by becoming the first cholera patient ever to die peacefully asleep in a deck chair. Similarly, the film director Luchino Visconti did not allow cholera's gross pathology to intrude upon his cinematically beautiful portrait of Venice in his 1971 movie version of the novella. Cholera is just too disgusting to depict.

Treatments

Throughout the nineteenth century physicians were overwhelmed by the sudden onset, severity, and rapid course of cholera. Nothing in their armamentarium made any significant difference in relieving suffering or prolonging life. In desperate attempts to save their patients, doctors often had to rely on the most experimental and invasive interventions—all to no avail.

Treatment was initially based on humoral principles, especially the concept of the healing power of nature (*vis medicatrix naturae*) and its companion doctrine that the symptoms of a disease were nature's therapeutics. Hippocratic and Galenic teachings suggested that the violent retching and diarrhea that characterized cholera were the body's strategy to rid itself of a systemic poison. To assist nature in this effort, physicians administered the most powerful emetics and cathartics known, including ipecac among the former and aloe, senna, cascara, and castor oil among the latter.

Pursuing this medical philosophy, doctors deployed the signature procedure of all orthodox physicians—venesection. Bloodletting then had numerous advantages. It was systemic, the strategy behind it was clear, it allowed the physician powerful control, and it had the sanction of two millennia of medical practice. A problem with it in treating cholera patients, however, was that they were already losing blood plasma at such a high rate that blood barely flowed and veins were difficult to locate. It therefore became necessary to open major blood vessels, arteries as well as veins.

Since bloodletting in such circumstances proved unproductive, doctors resorted to an array of desperate and experimental procedures. One of these, developed in the 1830s, was promising and, ironically, ultimately provided the basis for modern cholera therapy. This was rehydration. Since it was obvious that cholera patients were losing fluids at a lethal rate, many doctors reversed orthodoxy by replacing fluids rather than augmenting their flow. But simply giving copious drinking water to already exhausted patients caused them to wretch even more violently. Thus its main effect was to hasten death.

A more invasive alternative was what one might conceptualize as reverse phlebotomy—not the draining of fluid from the veins but the administration of water through them. This early form of rehydration, however, did nothing to alter patients' prognoses. Not knowing how much fluid to administer, physicians often overdosed, causing heart failure. Furthermore, since they were unaware of germs and of the need for the water to be sterile, they also induced septicemia.

An additional problem was that of salinity. The natural impulse was to replace like with like by administering fluid that was isotonic—that is, of the same saline level as blood. Unfortunately, however, body tissues absorb fluid only if it is hypertonic—of a much higher salt concentration. The often repeated result throughout the century was that the fluid injected into the veins simply drained into the bowel, enhancing the already profuse rice-water flow.

It was only in 1908 that the British physician Leonard Rogers initiated two additional procedures. The first was constructing a "cholera bed," which had a large hole in the center and a bucket beneath it to catch the rice-water evacuations in order to measure the loss. He was then able to administer the proper amount of fluid in a gradual drip that avoided sudden cardiac arrest. Equally important was Rogers's second innovation—to use a hypertonic saline solution in distilled water. The body retained this solution, and it did not cause septicemia. These developments halved the CFR of cholera—to 25 percent. A later refinement was the development of oral rehydration by means of packets of salt, sugar, and electrolytes that can be dissolved in a bucket of clean water. Oral hydration is effective because the glucose in the solution increases the capacity of the intestines to absorb salt and, with it, water. This strategy, which is simple, inexpensive, and easy to administer, lowers mortality still further and has been the sheet anchor of cholera therapeutics since it was widely adopted in the 1970s.

A puzzling question is why rehydration experiments persisted throughout a century of unalloyed failure. The reason is that often unsuccessful rehydration produced seemingly miraculous and tantalizing short-term effects. Even critically ill patients seemed to recover. The agonizing cramps, the sense of impending suffocation, and the loss of body heat all suddenly receded. They sat up in their beds, chatted, and had time to rewrite their wills before the inevitable relapse occurred and the disease resumed its course.

Of all nineteenth-century treatments for epidemic cholera, however, perhaps the most painful was the acid enema, which physicians administered in the 1880s in a burst of excessive optimism after Robert Koch's discovery of *V. cholerae*. Optimistic doctors reasoned that since they at last knew what the enemy was and where it was lodged in the body, and since they also understood that bacteria are vulnerable to acid, as Lister had demonstrated, all they needed to destroy the invader and restore patients' health was to suffuse their bowels with carbolic acid. Even though neither Koch nor Lister ever sanctioned such a procedure, some of their Italian followers nevertheless attempted this treatment during the epidemic of 1884–1885. The acid enema was an experimental intervention that, in their view, followed the logic of Koch's discoveries and Lister's practice. The results, however, were maximally discouraging, and they demonstrate that scientific advancements like the germ theory of disease can have lethal consequences if they are hastily and incautiously translated from the laboratory to the clinic.

Epidemiology and the Example of Naples

Vibrio cholerae, which did not live freely in the environment and had no natural reservoirs other than human beings, depended on the steamship and large-scale movements of people to arrive in the West. It was transported in the guts of crew members and passengers; on their bed linens, clothing, and personal effects; and in their waste. Arriving at a European or North American port city, the microbe was carried ashore or discharged into the waters of the harbor. The first cases broke out typically among people who were exposed to the harbor and those who frequented it, especially during warm weather. These first victims were consumers of raw shellfish that had feasted on untreated sewage, laundrywomen who washed the clothes of sailors and passengers, and the keepers of inns and restaurants in the city districts near the port. Their misfortune marked the onset of what was termed "sporadic cholera," which did not denote an epidemic but rather a trickle of individual cases in a neighborhood, family, or house in which the disease was spread directly from person to person. Venice in 1885 provides an illustration of the manner in which such an outbreak could emerge and persist. There it occurred in a single street, where an innkeeper prepared and served food on the ground floor but tended her son upstairs, who had fallen ill with a severe bout of diarrhea and vomiting. Afterwards she returned downstairs, without washing her hands, and filled orders from her customers. In fortunate cases, and Venice in 1885 was just such a place, the outbreak gradually faded away and the city as a whole escaped a large-scale disaster.

Less happily, the disease could move farther afield in spreading chains of transmission when sanitary conditions permitted. Squalid and congested slums provided a welcoming environment. Here the example of nineteenth-century Naples is especially illuminating because the great Italian port was the European city most frequently and heavily scourged by the unwelcome invader from Bengal. Naples, Italy's largest city, had a population approaching half a million by 1880. Crucially, by that date the city had not yet accomplished its own sanitary reform, and in 1884 it was on the verge of experiencing the most famous of the eight cholera epidemics that devastated it during the "long nineteenth century."

Occupying a site that resembled an amphitheater fronting the Bay of Naples, the city was divided in two: the Lower City built on the sea-level plain that formed the "stage" of the amphitheater, and the Upper City constructed on the semicircular hills that rose behind it. The Upper City housed the prosperous strata of Neapolitan society in salubrious quarters that were perennially free from the ravages of epidemic disease. Local doctors regarded it as a city apart.

Infectious diseases overwhelmingly afflicted the Lower City, which encompassed four of the twelve administrative boroughs, or *sezioni,* of the city that were notorious for poverty and unsanitary conditions—Mercato, Pendino, Porto, and Vicaria. There, 4,567 buildings housed 300,000 people in a dense maze of 598 streets. Not without reason, the local member of parliament Renzo De Zerbi termed this the "death zone," where cholera concentrated its fury during every visitation, with a striking overrepresentation of the population of the four boroughs among the dead. In 1837, for example, the death rate from cholera per thousand inhabitants was 8 for Naples as a whole but 30.6 in the borough of Porto. Identical patterns reappeared in 1854, 1865–1866, 1873, 1884, and 1910–1911.

Cholera, in other words, was unlike diseases such as tuberculosis, syphilis, influenza, and plague, which afflict all classes of society. Cholera's mode of transmission via the oral-fecal route marked it as a classic example of a "social disease," with a preference for inflicting the poor in environments with substandard housing, insecure water supplies, congestion, unwashed hands, malnutrition, and societal neglect.

Of these urban maladies, overcrowding was the most evident in Naples. Part of an ancient city tightly bounded by hills, marsh, and sea, Lower Naples by the nineteenth century had no further room for expansion. Furthermore, the impact of the sheer press of population numbers was greatly augmented by a politics of laissez-faire. The city had no growth plan, housing code, or sanitary regulations. Thus, caught in a building mania, the Lower City witnessed its gardens, parks, and open spaces disappear beneath an avalanche of stone. Towering tenements pressed so tightly together on both sides of narrow streets that three people could barely walk abreast and the sun never reached ground level. Hastily constructed, many buildings were in such advanced disrepair that they resembled heaps of broken masonry: "A typical example was the Vico Fico in the borough of Mercato. Permanently festooned with drying linen that dripped on pedestrians below, it measured fifty meters in length and three in breadth, with buildings thirty meters tall. Even in mid-summer, mud and moisture were omnipresent, and an evil black stream wound its way slowly down the center of a lane that no official of the city ever visited."[2] Unsurprisingly, visitors described Naples as the most shocking city in Europe—a place of such height, Mark Twain wrote, that it resembled three American cities superimposed one on top of another. Everywhere there were "swarms", "throngs," "masses," and "multitudes," such that every alley resembled Broadway in New York.[3]

But more striking still than the outdoors were the building interiors. In the words of an article published in the *British Medical Journal* in 1884, these "squalid" and "obscene" slums were the worst in Europe—bearing comparison only with the foulest habitations of Cairo. So savage was the overcrowding that in the Vico Fico the average size of a room occupied by seven people was only five square meters, and the ceilings were so low that a tall person was barely able to stand erect. Two straw mattresses covered the floor in such apartments, each providing sleeping accommodation for several occupants. Frequently in such a rookery, the inhabitants shared their space with chickens that helped maintain their subsistence-level lives.

Clearly a space of this kind was fatal to health because the sickroom served simultaneously as bedroom, kitchen, pantry, and living space. Bacteria had ample opportunity to spread directly from person to person on unwashed hands, linens, and utensils. Furthermore, food was stored in the same cramped room and was readily contaminated by a patient's evacuations. Rice-water stools were not easily detectable on a surface, especially in the semidarkness of a Neapolitan tenement.

John Snow, who first deciphered the epidemiology of cholera in London in the 1850s (see Chapter 12), stressed the danger posed to the working classes who lived in cramped and unsanitary accommodations. Since cholera evacuations have no color or smell, and since light in tenements is dim, soiled linens inevitably lead to soiled hands that are seldom washed. Therefore, those who tend a cholera patient unknowingly swallow a portion of the dried stools, and they then pass the material to everything they touch and to those with whom they share food or utensils. Thus, Snow argued, a first case of cholera in such an impoverished family leads rapidly to other cases.

The lack of facilities for cleaning enhanced the danger. Living quarters in the Lower City lacked running water and connections to the sewer system, so that personal hygiene was much neglected and the rooms were covered in a patina of human and animal filth. And, in defiance of contemporary doctrines of miasma, slops were simply dumped into the alleyway outside, giving rise to a pervasive stench and clouds of flies that also conveyed bacteria from space to space. Inevitably, such dwellings were infested with rats and every species of vermin.

Most notorious of all were the tenements known as *fondachi* that were interspersed through the slums of the Lower City and housed some one hundred thousand people. To contemporary visitors, these represented the depths of human misery. The Swedish doctor Axel Munthe, who served as a volunteer in the epidemic of 1884 and possessed a thorough knowledge of the Lower City, described the *fondachi* as "the most ghastly human habitations on the face of the earth."[4] As if to confirm Munthe's assessment, the *Times* of London described a representative tenement thus:

> Imagine the doorway of a cave where on entering you must descend. Not a ray of light penetrates into it. . . . There, between four black, battered walls and upon a layer of filth mixed with putrid straw, two, three and four families vegetate together. The best side of the cave, namely that through which humidity filtrates the least, is occupied by a rack and manger to which animals of

> various kinds are tied. . . . On the opposite, a heap of boards and rags represents the beds. In one corner is the fireplace and the household utensils lie about the floor. This atrocious scene is animated by a swarm of half-naked, dishevelled women, of children entirely naked rolling about in the dirt, and of men stretched on the ground in the sleep of idiocy.[5]

Thus a variety of household factors enabled the vibrio to spread in the tenements. Additionally, the southern Italian port was vulnerable through the practice of sewage farming that made market produce a risk to life and health. As in many nineteenth-century European cities, gardeners in the Neapolitan hinterland made trips to the city in the early morning hours to collect human and animal waste from the streets to use as fertilizer. As a result, the vegetables later sold at market stalls grew in untreated human waste and conveyed *V. cholerae* on their return trip to the city in the gardeners' carts. A further grower's artifice increased the danger. A common technique was to dip lettuce and other leafy produce in open sewers on the way to market as the ammonia in urine freshened the leaves for market. Those who consumed such produce were highly vulnerable to infectious diseases.

The notorious overpopulation of the city relative to resources had the additional result of severely depressing wages and destroying the bargaining power necessary for workers to establish trade unions. The poverty of southern Italian agriculture yielded a continuous stream of immigrants to the city, and every poor harvest or fall in the price of wheat quickened the flow. Thus, despite the near absence of an industrial base or secure prospects for employment, Naples was a never failing magnet for the destitute of the countryside. In the words of Girolamo Giusso, director of the Bank of Naples and later minister of public works: "Although Naples is the largest city in Italy, its productive capacity bears no direct relation to the number of its inhabitants. Naples is a center for consumption rather than for production. This is the principal reason for the misery of the populace, and for the slow, almost insensible, increase of that misery over time."[6]

According to assessments by the US Department of State, average Italian wages were the most meager in Europe, and Neapolitan wages were the lowest among major Italian cities. An average industrial worker's earnings were not sufficient, according to the American Consul, to permit the annual purchase of a change of clothing. During the normal ten- to twelve-hour workday, an unskilled male worker earned no more than the cost of four kilograms of macaroni. Four days of work were needed to purchase a pair of shoes. But such workers held relatively privileged positions—as dockers, engineering workers, metallurgical workers, and employees of the railroads and trams. They were fortunate in having stable employment.

Descending farther down in the social pyramid, one encountered a multitude of men and women occupied in an array of tiny, undercapitalized workshops that were caught in a downward spiral of decline as a result of competition from machinery and from the influx of migrants. In their ranks were cobblers, seamstresses, blacksmiths, bakers, porters, tanners, and hatters who lived from hand to mouth. There were also domestic servants, porters, and fishermen—too numerous for a polluted bay that had been overfished for generations by means of nets with a fine mesh.

More straitened still were the circumstances of those who earned their livings directly in the streets—hawkers of newspapers; peddlers of chestnuts, candy, matches, and shoelaces; messengers; washerwomen; water carriers; and private trashmen who emptied cesspits and removed domestic waste for a fee. They eked out a subsistence by peddling wares and their labor in a bewildering variety of roles that baffled specification. They were termed "micro-entrepreneurs" by some economists, and they formed a large floating population that was a distinctive feature of the city. It was they who provided Naples with its appearance of being a vast emporium. Frequently they lived seminomadic lives, roaming the city in a constant process of reinventing themselves and their jobs.

Most indicative of the degree of Neapolitan poverty, however, was the largest and most abject social stratum of all: the permanently unemployed. Members of the city council noted that of a population of half a million people, two hundred thousand (40 percent) were out of work. They woke in the morning not knowing when they would next eat. It was their misfortune to furnish Naples disproportionately with its beggars, prostitutes, and criminals—and victims of disease. Their poverty was directly linked with a vulnerability to cholera because malnutrition and undernutrition contribute to infectious disease by lowering resistance and undermining the immune system.

With specific regard to cholera, an important factor was that the poor lived in good part on a diet of overripe fruit and rotting vegetables as the cheapest produce available. But such a diet also caused gastric disturbances and diarrhea, which predisposed their sufferers to cholera by reducing the time of digestion. This was critical because the acidic environment of the stomach destroys bacteria and is the body's first line of defense against *V. cholerae.* As we have seen, preexisting digestive problems allowed the vibrio to survive passage through the stomach and to enter the small intestine still alive.

Clearly, once having gained access to the impoverished boroughs of the Lower City, cholera had ample means to spread from person to person and from tenement to tenement in clusters of "sporadic cholera" cases. What transformed such sporadic appearances of the disease into an explosive, generalized epidemic was the contamination of water supplies. Here too there was a major difference between the Upper City and the Lower City, as residents from the two areas drank from different sources. In the Upper City, the population relied on rainwater collected and stored in well-maintained private tanks. In the Lower City, people turned to more hazardous supplies. Most important were three aqueducts that delivered forty-five thousand cubic meters a day of severely polluted water. The Carmignano aqueduct illustrates the perils. It conveyed its supply of water forty-three kilometers from springs at Montesarchio. This water flowed across the countryside in an open canal, traversed the town of Acerra, and then flowed again across the fields to the gates of Naples. There it divided into branches that fed the two thousand cisterns of the Lower City. By lowering buckets into the tanks, which were usually located in courtyards, the inhabitants quenched their thirst and met domestic needs.

But every point from source to storage tank provided opportunities for contamination. Leaves, insects, and debris were blown into the uncovered canal. Peasants retted hemp in the stream, laundered garments, and disposed of household waste and the carcasses of farm animals. Meanwhile, mud and manure from the fields seeped through the porous stone or drained directly into the canal after a rain. Worse still, the aqueduct branch lines and the sewer mains ran parallel beneath the city. Since both were made of porous limestone, their contents slowly intermingled. Often the water arrived brown and thick with sludge that finally settled at the bottom of courtyard cisterns.

There, the contamination continued. Built of limestone like the aqueduct, the water tanks allowed surface water to percolate through the soil and into them. In the many buildings where cesspits were placed adjacent to the tanks, the bacterial exchange was immediate and overwhelming. Tenants, in addition, were not always careful when emptying their slops. Always poorly maintained, the cesspits frequently overflowed, allowing their contents to join the water already in the tanks. Even the final collection of water in household buckets had medical relevance. Rarely washed, they lowered their cargo of germs into the supply even more readily than they raised the waiting water, which was brown, malodorous, and teeming with surprises.

For all of these reasons, water from the cisterns of the Lower City had an evil reputation, which was augmented by its discouraging smell and visible contents. Ground-floor residents, for whom it was easiest to transport water, often reserved the cisterns for household purposes and fetched drinking water from more distant and less suspect public fountains. Residents of the upper stories, for whom carrying buckets of water was an onerous chore, made more extensive use of the cisterns and paid a heavier tribute in death and disease.

In the Neapolitan urban environment, which differed only in degree from the norm of major European cities in the nineteenth century before the sanitary revolution, cholera followed a typical pattern. After arriving in the warm spring or summer, it first broke out in scattered clusters of sporadic cholera that were often unnoticed by the authorities of a city where few inhabitants ever saw a doctor or a municipal health official. Thus the disease spread, perhaps for days, in neighborhood chains of transmission, until the vibrio gained access to the water supply. Then, as John Snow demonstrated in London, a major epidemic could erupt violently, claiming its victims preferentially among the impoverished and unsanitary boroughs of the Lower City. At the height of the most savage epidemics, such as those of 1837 and 1884, as many as five hundred people died of this one cause in a single day, invariably in the heat of summer when people drank most plentifully to quench their thirst.

Why the disease then slowly declined after a period of weeks and finally faded away is not fully understood and remains what Robert Koch termed one of the many "cholera mysteries." Certain factors are likely to have played a significant role, however. One is seasonality. Like other gastrointestinal pathogens, *V. cholerae* is not hardy in cool temperatures, and people drink less copiously in autumn than in summer. In addition, the vibrio can survive for an extended period in perspiration, so a change in the weather can make an appreciable difference in the number of cases. In the autumn people also consume less overripe fruit. And as we have stressed, the whole of the population is not equally vulnerable. The peak of an epidemic coincides with the affliction of those who—by reason of occupation, housing conditions, the presence of prior medical conditions, dietary habits, and physical constitution—are most at risk. Those who remain unaffected possess some degree of resistance. Having attacked people who are most vulnerable and most exposed to danger, cholera then behaves like a fire that has run short of fuel.

In addition to such "spontaneous" reasons for the decline of a cholera epidemic in a community, the health policies adopted by municipal and national authorities during the nineteenth century probably also played an appreciable role, sometimes positive but often not. In the first pandemic to reach the West in the 1830s, the first reaction of states was to impose coercive plague measures of self-defense by means of sanitary cordons and quarantine. The Italian government even resorted to such policies in 1884 in a desperate bid to seal its borders against microbes. Both in the 1830s and the 1880s, such a strategy proved counterproductive as it served only to spread both cholera and social disorder by inducing large-scale flight and devastating the economy. Antiplague measures were therefore abandoned in the face of cholera.

Other policies adopted during the course of cholera epidemics were more positive in their effect. Voluntary and municipal services, for instance, were instituted in some places to distribute assistance in the form of blankets, food, and medication. In addition, municipalities banned mass assemblages of people, regulated burials, organized campaigns to clean up streets and empty cesspits, lit sulfur bonfires, closed workshops that produced noxious odors, and opened isolation facilities for the stricken.

Which factors were decisive, and to what degree, is unknowable. Cholera epidemics, however, proved self-limiting, and after weeks of high drama and suffering, they ebbed and faded away. In Naples, the first cases of the epidemic of 1884 probably occurred in the second half of August; the disease flared up in full force at the beginning of September; and the last victim was interred on November 15. Besieging the city for two and a half months, the disease claimed approximately seven thousand lives and sickened fourteen thousand people.

The Terror of Cholera: Social and Class Tensions

The passage of cholera through a city heightened social tensions and frequently triggered violence and revolt. One reason was the striking inequality of the burden of suffering and death, which, as we have seen, fell overwhelmingly on the poor. To many, however, this apparent immunity of the prosperous seemed mysterious and suspect. Why were doctors, priests, and municipal authorities able to move with impunity while visiting patients and enforcing protective health regulations in neighborhoods where so many around them suffered and died?

Sound epidemiological reasons explain such invulnerability. Those who intervened from outside in the midst of the emergency did not in fact share the conditions of those who lived in the area. Cholera is not transmissible except in a very specific manner that rarely exposed visiting members of the municipal government, the medical profession, and the church to serious risk. Those officials did not live in the tenements they visited. They did not take meals or sleep in the sickroom, or drink from cisterns in the courtyards. In addition, they ate a diet that did not include fruit, vegetables, or seafood from local markets; and they washed their hands.

To already frightened residents, however, much seemed suspect, and rumors easily spread that the visitors were the agents of a diabolical plot to destroy the poor in a literal class war. The symptoms of cholera and its sudden onset, after all, suggested poisoning, as did strict burial regulations that excluded the community and caused the bodies of family members to be taken unceremoniously away while their limbs still stirred with apparent signs of life. Why, in any case, were outsiders suddenly interested in the health and habits of the poor when they had never been seen before? The concept was baffling. Furthermore, as the disease suddenly surged in a community, it seemed that the more frequently outsiders intervened, the more intense the epidemic became. Causation and correlation were conflated in the minds of those who suffered the most.

But in Naples in 1884, perhaps the most important lesson in the plausibility of popular conspiracy theories was the behavior of the municipal officials during the early weeks of the outbreak. The city sent teams of people who would now be termed health-care workers and disinfection squads, but they conducted themselves almost like an army in enemy territory. Making a great show of force, they arrived at tenements with weapons drawn, sometimes at night, and ordered distraught tenants to surrender critically ill relatives to undergo isolation and treatment at a distant hospital that was rumored to be a death house. The teams also compelled families to surrender precious linens and clothing to be burned and to allow disinfectors to fumigate and cleanse the premises. The behavior of city personnel was so unnecessarily high-handed that it was universally condemned by the press, which admired their zeal but criticized their methods. As the mayor later admitted, the procedures he had approved served above all to spread distrust and resistance.

It seemed implausible to residents that the city was acting for their benefit. The population drew its own conclusions, and it is hardly surprising that the press reported that the Lower City seethed with "class hatred" and that officials complained about what they saw as "the indescribable resistance opposed by the lower classes to the measures intended for their salvation."[7] An important and recurrent expression of popular resistance was ritualized public defiance, which took a variety of forms. One was dietary. The city had posted notices giving advice to avoid unripe or overripe fruit. It then attempted to enforce this admonition by a rigorous inspection of market stalls and the confiscation and destruction of suspect produce. Neapolitans responded by holding a series of public dietary demonstrations. Crowds gathered in front of city hall, where they set baskets of figs, melons, and assorted fruit on the ground. Demonstrators then proceeded to consume the forbidden fruit in enormous quantities while those who watched applauded and bet on which binger would eat the most. All the while they shouted defiant epithets at officialdom.

Alternatively, people obstructed municipal workers as they attempted to combat the epidemic by means of purifying sulfur bonfires. To the population the fires were abhorrent because the pungent fumes were irritating and they drove armies of sewer rats into the streets. A French volunteer during the emergency wrote:

> I shall never forget . . . the famous sulphur bonfires. In ordinary times fresh air is in short supply in the Lower City; with the arrival of cholera, it became impossible to breathe even on the highest point of Upper Naples. As soon as evening fell, sulphur was burned everywhere—in all the streets, lanes, and passages, in the middle of the public squares. How I hated those sulphur fumes! The sulphuric acid cauterized your nose and throat, burned your eyes, and desiccated your lungs.[8]

Citizens therefore gathered to prevent the workers from carrying out their assignments and to extinguish the flames as soon as they were lit.

Similarly, the residents of the Lower City flouted municipal regulations banning public assemblies. Processions of hundreds of penitents wearing crowns of thorns and bearing images of saints marched through the borough streets and refused police orders to disband. When traditional festival days of local saints arrived in September, the people observed the occasions in the usual manner marked by crowds and excesses of fruit and wine.

More conflictual was the opposition to the public health regulation that all cases of gastrointestinal disorder, that is, diarrhea, be reported to city hall. Here municipal and borough authorities met a resolute barrier of noncompliance. An epidemic of concealment confronted the epidemic of cholera. Everywhere in the Lower City families refused to inform city hall that a case of the disease had broken out in their midst, and they protected their possessions from the purifying flames. When physicians and their armed guards nonetheless appeared unasked, tenants denied them entry and barricaded themselves in their rooms. People preferred to die where they were rather than to entrust themselves to the care of strangers. On numerous occasions a crowd gathered and forced unlucky physicians to open their phials of laudanum and castor oil and to swallow the contents that residents thought were poisons.

Physical violence often followed. The arrival of physicians and armed escorts easily unleashed a fracas. Hostile crowds assembled to confront the outsiders, hurling epithets and accusing them of murder. Sometimes doctors and stretcher-bearers were handled roughly—pushed down stairs, beaten, or stoned. Local newspapers were filled during the emergency with accounts of "tumults," "mutinies," and "rebellions"; the population of the Lower City were called "beasts," "rabble," "idiot plebeians," and "mobs." The mayor noted that many guards and doctors were injured by missiles hurled by angry residents.

On some occasions unwanted medical interventions ignited full-scale rebellions. One example occurred in the borough of Mercato when the outbreak was just gaining force. On August 26 the municipal doctor Antonio Rubino and a police escort were sent to examine a stricken child in a notoriously unhealthy tenement. There a crowd armed with cobblestones greeted them and shouted, "Get them! Get them! They've come to kill us!"

Rubino and his guard were rescued by the arrival of military police summoned by a passing street-sweeper. By then the swelling crowd numbered several hundred, and the arrival of the troops only inflamed the situation. People who had come to assault the doctor now vented their fury by stoning the soldiers. A street battle unfolded, and the troops gained control only by drawing their weapons and opening fire.

An equally revealing and fortunately bloodless event happened a day later when a grocer named Cervinara lost his young son to cholera at the Conocchia Hospital. Grief-stricken and convinced that the child had been murdered, Cervinara and his brothers took up arms and forced their way onto the wards with the intention of killing the attending physician. The intervention of the chaplain saved the situation. He calmed the brothers and talked them into surrendering their weapons to the orderlies. Their action ended the first of several invasions of the cholera wards intended to liberate relatives who were still alive or to avenge the deaths of those who had perished.

Cholera hospitals rapidly became the scenes of full-scale riots as the epidemic reached its peak in September. By that time the Conocchia Hospital was filled to capacity and the city had opened two new facilities—the Piedigrotta and the Maddalena. These institutions were anathema to the neighborhoods in which they were located. Cholera hospitals were the most dreaded feature of the municipal campaign to contain the epidemic. Neapolitans regarded them as places of death and horror from which no one ever returned. At a time when miasmatic theories still prevailed, these institutions were considered a danger to everyone because of the lethal effluvia that emanated from them. It seemed clearly malevolent to place sources of poisonous gas in the midst of densely populated neighborhoods.

At the Piedigrotta, a former military hospital converted to treat infectious diseases, opening day on September 9 sparked an insurrection. A crowd gathered around the facility and forced the first stretcher-bearers to drop their charges on the ground and withdraw in haste. Encouraged by initial successes, the throng grew in size and purpose. Intent on frustrating the plans of city hall, residents hastily erected barricades and armed themselves with paving stones, sticks, and firearms. When the police arrived in force, a pitched battle ensued until, having sustained heavy injuries, they withdrew and were replaced by mounted police who attempted to force a passage and were also beaten back. Calm was restored only when parish priests intervened as mediators, bearing the welcome news that the people were victorious and the city would not force the issue by opening the hospital.

Identical scenes followed a week later with the opening of the Maddalena Hospital. Stretcher-bearers were met with violent opposition, and residents gathered at the windows of the upper stories and hurled tables, chairs, mattresses, and stones at the hapless porters below, some of whom were seriously injured. The crowd again barricaded the streets and forced the city to cancel the hospital's opening.

Meanwhile, mutinies broke out in both of the city's prisons, where the inmates concluded that they had been left to die. Detainees attacked the guards and seized the wardens in separate and uncoordinated but almost simultaneous riots. Prisoners attempted to force the gates and opened fire from the roofs. Only the full-scale intervention of the army in battle gear restored authority and order.

Such pervasive and violent opposition to the city's public health program prevented the plan from being implemented as every municipal initiative provoked a violent backlash and drove Naples to a state of near anarchy. The *Times* of London commented that the port was afflicted with something worse than cholera—"medieval ignorance and superstition."[9] In mid-September the secular mayor of the largest city in a kingdom that the papacy had condemned appealed to the cardinal archbishop of Naples for help. Through this unprecedented cooperation of church and state in Italy, a public health plan was slowly deployed—under new auspices and without further displays of force.

The city was also assisted by the arrival of a thousand volunteer health workers organized by a nongovernmental organization founded by private donors and philanthropists to combat the Neapolitan emergency. The White Cross functioned as a nineteenth-century precursor to the present-day Doctors without Borders. Its mission was that of emergency response—to treat patients in their homes and to provide assistance to their families. An international agency with no ties to Neapolitan or Italian authorities, it rapidly gained the trust of the population, calmed social tensions, and cared for a significant proportion of the victims of the 1884 calamity.

Many European cities struck with cholera suffered events similar to the tumults in Naples. Suspicions did not run solely in one direction, however; the "rebellious" century was a time when fear among the social and economic elite was widespread concerning the "dangerous classes." The poor and working classes had already demonstrated that they were dangerous politically and morally; now cholera revealed that they presented a medical peril as well.

It is thus possible to speculate that cholera formed part of the background to the century's two most egregious examples of extreme class repression, both of which occurred in Paris, where social tensions ran highest. These were the violent crushing of the 1848 revolution by the army commanded by General Louis-Eugène Cavaignac, and the "Bloody Week" in 1871, when Adolphe Thiers destroyed the Paris Commune in a frenzy of bloodshed. Cholera was by no means a direct cause of those reactionary excesses, but at the time, the disease had recently passed, it was deeply dreaded, and it had revealed yet another reason to fear the dangerous classes. Perhaps this revulsion played some part in the buildup of tensions that exploded in the repression carried out by Cavaignac and Thiers, who regarded the working classes of the French capital as enemies to be destroyed.

Sanitation and Cholera: Transforming Cities

Although antiplague measures were largely discarded early in the nineteenth century as defenses against cholera, the disease did give rise to enduring public health strategies. Of these the most influential and effective was the British sanitary movement theorized in the 1830s and progressively implemented in stages until World War I. Cholera, as we saw in Chapter 11, was not the only disease that contributed to the sanitary idea, but it was an important factor that Edwin Chadwick carefully pondered. The goal of preventing its return was a central consideration in the drafting of the *Sanitary Report* and the urban reform movement that followed.

Elsewhere in Europe the sanitary idea took root as well in the aftermath of cholera visitations, but sometimes in very different forms. Of these the most dramatic and far-reaching were the efforts not to retrofit the cities of Paris and Naples, but rather to rebuild them comprehensively.

The Haussmanization of Paris

Modern Paris is part of the lasting legacy of cholera. Under the Second Empire, Napoleon III charged the prefect of the Seine, Baron Georges-Eugène Haussmann, with the project of demolishing the squalid slums in the center of Paris in order to erect a salubrious modern city worthy of being an imperial capital. This vast project involved the digging of underground sewers while creating an aboveground network of broad boulevards, an array of gardens and parks, bridges, and straight lines of imposing modern buildings. The displaced population was exported, together with its urban problems, to the periphery.

Many of the motives for the reconstruction had nothing to do with disease. Haussmann acted in part to create—through the "Grands Travaux" that he oversaw—an enormous public works project that would provide employment for thousands of workers. He also intended to bring the long history of revolutions in the city to an end because his boulevards would facilitate the movement of troops across the city while also making the construction of barricades impossible. The new city would display the power and grandeur worthy of an imperial regime. As Haussmann makes clear in his memoirs, however, a crucial motivating force was the desire to stop the return of cholera. The disease was intolerable under the Second Empire because it suggested disorder, incivility, and Asia.

Rebuilding Naples: Risanamento

If considerations of public health were one component in the reconstruction of Paris, the renovation of Naples—a project known as Risanamento—was conceived entirely as a means to make the city cholera-proof. Naples was unique in being a city that was rebuilt for a single medical purpose and in accord with a specific medical philosophy—the miasmatic doctrine of the Bavarian hygienist Max von Pettenkofer.

By the time of the major 1884 epidemic of cholera in Naples, advances in sanitation and hygiene had nearly banished the disease from the industrial world. To suffer a return of the unwelcome visitor was therefore to admit conditions of sanitation, housing, diet, and income that were shaming. Embarrassment was all the greater because Naples in that year was the only major European city to be so severely scourged. For that reason Italy's great seaport attracted the full attention of the international press, which carefully documented the social conditions prevailing in the Lower City.

Even while the epidemic raged, King Umberto I visited the infamous tenements of the Lower City and called for a "surgical" intervention, or *sventramento*. In his lexicon, an excision was required that would remove the hovels where *V. cholerae* had nestled. Making good on the sovereign's promise, the Italian parliament passed a bill in 1885 for the city's Risanamento, or restoration to health. Then in 1889 pickaxes struck the first blows to initiate the project, and work progressed fitfully for almost thirty years—until 1918.

According to von Pettenkofer's miasmatic philosophy, the danger posed by *V. cholerae* to Naples was not the contamination of the city's drinking water in the manner John Snow and Robert Koch suggested. In Pettenkofer's system, disease did not began when vibrios entered the guts of the population, but when they reached the groundwater beneath the city. There, under the right conditions of temperature and humidity, they fermented and emitted poisonous effluvia that the population inhaled. Susceptible individuals among them succumbed, and half of them died.

Following this miasmatic analysis, the Neapolitan planners imagined a scheme that would make the city forever free of cholera. The first coefficient of health was to deliver clean water in abundance—not for the purpose of drinking, but to protect the soil beneath the city from contamination with sewage and to flush streets, drains, and sewers so that all excremental matter bearing the vibrios would be borne harmlessly away. Having supplied clean water, the engineers then built a great network of sewers and drains below the streets to dispose of the flow.

Having constructed waterworks belowground, the planners turned their attention aboveground. There, the first task was to demolish the overcrowded tenements, "thinning out" the population and making use of the rubble to raise the level of the streets from the first or ground floor to the second. The idea was to place a layer of insulation between the soil and the residents, trapping vapors belowground so that they would be unable to poison the atmosphere.

Finally, Risanamento involved bisecting the city with its narrow, winding lanes with a broad major boulevard, known as the Rettifilo, built in the direction of the prevailing wind. Crossing the boulevard at right angles the planners envisaged a series of wide streets with well-separated buildings. Together, these roads constituted the city's "bellows of health." They would admit air and sunlight into the heart of the city, drying up the soil and wafting all foul effluvia harmlessly away.

Thus protected both belowground and above, renovated Naples would be impervious to the threat of a return of Asiatic cholera and would present itself as the only major city ever rebuilt for the purpose of defeating a single infectious disease.

Rebuilding: An Assessment

Inevitably, the question arises of how effective rebuilding was in achieving its great anticholera objective. In the case of both Paris and Naples, the results were mixed. The public health defenses of Paris were tested by the arrival of the fifth pandemic in 1892, the year of a major outbreak at Hamburg, the only European city to have suffered severely during that year. Paris was more fortunate than Hamburg, but it did not escape as fully as other urban centers of industrial northern Europe. The rebuilt center of Paris experienced nothing more serious than a scattering of isolated cases. On the other hand, a major limit of Haussmann's project was that it gave little thought to the fate, sanitary or otherwise, of the workers who had been resettled in the *banlieue,* or suburbs. There, problems of overcrowding, substandard housing, and poverty persisted, and cholera returned in 1892 in an epidemic substantial enough to qualify the grand claims of the Second Empire. Cholera had been banished from the center but exported to the periphery.

In the case of Naples, a final judgment is more severe. The test occurred in 1911 during the sixth pandemic, when Risanamento revealed flaws of both conception and execution. In conception the rebuilding program was anachronistic. When the pickaxe struck its first blow, the medical philosophy on which the plan was based was no longer widely accepted by the international medical profession. By 1889 Pettenkofer's miasmatic philosophy had been largely supplanted by the germ theory propounded by Pasteur and Koch. Given the new understandings, the attention devoted by Risanamento to groundwater, soil, and effluvia on one hand and the lack of concern for the purity of drinking water on the other seemed out of date. Furthermore, the plan was flawed in execution because it was never completed. Over time significant portions of the funding for the project disappeared into mysterious channels, and *stralci,* or cuts, were made, undermining the ambitious original project.

In 1911, therefore, Naples resisted cholera less well than Paris had in 1892, succumbing to the last significant epidemic in western Europe. Unfortunately, the exact profile of the disaster is complicated by the fact that Italian authorities at both national and municipal levels adopted a policy of concealing and then denying the existence of the outbreak. To admit an outbreak of Asiatic cholera was too dishonorable a confession of backwardness. It also entailed acknowledging that the funds allocated for the rebuilding project had been partially misappropriated, thereby legitimizing claims of corruption advanced by opposition political parties—anarchists, republicans, and socialists. Furthermore, such acknowledgment would have exposed Italy to major economic losses through the closure of emigration routes to Italian migrants and through the loss of tourism to attend the festivities marking the fiftieth anniversary of Italian unification. Rather than risk any of these consequences, Prime Minister Giovanni Giolitti ordered a policy of silence and statistical fabrication. Cholera circulated once more, but silently.

A New Biotype: *Vibrio El Tor*

In 1905 a new cholera biotype was isolated in Egypt and named for the quarantine station—El Tor—where it was discovered in the guts of pilgrims returning from Mecca. Gradually, this biotype has taken over the ecological niche formerly occupied by the classical *V. cholerae* that caused six pandemics. The seventh pandemic, which began in 1961 and is ongoing, is caused by *Vibrio cholerae* O1 El Tor, which has transformed the nature of the disease. For the purpose of making the distinction between the two radically different biotypes, we refer to the nineteenth-century bacterium as *V. cholerae* and to the twentieth- and twenty-first-century pathogen responsible for the seventh pandemic as *V. El Tor.*

A number of features give the new strain an evolutionary advantage. Most important, it is much less virulent than classic *V. cholerae.* Comparisons of cases during the sixth pandemic of 1899–1923 (the last caused by classic cholera) and the seventh since 1961 (the first caused by *V. El Tor*) indicate that severe disease, or *cholera gravis*, affected 11 percent of patients in the former but just 2 percent in the latter. Its presenting symptoms are recognizably similar to the nineteenth-century disease, including rice-water stools, vomiting, and severe cramping. But the El Tor disease is less fulminant: its onset is less sudden, its symptoms are less florid, it gives rise to fewer complications, and its sufferers receive a far more favorable prognosis. Furthermore, transmission is enhanced by large numbers of asymptomatic carriers and by recovered patients who continue to shed bacteria for months.

In the expression of Myron Echenberg in his 2011 book *Africa in the Time of Cholera*, the disease had "changed its face," and to such an extent that at first it was considered "para-cholera" and not cholera at all.

In the Peruvian epidemic that erupted in 1991, epidemiologists concluded that three-fourths of infected people were asymptomatic. This high incidence of inapparent cases was a prominent characteristic of the disease, and it undermined efforts at containment. Quarantine and isolation were useless when most infectious people were apparently well and detectable only by laboratory examination of their stools. Cholera El Tor was accordingly dubbed the "iceberg disease," as diagnosed cases were only the tip of a much larger but hidden mass of cases too mild to be noticed.

All of these features dramatically increased the transmissibility of the disease by comparison with nineteenth-century cholera, whose victims were rapidly immobilized, perished in 50 percent of cases, and were no longer infectious after recovery. *Vibrio cholerae* was too lethal for the good of a microbe diffused by the oral-fecal route. There is no invariable tendency in nature for microbes to decrease in virulence, but the oral-fecal mode of transmission relies on the mobility of the host. It therefore places severe selective pressure on pathogens that immobilize and kill too rapidly. The mutations that yielded *V. El Tor* have enabled it to successfully fill the evolutionary niche once occupied by *V. cholerae*.

The Onset of the Seventh Cholera Pandemic

By 1935 *V. El Tor* had lodged itself in Indonesia as an endemic pathogen on Celebes (now Sulawesi) Island. Only in 1961, however, did it produce outbreaks of disease further afield. In that year El Tor cholera began a slow but relentless journey westward that ultimately circled the world. In the early 1960s epidemics of the strain broke out in China, Taiwan, Korea, Malaysia, East Pakistan (now Bangladesh), and all of Southeast Asia. By 1965 cholera reached the Volga Basin and the shores of the Caspian and Black Seas, particularly afflicting the Soviet city of Astrakhan. The disease then engulfed Iraq, Afghanistan, Iran, and Syria and spread across the Middle East, striking Turkey, Jordan, and Lebanon, and moving on to North Africa, where major outbreaks occurred in Egypt, Libya, and Tunisia.

In 1970 the World Health Organization (WHO) reported that twenty-seven nations had been stricken by *V. El Tor.* Accurate statistics of morbidity and mortality were impossible to produce, however, because many nations violated international health conventions by concealing the disease for fear of international stigma, an embargo on their exports, and a chilling impact on tourism. A widely accepted estimate, however, was that 3 to 5 million cases occurred, and tens of thousands of deaths.

During the first two decades of the El Tor pandemic, three momentous events occurred. One was positive—the development of oral rehydration therapy. In the early twentieth century Leonard Rogers had demonstrated the potential of rehydration as a life-saving procedure. Unfortunately, however, since intravenous administration of hypertonic saline solution requires trained medical personnel, who are scarce in impoverished nations, it was not available to most cholera victims. In 1963, medical scientists in Pakistan perfected the alternative methodology of administering an oral rehydration solution of salts and dextrose. The cost was minimal, and anyone could prepare and administer it, with only clean water needed. Physicians in Pakistan concluded that no cholera patient with a beating heart and the ability to mix the packet of salt and sugar in clean water would ever again die.

By contrast, two other early events in the pandemic were negative. One was that in 1971 cholera reached sub-Saharan Africa for the first time. The epidemic struck Guinea and then swept across West Africa, causing ten thousand cases and hundreds of deaths in that year alone. As public health officials immediately realized, social and economic conditions in the affected countries would enable El Tor cholera to thrive, to spread throughout the continent, and to become endemic. With considerable prescience, they also speculated that the spread of the disease across Africa posed a threat to Latin America—the second major negative event.

The epidemics unleashed by *V. El Tor* were less violent than those triggered by wandering *V. cholerae;* but since poverty, overcrowding, lack of sanitation, and unsafe water supplies persist, cholera outbreaks have become more frequent across the planet. Ever a "social disease," cholera now erupts solely in conditions of societal neglect, especially in the presence of political crises that WHO labels "complex emergencies."

Rita Colwell and the Discovery of the Environmental Bases of Cholera

The most striking development of the seventh pandemic is the transformation of cholera from an invader into an endemic disease. After 1970 in both Asia and Africa, El Tor cholera did not fade away. No longer "Asiatic" in any meaningful sense, *V. El Tor* became indigenous throughout the developing world.

Beginning in the late 1960s the mechanisms of El Tor cholera were gradually unraveled by the microbiologist Rita Colwell and her co-workers. Colwell discovered that attenuated virulence is not the only evolutionary advantage possessed by *V. El Tor.* An important factor is its ability to persist in the environment—unlike *V. cholerae,* whose sole reservoir is the human gut. The El Tor strain of cholera can inhabit aquatic environments, both brackish and freshwater—offshore waters, estuaries, rivers, lakes, and ponds. As Colwell demonstrated, *V. El Tor* spreads by what she called "human-environment-human transmission." This mode complements, but does not entirely supersede, the well-known oral-fecal pathway that is the sole transmission mode of classic cholera.

Ever present in the environment, cholera is permanently capable of "spilling over" into human populations. This occurs whenever social, climatic, and sanitary conditions enable the disease to resume its older mode of transmission—from human to human via fecally contaminated food and drink.

In temperate and tropical climates, when contaminated fecal matter gains access to aquatic environments, usually through the discharge of raw sewage into rivers and coastal waters, the cholera bacteria successfully multiply on and within the bodies of amoebae, zooplankton, and copepods with which it develops a commensal relationship. Aquatic flora and fauna do not succumb to disease, as cholera El Tor is an affliction only of human beings.

When plankton and algae, the preferred reservoirs of the vibrio, are eaten by bivalves or fish, and by aquatic birds such as gulls, cormorants, and herons, the bacteria are transmitted to them as alternative hosts and disseminators. Birds in particular carry the microbes to inland rivers and lakes, where they shed them from their feet, wings, and feces, enabling the vibrio to create reservoirs far from the coast. Duckweed, algae, and water hyacinth are freshwater plants that figure prominently in the bacterium's life cycle although the vibrio can also thrive in sediment.

Spillover to human populations depends on climatic factors, and on human activities. Warm weather, rising sea temperatures, and algae blooms promote the multiplication of bacteria, which can then be ingested by people when they eat fish and crustaceans or when they bathe in or drink contaminated water. Climate change, summer, and such weather determinants as El Niño, which circulates warm surface water and currents that bring nutrients in abundance, promote a profusion of vibrios, which are then available for human consumption.

Modern Cholera in Peru

The mechanisms of cholera El Tor were carefully documented during one of its most important outbreaks—the Latin American epidemic that began in 1991, as epidemiologists had anticipated when they witnessed the spread of the seventh pandemic to Africa during the 1970s. Peru was its epicenter. After ravaging that country, cholera then spread, less intensively, across the whole of South America. This epidemic was unexpected because the forebodings of the 1970s had been long forgotten amidst other pressing public health issues and the knowledge that cholera had not broken out in Latin America in a century.

Peruvian doctors were thus caught unprepared, knowing of cholera only what they remembered from medical school textbooks. On January 22, 1991, the Peruvian physician Walter Ortiz treated Daniel Caqui, a young and seriously ill farmworker. Caqui staggered into the Chancay hospital with puzzling symptoms that might have suggested food poisoning, a spider bite, or pneumonia. Ortiz, the first South American physician to confront *V. El Tor,* said: "I tried all the usual treatment and he just got worse. I had no idea what it was."[10] A week later, after thousands of Peruvians had fallen ill, the Ministry of Health announced the presence of epidemic cholera. Daniel Caqui was its first official victim.

Facing the outbreak, epidemiologists initially thought in traditional terms of an importation of cholera from abroad and focused on the possibility of the discharge of contaminated bilgewater from a ship that had sailed from Asia. No ship, however, was ever identified as having reintroduced cholera to the Americas. Indeed, epidemiology undermined all possibility that a single event was responsible. Instead of flaring up at a single place in Peru and then spreading in the manner of classical cholera, the El Tor epidemic started simultaneously at six different cities spread across six hundred miles of Peruvian coastline—Piura, Chiclayo, Trujillo, Chimbote, Chancay, and Lima.

Subsequent genomic investigation determined that the pathogen in Peru had originated in Africa and dated from the large-scale immigration of Africans during the 1970s. The presence of African isolates of *V. El Tor* in the Peruvian aquatic environment was also confirmed. Persistence in the environment rather than importation laid the bases for the Latin American microbial ordeal.

In January 1991 climatic conditions in Peruvian waters were ideal for the vibrio, as summer arrived and El Niño persisted. Vibrios multiply most readily in the summer heat, and people drink water, safe or unsafe, copiously to quench their thirst. Spillover occurred when fishermen marketed their catch ashore. Having completed the human-environment-human circuit, the vibrio reached town in the bodies of mollusks and fish. It then pursued the alternative mode of transmission from human to human. Raw fish and seafood were the Trojan horses in whose guts the vibrio was carried to the population of Lima and other cities. Especially suspect at the time was ceviche, a dish of raw fish marinated in lime juice, onion, chilies, and herbs. Ceviche was a staple of the diet because fish—available at a fraction of the cost of meat and chicken—was the chief affordable source of protein. The Ministry of Health banned the sale of ceviche and sought to persuade Peruvians to avoid it in their homes.

During the following year, three hundred thousand people in a population of 22 million contracted cholera even though the disease was well understood and it was both preventable and treatable by oral rehydration. The reason was that Peruvian cities presented conditions of poverty, density, and sanitary neglect that were reminiscent of nineteenth-century European cities such as London, Paris, and Naples that had been so favorable for "classic" *V. cholerae.* Indeed, *Caretas,* the Lima news magazine, explicitly invoked the comparison in March 1991 when fifteen hundred people were being stricken daily. "Terrifyingly," it explained, "health conditions in nineteenth-century London were similar to those of Lima today."[11]

The infrastructure of the Peruvian capital had been overwhelmed by urban growth. Lima's population had swollen to 7 million people, 4 million of whom lived in shantytowns lacking space, water, and sewerage. As the *New York Times* described conditions in the Lima slums, or *pueblos jóvenes,* habitations were "no more than rows of cardboard shacks, with no roofs, dirty floors and no electricity or running water. Children play in the streets alongside garbage. Dogs forage among piles of trash."[12] Two million residents had access to water only from the river in which raw sewage was dumped or from delivery trucks. They then stored it in filthy cisterns and tanks.

Unfortunately, poverty in Lima had grown exponentially throughout the 1980s while the Peruvian economy plummeted in what the press described as a "free fall." Industrial and agricultural production declined; unemployment and underemployment mushroomed to encompass 80 percent of the workforce; hyperinflation galloped at 400 percent a year; and real wages were halved. Malnutrition and hunger spread widely, and the incidence of infectious diseases such as tuberculosis and gastroenteritis spiked.

Conditions were also starkly worsened by ongoing guerrilla warfare, which was stoked by the economic crisis. The war, conducted amidst paroxysms of violence on both sides, was waged by the Peruvian Army against two well-organized but mutually antagonistic revolutionary forces—the Maoist party Shining Path (*Sendero Luminoso*), founded in 1980, and the pro-Russian movement named after the last Incan monarch, Túpac Amaru. As many as twenty thousand people died in the conflict, and the lack of security in the countryside decimated agricultural production and drove migration to cities that were already overcrowded.

Chairing a congressional investigation into the Latin American epidemic, Democratic representative Robert Torricelli concluded that there was no doubt about the causes that stoked the Peruvian medical crisis. Cholera, he explained, was the classic "social disease" of poverty. Throughout the 1980s, he noted, the World Bank, the International Monetary Fund, and the US Agency for International Development (USAID) had preached the doctrine of "structural adjustment," by which Third World countries were pressed to adopt laissez-faire economic policies. Thus prodded, Latin American governments favored unfettered market forces while slashing public spending on health, housing, and education. Cholera flourished when poverty rather than economic growth ensued. "The cholera epidemic," Torricelli explained,

> is the human face of the debt problem, a repudiation of the trickle-down theories that have guided development policy for more than a decade. . . .
>
> It occurs solely because of inadequate sanitation, because people lack access to clean water. It is a consequence of Latin America's lost decade.[13]

All that Torricelli omitted was that on the eve of the cholera explosion, Peruvian president Alberto Fujimori brought the economic crisis to a final crescendo by implementing a program of drastic fiscal austerity measures popularly known as "Fujishock." Even Hernando de Soto Polar, Fujimori's economic policy advisor, bemoaned the consequences, declaring starkly in February 1991, "This society is collapsing, without a doubt."[14]

Latin America in the 1990s also clearly revealed differences that distinguish "modern" from "classical" cholera. Since epidemics caused by *V. El Tor* are far less lethal than those epidemics caused by *V. cholerae*, they do not give rise to similar apocalyptic fears, riots, and mass flight. During the decades since the 1960s, Latin America, Africa, and Asia have not reproduced the social upheavals that marked the passage of Asiatic cholera through European cities in the nineteenth century.

On the other hand, modern cholera leaves a lasting and painful legacy. Most immediately and obviously, it produces widespread suffering as well as a not insignificant mortality. In addition, it overburdens fragile health-care systems, diverts scarce resources from other purposes, and decimates economies by its impact on trade, investment, tourism, employment, and public health. In Christian contexts, it also leaves an upsurge in religiosity. In Peru, for example, the passage of cholera through the country was widely experienced as a "divine curse" on the people. Journalists reported that it was popularly regarded as "something from the Bible," a modern equivalent of the plagues of Egypt. Indeed, the magazine *Quehacer* made the point clearly when it noted, "We are experiencing something that the Pharaoh of Egypt, with his seven plagues, would not envy."[15]

But the most enduring legacy is the threat of recurrence. Unlike classic cholera, the seventh pandemic has already lasted for more than half a century and, showing no signs of abatement, is capable of causing outbreaks wherever poverty, insecure water supplies, and unsanitary conditions persist. Indeed, in 2018 WHO reported that "every year there are roughly 1.3 to 4.0 million cases, and 21,000 to 143,000 deaths worldwide."[16]

Such an ongoing threat, concentrated in Asia, Africa, and Latin America, carries additional risk related to climate change. Absent technological interventions such as a successful vaccine, or major programs of social uplift and sanitary reform in vulnerable countries, global warming is certain to promote the survival of *V. El Tor* and to enhance the number and severity of outbreaks. Far from vanquished, modern cholera is what might be termed a perpetually reemerging disease.

Haiti since 2010

Cholera in Haiti, which began in 2010 and continues in mid-2018, is an important reminder of the threat of modern cholera. Nepalese soldiers, arriving on Hispaniola as part of the UN peacekeeping mission, brought cholera to the Caribbean. As the United Nations—fearful of the political and financial implications of admitting responsibility—belatedly acknowledged, some of the Nepalese contingent suffered from mild cases of the disease. Unfortunately, the peacekeeping force improperly disposed of their excrement by dumping it into the Arbonite River. This waterway serves the domestic, agricultural, and drinking needs of those inhabiting the valley through which it flows. In the words of observers, contaminating the Arbonite River in this manner was just as certain to trigger a conflagration as lighting a flame in a dry forest.

When the match was lit in October 2010, Haiti, the poorest nation in the Western Hemisphere, provided an ideal setting for cholera to take hold. Its citizens lacked all of the prerequisites of a healthy existence. They were hungry and malnourished; they already suffered disproportionately from a long list of diseases associated with poverty; the country had only an embryonic medical and sanitary infrastructure; and the people possessed no "herd immunity" to resist a disease that had been unknown in the hemisphere for more than a century.

A major earthquake in October 2010 prepared the way for the transmission of bacteria. Seismic shocks displaced people, shattered an antiquated system of water supply and sewerage, and damaged the fragile Haitian network of clinics and hospitals. Thus introduced and welcomed, *V. El Tor* multiplied exponentially in its new human hosts. Within weeks of the earthquake, every province of the country was infected, while 150,000 cases and 3,500 deaths had occurred.

The epidemic thereafter proved refractory to all efforts to contain it. For most of the population, safe water was unavailable, and clinics rapidly ran out of oral rehydration packets, intravenous rehydration solutions, and antibiotics. Meanwhile, the international public health community indulged in endless debate over the ethics and efficacy of an anticholera vaccine that in any case had not been approved for distribution and was in such short supply as to be useless. Lack of physical security also severely hampered relief efforts.

Natural disasters other than the earthquake also played a lethal role in the health crisis. Unusually heavy rains flooded water sources with runoff, and then in October 2016 Hurricane Matthew, a category 4 storm, made landfall and left a trail of devastation. It rendered hundreds of thousands of people homeless, wrecked health-care facilities, and ruined existing networks of sewerage and drainage. Meanwhile, as the epidemic held firm year after year, cholera faded from the international news, donor fatigue overcame philanthropic intentions, and charitable NGOs withdrew. By 2018, Haiti, a nation of nearly 10 million people, had already suffered a million cases by conservative estimates and ten thousand deaths. But these totals, dramatic as they are, are only provisional because the epidemic still prevails. The most recent available figures at the time of writing in spring 2018 indicate 249 new cases in February and 290 in March.[17]

But the Haitian epidemic also raises anxiety about the attenuated virulence of *V. El Tor.* Although a low level of virulence provides selective Darwinian advantage, there is no assurance that future epidemics will not again become more virulent. In Haiti the percentage of mild and asymptomatic cases was significantly lower than in previous outbreaks during the seventh pandemic, and physicians noted that the incidence of severe rehydration, or *cholera gravis*, was substantially higher. As recent studies indicate, a feature of the cholera genome is that it "readily undergoes change, with extensive genetic recombination . . . , resulting in what have been termed shifts and drifts in the genome sequence."[18] With such a high level of plasticity, *V. El Tor* is unpredictable, and one cannot assume that future epidemics will not be characterized by a renewed violence reminiscent of the nineteenth century.

CHAPTER 14

Tuberculosis in the Romantic Era of Consumption

Tuberculosis (TB), caused by the bacterium *Mycobacterium tuberculosis,* is one of the oldest of human afflictions. Since diseases of the genus *Mycobacterium* also affect animals, current thinking is that at an early date in human evolution a member of the genus crossed the species barrier from animals to humans, thereby launching its uninterrupted career as a disease of mankind. This idea of an early transmission of the disease to humanity is supported by the ever-growing evidence of tuberculosis among *Homo sapiens* during the Neolithic period (which started about 10,000 BCE), and there is robust evidence of its prevalence in ancient Egypt and Nubia, where its passage is marked by DNA, artwork, and the skeletal remains of mummies. From antiquity there is then an unbroken record of its constant presence wherever people have clustered in substantial settlements. It was a scourge of ancient Greece and Rome, of the Arab World, of the Far East, and of medieval and early modern Europe.

In the West, TB achieved a peak of ferocity during the eighteenth and nineteenth centuries in the wake of the Industrial Revolution and the massive urbanization that accompanied it. It is principally a respiratory disease spread through the air. The crowded slums of northwest Europe and North America during the "long nineteenth century" between 1789 and 1914 provided ideal conditions for a respiratory epidemic. These conditions included crowded tenements, sweatshops, poor ventilation, air thick with particulate matter, poor hygiene, and bodies whose powers of resistance were compromised by poverty, malnutrition, and preexisting diseases.

In such circumstances, morbidity and mortality from pulmonary tuberculosis rocketed. Contemporary physicians, relying above all on evidence from autopsies, believed that virtually everyone was exposed to the disease and had pulmonary lesions—even though in most cases the lesions were contained by the immune response of the body and did not progress to active disease. At its peak, tuberculosis may have infected 90 percent or more of the population in such industrializing countries as England, France, Germany, Belgium, the Netherlands, and the United States. For that reason, whole nations were said to be under the influence of the "frightful tubercularization of humanity."[1]

Inevitably, such widespread infection also led these industrializing nations to devastating levels of morbidity and mortality. In 1900, for example, TB was the leading cause of death in the United States. At that time approximately seventy-five thousand people died of TB per year—an annual mortality rate of 201 per 100,000 inhabitants. The close relationship between industrialization and tuberculosis led to the widespread nineteenth-century perception that tuberculosis was intrinsically a "disease of civilization." Throughout this period in the West, tuberculosis became, among all diseases, "the captain of all these men of death," as the writer and preacher John Bunyan dubbed it. It was consumption that took Mr. Badman to his grave as the wages of an ill-spent life.[2]

This upsurge in tuberculosis across the industrializing world over centuries raises the question of whether it is better understood as an endemic or epidemic disease. From the standpoint of any single generation, the disease was grimly present year after year with none of the characteristic ebbs and flows that mark the passage of epidemics through societies. Furthermore, even in the body of an individual patient, the progress of tuberculosis was capricious. It frequently lingered for decades rather than causing a sudden and dramatic illness in the manner of bubonic plague, influenza, yellow fever, or cholera. Indeed, patients often experienced tuberculosis as a chronic affliction, rearranging their lives in order to deal with their illness and devoting themselves entirely to the goal of regaining their shattered health.

From a longer time perspective, however, TB fits the epidemic model, albeit with allowance for its lengthy course. On the level of the individual patient, tuberculosis is now understood to be an infectious disease, but one that spreads unhurriedly from person to person, usually after a prolonged period of contact, and that characteristically attacks the body in a protracted siege. Similarly, in society as a whole, TB can be understood as an epidemic in slow motion, lasting even for centuries in a single place and then retreating gradually and mysteriously over the course of generations. Those who lived through the nineteenth-century peak of tubercularization regarded the disease as an epidemic. Indeed, they invoked the analogy of bubonic plague by calling it the "white plague."

Industrialization and the dislocations it brought in its train go a long way toward explaining the chronology of modern tuberculosis. In England, the "first industrial nation," the tuberculosis epidemic wave crested from the late 1700s through the 1830s and then began a slow but steady decline thereafter as conditions of housing, wages, diet, and sanitation began to improve. For countries that began to industrialize later, such as France, Germany, and Italy, the upsurge of tuberculosis began later, and the decline set in only after the turn of the twentieth century when industrial takeoff gave way to economic modernity.

Further confirming the relationship between pulmonary tuberculosis and economic transformation, the disease became a major killer above all in the industrial nations of northern Europe but played a less important part in the public health of countries that were still largely agricultural, such as Italy and Spain. Indeed, the geographical incidence of TB within countries followed the pattern of economic development. In Italy, for example, the affliction primarily affected the industrial cities of the north such as Milan and Turin while the peasant society of the south, where outdoor labor on the land prevailed, suffered far less. In the words of Maurice Fishbert, a leading American authority writing in 1922, "The frequency of infection goes hand-in-hand with civilization, or contact of primitive peoples with civilized humanity. . . . The only regions free from tuberculosis appear to be those inhabited by primitive peoples who have not come in contact with civilization."[3]

There are two eras in the modern history of tuberculosis—before and after Robert Koch's 1882 discovery of *Mycobacterium tuberculosis,* which successfully established the germ theory of disease and demonstrated that tuberculosis is contagious. These two periods are profoundly different in terms of the ways in which the disease was understood and experienced, in terms of the impact it had upon society, and in terms of the public health responses it inspired. This chapter is devoted to a consideration of the "romantic" era of "consumption" from 1750 to 1882; the next chapter will discuss modern tuberculosis. But first we consider the etiology and symptoms of the disease.

Etiology

Mycobacterium tuberculosis was discovered by Robert Koch in 1882, and for that reason it is also known as "Koch's bacillus" or the "tubercle bacillus." Humans contract it through four disparate and highly unequal modes of transmission. Of these, three are relatively infrequent and have played only an auxiliary role in the modern epidemic: (1) transplacental transmission from mother to fetus, (2) inoculation of bacteria through an abrasion or shared needles, and (3) ingestion of bacilli through infected milk or meat.

The fourth mode of transmission stands apart as overwhelmingly important—the inhalation of *M. tuberculosis* in contaminated airborne droplets expelled by sufferers as they cough, sneeze, or talk. After being inhaled, Koch's bacilli are implanted in the bronchioles and alveoli of the lungs—the tiny passageways and sacs at the end of the respiratory tract that receive air with every inhalation.

Drawn with the breath into the deepest tissues of the lungs, *M. tuberculosis* gives rise, in the vast majority of cases, to primary pulmonary infection. Infection, however, is not synonymous with disease. In most healthy people the body mounts an effective cell-mediated immune response. Activated macrophages—motile leukocytes recruited from the bloodstream to the site of the initial lesion—ingest the invading organisms, giving rise to a nodule, or granuloma. Meanwhile, other phagocytes known as epithelioid cells surround the granuloma, walling it off within a palisade. Within weeks or a few months, healing results in 90 percent of cases. The infection is contained and there is no further progression to active disease and the onset of symptoms. The individual is unaware of the lingering presence of the successfully contained lesion, but the bacilli within the granuloma are sequestered rather than destroyed. They remain alive and are capable of giving rise to disease at any later time when the immune system is severely weakened. Until reactivation or reinfection, the arrested disease is characterized by physicians as "abortive," and the patient is said to have "latent TB."

Within five years of the initial invasion, 10 percent of cases will progress to clinically significant disease. In those cases, the tubercle bacilli successfully counter the macrophages and evade the encompassing phagocytes. "Active tuberculosis" therefore ensues, with its train of symptoms and complications. This outcome is the consequence of several possible contingencies. The invading strain of *M. tuberculosis* may be highly virulent, or the number of invading bacteria may be overwhelming. Alternatively, the body's immune system may be compromised by the presence of a variety of immunosuppressive conditions, such as malnutrition, diabetes, HIV infection, alcoholism, malaria, drug abuse, silicosis, or chemotherapy. In these cases myriads of tubercle bacilli invade surrounding tissues in the lungs or seep into the lymph system or the bloodstream to be carried to any part of the body. This is a process of dissemination known as "miliary" tuberculosis—so named as an analogy with the broadcasting of millet seeds across a plowed field. This dispersion, which is almost invariably fatal, may occur immediately after a primary infection that is not successfully healed, or it may occur in the body of a previously abortive case whose immunity is subsequently impaired, allowing the reactivation of the original lesion. Alternatively, an abortive case can be reinfected since the original infection leaves no acquired immunity in its wake.

Symptoms and Stages

In miliary tuberculosis, Koch's invading bacilli can metastasize throughout the body from an initial lesion in the lungs. But since Mycobacteria can sometimes enter the body via the different pathways of ingestion, inoculation, or vertical transmission, they can establish the original infection in a locus other than the lungs. This multiplicity of possible sites of infection makes tuberculosis one of the most polymorphous of all diseases, capable of attacking any tissue or organ—skin, heart, central nervous system, meninges of the brain, intestines, bone marrow, joints, larynx, spleen, kidney, liver, thyroid, and genitals. Potentially, therefore, TB can appear in a wide array of guises, enabling it to mimic other diseases and making physical diagnosis notoriously difficult. Until the introduction of reliable diagnostic tests in the twentieth century, puzzled physicians often mistook TB for typhoid fever, bronchopneumonia, cholera, bronchitis, malaria, septicemia, meningitis, and other ailments.

Since pulmonary tuberculosis, or phthisis, is the most common and important form of the disease, however, we will limit ourselves to examining it alone in detail. A first feature of the disease is the extreme capriciousness—for reasons that are still mysterious—of its progress in the human body. At one extreme, its course can be fulminant, leading to death within months after the onset of symptoms. This was known in the nineteenth century as "hasty," "malignant," or "galloping" consumption.

Alternatively, tuberculosis can establish itself as a chronic illness progressing slowly over a period of decades punctuated by remissions and even apparent recoveries followed by mysterious relapses and the inexorable advance of the disease. Before antibiotics, some 80 percent of cases were estimated to end in death over a span of time that ranged from one to twenty years, although in every age some people recovered spontaneously or were apparently cured.

The range of possible experiences for sufferers is exemplified by the starkly contrasting medical careers of two major nineteenth-century British writers who had tuberculosis. One is John Keats, whose illness epitomizes the progression of galloping consumption. Keats fell ill in February 1820 and died exactly a year later at the age of twenty-six. This Romantic poet became an icon of the relationship of consumption to the arts and to genius. An entire century knew of his contracting the disease while tending his dying brother; of his desperate departure from England for the more health-inducing climate of Rome; of his separation from his beloved Fanny Brawne; and of his death and burial in Rome after a final period of poetic inspiration regarded by many as his most brilliant. His brief life—often compared to that of a meteor, comet, or candle that burns itself out—became a focal point for the social understanding of tuberculosis in the mid-nineteenth century. Consumption was a marker of high culture, sensibility, and genius, and Keats was its poster child.

At the other end of the tuberculosis spectrum was the Scot Robert Louis Stevenson. In contrast to Keats, Stevenson battled his illness for decades, moving in and out of health spas and sanatoria and traveling across the world in a never-ending attempt to restore his health. When he died at the more advanced age of forty-four in 1894, he had led a comparatively extended and productive life that was in all probability terminated not by tuberculosis but by a stroke.

Nineteenth-century physicians regarded the baffling progress of pulmonary tuberculosis as passing through three successive stages, although the passage from one to the next was often imperceptible, the symptoms overlapped, and—until the development of reliable diagnostic tools in the form of the sputum test, the tuberculin skin test, and the x-ray—diagnosis was uncertain until the disease was far advanced. Furthermore, it was not uncommon for patients to die before they reached the third stage or to recover even with an advanced case. As with all features of tuberculosis, the duration of each stage was unpredictable, and the advance from one stage to the next was not inevitable. Lung infections, moreover, could be bilateral or unilateral, although unilateral infections were more frequent, and the left lung was far more commonly involved than the right. In the pulmonologist Fishbert's words describing the variability of the disease:

> A continuous course from bad to worse till the patient dies, or with improvement till he recovers, is uncommon in chronic phthisis. It is characteristic of either the abortive form of phthisis, on the one hand, or of acute galloping phthisis, on the other. But the usual case of chronic phthisis pursues a discontinuous paroxysmal, I may say a capricious course, marked by periods of acute or subacute exacerbations of the symptoms, and periods of remission during which the patient is more or less free from the troublesome symptoms, or he may even feel comparatively well.[4]

The stages are therefore fluid and somewhat arbitrary and should be understood as suggestive rather than neatly demarcated.

First Stage: Incipient Tuberculosis

The onset of active tuberculosis is usually gradual, but a classic early symptom, widely mistaken for the common cold, is a dry, hacking cough that disturbs patients above all at bedtime and is accompanied by clearing of the throat. The cough may then disappear during the night, only to return at dawn—sometimes in racking paroxysms and to continue until the chest is cleared. If the cough is entirely unproductive, it is likely to persist in fits throughout the day, only to increase in violence toward evening, waking sufferers and leading to insomnia, exhaustion, and pain in the chest and throat. Not infrequently the cough is also emetic, leading to vomiting of various degrees of intensity. At this point in the disease, patients also present with shortness of breath after even moderate exercise, loss of weight, increasing pallor, declining productivity at work or study, diminished appetite, and swollen lymph nodes.

Persistent languor is a leading clinical manifestation of the disease from the onset and is often the first symptom to attract the attention of patients. The fatigue occurs without apparent explanation and is succinctly described by the early twentieth-century "tubercologist" Charles Minor: "The whole body seems filled with 'tiredness'; even to breathe is an effort, and if the patient lies down to rest, weariness seems to run through his limbs.

They ache with fatigue and . . . on waking, a hitherto active man will find himself not rested or refreshed and with no ambition for work."[5]

Second Stage: Moderately Advanced Tuberculosis

There is no clearly defined boundary between the first and second stages of TB. In the latter, however, the cough becomes more frequent and tormenting. In the lungs the tubercles multiply and progressively hollow out cavities that become inflamed. These repeatedly fill with phlegm that patients cough up in ever greater quantities of thick, greenish, and fetid mucous. The amount of sputum varies widely but can total as much as a pint in a single day. Clearing this phlegm provides relief for a time, but always the cough returns, ever more tormenting and frequent.

Other signs of the progress of the disease include "hectic fever"—a temperature that spikes once or twice daily at 103°F to 104°F and is accompanied by chills and profuse night sweats. These sweats are a troublesome feature of the disease, leaving sufferers drenched and exhausted and repeatedly disturbing sleep. The fever curve, however, is not uniform in all cases. The usual pattern is intermittent, but it can be continuous, which indicates an unfavorable prognosis. Alternatively, fever can be absent altogether.

In addition, patients experience tachycardia, with pulse counts of 120 beats per minute or more, even in the absence of fever; severe exhaustion that is not relieved by sleep; hoarseness that makes it impossible to speak above a whisper; pain in the joints and the chest; vertigo; headaches; respiratory distress after minor exertion; and, among women, menstrual disorders such as absence of periods (amenorrhea) and acutely painful cramping (dysmenorrhea). Typically, patients also cough up blood (hemoptysis), spitting out mouthfuls of bright red, frothy blood, especially after exertion or strong emotion. This was the most dreaded symptom of all for patients and their families because it clinched the diagnosis and presaged an evil outcome. It was said that in sanatoria the anxiety induced by the sight of one tuberculous patient hemorrhaging would cause his or her neighbors to bleed as well.

Finally, this stage of TB was thought to affect the personality. Moderately advanced consumptives typically grow unrealistically sanguine about their condition, euphoric about life, ambitious regarding the future, and libidinous. Taken together, these traits form an outlook that is diagnostically relevant—the unfounded optimism, or *spes phthisica*, of the so-called tuberculous personality.

Third Stage: Advanced Tuberculosis

Before the development of reliable tests, it was only in the third stage of TB that a dependable diagnosis could be made. By then patients bore all of the stigmata of the disease, which was easily recognized. The body was severely emaciated—a feature so universal that it led to the two terms by which the disease was most commonly known—"consumption" and its synonym "phthisis," which is derived from Greek roots meaning "wasting away." In advanced tuberculosis the body progressively loses flesh and muscle tone and is finally reduced to a mere skeleton. This wasting process is reinforced by other characteristic complications of the disease that make proper nutrition impossible—anorexia, diarrhea, and often the involvement of the larynx in the spreading infection so that swallowing becomes painful and almost impossible—a development known as dysphagia. Conversely, increasing weight in patients is one of the most favorable prognostic signs.

Typically, the entire physiognomy takes on what was known as the "consumptive look": sunken cheeks, hollow eyes, elongated neck, atrophied facial muscles, rounded shoulders, and a pallid facial color. At the same time the chest acquires a series of deformities. Stressed by every breath and the effort to force blood to flow through diseased pulmonary tissues, the lungs, trachea, and heart are displaced within the thoracic cavity, which therefore loses its symmetry. Meanwhile, the spaces between the ribs are deeply indented, the clavicles protrude prominently, and the scapulae assume a decidedly winged appearance. Edema from advancing circulatory failure renders the extremities swollen and chilly, and the heart is dilated, especially the deeply taxed right side that supplies blood to the lungs.

Disease processes in tuberculosis therefore leave a clearly visible imprint on the external shape and appearance of sufferers. At the same time those processes are distinctly audible when doctors listen to a patient's chest with a stethoscope—a procedure called "mediate auscultation"—a technique that René Laënnec developed into an art form after inventing the monaural stethoscope in 1816. He systematized his methodology in his famous text of 1819 *A Treatise on Mediate Auscultation*. Devising a lexicon to describe the sounds they heard in the diseased thoraxes of their patients, Laënnec and his followers wrote of crackling, dry crackling, clicking, gurgling, wheezing, and whining; of cavernous respiration; of cracked-pipkin sounds; of ronchi; and of cog-wheel breathing. But above all they described "rales" in all their varieties—fine, medium, or large; moist, coarse, bubbling, consonating, provoked, crepitant, subcrepitant, sibilant, and sonorous. Such subtle differences gave an advantage to the diagnostician with a musical training and an acute sense of pitch. As the Progressive era tuberculosis expert Francis Pottenger wrote of the range of tuberculous sounds he heard: "Sometimes they resemble the croak of a frog, the rattle of musketry, the purr of a cat, the whine of a puppy or the low note of a bass viola. The higher toned ones originate in the smaller passages and the lower ones in the larger passages."[6]

With lungs so audibly ravaged, the most distressing symptom of all was the oppressive, and unrelenting hunger for air now known as acute respiratory distress syndrome (ARDS). This "suffocative" syndrome occurs with particular frequency among patients with bilateral pulmonary tuberculosis and among those who suffer from an associated inflammation of the trachea or larynx. In the latter complication, the invasion of the trachea by tubercle bacilli gives rise to stenosis—a narrowing of the airway that physically impedes respiration. "It does not," wrote one physician in 1875, "consist so much in difficulty or obstruction of breathing as it does in the imperative demand for air."[7] Vividly describing the death of a patient suffering respiratory distress, Fishbert wrote:

> Suddenly, like a thunderbolt out of a clear sky, after a fit of coughing . . . , he is seized with a sharp agonizing pain in the chest, he feels as if "something has given way," or as if something cold is trickling down his side. He at once sits up in bed holding his hand over the affected side, gasping for breath. Acute, distressing dyspnea, cyanosis, a small rapid and feeble pulse, cold, clammy extremities and other phenomena of collapse make their appearance. The facial expression is that of profound agony, the eyes prominent, the lips livid, and the forehead clammy.[8]

Death, which is the outcome in more than 50 percent of untreated cases of tuberculosis, can result directly from such asphyxia as patients choke on the fluids accumulating in their chests. But death in advanced pulmonary tuberculosis also arises from other closely related causes. Of these the most prominent are heart failure and paroxysmal tachycardia, when the pulse rate rises to two hundred beats per minute; hemoptysis, when large pulmonary blood vessels are affected, causing massive hemorrhaging and/or an aneurysm that drowns sufferers in blood; and sudden pneumothorax, or lung collapse, leading to suffocation. The denouement of terminal phthisis is invariably gruesome; it usually involves asphyxiation; and it often occurs suddenly after a widely variable but always protracted period of exquisite suffering.

The Medical Theory of Consumption

The most comprehensive expression of the "romantic" interpretation of consumption was the work of Laënnec, the foremost authority of the era. Ultimately dying of tuberculosis himself in 1826, he spent his foreshortened life listening to the tuberculous chests of patients he attended on the wards of the Necker Hospital and correlating them with the lesions he observed postmortem on the autopsy table. His major contributions to understanding tuberculosis were multiple, including the development of mediate auscultation and the conviction, developed through tracing and carefully dissecting tubercles throughout the body, that the apparently disparate afflictions of different organs were manifestations of a single specific disease. By recognizing the primacy of the tubercle in the course of the disease, Laënnec united a series of pathologies that had been previously regarded as unrelated. Thus he became the "father" of the unitary theory of tuberculosis. Most influential of all, however, was his medical philosophy—the "essentialist theory," which dominated both professional and popular thinking about tuberculosis until Koch's transformative discovery. It was so named because it located the underlying cause of the disease in the nature—the "essence"—of the body itself.

Laënnec was an ardent anticontagionist. Consumption, he argued, was hereditary and developed from causes internal to the body and attributable to its "constitution." In his own jargon, this was the individual's innate *diathesis*—a Greek word meaning predisposition. This inherited or inborn defect was the ultimate cause that inclined individuals to the disease. It rendered them vulnerable to environmental influences—the proximate, or "exciting," causes—that precipitated actual illness in susceptible bodies. Consumption, therefore, was destiny: it was an individual's fate, inscribed in his or her body at birth. Its victims, therefore, were blameless, and—unlike sufferers of syphilis or smallpox—they were not dangerous to their fellows.

Believing in spontaneous generation, Laënnec thought that a body's diathesis could even lead directly to disease from internal causes. Primarily, however, he emphasized the role of immediate causes that were external and secondary. If diathesis was ultimately responsible for tuberculosis, the exciting causes were varied. Laënnec stressed the importance of emotional shocks and "sad passions" (*passions tristes*)—grief, disappointed hope, religious zealotry, and unrequited love—that depress the body's "animal energy." In the view of the medical profession, similar consequences could also ensue from abuse of the intellectual faculties through overwork and "unholy ambition."

Alternatively, physical factors could be decisive. Sententious doctors particularly lamented the role of sexual excess, which, they taught, produced a loss of essential bodily fluids, thereby weakening the organism and opening the way to disease. Above all, masturbation attracted suspicion in this regard. "Onanism," thundered the pulmonologist Addison Dutcher, was "self-pollution" and "the great bane of the human family . . . that rivals war, intemperance, pestilence and famine" in its baleful consequences for health.[9]

"Tubercologists," as some specialists called themselves, also underlined the nefarious role of liquor in the modern upsurge of the disease. The powerful temperance movements of the nineteenth century arose not only on the strength of moral and religious precepts, but also on the medical belief that alcohol fueled the white plague. Indeed, the English physician and novelist Benjamin Ward Richardson imagined a medical utopia in his 1876 book *Hygeia: A City of Health*, where both tuberculosis and the public houses that promoted it had been eradicated. The link between the disease and drink was given compelling statistical support by the French doctor Jacques Bertillon, who demonstrated that tuberculosis was an occupational hazard of publicans and mailmen, who were both well known to drink to "saturation."

Consumption, Class, and Gender

Since consumption was considered to be both hereditary and noncontagious, it did not give rise to stigmatization. The sufferers were not at fault, and they were not thought to spread the malady. Furthermore, phthisis was not associated with a particular social class or ethnic group. Although in modern epidemiological terms tuberculosis is well known to afflict the poor, who disproportionately live and work in crowded and unsanitary conditions, throughout the eighteenth and nineteenth centuries it also claimed large numbers of victims among the social, cultural, and economic elite. The list of its nineteenth-century victims includes not only Keats and Stevenson, but other luminaries such as Friedrich Schiller, Anton Chekov, the Brontë sisters, Edgar Allan Poe, Honoré de Balzac, Frédéric Chopin, Percy Bysshe Shelley, Eugène Delacroix, and Niccolò Paganini. The mode of transmission played a critical part in the social construction of consumption because a disease disseminated through the air is inevitably "democratic" in its reach and more likely to touch the lives of the privileged than, for example, social diseases such as cholera and typhoid, which are transmitted by the oral-fecal route and are visibly linked with poverty.

Indeed, one of the ironies of a disease that was painful, deforming, and lethal was that it was popularly deemed not only to affect men and women of high social standing, ability, and refinement, but also to augment and elevate their beauty, genius, and sex appeal. This understanding was highly gendered, and it had widely divergent implications for both women and men. For women, pulmonary consumption promoted a new anemic ideal of female beauty in its own image—thin, pale, delicate, elongated, and almost diaphanous. Henri de Toulouse-Lautrec perfectly captured this tuberculous somatic ideal in his 1887 painting *Poudre de riz* ("Rice Powder," also known as *Young Woman at a Table*). Toulouse-Lautrec depicts a slender woman—probably his mistress Suzanne Valadon—at her toilette before a jar of rice powder intended to whiten her face to a consumptive pallor according to the fashion of the age (fig. 14.1). Similarly, the Pre-Raphaelites specifically chose models who

Figure 14.1. Henri de Toulouse-Lautrec, *Young Woman at a Table, "Poudre de riz"* (1887). Some nineteenth-century women applied perfumed rice powder to their faces to mimic a fashionable consumptive pallor. Van Gogh Museum, Amsterdam (Van Gogh Foundation).

suffered from the disease, such as Elizabeth Siddal, who became Dante Gabriele Rossetti's favorite model and then his wife, and Jane Burden, whom William Morris frequently painted and later married. This tubercular aesthetic persisted into the early twentieth century in the gossamer frames, pale faces, and swanlike necks of the women in Amedeo Modigliani's portraits.

So too in his ballad "La Belle Dame sans Merci" (beautiful lady with no mercy), John Keats (1795–1821) portrays consumption as the ultimate *femme fatale*—enchanting, "full beautiful," and irresistible. Nor does the poet resist, but yields to her charms and "sweet moan." Only after "she lullèd me asleep" does a dream reveal that he has fallen irretrievably—"Ah! woe betide!"—into her "thrall," together with a vast company of "pale kings and princes too, / Pale warriors, death-pale were they all."[10]

Opera and literature of the period also extolled the haunting, ethereal allure of such tuberculous heroines as Mimì, the seamstress of Puccini's opera *La Bohème,* who was an adaptation of Musette in Henri Murger's story *La vie de bohème;* Marguerite Gautier, the courtesan in the novel *La dame aux camélias* by Alexandre Dumas; and Violetta, Marguerite's counterpart whom Giuseppe Verdi transposed to the operatic stage in *La traviata*. It is probable that anorexia nervosa, which became an established term with an 1873 paper by the physician William Gull, increased among young women as a result of tuberculosis. They measured themselves against an ideal of feminine beauty "advertised" in poems, paintings, plays, novels, and operas about ethereal tuberculous heroines.

Fashion also powerfully promoted this ideal of feminine beauty modeled on the emaciated figure of the consumptive. This trend—implemented in diet, comportment, dress, and beauty practices—is what historian Carolyn Day aptly terms "the consumptive chic," in which the affectation of illness became a mark of social standing while corpulence and strength were vulgar. Day comments, "Since health was out of style, if an illness did not occur naturally, a number of ladies affected the trappings of sickness." Tuberculosis, in other words, spawned a "rage for illness."[11] The medical doctrine of essentialism, asserting that consumption was hereditary, increased the appeal of appearing to suffer the disease as it proclaimed for all to see that a woman could claim a place in the upper ranks of society.

Fashion, of course, is suggestive rather than literal. It was by no means contradictory to hope to avoid the disease while making an outward show of an ennobling predisposition to it. To approach the necessary standard, healthy women of means cultivated a gaunt, fragile figure suggestive of an inner state of sensibility, intelligence, and refinement. To that end, they avoided exercise, physical exertion, and zeal at the table while learning to lisp and practicing a tottering gait—in imitation of the poor appetite and inactivity of the tuberculous. Fashionable women were also careful with their toilette. Many followed the advice of society magazines that a touch of belladonna on the eyelids would dilate the pupils in order to give the eyes the wide-eyed consumptive look that was a mark of beauty, and that a hint of elderberry juice rubbed into the eyelids to darken them would draw attention to the eyes and skillfully frame them. Meanwhile, rice powder, as we have seen, could mimic the translucent pallor of the skin that was de rigueur; and a thin application of red paint to the lips would replicate the effect of the so-called hectic or tuberculous fever while emphasizing the pale appearance of the cheeks.

Elegant women then turned to the requirements of dress. Physicians wrote that, among its visible diagnostic signs, consumption produced general wasting, a flat chest and abdomen, a thin waist, a stooped neck, and prominent shoulder blades. Thus, women's clothing was enlisted to simulate these symptoms. The ever lower backline of gowns compressed the shoulders and exposed the scapulae, which assumed "the form of wings . . . just raised from the body and about to expand for flight."[12] Some gowns exhibited an artificial, slightly raised hump in the back to give the wearer the appearance of stooping forward. Heavily laced and elongated corsets with broad stays reshaped the torso. And a V-shaped bodice allied with wide skirts and enormous sleeves enhanced the smallness of the waist.

There is no more powerful illustration of the beautiful and uplifting death of a romantic literary heroine than that of Little Eva in Harriet Beecher Stowe's 1852 novel *Uncle Tom's Cabin,* which is a narrative not only of slavery, but also of disease. Little Eva is terminally afflicted with consumption, but she dies a thoroughly edifying and lovely death that is at polar opposites from the terrifying suffocation described by the modern physician Fishbert. Little Eva's death made cultural sense to a reader of the time—but in a way that would puzzle a present-day clinician. Repeatedly stressing Little Eva's beauty at the climactic moment of her brief life, Stowe writes:

> For so bright and placid was the farewell voyage of the little spirit,—by such sweet and fragrant breezes was the small bark borne towards the heavenly shores—that it was impossible to realize that it was death that was approaching. The child felt no pain—only a tranquil, soft weakness, daily and almost insensibly increasing; and she was so beautiful, so loving, so trustful,

> so happy, that one could not resist the soothing influence of that air of innocence and peace which seemed to breathe around her. St. Clare [her father] found a strange calm coming over him. It was not hope,—that was impossible; it was not resignation; it was only a calm resting in the present, which seemed so beautiful that he wished to think of no future. It was like that hush of spirit which we feel amid the bright, mild woods of autumn, when the bright hectic flush is on the trees, and the last lingering flowers by the brook; and we joy in it all the more, because we know that soon, very soon, it will all pass away.[13]

If women were held to standards of beauty epitomized by fictive heroines, the culturally expected impact of tuberculosis on men was to bring their creative powers to new heights. In this regard, as we have seen, Keats served as the ideal type of the male artist whose full creativity was said to have reached its fruition only in the crucible of his fevered final year in Rome. There, fever consumed his body, allowing his mind and soul to soar to new heights that, free of disease, they would never have attained. An apocryphal story about the French novelist Victor Hugo conveys the point: it is said that Hugo's friends often chided him for not having contracted consumption; had he done so, they suggested, he would have become an even greater writer.

For the same reason, as the tide of the tuberculosis epidemic in the United States ebbed in the early twentieth century, the Brooklyn physician Arthur C. Jacobson fretted in his work *Tuberculosis and the Creative Mind* (1909) that the quality of American literature would inexorably decline. The physical suffering induced by phthisis, he wrote, had long been compensated by the artistic boon it conferred—its ability to serve as "a means of quickening genius, a fact wherefrom have flowed benefits that concern the whole world of intellect." Jacobson explained the unexcelled intellectual productivity of a galaxy of tuberculous men as based on the specific clinical trait of the *spes phthisica,* which conveyed an inner sense of unbounded mental power and optimism. The lives of consumptives, he stated, "are shortened physically, but quickened psychically in a ratio inversely as the shortening." This is, he wrote, a "divine compensation."[14]

The only stigmatizing aspect of the essentialist theory of consumption arose in the marriage market. Since the disease was deemed to be hereditary, it was a wise precaution for physicians to discourage consumptives from marrying and possibly passing on their tainted constitutions to the next generation. The most dangerous possibility of all was the intermarriage of two tainted families. Addison Dutcher, for example, regarded it as his duty in the 1870s to explain the implications of the diathesis to his consumptive patients. His hope was thereby to avoid "perpetuating a malady that blights the fairest prospects of the race and consigns so many to a premature grave."[15] By the eve of Koch's discovery, essentialism thus led some to envisage a public policy that regulated marriage and fostered eugenics.

Consumption and Race

The stress on the relationship between consumption and mental acuity was inflected with race as well as class and gender. The idea that consumption was a disease of civilization supported two major tenets of the racialized medicine of the era. The first was that the various races of humankind were so biologically distinct that they succumbed to separate diseases. Since consumption was a marker of intellectual superiority, it was understood to affect only the white race. Here was one meaning of the designations of TB as the "white plague" and, even more tellingly, the "white man's scourge." In the United States, the prevailing wisdom was that African Americans contracted a different disease. The disinclination even to give it a name speaks volumes with regard to the prevailing racial hierarchy and the lack of access to medical care by people of color. Consumption as a diagnosis was the ennobling monopoly of whites. Dr. Samuel Cartwright (1793–1863), a prominent chest specialist who practiced in New Orleans and Jackson, Mississippi, staunchly defended Negro slavery as an institution ordained by God. Enunciating the prevailing view in the antebellum South, he pointed out that

> Negroes are sometimes, though rarely, afflicted with . . . phthisis. . . . Phthisis is, par excellence, a disease of the sanguineous temperament, fair complexion, red or flaxen hair, blue eyes, large blood vessels, and a bony encasement too small to admit the full and free expansion of the lungs. . . . Phthisis is a disease of the master race and not of the slave race—it is the bane of the master race of men, known by an active haematosis, by the brain receiving a larger quantity of aerated blood than it is entitled to, by the strong development of the circulatory system, by the energy of intellect, vivid imagination and an indomitable will and love of freedom. The Negro constitution . . . , being the opposite of all this, is not subject to phthisis.[16]

In strict logic, the second tenet relating race and consumption in the United States contradicted the idea of the disease as exclusively an ailment of whites. This second view agreed that blacks were free from tuberculosis, but it pointed to social rather than biological explanations. The prevailing racialized medicine buttressed the slaveholders' cause by its argument that blacks had long been sheltered by slavery from exposure to tuberculosis and from the stresses of modern life that were its exciting causes. African Americans' alleged freedom from tuberculosis in the antebellum South, in other words, was not biologically determined. It was instead a testament to the benevolence of the "peculiar institution," which met the needs of an inferior people. Against abolitionism, this argument held that to destroy slavery was to destroy the Negro race, which would be ravaged by tuberculosis without the protection of the white man.

Romanticism

An aspect of the cultural resonance of consumption was its contribution to the sensibilities, metaphors, and iconography of Romanticism. Not every major epidemic disease has a significant effect on culture and the arts. Influenza and Asiatic cholera, for example, had a limited cultural impact while bubonic plague had a transformative role, as we have seen. Tuberculosis is a further example of a disease that played a salient artistic role, but one that was very different from that of the plague. In the European imagination, pestilence raised above all the specter of sudden and agonizing mass death.

With tuberculosis, however, there was no possibility that victims would be caught unaware and lack time to put their affairs and their souls in order. Consumption therefore evoked something different from *mors repentina* and its terror. It evoked the idea of sadness, of lives foreshortened, when the most artistic and creative individuals were cut off in their prime. Consumption, unlike plague, was uplifting because it pointed to the spiritual realm, warning sufferers of death in ample time to work out their relationships with God and their communities. Keats himself expressed the melancholy of abbreviated life in a celebrated sonnet:

When I have fears that I may cease to be
  Before my pen has glean'd my teeming brain,
Before high piled books, in charactery,
  Hold like rich garners the full-ripen'd grain;

When I behold, upon the night's starr'd face,
    Huge cloudy symbols of a high romance,
And think that I may never live to trace
    Their shadows, with the magic hand of chance;
And when I feel, fair creature of an hour,
    That I shall never look upon thee more,
Never have relish in the faery power
    Of unreflecting love; then on the shore
Of the wide world I stand alone, and think
    Till love and fame to nothingness do sink.[17]

Many of the leading themes of Romantic literature express sensibilities close to a consumptive vision of the world: a deep awareness of the transience of youth; a pervasive sense of sorrow; nostalgic longing for the past and what has been lost; the search for the sublime and the transcendent; the cult of genius and the heroic individual; and a preoccupation with the inner self and its spiritual state—Laënnec's "essence"—cleansed of what is material, gross, and corrupt. Autumn is a recurring and revealing trope—autumn, that is, no longer as a time of harvest and bounty but as the season of falling leaves and flowers wilting as early death sets in.

In his elegy "Adonais," Shelley therefore mourned Keats as a "pale flower"—"the bloom whose petals, nipp'd before they blew / Died on the promise of the fruit." With this view of the sad esthetics of consumption, Romantic artists placed consumptives at the center of their work. In turn, the Romantic elevation of sublime imagination over sordid fact was a feature in the social construction of consumption in defiance of what appears to the empirical modern mind as its abhorrent and degrading symptomatology.

The Impact of Consumption on Society

Comparing the two great infectious diseases of different centuries—bubonic plague and tuberculosis—clearly illustrates the point that epidemic diseases are not simply interchangeable causes of death. On the contrary, each generates its own distinctive societal responses. From its first visitation in western Europe in 1437 until its last significant outbreaks at Marseille during 1720–1722 and at Messina in 1743, plague, as we have seen, was synonymous with mass hysteria, scapegoating, flight, economic collapse, and social disorder.

Tuberculosis, by contrast, led to none of these phenomena. A disease that was ever present and glacially slow in its pace, and that never gave rise to a sudden spike in mortality, consumption never produced the terror associated with a sudden invasion from outside. Flight and the imposition of quarantine in any case were pointless since the victims were deemed to be harmless, and the disease itself was the result of one's fate—the product of an inner hereditary taint. In a city beset by the white plague rather than the Black Death, the authorities therefore remained at their posts, trade and commerce were unaffected, and public life continued in its quotidian routines. Consumption produced profound societal effects, but not a reenactment of the high drama of a city besieged by plague. Consumption gave rise to private dread, not public terror. In the words of the historian Katherine Ott, "The cumulative morbidity and mortality figures for consumption were higher than for any epidemic, but since civic affairs remained intact, few people became alarmed."[18]

Consumption also failed to alarm because of simple comparison. Death from tuberculosis was "beautiful," if not absolutely, at least relative to other contemporary epidemic diseases. Pulmonary tuberculosis did not hideously disfigure its victims in the manner of smallpox; and its symptoms, though painful, were less degrading than the profuse diarrhea of Asiatic cholera. Lungs are more ethereal than bowels.

Invalidism

Of all the effects of consumption on society, the most visible and pervasive was invalidism. At a stage in what Abdel Omran in the 1970s famously dubbed the "epidemiological transition," or "health transition," when chronic diseases were exceptional and infectious diseases the rule, protracted illnesses were also unusual, apart from tuberculosis. Consumption thus set a new standard for long-term maladies as the disease became the occupation of a lifetime. After diagnosis, the rest of a patient's future suddenly became unknowable. Sufferers had to confront painful decisions about career, marriage, and family. Consumptives put normal responsibilities, friendships, and aspirations on hold in order to undertake in their place the new and all-consuming task of either restoring health or learning to accept an untimely death.

Highly suggestive of the nature of life as a consumptive are the plays of Anton Chekov (1860–1904). The Russian writer, who was a physician as well as a playwright, fell victim to phthisis. After falling ill, he abandoned his theatrical career in Moscow and departed for the Crimea in an ultimately vain attempt to recover his health on the milder shores of the Black Sea. Chekov wrote all five of his most famous plays—*Ivanov* (1887–1889), *The Seagull* (1896), *Uncle Vanya* (1897), *The Three Sisters* (1901), and *The Cherry Orchard* (1904)—during the period of his illness. Only the first, *Ivanov,* deals explicitly with consumption; but all of them have consumptive invalidism as their hidden but unmentioned autobiographical theme. It is no coincidence that all five portray protagonists who, like consumptives, are unable to act, as they find themselves trapped, waiting and waiting for the outcome of events beyond their control.

In *The Cherry Orchard* of 1904, the year of his death, Chekov examines the fate of characters whose lives are all mysteriously and unalterably paralyzed. Peter Trofimov is the eternal student incapable of completing his degree; Ermolai Lopakhin is a merchant unable to propose to his beloved; Lubov Ranevsky is a landlady powerless to save her estate from destruction as her resources are devoured by a ne'er-do-well lover; and Boris Simeonov-Pischin is a landowner unwilling to implement a plan to preserve his property from being consumed by debt. Speaking both for them all and for the invalid Chekhov, Simeonov-Pischin already declares in Act I: "I thought all was lost once and I was done in, but then they built a railway across my land . . . and paid me. Something else will come up if not today, tomorrow. Dashenka will win two hundred thousand. She's got a lottery ticket."[19]

Chekhov's career as a patient is typical of middle- and upper-class consumptives during the "long nineteenth century." Consumption set in motion one of the great population movements of the period—a movement known as "taking the cure." A change of environment as a therapeutic intervention had an honored place in the medical armamentarium since Hippocrates's famous work *On Airs, Waters, and Places.* Thus a consistent recommendation of physicians in dealing with consumption was "climate therapy"—the advice to travel to a salubrious destination.

Medical opinions differed with regard to the particular climate to be endorsed and the reasoning underlying the recommendation. Frequently, physicians urged their tuberculous patients to travel to the mountains, where breathing becomes deeper with a longer inspiration and more complete expiration, where the thinner air allows the sun's rays to penetrate more easily to tan the skin and quicken the circulation, and where "the glorious sunshine and the grandeur of the mountain scenery infuse new hope and courage."[20] Mountain air was also said to stimulate the appetite and combat the consumptive's fearful emaciation. Other doctors spoke in favor of warm, dry weather at sea level; and some embraced the idea of a mild but uniform climate. One current of opinion suggested that climate change constituted a specific remedy for consumption; another held that it merely provided an adjuvant remedy. Physicians also varied the suggested destination in accordance with the stage of the disease and the age of the patient. For a number of doctors it was not the destination but the journey that possessed curative powers. In their view, an ocean voyage would provide bracing "pulmonary hyper-aeration" while seasickness would purge the body of corrupted humors. Even "long-continued" horseback riding was deemed beneficial.

The underlying idea was that epidemics are by nature "inflammatory" or "sthenic" so that the proper remedy was a regimen of air and diet that was a depleting "counterirritant." Thus it was that European consumptives with means journeyed to the Alps, the French Riviera, Italy, and the Crimea. Keats and Shelley went to Rome, Tobias Smollett to Nice, Elizabeth Barrett and Robert Browning to Florence, Chopin to Majorca, Paul Ehrlich to Egypt, and Chekov to the Crimea. This migration in search of the "cure" was stimulated by a flood of medical books, by rumor and anecdote, by "chain migration," and by brochures prepared by such interested parties as railroad and steamship companies.

In the United States the migration of consumptives was so intense that it produced a new, medical version of the "frontier thesis" in US history, especially after the completion of the transcontinental railroad in the 1870s transformed the current of health seekers—"interstate migrants"—into a flood. Entire communities were founded by and for the tuberculous, including most notably Colorado Springs and Pasadena. Southern California, a Mecca for patients hopeful of a cure, came to be known as "Nature's Great Sanitarium" and the "Land of New Lungs."

One of the most famous consumptives to "go West" was the folkloric hero of the gunfight at OK Corral and friend of the lawman Wyatt Earp—John Henry "Doc" Holliday. Originally a dentist in Georgia, Holliday moved to Dodge City, Kansas, and then Tombstone, Arizona, after his persistent cough was diagnosed as pulmonary tuberculosis. The move was an attempt—ultimately unsuccessful—to preserve his life. Once settled in the Southwest, he gave up dentistry in favor of gambling and gunfighting because potential patients were repelled by his cough. Holliday succumbed to TB, together with the alcohol and laudanum with which he self-medicated, at the age of thirty-six in 1887.

For consumptives of more limited means, alternatives to "taking the cure" could be found closer to hand. One of these was "inhalation therapy," which was devised as a means of bringing a vital element from the distant climate to the physician's office or the home. Inhalers, atomizers, and vaporizers enabled doctors to apply a spray, fume, or stream to the nose, lungs, and throat in order to treat the disease at its seat. Just as the destination recommended in climate therapy varied, so too did the active element to be inhaled—creosote, chloroform, iodine, turpentine, carbolic acid, and a variety of mercurials. A different but more exotic substitute for a journey was "altitude therapy," achieved by going aloft in a basket beneath a hot air balloon and thereby receiving many of the benefits of mountain air while avoiding the expense and inconvenience of travel.

A facetious conjecture is that the rigor of the treatments available at home may have been a factor that motivated patients to travel. Inhalation therapy with acidic sprays was painful and, apart from hope, provided little relief. Other standard nineteenth-century home remedies included the holistic humoral strategies of purging by means of venesection, cupping, and emetics; of adopting an antiphlogistic, or anti-inflammatory, diet based on vegetables, fish, and cold soups while severely limiting animal flesh because of its stimulating influence; and of reducing exercise and stress to a minimum. Creosote, hydrochloric acid, ox-gall, and pepsin were regarded as potent stimulants to the appetite and were administered internally in an effort to promote weight gain and combat the flaccid muscle tone of patients. Even as humoral theory lost its intellectual hold, physicians had few options in practice to replace time-honored remedies. Meanwhile, doctors also adopted symptomatic approaches, treating pain with morphine and opium; fever with quinine, strychnia, and atropine; and hemoptysis with opium or buglewood tea.

CHAPTER 15

Tuberculosis in the Unromantic Era of Contagion

The medical and social meanings of phthisis were transformed between the 1860s and the turn of the twentieth century in what one might term a passage from the "age of consumption" to the "age of tuberculosis." Consumption was a romantic and glamorous hereditary affliction of beautiful and creative elites; tuberculosis was a vile, contagious, and stigmatizing disease of the poor and the filthy. As we have seen, Little Eva's death in *Uncle Tom's Cabin* perfectly expressed the conception of consumption as an ineffably lovely and ethereal disease that ennobled patients and charmed visitors to the sickroom. By contrast, André Gide took a decidedly positivist view of tuberculosis as a painful, disgusting, and dangerous affliction. In the 1902 novel *The Immoralist*, the tuberculous protagonist Michel found his condition abhorrent. In a counterpoint to the description of Little Eva's uplifting demise, Michel looked back without a trace of romanticism on his own distressing career as a patient. In words that neither Stowe's Eva nor Puccini's Mimì ever would have uttered, he declared:

> Why speak of the first days? What is left of them? Their hideous memory is mute. I no longer knew who, or where, I was. . . . Death had brushed me, as the saying goes, with its wings. . . .
>
> A few hours later, I had a hemorrhage. It happened while I was walking laboriously on the veranda. . . . Feeling out of breath,

> I inhaled more deeply than usual, and suddenly it came. It filled my mouth . . . but it wasn't a flow of bright blood now, like the other hemorrhages; it was a thick, hideous clot I spat onto the floor with disgust.
>
> I turned back, bent down, took a straw and raising the clot of spittle, laid it on a handkerchief. I stared at it. The blood was ugly, blackish—something slimly, hideous.[1]

The accounts written by Stowe and Gide encapsulate the distinction between edifying consumption, which seemed to exalt patients, and vile tuberculosis, which collapsed and destroyed lungs and lives (figs. 15.1 and 15.2). The two accounts were separated by several factors that together fostered new attitudes toward both medical treatment and public health regarding the disease.

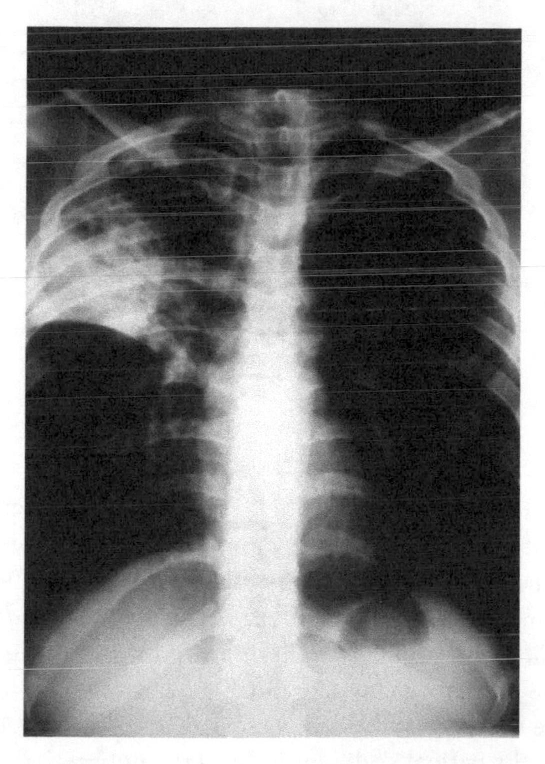

Figure 15.1. Chest x-ray of a person with tuberculosis shows a mild collapse of the right upper lobe of the lung. (Wellcome Collection, London. CC BY 4.0.)

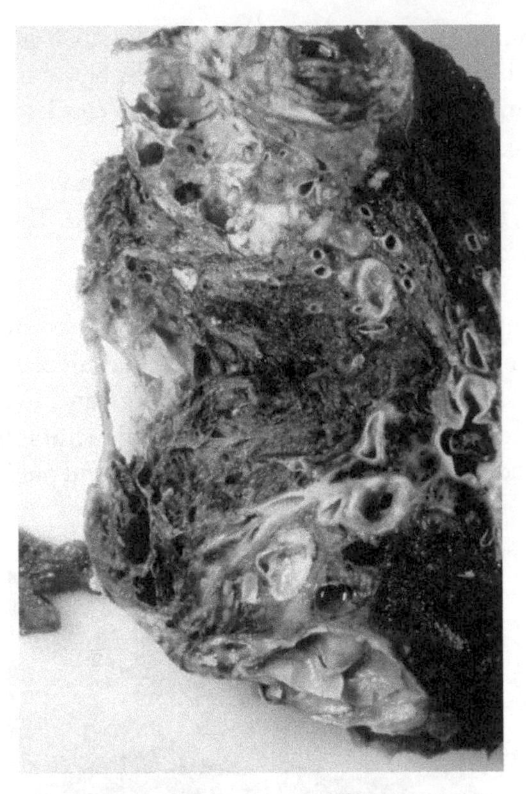

Figure 15.2. Specimen of a lung with chronic fibrotic pulmonary tuberculosis. (Wellcome Collection, London. CC0.)

Contagionism

Consumption lost its allure with the demonstration that it was not a hereditary but an infectious disease. A first major step in this direction was the work of the French military physician Jean Antoine Villemin (1827–1892). In his landmark works *Études sur la tuberculose* (*Studies on Tuberculosis*) and *De la propagation de la phthisie* (*On the Transmission of Phthisis*) of 1868 and 1869, respectively, Villemin challenged the two leading medical theories of the time regarding tuberculosis: the essentialist doctrine that it was caused by the patient's "diathesis," and the notion that it was a hereditary disease. Both, he argued, were logically flawed examples of circular reasoning that impeded the advance of scientific understanding and preventive medicine. In different guises the two theories had in common the assertion that tuberculosis was caused by a person's predisposition to the disease—a conclusion that left physicians powerless to seek means to combat the scourge that killed, according to the available figures, 160,000 French people every year. Diathesis and heredity, moreover, raised the specter of spontaneous generation in their claim that the disease came from within the body that mysteriously generated tubercles where none had been before. Villemin was convinced that tuberculosis came from without—from a contagion that gained access to the body and then diffused in the form of tubercles.

In addition to being logically circular and medically disempowering, neither diathesis nor heredity could account for the epidemiology of tuberculosis in the environment he knew best—the army. Villemin noted that French soldiers in the mid-nineteenth century were overwhelmingly of peasant stock, and yet they paid a heavy tribute in death and disease to tuberculosis. This fact posed the serious etiological problem of how the two prevailing theories could explain the onset of phthisis in a population of young men whose families did not suffer from the disease and who were vigorously healthy until they were suddenly confined together in congested military barracks. To Villemin, it seemed apparent that neither the constitution of the troops nor their heredity could solve this riddle, and that only the competing idea of contagion provided a plausible answer. In his view, the army exposed healthy young men to an infectious agent while housing them in an environment that favored its spread from body to body.

To buttress his hypothesis, the French doctor conducted a series of laboratory experiments to establish whether consumption is communicable. He inoculated rabbits with matter taken from the tubercles of consumptive human beings and cattle; the rabbits sickened as a result. Then he repeated the process—but this time inoculating healthy rabbits with matter taken from the consumptive ones. Those rabbits sickened as well. Villemin thought that these results confirmed that tuberculosis was indeed a contagious disease and that an invisible "germ" (or, in his terminology, a "virus") was responsible.

Villemin's position, supported by theory, epidemiological evidence, and laboratory tests, was highly suggestive, and his work marked an important and influential step in the emergence of the modern germ theory of disease and the victory of contagionism. Like John Snow in England, however, Villemin was unable to clinch the argument because he could not identify the responsible pathogen. That proof required the further development of microscopy, and the work of Robert Koch.

As we have seen, the decisive intervention was that of the microbiologist Robert Koch, a pioneering figure of the modern germ theory of disease (Chapter 12). In 1882 he identified *Mycobacterium tuberculosis,* and he established rigorous proof—in compliance with his own famous postulates—of its causative role in the disease. This discovery was momentous because it signified that the pathogen responsible for the most important infectious disease of the century was now known and because it marked the victory of the contagionists such as Koch and Louis Pasteur over the anticontagionists such as Max von Pettenkofer.

Inevitably, the new understanding took root over an extended period rather than overnight. Even in 1914, twelve years after *The Immoralist,* Thomas Mann could still write *The Magic Mountain,* his influential depiction of consumption in Romantic guise. In this work, the elegant elite still took the cure at Davos in the Swiss Alps and whiled away the time in improving conversation. Furthermore, as late as 1922 Somerset Maugham's short story "Sanatorium" conveyed a similarly "essentialist" view in a Scottish context.

Victory for contagionism—and the view of tuberculosis as a disease of poverty—moved all the more slowly because the discovery of the tubercle bacillus did not lead to the expected gains in prevention and cure. Koch himself inadvertently fostered skepticism. In a premature rush of optimism, he announced mistakenly in 1890 that armed with a knowledge of the cause of TB, he had also discovered a specific remedy in the form of tuberculin—"Koch's lymph." A derivative of *M. tuberculosis,* tuberculin dramatically failed as a therapeutic agent, causing widespread disillusionment. Employed as a specific, Koch's preparation caused painful side effects and even fatalities. The fact that tuberculin later provided the basis for an effective diagnostic test did little to restore confidence. Until the arrival of the antibiotic era after the Second World War when streptomycin was discovered, physicians remained as helpless in the battle against tuberculosis as they had been before Villemin's and Koch's discoveries.

Important features of the etiology of tuberculosis masked its mode of transmission and made the notion that it was an infectious disease difficult to accept. The prolonged latent period, when patients were asymptomatic, concealed the link between exposure to bacilli and the onset of active disease. An asymptomatic latent period that could last for months or even years was unprecedented, and it complicated acceptance of the germ theory.

Holistic and humoral theories persisted. Even physicians who accepted the germ theory of disease sometimes integrated it into older views, theorizing that Koch's bacillus was simply one more "exciting cause" of tuberculosis. Such a stance was especially likely among older doctors. They had no training in bacteriology and microscopy, or contact with scientific laboratories, and they noted that bacteriology had not generated any useful therapies. Doctors who clung to tradition were also disquieted by the association of the new doctrines with quantification, and with the new and—for some—disconcerting technologies of microscopes, staining methods, slide covers, and agar. Hippocratic doctrines were not suddenly overthrown but progressively eroded.

If the slow but steady collapse of the medical philosophy of essentialism and heredity in favor of a scientifically robust doctrine of contagionism was a major factor in producing new attitudes toward tuberculosis, another was the result of epidemiological and statistical studies of the disease and its social profile. Such studies demonstrated that although tuberculosis did affect the elite of society, it was above all a social disease that disproportionately afflicted the "dangerous classes"—working men and women and the urban poor. In Hamburg, Germany, for example, it was reported in 1922 that tuberculosis death rates were inversely proportional to the amount of income tax paid; and it was noted in Paris that mortality from phthisis was most pronounced in the impoverished Twentieth Arrondissement and lowest in the wealthy districts of the city. A typical comment of the early years of the new century echoed Gide and broke decisively with essentialism and the nineteenth-century understanding of consumption, calling TB "a coarse, common disease, bred in foul breath, in dirt, in squalor. . . . The beautiful and the rich receive it from the unbeautiful and the poor."[2]

Similarly, in New York City pulmonary disease was so intense in certain tenements—whose squalid conditions are vividly captured in the photojournalism of Jacob Riis, especially his work *How the Other Half Lives* (1890)—that they were known as "lung blocks" (fig. 15.3). In 1908 public health advocates in the city sponsored a "tuberculosis exhibit" that was attended by 3 million people. Its purpose was to illustrate the dirt, crowding, dark, and poor ventilation that had given rise to tuberculosis in thirty thousand tenement dwellers in the city and had led to the "horrible examples" of lungs ravaged by tuberculosis and preserved in alcohol for the public to examine. Meanwhile, the exhibitors distributed six hundred thousand pink "Don't Cards" that imparted two central tenets of the new understanding Koch had promoted: "Don't give consumption to others; don't let others give consumption to you." The only distinctive feature separating New York and other major US cities from their European counterparts in this regard was that

Figure 15.3. Jacob Riis, "Lodgers in a Crowded Bayard Street Tenement," from *How the Other Half Lives: Studies among the Tenements of New York* (1890).

the relationship between poverty and disease had a pronounced ethnic inflection: Irish and Italian immigrants were overrepresented among New York's victims.

Another factor promoting a profoundly new social construction of tuberculosis after Koch's scientific bombshell of 1882 was the international climate of gathering tensions among the European powers. These were years of Social Darwinism; of imperial rivalry in the "scramble for Africa"; of nationalistic economic competition; of enmity between France and newly unified Germany, marked by the French loss of Alsace and Lorraine; of the Anglo-German arms race; and of the crystallization of clashing insecurities in the form of two opposing blocs—the Triple Entente of the Russian Empire, France, and Great Britain and the Triple Alliance of Germany, Austria-Hungary, and Italy. The result was a deep awareness of national vulnerabilities and of the need for preparedness or, in the English phrase, "national efficiency." Pulmonary disease, however, was a major liability. It lowered the birthrate, compromised productivity, undermined military power, and diverted precious resources. Consumptives, therefore, were a threat not just to themselves and their communities, but also to economic and demographic growth. They undermined empire and even threatened national survival.

The results for sufferers were severe. Fear ushered in a time of stigmatization and shunning for anyone with a diagnosis of TB or a telltale persistent cough. American newspapers and magazines reported a rising tide of what they called "phthisiphobia" and "tuberculophobia" that was stoked by the ubiquitous messages that public health authorities disseminated. Pamphlets and posters warned of the dangers presented by consumptives, and doctors and nurses reinforced the message during consultations in the clinic.

In the new understanding of tuberculosis as contagious, the general public viewed persistent coughers as dangerous and even unpatriotic. Accordingly, tuberculosis sufferers were shunned. They found it difficult to obtain lodgings, employment, or insurance, and their condition was a serious barrier to marriage. Parents of schoolchildren demanded that pupils be tested for fever as they entered school and that any child with a reading above 98.6°F be sent home.

There was also a related hysteria. People panicked about the dire consequences of licking postage stamps. In many cities residents viewed library books with suspicion as possibly bearing the lethal tuberculous bacilli from a previous reader. They demanded that all books be fumigated before they were recycled. Accordingly, the New York Public Library sent its returned books to the Board of Health, which "disinfects books by means of formaldehyde gas under pressure. The books are placed in an airtight compartment and suspended in such a way that the leaves will hang loosely and be exposed to the action of the gas."[3] For the same reason, pressured by "Clean Money Clubs," banks sterilized coins and the Treasury Department retired old bills and issued uncontaminated replacements. According to the Research Laboratory of New York, tests determined that dirty pennies bore an average of twenty-six living bacteria each, and dirty bills seventy-three thousand.

Beards and mustaches fell out of favor after being fashionable for most of the second half of the nineteenth century. Bacteria could nestle amidst the whiskers, only to fall into other people's food or onto their lips during a kiss. Indeed, some public health authorities advised that kissing was excessively dangerous and should be avoided altogether. The *Atlanta Constitution* newspaper conducted an informal survey of male pedestrians on the city's streets in 1902. Noting that only 5 percent sported beards, "in contrast to some years ago when every third man on Broadway wore a beard," the paper fretted, "It will not be long before we will be as whiskerless a race as in the days of Napoleon."[4]

Fear of the tubercle bacillus also invaded churches, leading to protests against common communion cups and the promiscuous sprinkling of holy water. Similarly, there were campaigns against the use of a single metal drinking cup at water fountains and against the practice by ice cream parlors of recycling glass and metal cups. At the same time residents in many cities petitioned against opening pulmonary disease wards and dispensaries in their neighborhoods. They feared that ambulatory patients and their families would deposit Koch's bacillus on the straps, handles, and floors of the buses and trams that bore them to and from these facilities. Real estate prices plummeted when a neighborhood was threatened with such dangers.

Most clearly demarcating the boundary between consumption and tuberculosis was the decision taken by fashionable hotel keepers along the French and Italian rivieras that were destinations of choice for wealthy consumptives taking the cure. In the aftermath of Koch's discoveries, hoteliers announced that the tuberculous were no longer welcome at their establishments. Coughers, they argued, frightened away other guests and threatened the health of their employees. Clearly, consumption had ceased to be romantic. The *New York Tribune,* however, argued in 1901 that things had gone too far:

> The American people and their officials, animated with zeal not according to knowledge, are in danger of going to senseless and cruel extremes in hunting down consumptives. There is a tendency on the part of people who have grasped the idea of the infectious nature of this disease to become panic-stricken and act as badly as we from time to time see communities doing when they burn down contagious disease hospitals. . . .
>
> In California and Colorado talk of barring invalids from other states has been heard and there is danger that the common and natural anxiety to guard against consumption may be indulged with a heartlessness more characteristic of the Middle Ages.[5]

Nonetheless, this new and pervasive sense that tuberculosis threatened the health and well-being of society as a whole persisted. In 1908 Dr. Thomas Darlington, the commissioner of health for New York City, described the disease as a great destructive force responsible for four hundred deaths a day in the United States and an annual expenditure of $300 million for its control and treatment. Darlington argued that major natural disasters, such as the devastating earthquake at Valparaiso in 1906, paled by comparison with the havoc wrought by tuberculosis, although fortunately the general public was finally waking up to the danger.

Propelled by this sense that tuberculosis was a national emergency on grounds that were at once humanitarian, sanitarian, patriotic, and economic, powerful interests throughout the industrial world launched a series of "wars" against tuberculosis during the late nineteenth century and into the early twentieth. Taken together, these efforts arguably constituted the most powerful movement ever directed against a single disease until that time. The mix of interested parties involved in these campaigns varied from country to country, but generally it included charities, medical societies and associations, chambers of commerce, public health officials, educators, and the state at national, regional (or state in the US), and local levels.

Everywhere, reflecting the international tensions of the era, the effort employed military metaphors to describe its efforts. Terms such as "war," "campaign," "weapon," and "combat" dominated the lexicon, as Susan Sontag discusses in her 1979 book *Illness as Metaphor.* Similarly, images of bayonets, knives, and guns wielded against the evil dragon TB prevailed in posters. In France, tuberculosis was often personified, after 1914, in the form of the national enemy derisively termed *le boche,* or "the German."

War on Tuberculosis

The war on tuberculosis operated throughout western Europe and North America from the late nineteenth century to the advent of streptomycin after World War II. Organizational methods, levels of funding, and strategies differed from country to country and over time; however, major similarities unified the movement, partly because everywhere the problems were the same. Another factor promoting convergence is that medical and public health are international disciplines that operate on the basis of common scientific understandings. In addition, the various national campaigns emulated the "best practices" of other countries, and in 1905 a unifying series of International Congresses on Tuberculosis began. The first congress opened in Paris with the specific intention of promoting shared experiences, research, and institutional arrangements.

Developments in the United States were in the forefront of the international movement, and they illustrate the organizational trajectory common to the movement as a whole. The initiative began at the local level through medical associations in New York, Philadelphia, Chicago, and Boston. Sigard

Adolphus Knopf, a leading figure of the National Tuberculosis Association, dated the antituberculosis crusade from 1889. At that time three New York physicians, led by Hermann Biggs, presented the municipal health department with a set of recommendations to control the spread of the disease. Symbolically, this initial step was significant, but the campaign did not acquire a stable organizational structure until the final decade of the century.

In 1892 the first antitubercular association was established in Philadelphia as the Pennsylvania Society for the Prevention of Tuberculosis. The creation of this society was a decisive event because it was the first institution to be established for the specific purpose of preventing TB, because it served as a model for two further critical local associations, and because it provided a base for the creation of a national organization in 1904. Two additional local associations followed the Philadelphia model—the Tuberculosis Committee of New York City in 1902 and the Chicago Tuberculosis Institute in 1906.

The National Tuberculosis Association, with Edward Livingston Trudeau as its president, arose from the three local initiatives, and its purpose was to provide leadership, coherence, and stimulus for the movement throughout the country. By 1920 affiliated tuberculosis associations existed in every state and the District of Columbia.

The campaign sponsored by the National Tuberculosis Association in the United States relied on instruments that were soon adopted throughout the industrial world: the sanatorium, the dispensary, and health education. These were the weapons of choice in Britain, France, Germany, Belgium, Portugal, Canada, Denmark, Sweden, Russia, Japan, Norway, Australia, and the United States.

The Sanatorium

From Görbersdorf to Saranac Lake

Of all the measures devised to combat tuberculosis, the sanatorium was the most distinctive and most important. The world's first tuberculosis sanatorium was the idea of the German physician Hermann Brehmer (1826–1889), who established a prototype at Görbersdorf in Silesia in 1859. As a medical student at the University of Berlin at midcentury, he contracted tuberculosis and received the standard advice from his doctors—to attempt to restore his health by taking the cure. Pessimistic about his prognosis, Brehmer elected altitude therapy in the Himalayas. There, to his surprise, his condition improved. He attributed this recovery to life in the open air and reasoned that his experience could be universalized. Returning to Berlin, he completed his medical studies with a dissertation whose title proclaimed a disarmingly optimistic extrapolation from his own case: "Tuberculosis Is a Curable Disease."

Settling in Silesia, Brehmer decided to put his theory to the test by creating an institution dedicated to providing consumptives with the three specific remedies he had discovered in the mountains of India: an outdoor life, total rest, and good nutrition. To implement this rigorous threefold program, Brehmer set up the Görbersdorf Sanatorium—a small group of cottages to accommodate several hundred patients. Treatment followed the guidelines he had discovered in India. Brehmer's disciple and former patient Peter Dettweiler opened a companion sanatorium run along identical lines at Falkenberg in 1876.

Despite the presence of two pioneering institutions in Germany, the sanatorium movement did not become a major factor in public health until the work of the American doctor Edward Livingston Trudeau (1848–1915). By the time of Trudeau's venture, the new knowledge that TB was a contagious disease and a peril to the well-being of a nation provided impetus to the movement. A generation after the isolated initiatives at Görbersdorf and Falkenberg, combating tuberculosis had become an imperative. Furthermore, there was a strong belief that, since the formerly unknown enemy was now revealed to be Koch's bacillus, it was vulnerable to attack—if the proper weapons could be deployed. Nations and their health officials were no longer powerless. In this resolute climate, Trudeau became Brehmer's most influential follower.

Like Brehmer and so many tuberculosis patients, Trudeau experienced a life-changing health crisis. After receiving his medical degree at Columbia University, he tended his brother, who was dying of consumption, and he himself was diagnosed with the disease in the 1870s. Expecting to die, he experimented, like Brehmer, with an outdoor rest cure in the wilderness at Saranac Lake in the Adirondack Mountains. There he spent his time resting in the open air and hunting from a canoe on the lake. After a time, he felt better and recovered.

Devoting himself to the cause of combating tuberculosis, Trudeau learned about both Brehmer's pioneering sanatorium and Koch's discovery. Armed with this knowledge, he resolved to apply Brehmer's approach to the disease among the urban poor, who were now known to be its principal victims. Having raised money from philanthropists to support the venture, the New York doctor opened a cottage-based sanatorium for impecunious consumptives in 1884. Patients at Saranac Lake were heavily subsidized. Those who were able, paid half the cost of their maintenance; the severely impoverished were funded by the institution's endowment.

Brehmer was largely ignored beyond the narrow circle of his disciples. Trudeau, however, was not only a physician and humanitarian, but also a skilled and determined publicist for his therapeutic views. From the outset he devised Saranac Lake as a showcase for the sanatorium idea. Its establishment marked the beginning of the sanatorium movement as a powerful new tool in the battle against tuberculosis both in the United States and abroad. Already by 1922 there were seven hundred sanatoria in the United States alone with a capacity of more than one hundred thousand beds.

Organizationally, these establishments were highly diverse, as *A Directory of Sanatoria, Hospitals, Day Camps and Preventoria for the Treatment of Tuberculosis in the United States*, published by the National Tuberculosis Association, clearly demonstrated.[6] Some were private institutions; others were run by the federal government or a state, county, or municipality. A number restricted their admissions to those who could afford private care, but most made charitable provision for the poor or charged according to a person's ability to pay, with fees in 1930 ranging from gratis to the $150 a week charged at Pamsetgaaf, a private sanatorium outside of Prescott, Arizona. The majority were open to the general public within a county or state, but many restricted admission by race, gender, or age. It was common, for example, to exclude African Americans entirely or to provide "separate but equal" provision in segregated buildings, or in separate wings or pavilions. A few states and counties opened sanatoria, such as the Maryland State Sanatorium, Colored Branch, at Henryton, dedicated solely to the treatment of "Negroes."

Many sanatoria catered to specific sectors of the population—immigrants, children, veterans, Native Americans, Jews—or to members of a particular occupational group, trade union, or Christian denomination. Some establishments were dedicated to the employees of specific companies, such as the insurance giant Metropolitan Life; to moving picture artists; or, in the paternalistic language of the Night and Day Rest Camp at St. Louis, to "working girls in a run-down condition." Usually sanatoria were free-standing institutions, but a significant number occupied a wing or pavilion of a general hospital, a prison, or an insane asylum. A few were located in cities, but the typical sanatorium occupied hundreds of acres in the countryside, preferably on elevated terrain and with easy access to a railroad station.

Patient capacity was also a significant differentiating feature. At one extreme, the two largest US sanatoria in 1931—the Hennepin County Sanatorium in Minnesota and the Detroit Municipal Sanatorium at Northville, Michigan—could accommodate 704 and 837 tubercular patients, respectively. At the other extreme, small institutions such as the Tuberculosis Home for the Colored at West Palm Beach, Florida, had twelve available beds, and the Alpine Sanatorium near San Diego, California, only twenty. Several states, such as Colorado, did not fund public sanatoria at all, aside from the facilities provided for tuberculous patients in penal institutions, but they did refer sufferers to an array of private sanatoria and designated boarding houses.

Finally, there were many admissions requirements related to the form and stage of the disease. Most institutions accommodated only pulmonary tuberculosis in its incipient or "minimal" stage in the manner of Saranac Lake. But some sanatoria, such as the Olive View Sanatorium of Los Angeles County and the Pottenger Sanatorium at Monrovia, California, provided for "all stages of pulmonary, genitourinary, laryngeal, and intestinal tuberculosis."

The entry in the *Directory* outlining tubercular care in Minnesota provides a clear idea of the variety of institutions available:

> [Minnesota] has 2463 beds for tuberculosis, exclusive of beds in federal hospitals. There are 16 public institutions, including a state sanatorium, 14 county sanatoria, and a municipal preventorium school. There are six private and semi-private sanatoria and a boarding home for inactive tuberculous patients. The federal government operates a hospital for ex-service men and a sanatorium for Indians. In addition, provision is made for 169 tuberculous patients in state hospitals for the insane, and for 30 in institutions for epileptics and the feeble-minded. State penal institutions make provision for 19 tuberculosis cases.

All sanatoria in the United States, however, made it their mission to follow the general outline of the "sanatorium treatment" featured at Saranac Lake.

Education for Prevention

In Trudeau's understanding, Saranac Lake simultaneously fostered both preventive and curative objectives. Preventively, it removed impoverished consumptives from crowded tenements and workshops where they spread disease through the bacteria that they coughed, spat, and breathed into the atmosphere. Trudeau calculated that, in the course of a year, an average consumptive in New York City infected twenty other people. By removing people with the disease from the city to the wilderness of the Adirondack Mountains, Saranac Lake Sanatorium interrupted the chains of transmission. It therefore served a prophylactic function analogous to quarantine with respect to plague.

In addition, Saranac Lake attempted to lower the incidence of tuberculosis by educating its patients in matters of "tuberculosis etiquette" that would reduce their likelihood of infecting others after discharge from the sanatorium. Patients invariably stayed in the Adirondacks for more than six months, and frequently they stayed many years. During that time the sanatorium had every opportunity to teach sanitary rituals to be practiced by patients for the rest of their lives. They were schooled, for instance, to suppress their coughs when they could. When coughs became too powerful, they learned to cough into handkerchiefs that they always carried with them.

Equal thought was given to expectoration, which was one of the major preoccupations of the crusade. Spitting, patients were taught, always presents a danger of contamination to others. An international "dust scare" arose on the strength of the belief that floating bacilli were a major factor in the ongoing epidemic. In the words of Sigard Adolphus Knopf in 1899:

> As long as the expectorations remain in the liquid state there is less danger from them, but matter expectorated on the floor, in the street, or in a handkerchief usually dries very rapidly, and, becoming pulverized, finds its way into the respiratory tract of any one who chances to inhale the air in which is floating this dust from expectorations, laden with many kinds of bacteria. The most dangerous of them all is the tubercle bacillus, which retains its virulence in the dried state for several months.[7]

It was therefore essential for sufferers scrupulously to avoid "promiscuous spitting." The sanatorium instructed them always to carry a cardboard sputum cup in a pocket or handbag as the exclusive recipient of their expectorations. At the end of each day patients were required to burn the flammable spittoon, thereby destroying not just the cuspidor, but also its accumulated microbial load.

Because of the danger of dust-borne bacilli, consumptives needed to learn new domestic sanitary practices and to teach them to their families, tenants, and fellow residents after their eventual return to communities. Foremost among domestic admonitions was the danger of the traditional practice of sweeping floors with a broom. Sweeping, it was taught, could be fatal because it lifted contaminated dust into the air. A fundamental principle of sanitary wisdom, therefore, was to replace brooms with mops. In myriad ways, Saranac Lake patients learned, practiced, and taught techniques intended to break the cycle of transmission.

The Curative Regimen

Meanwhile, Trudeau believed that the sanatorium served a major therapeutic purpose. Here a compelling case based on robust data cannot be made. Saranac Lake produced statistics indicating a significant "cure rate" of more than 30 percent for active consumptives. This compared favorably with the general tuberculous population outside the sanatorium, where an active case was thought to be almost universally fatal, though often at a distant time. Saranac Lake's reassuring figures are seriously misleading, however, because the sanatorium intentionally admitted only mild cases still in the incipient stage of the disease. Severe and advanced cases were rejected on the grounds that the sanatorium should prioritize aiding those patients for whom it believed it could make a difference.

As a result, the patients under Trudeau's care were not representative of the general population of the tuberculous. After such rigorous triage it is impossible to determine whether Saranac Lake's positive findings reflected the efficacy of its treatment or the skill of its admissions office in eliminating the incurable. Trudeau himself recognized that the first ambition of Saranac Lake was not to cure but to offer hope at a time when a diagnosis of tuberculosis was regarded as a death warrant. This approach was embodied in the motto of his institution, "To cure sometimes, to relieve often, to comfort always."

On the other hand, Trudeau, his followers, and the wider medical profession regarded Saranac Lake and all sanatoria as therapeutic institutions. Indeed by the end of the nineteenth century many pulmonary specialists considered TB a curable affliction provided that it was diagnosed early and an appropriately rigorous therapeutic regimen was adopted. The American authority Knopf declared in 1899 that TB was "one of the most curable and frequently cured of all diseases."[8] And the ideal treatment was that administered at sanatoria in accord with Brehmer's original vision and Trudeau's systematic application of it.

Therapy for tuberculous patients in other settings—on chest wards in hospitals, in dispensaries, and at home—followed as far as possible the sanatorium model, but with a universal recognition that it was inevitably flawed and inferior. For the first half of the twentieth century, sanatoria were the backbone of phthisiotherapy and everywhere approached the treatment of tuberculosis on the basis of four central principles—the outdoor life, rest supplemented by graduated exercise, a robust diet, and total control of the patients by the medical staff. There were, however, medical fashions that added variations to the essentials.

THE OUTDOOR LIFE

The bedrock of sanatorium therapeutics was the so-called wilderness cure originally preached by Brehmer. While developing this treatment strategy, Trudeau conducted an experiment on a tiny island—subsequently dubbed "Rabbit Island"—in the middle of Saranac Lake. There he confined two groups of rabbits in radically contrasting enclosures. In the first he reproduced what he regarded as the defining features of urban tenements—filth, severe congestion, and poor ventilation. In the second he allowed the rabbits to experience the unaltered outdoor environment of Rabbit Island itself. When the "slum" rabbits died and the "outdoor" rabbits survived, he drew the appropriate therapeutic conclusion. The rabbit experiment was not scientifically convincing, but it persuaded Trudeau of a conclusion he had already drawn, and it provided him with a vivid example to illustrate it.

At Saranac Lake, and at later sanatoria on every continent, the first principle was that patients should live the outdoor life in all weather. Most institutions followed two contrasting architectural models—the cottage and the pavilion systems. Saranac Lake and its sister private establishment—the Loomis Sanatorium, opened in 1896 at Liberty, New York—typified the cottage pattern. A cottage sanatorium normally consisted of twenty to thirty cottages built one hundred feet apart, with four to eight patients assigned to each. There patients spent nearly all of their waking hours reclining in steamer chairs on small porches, either singly or in small groups of three or four. They were sheltered from rain and snow and wrapped in blankets against the cold, but otherwise they were exposed to bracing outdoor conditions.

Under the pavilion system, seventy-five to one hundred patients lived under a single roof, but a long verandah ran the length of the building. Patients spent their days stretched out on deck chairs in this common space (fig. 15.4). Sometimes several pavilions were built and linked by covered galleries that allowed for promenades in inclement weather.

Figure 15.4. The Stannington Sanatorium for children with tuberculosis opened in 1907 in the United Kingdom. This photo shows the outdoor pavilion where patients spent much of their time. (Wellcome Collection, London. CC BY 4.0.)

Globally, pavilions predominated because they were far less expensive to build than a network of cottages, but the essential characteristic of both types was life outside, whether it was in semi-isolation or as a member of a large group. The pamphlet "Rules for Patients" at Saranac Lake was typical in its specifications: "Patients are expected to lead an out-of-door life—that is, to remain eight to ten hours in the open air each day. . . . Each patient is required to be out of doors from 9 a.m. to 12:45 p.m. and from 2 p.m. to 5:45 p.m. Sleeping out is considered in no way to affect this requirement." Furthermore, even indoors patients slept with their windows open, "rain or shine, warm or cold" because "a consumptive, if he wishes to get well, should live every moment of his existence in the purest and freshest air possible."[9] In the succinct statement of two British sanatorium medical directors in 1902, "TB is vanquished by the pure air of Heaven."[10] In the interiors, whether of cottages or of pavilions, the concern for dust dictated structure and decor. All angles were rounded to prevent the accumulation of dust; walls were painted rather than papered to facilitate washing; heavy furnishings and carpets were banned; and floors were made of hardwood that was easily mopped. Sweeping was absolutely forbidden.

REST AND GRADUATED EXERCISE

A second pivotal feature of sanatorium life was the amount and type of exercise. Here national fashions predominated. In the United States the norm, as at Saranac Lake, was absolute rest with no exertion at all for patients with a temperature above 99.5°F or a pulse rate above one hundred beats per minutes. Patients with less than a degree of fever were allowed half an hour of exercise per day, but the simplest movements of daily life—walking to the refectory for meals, getting in and out of bed, and standing—were deducted from the half hour.

In Britain, the prevailing thought was that habituating the working classes—who predominated among the tuberculous—to absolute rest would ruin their moral fiber and render them unfit for a productive later life. This standard was exemplified by Brompton Sanatorium, which featured a program of "graduated exercise." Exertions were increased gradually in accordance with improvement in a patient's condition as measured by temperature readings. Paths were laid out that were equipped with benches for rest stops and were color-coded according to degree of difficulty. As patients "graduated" from green to blue to red, inclines sloped more steeply.

DIET

A third therapeutic tenet of the sanatorium movement was the value of a hardy diet to counteract consumptive emaciation and to build the patient's resistance. At Saranac Lake, for example, patients were encouraged to eat four full meals a day, with a glass of milk in between each. The meals emphasized beef and carbohydrates with the goal of persuading patients, many of whom found food abhorrent, to consume thirty-five hundred to four thousand calories daily.

Diet is one of the oldest therapeutic interventions known to medicine, and it had been systematically deployed by Hippocrates and Galen. What was novel in the use of diet to treat tuberculosis was that the strategy involved no humoral philosophy or the search for foods that were hot, cold, moist, or dry to reverse imbalances or dyscrasia. Therapeutics in the age of Trudeau adopted instead an approach that was essentially symptomatic. The strategy was to augment caloric intake in order to combat the anorexia that seemed to weaken consumptives by depleting their vital energy and recuperative powers. An important aspect of medical supervision, however, was to resist the widespread popular misconception that the best advice for consumptive patients was to gorge themselves continuously. The sanatorium insisted that that two of its principal functions were to instruct patients instead that the proper selection of foods appropriate for their condition was even more important than sheer quantity, and then to exercise the necessary surveillance to ensure compliance.

CLOSED INSTITUTIONS

A further feature underlying therapeutics at sanatoria is that they were "closed" institutions where patients were subject throughout their stay to constant surveillance and supervision by the medical staff. This was to ensure compliance with all facets of the treatment regimen that dictated every aspect of a patient's life. Recovering from tuberculosis was considered a full-time commitment, and any deviations were viewed as life-threatening. Sanatoria therefore developed complex and all-encompassing regulations to promote the patients' physical health while also sheltering them from emotional shocks from the outside world or depressing medical news from fellow patients that would dampen their spirits.

Accordingly, patients were strictly forbidden to leave the premises, and visits from outsiders were carefully monitored. Patients' mail was censored to protect them from upsetting news, and the reading material in the sanatorium libraries was carefully chosen as a part of what was termed "brain cure"—a uniformly upbeat and optimistic outlook on life. For the same reason, conversations among patients about the progress of their disease were forbidden, and socializing was limited to meals and an hour of sanctioned conversation during the day.

To prevent emotional stress and physical excess, the sexes were segregated and patient attempts to form emotional and sexual relationships were discouraged. In addition, there were strict rules against gambling, profanity, and the use of tobacco. In implementing this almost monastic way of life, the rest cure performed a disciplinary as well as a medical function. The long hours spent "in the horizontal" on the porches and verandahs kept patients as constantly in view as the prisoners in Jeremy Bentham's Panopticon penitentiary. Sanctions were also severe: infringement of the regulations was punishable by expulsion. Explaining the rationale for such measures, the American pulmonologist Francis Pottenger wrote:

> The sanatorium is the agency through which the open-air hygienic, dietetic, and scientific treatment of tuberculosis can be carried out to best advantage. . . . While it is possible to obtain excellent results outside of sanatoria, yet, it is impossible to have that same absolute control over a patient and all his actions, that

> close feeling of mutual interest and co-operation between patient and physician and that cheerful helpfulness which comes from the association of many who are making the same sacrifices, struggling toward the same end and constantly finding themselves and their associates making steady advances toward recovery. The psychic effect of a sanatorium is one that cannot be measured.[11]

The Appeal of the Sanatorium

Clearly, sanatoria were intentionally paternalistic and hierarchical, with enormous authority vested in the medical staff entrusted with efficiently administering the therapeutic and educational visions propounded by Brehmer and Trudeau. But for a large proportion of patients, such a way of life had enormous appeal. In the era before antibiotics it offered the only widely accepted hope of recovery from a deadly and dreaded disease. It also provided the impoverished, who formed the majority of consumptive patients, with a place where they were secure, well fed, and sheltered from unwelcome news. Furthermore, sanatoria usually gave thought to the economic future of patients after their discharge. They provided advice about the activities patients would still be able to perform in the labor market, and they held classes for more advanced convalescents in useful skills. On occasion the sanatorium could even offer employment with philanthropic companies such as the Reco Manufacturing Company and the Altro Manufacturing Company that specialized in hiring "arrested" consumptives for limited hours at such relatively light tasks as making garments, watches, and jewelry.

Not surprisingly, the sanatorium at Saranac Lake was overwhelmed with applications, with a rate by 1920 of twenty applicants per place at the institution. Indeed, the demand was so great that the town beyond the sanatorium experienced a significant developmental boom as commercially run "rest cottages" sprang up to accommodate the overflow of consumptives rejected by the sanatorium. These institutions advertised a diluted version of the rest cure, and their practices followed the advice and supervision of Trudeau's medical staff, and even of Trudeau himself, who served as mayor of the town. Often, rest cottages specialized in their provision; some, for instance, were designed for advanced cases, for Italians, or for women.

Far from having a reputation as a fearful institution, Saranac Lake was known for its optimism, its helpfulness, and Trudeau's personal kindness. Its appeal was evident in the fact that many ex-patients sought readmission in later life. Often, patients became dependent on the institution and its routine and tried to delay or even to avoid discharge entirely. A problem for the staff was to distinguish between the physical disease of phthisis and the psychological condition then diagnosed as "neurasthenia." The latter affected "equivocal consumptives" and mimicked many symptoms of tuberculosis, especially those of the "tuberculous personality"—headache, fatigability, insomnia, languor, and irritability. Neurasthenics sought to remain in care although the institution found no physical disease to treat. The problem was sufficiently widespread that some authorities recommended that care be taken to construct sanatoria in a spartan and economical manner in order that they not be too comfortable and likely to make patients "unduly discontented with the environment and life to which they belong."[12]

Such positive responses to sanatoria are difficult to reconcile with a recent trend in the historical literature. Some scholars regard sanatoria as coercive for ulterior Foucaultian motives—not of promoting health but of exercising social control, easing the job of the medical staff, and adapting the patient to social hierarchy. Most influentially, Erving Goffman, in his 1961 study *Asylums,* suggested that sanatoria should be seen as "total institutions" analogous in their methodologies of discipline and control to prisons, concentration camps, prisoner of war camps, and mental asylums. Such interpretations seem to be based on a resolutely negative and political reading of medical commentaries and patients' letters. Furthermore, they overlook the pivotal distinction that unlike patients in other total institutions, adult patients in all sanatoria—apart from the few who were in prisons or asylums—always had the freedom to discharge themselves and return home. Their presence was voluntary.

Two examples of works that can be easily misread are those by the tuberculosis authorities Pottenger and Knopf. Both accent the need in sanatoria for physicians to have enormous authority over their patients, and Pottenger even suggests "absolute control over a patient and all his actions." The context, however, suggests that such control is intended only in pursuit of therapeutic goals and should be accompanied by "cheerful helpfulness" and a "close feeling of mutual interest and co-operation between patient and physician."[13] Meanwhile, Knopf argues that discipline in a sanatorium "need not be too severe" but should be confined to the enforcement of regulations necessary for the patient's health.[14] Founded at the peak of the tuberculosis epidemic, the sanatorium movement regarded cure as possible, but only under the most carefully regulated conditions. Extensive authority to govern the activities of the patients was not a means of social control but an issue of life and death. In the words of an early analyst of the movement and himself a sanatorium doctor, recovery from TB "is a sufficiently difficult task, to which everything else must be subordinated."[15]

Additional Therapeutics

Although the outdoor life, complete rest, and a robust diet made up the therapeutic triad of sanatorium therapy, additional interventions became fashionable at particular institutions, over specific time periods, or in defined national contexts. A feature of the sanatorium was its specialization in the exclusive care of consumption, which made the institution a focus for pulmonary specialists and a site for the introduction of experimental new methodologies.

Of these, some were minimally invasive, such as aerotherapy, which involved "thoracic gymnastics" intended to teach deeper breathing to oxygenate the blood and stimulate pulmonary function. There was even a vogue in "lung cabinets" in which a patient sat for a period of two to eight minutes while a partial vacuum was created to encourage deeper contraction and expansion of the lungs. Similarly, there were fashions in hydrotherapy, which entailed having the consumptive sponged with cold water to stimulate energy and resistance, and heliotherapy, which involved extensive sunbathing in fine weather.

The United States during the interwar period witnessed the heyday of more heroic surgical interventions. The fact that the standard medical procedures were not yielding increasing cure rates led one surgeon to conclude that since physicians had treated the disease for thousands of years without finding a cure, it was time for surgeons to take over the job. Most popular among surgical treatments for pulmonary tuberculosis was artificial pneumothorax, or "pneumo," which involved the injection of air or nitrogen into the pleural cavity to collapse the lung by subjecting it to external pressure. Carlo Forlanini developed the procedure in Italy during the 1890s, but it was widely adopted only from the 1920s with the advance of faith in technology, especially in North America. The theory behind the procedure was that it permitted total rest of the affected lung just as a cast would for a broken limb. It was a localized application of the holistic strategy of rest therapy. Some surgeons went a step farther, making the collapse permanent by removing ribs to paralyze the diaphragm or by severing the phrenic nerve and excising a section of it. Equally daring was bilateral pneumothorax, in which both lungs were partially collapsed.

Surgeons at some institutions even carried out lobectomies, removing portions of a diseased lung or even an entire lung. Surgically reducing the bacterial load in the chest, it was thought, would provide an adjuvant therapy and enhance the efficacy of the concurrent medical treatment. Unfortunately, this procedure yielded a high rate of complications and mortality, and surgical therapies for tuberculosis were abandoned by 1940 as tempting in theory but ineffective or lethal in practice.

Of all the accounts of sanatorium life, the most clearly negative is the autobiographical novel *The Rack* (1958) by A. E. Ellis (the pen name for Derek Lindsay), a work that he composed as a bitter reply to the romantic and edifying portrayal of *The Magic Mountain* by Thomas Mann. As the title suggests, Ellis considered his prolonged stay at Brisset in the French Alps as bearing comparison only to an instrument of torture. As the medical director, Dr. Bruneau, says to the protagonist, Paul, just before Paul attempts suicide, "Consider yourself an experiment of the gods in what a man can endure."[16] What Paul and his fellow inmates endure is an unending round of agonizing surgical procedures—pneumothorax, thoroscopy, thoracoplasty, puncture, and lobectomy—that produces only pain, pus, and stench without hope and without end. In Ellis's description of Paul's time "on the rack,"

> *Ponction* succeeded *ponction*, transfusion succeeded transfusion. The increasing pressure within Paul's chest, caused by the continuous secretion of the pus, necessitated regular *exsufflations*, or extractions of air. In between . . . *Soeur* Miriam would come to make intravenous injections or to drain off five ccs. of blood in order to measure the rate of its sedimentation. . . . A dozen intramuscular injections each day, and his buttocks and thighs became so painful that he felt that he was lying on a heap of embers. . . .
>
> Day and night represented no more than segments in the over-all cycle of his fever. Though his lucidity remained unimpaired, his existence was purely physical, an agglomeration of aching, burning flesh, he felt himself to be no more than the sum of its functions and sensations.[17]

In considering this catalogue of agonies, however, one should recall that Ellis does not recount life as it had unfolded in the traditional sanatorium known to Brehmer and Trudeau, when treatment was purely medical and was founded on the principle of rest. Ellis instead describes the waning years of sanatoria when many institutions delegated care to surgeons and promoted the operating theater rather than the cure porch as their centerpiece. It is important to note that, in real life, Derek Lindsay began his fearful stay at Brisset only after discharge from the army in 1946.

The Dispensary

After the sanatorium movement, the second leading feature in the war on tuberculosis was the establishment of dispensaries. These institutions could be defined as specialized tuberculosis health stations and were established simultaneously with the network of sanatoria as a means of completing their mission. Located in the countryside, sanatoria functioned by removing working people from the congested neighborhoods where they had fallen ill. Dispensaries, by contrast, brought specialized clinical services—diagnostic, therapeutic, and preventive—into the urban centers where the disease prevailed.

The world's first specialized tuberculosis dispensary—the Victoria Dispensary for Consumption and Diseases of the Chest—opened in 1887 at Edinburgh. It differed fundamentally from earlier nineteenth-century institutions also called "dispensaries" that were all-purpose outpatient clinics that "dispensed" medication and played no part in the war on TB. The Victoria Dispensary at Edinburgh, by contrast, was the work of the consumption expert Robert William Philip (1857–1939), who intended it to assume a targeted and carefully defined role in the larger antituberculosis campaign. Philip also played a major part a few years later in setting up a sanatorium, the Victoria Hospital for Consumption, as he always regarded the two institutions—the TB dispensary and the sanatorium—as interlocked. After this Scottish precedent, the first US dispensary opened in 1896 in New York, and by 1911 there were more than five hundred in cities across the country following a period of mushrooming growth.

Diagnostically, the dispensary provided the means to identify incipient cases of tuberculosis well before patients were aware of their condition. One way to accomplish that goal was to offer free and accessible walk-in testing during hours convenient for working people. The dispensary took case histories of everyone who came forward, examined their sputum under a microscope, carried out physical exams, and administered tuberculin tests and x-rays in order to diagnose the disease (fig. 15.5). Rather than passively waiting for residents to seek their services, however, the dispensary also employed visiting nurses to call at the homes of all known local cases of the disease. The goal was to persuade patients' family members to attend the clinic even if they felt entirely well. Since it was universally accepted that only

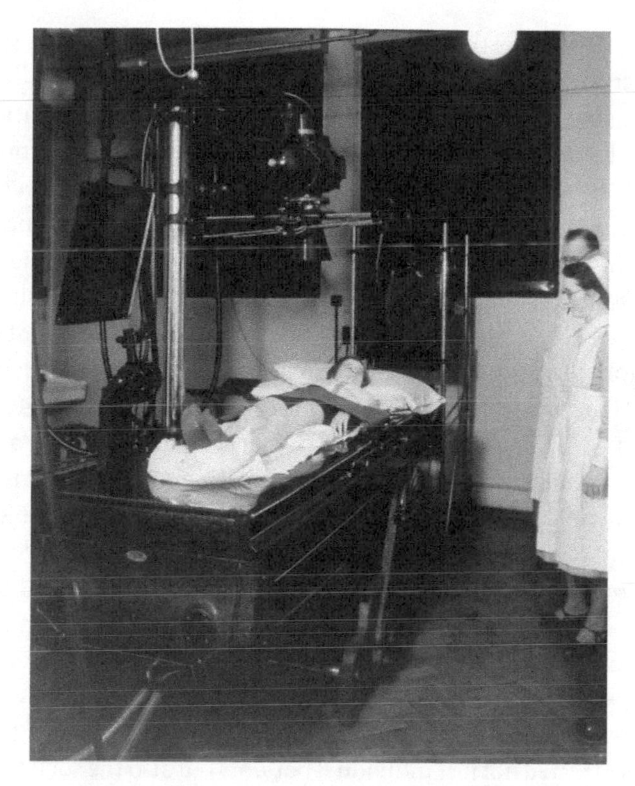

Figure 15.5. X-rays were an important part in the diagnosis of tuberculosis. ("West Midlands Tuberculosis Sanatoria and Public Information," by Adrian Wressell. Heart of England NHS Foundation Trust. CC BY 4.0.)

patients in the earliest stages of tuberculosis were likely to benefit from sanatorium treatment, the dispensary served as an instrument of triage and referral. So critical was this function to the war on tuberculosis that the New York City campaigner Elizabeth Crowell asserted that "sanatoria would go empty if it were not for the persuasive powers of the nurse."[18]

Unfortunately, most tuberculosis patients were ineligible for admission to a sanatorium. The many categories of patients officially barred included those whose disease had been diagnosed when it was already too far advanced, those who had been discharged as untreatable, and those who had returned home from a sanatorium with an "arrested" disease that required only follow-up care. The dispensary rather than the sanatorium provided therapeutic services for such cases. For their benefit the campaign carried out a physical examination, diagnosed the stage of the disease, and admitted the sufferers to the dispensary and its range of medical and social services.

A first step in drawing up a therapeutic plan for each patient was a lengthy case history that was profoundly different in emphasis from the case histories taken by contemporary hospitals and clinics. The dispensary history stressed not only patients' symptoms but also their personal circumstances with regard to housing, rent, family size, occupation, salary, indebtedness, diet, sanitation, and household ventilation. A visiting nurse, who was invariably a woman in this phase of the development of the nursing profession, then supplemented these data with a home inspection to assess congestion, the health of all residents, and their financial situation.

Armed with this information, the dispensary then drew up a therapeutic regimen for the newly admitted patient. As always in the treatment of consumptive patients before the age of antibiotics in the 1940s, the objective was to reproduce the main features of sanatorium care as far as possible in densely populated and impoverished settings. To that end, it was essential to practice what dispensaries termed "social treatment." This was founded on the principle that to return patients to the same unsanitary conditions in which they had originally fallen ill was to issue a death sentence. Consciously evoking the earlier nineteenth-century medical doctrine of "social medicine," the dispensary therefore treated not just individual patients but also the social, economic, and physical environments in which they were immersed.

The dispensary, therefore, attempted to ensure that even in the most crowded tenement, each tuberculous patient had a private sickroom with no other occupants. The designated room was then cleared of all dust-collecting furnishings and frequently disinfected, while the sickbed was moved to the center for the convenience of caregivers. Most important, the window—if there was one—was opened wide to admit light and air. This constituted an urban version of the wilderness cure, and the dispensary provided adequate bed linen for winter weather. At the same time social treatment implied a disposable cuspidor for all expectoration, admonition against open-faced coughing, rules limiting visitors and exertion, and careful instruction on the need for prolonged and total rest in the recumbent position.

If the crowding or the floor plan of a tenement made such conditions impossible to implement, the dispensary sought more suitable accommodation so that patients would have at least a chance of restoring their shattered health. Similarly, if a family's economic circumstances made adherence to the prescribed regimen problematic, the staff sought donations from philanthropists in order to subsidize rent, redeem pawned clothing or furniture, and settle debts. Dispensary staff also used their influence to find suitable employment for tuberculous family members and to prevent their being dismissed because of their disability. Visiting nurses also supervised diet, ensured that premises were regularly cleaned, and made house calls to monitor patients' temperature and pulse.

In addition, social treatment entailed health education so that patients and their families could protect themselves. Dispensary nurses instructed all residents of an infected household in the basics of tuberculosis as a disease, stressing the perils of expectoration and dust. They ensured that no family members shared the sickroom; and explained that, since the family of a tuberculous patient was at serious risk, relatives needed to pay periodic visits to the dispensary for monitoring. Nurses also distributed posters and informed patients of forthcoming lectures and exhibits sponsored by the war on tuberculosis as well as relevant events promoted by health departments and medical societies.

The Preventorium

In the early twentieth century dispensaries promoted a new institution known as a "preventorium" to complete their program of outreach into impoverished neighborhoods. The medical basis for the preventorium was a new understanding of the epidemiology of tuberculosis made possible by Koch's tuberculin skin test. Although discarded, in the midst of wide controversy, as a remedy for TB, tuberculin during the Progressive Era became the standard means of detecting "latent" tuberculosis. A positive reaction to the test was a reliable means of determining exposure and "abortive" but uncured disease. Extensive administration of the test revealed that latent disease of this kind was unexpectedly prevalent among children and that the onset of active disease among adults frequently resulted not from a new primary infection, but rather from the triggering of disease unknowingly contracted during childhood. In the concise expression of the Nobel Laureate Emil von Behring (1854–1917), "Phthisis in the adult is but the last verse of the song, the first verse of which was sung to the infant at its cradle."[19]

If childhood lesions were therefore the source of the worldwide pandemic of tuberculosis, pediatricians theorized, an effective antituberculosis campaign strategy—one that might even lead to the high road of eradication—was to prevent primary infections among children. By contemporary thinking, there were two means to accomplish this objective. One was to minimize the contact of vulnerable children with tuberculous family members by removing them to the care of nurses and educators and to the company of other children who had been screened by a period of quarantine to determine that they were not infectious. The other means to protect children was to improve their resistance by strengthening their immunity, employing all the lessons learned from sanatoria. Thus, children would be placed in hygienic surroundings, given an ample and nutritious diet, go to bed early, receive a carefully regulated exercise regime, and undertake an outdoor life achieved by schooling in fresh-air classrooms, sleeping on porches in all weather, and participating in outdoor sporting activities. During the children's absence from their homes, moreover, nurses were employed to oversee the sanitary improvement of those homes.

Precursors to the fully developed preventoria were set up in various national contexts. Among the earliest was the French Society for Tubercular Children (Oeuvre des Enfants Tuberculeux) founded in 1888, which pioneered the concept of pediatric tuberculosis facilities. Another was the Ste. Agathe des Monts Sanatorium, popularly known as the Brehmer Rest Preventorium, which opened its doors in Canada in 1905. Both of these pioneering institutions began with the notion that all severe illnesses rendered children susceptible to tuberculosis; therefore, the only way to rescue "sickly children" from TB was to shelter them during a lengthy convalescence from all other severe infections while they slowly rebuilt their strength. Neither the French nor the Canadian institution attempted a consistent prophylactic regimen or followed a precise medical philosophy.

For tubercologists, however, the novel lesson from both prototypes was that preventive measures for children were key tools for combating consumption. Here the decisive initiative was taken in the United States in 1909 when an initial group of ninety-two children arrived at a newly constructed preventorium—"an absolutely unique institution," according to the *Nashville Tennessean*—at Lakewood, New Jersey. The venture began in a blaze of nationwide publicity due to the array of luminaries who endorsed it: philanthropist Nathan Straus, industrial titan Andrew Carnegie, social reformer and photojournalist Jacob Riis, prominent physicians Abraham Jacobi and Herman Biggs, and virtually the entire New York press corps. Some supporters were motivated by the medical revelations of the tuberculin skin test; others by compassion for indigent children; and still others by the calculation that preventoria, if successful, would represent an enormous savings compared with the prodigiously expensive care provided to adults by sanatoria and dispensaries. All agreed, however, that the time had come to experiment with a fresh-air boarding school for impoverished children at risk.

After the opening of this first preventorium, which was relocated from Lakewood to Farmingdale, similar facilities were set up across the US and abroad. The war on tuberculosis thus came to embrace prevention as well as cure, and to add a new institution to the established institutions of sanatorium and dispensary. Furthermore, as the preventorium movement expanded, it diversified. Most preventoria were schools under strict medical direction, but some were summer camps; and New York City boasted floating day-care facilities such as the ferryboats that took Manhattan children for day-long outings in New York Harbor. These included both the Bellevue Day Camp Boat and Day Camp Manhattan, which operated from Pier Four to benefit children who were malnourished or had been exposed to active tuberculosis.

Health Education: Sanitary Consciousness

Education went beyond the dispensary and preventoria, however. The war on TB involved a massive, ongoing effort to provide the general population with a sanitary consciousness based on understanding the danger posed by germs. Here the tuberculosis associations, assisted by municipal health departments, took the leading role. They covered wall spaces—in railroad stations, post offices, factories, hospitals, and schools, together with the sides of buses and trams—with notices proclaiming the dangers of "promiscuous spitting" and the uncovered cough. Contemporary cultural practices provide the context for appreciating the emphasis on spittle. In an age when cigars and chewing tobacco were fashionable among men, sputum was ubiquitous, especially in the United States during the Victorian and Progressive eras. Charles Dickens, touring the United States in 1842, reported his revulsion upon discovering that yellow-tinged saliva was everywhere. "In all the public places of America," he wrote, "this filthy custom is recognized. In the courts of law the judge has his spittoon, the writer his, the witness his, and the prisoner his; while the jurymen and spectators are provided for, as so may men who in the course of nature must desire to spit incessantly."[20]

In the view of the Maryland Tuberculosis Commission, New York City was exemplary in the rigor of its measures to repress so grave and general a danger. There, a municipal ordinance forbade spitting "upon the floors of public buildings and elevated cars, in railway stations and on steps, and on the pavement."[21] To enforce the law, the city deployed teams of plainclothes police to observe violations and make arrests, and it empowered magistrates to impose fines of five hundred dollars and imprisonment for up to one year.

In addition to controlling sputum, the campaigners organized lectures, drafted health brochures, and printed articles in newspapers. Most innovatively, beginning in 1904, the National Tuberculosis Association and the International Congress on Tuberculosis organized exhibits, both fixed and traveling, on all major aspects of the disease—its history, varieties, cost, mode of transmission, prevention, epidemiology, and treatment. To drive the message home, the organizers arranged panels of expert or celebrity speakers and displayed lungs devastated by TB in glass bottles filled with formaldehyde. These exhibitions, which attracted millions of visitors in major centers, were among the most influential forms of persuasion produced during the "war."

As a further means of outreach, the National Tuberculosis Association published three important journals: the *Journal of Tuberculosis,* which appeared in 1899; the *Journal of the Outdoor Life,* which began in 1903; and the *American Review of Tuberculosis,* which started in 1917. The first two were aimed at informing the general public; the third was aimed at the medical profession. The association also produced a film—*The Temple of Moloch* (1914)—that explained the new understanding of TB. It did so in the form of an instructional melodrama recounting the misfortunes of a pottery worker and his family whose impoverished environment and failure to heed medical advice doomed them to become human sacrifice to the fearful god Moloch, whose insatiable appetite for human flesh represented TB.

Assessment of the "War"

What was the impact of the war on tuberculosis? In some aspects of society, its influence was decisive. The crusade persuaded the general public that TB was a dangerous, infectious disease transmitted principally by the poor. The new medical philosophy, as we have seen, radically altered literary representations of consumption, styles of dress and facial hair, interior decoration, modes of maintaining domestic hygiene, and the treatment of library books.

Unintentionally, however, this message generated a backlash of stigma and a morbid fear of people who suffered from the disease. The press described the result as reminiscent of medieval attitudes toward lepers and early modern views of the victims of plague. As a result, society further marginalized social groups—the poor and ethnic minorities—that were already social pariahs. The analogy with leprosy and plague drawn by contemporaries is, however, exaggerated in two decisive respects. Victims of TB were not forcibly banished for life to leprosaria, nor were they subjected to violence like witches, Jews, and foreigners during times of plague. Social attitudes hardened in ways that were often unpleasant and discriminatory, but social order was untroubled and the talk of witch hunts and riots was metaphorical rather than literal.

The new scientific understanding of tuberculosis due to Koch's discovery and trumpeted by the "war" also profoundly affected medical practice. Therapies based on Hippocratic understandings such as depletion by means of venesection and emetics or the advice to take the cure in distant locations gradually ceased to be administered. Patients instead received treatment delivered either in sanatoria or at home that was based on the therapeutic triad of air, rest, and diet. Furthermore, during the interwar years, the United States witnessed the use of heroic experiments in the surgical management of TB.

In terms of patient outcomes, however, there is no robust evidence that the therapeutics devised during the decades of the war on tuberculosis were more effective than traditional humoral approaches. Physicians and institutions experienced a surge of optimism regarding their curative powers, but their claims are anecdotal and lack firm statistical grounding. Furthermore, the physical mechanisms underlying the supposed advances were never delineated. Retrospective views by chest specialists after World War II regard the medical armamentarium deployed by the crusade as probably ineffectual but unlikely to have inflicted harm—with the exception of surgical interventions. Indeed, there are reasons to suspect that patients were likely to have benefited—at least psychologically—from the sense of optimism and empowerment that accompanied the rise of the sanatorium.

What changed more demonstrably was not medicine itself, but rather the doctor-patient relationship. The war on tuberculosis proclaimed a new imperative—that an unquestioned authority by physicians over their patients was essential to recovery. Only they, and not the patients, could properly read the signs of tuberculosis made visible by the microscope, the thermometer, or the x-ray and made audible by the stethoscope. Only the practitioner could therefore assess the progress of the disease and determine the proper course of treatment. This new assertion of authority was expressed in chest specialists' need for what they termed "absolute control." This aspiration found expression in the minute details of sanatorium regulations and the sanctions physicians wielded against noncompliance.

Undoubtedly the most important question, however, is the one raised by the debate over the McKeown thesis (see Chapter 11). To what extent did the war on tuberculosis contribute to the recession of the disease that was visible in the most advanced industrial powers such as Britain and the United States by the middle decades of the nineteenth century and nearly everywhere in western Europe by the early twentieth? It would be mistaken to regard the major decline of the tuberculosis epidemic that occurred across the industrial world as causally determined by the conscious policy of public health officials, campaigners, and medical practitioners. The beginning of the decline of TB predated the launching of the "war," and its swift and steady progress is difficult to explain in terms of the mechanisms available to the campaign. McKeown is undoubtedly correct to identify factors separate from, and more substantial than, sanatoria, dispensaries, and education. Improved diet, better housing, more sanitation, higher literacy rates, reduced working hours, increased legislation of child labor, higher wages, the development of trade unions—these factors significantly improved the living standards of working men and women, who were the principal sufferers from the epidemic, and they did so in ways that were important in significantly reducing the risk factors for TB.

At the same time, by modern epidemiological understandings, the "war" devoted much of its energy to issues that were largely irrelevant to the etiology of tuberculosis. An example is the preoccupation of the campaign with spitting, dust, and fomites, which are minor factors in the transmission of the disease in comparison with the dominant mode of transmission by the inhalation of airborne droplets after coughing, sneezing, talking, or simply breathing. A large proportion of the resources of the campaign were systematically devoted to preventing transmission in manners that, by later epidemiological understandings, were limited or negligible in their effect.

On the other hand, despite many caveats, it would be incorrect to maintain that the instruments of the "war" were entirely unrelated to the sharp reduction in mortality and morbidity from the white plague. Precision is impossible, as reliable statistics for the relevant variables do not exist; conclusions are necessarily conjectural. But the logic of the situation suggests that sanatoria, for example, played an unquantifiable but important part. The establishment of sanatoria removed consumptives from crowded urban slums, effectively quarantining them and preventing them from infecting others. Since the total sanatoria population of one hundred thousand in a country as populous as the United States—with 92 million inhabitants in 1910—was limited, the impact also was limited. Sanatoria, in other words, reinforced a trend already under way for other reasons.

On the other hand, the International Congress on Tuberculosis estimated in 1908 that at any one time there were approximately five hundred thousand cases of active TB in the total US population. By such a calculation, the possibility of isolating a fifth of that number in sanatoria made a sensible difference in the incidence of the disease, because each TB patient was thought to infect as many as twenty other people a year if no effective prophylactic or therapeutic measures were taken. Such an institution, therefore, would not halt the epidemic, but it would lower its impact by slowing transmission. The other instruments of the antituberculosis campaign are probably similar in their effect. Five hundred dispensaries, dozens of preventoria, and widespread health education are unlikely, even in combination with sanatoria, to have played more than a partial role in turning the tide of the great tuberculosis epidemic.

The war on tuberculosis, in other words played a significant part in lessening morbidity and mortality, continuing the decline that had begun with better sanitation, higher wages, improved housing, and increased education. To advance further, a new outside tool was required from the laboratory. This was the introduction of antibiotics, and especially streptomycin, in the 1940s that made possible the next major advance toward the control of the disease and even raised the tantalizing prospect of its eradication. No single factor brought about the retreat of TB. Social and sanitary reform; a health-care system providing access to care, case tracing, and isolation; and technological tools were all required to approach the final goal.

Nevertheless, the "war" had another lasting but unmeasurable public health legacy. As we have seen, it created a lexicon that gravitated toward the revealing descriptor "social"—social treatment, social work, social medicine, social problem, social care, social class, social disease, social outlook. Thus the International Congress on Tuberculosis meeting in London in 1901 provided a revealing sign of the times, as its president, the American Edward Thomas Devine (1867–1948), stressed. Unlike most medical conferences, this antituberculosis event placed "social" in the title of one of its sections: "The Hygienic, Social, Industrial, and Economic Aspects of Tuberculosis." It also selected Devine, who devoted his life to social work and social reform, to deliver the keynote address. As the American explained:

> It may be that we have this section . . . precisely because physicians and sanitarians have of late begun better to understand their business and to gauge the height and the depth and the multifarious aspects of this their ancient enemy. It may be that they have come to look beyond the patient to his family and his neighbor. . . . May it not be that the medical profession is here giving evidence, by establishing this section, . . . that it realizes at last that to overcome tuberculosis something more is needed than the

> treatment of the individual patient, something more than the enforcement of the most enlightened health regulations? . . . We shall ask high authorities in jurisprudence and in medicine to discuss the principles on which the state should exercise its police power for the protection of health. . . . Into these spheres also the campaign against tuberculosis must remorselessly extend.[22]

As it thus expanded its mandate by becoming social as well as medical, the "war" also provided hope at a time when there were no other known defenses against the most devastating medical problem of the age. Few doubted that the campaign was at least partially successful, and for half a century it tirelessly imparted the lesson that the most deadly and widespread disease of the era could be met headlong by preventive measures, access to care, and societal reform. The nurse Elizabeth Crowell, for example, explained that a dispensary nurse spent "very little" of her time and energy on "technical nursing": "By far the larger part of her work will be instructive and social. Infinite tact, patience, and infinite perseverance alone will accomplish the installation of the fundamental principles of hygiene and sanitation into unwilling, uninterested, prejudiced, and uneducated minds."[23]

It is tempting to conjecture that this long war on tuberculosis was one of the factors that prepared the way for the establishment of the "social state" that was constructed in western Europe after World War II. From that time European societies adopted the final goal toward which Devine and Crowell had pointed and that might even be defined as "tuberculosis care for all." Even the United States came close to establishing national health care of that sort at the same time but recoiled when President Harry Truman insisted that the country could not afford both health and Cold War rearmament.

The Postwar Era and Antibiotics

After World War II an era of unalloyed optimism prevailed in the world of TB control and lasted until the 1980s. This confidence was not founded on the traditional arsenal of the war on tuberculosis—sanatoria, dispensaries, and education. The structures of the "war" were instead rapidly dismantled as ineffectual and unnecessary in the light of two new scientific tools, one preventive and the other curative. Together, these instruments seemed to provide hope not simply for better control, but for the final eradication of tuberculosis from the earth.

The preventive weapon was a vaccine—the so-called bacillus Calmette-Guérin (BCG) vaccine, which had been developed in line with the methodology employed by Edward Jenner against smallpox and by Louis Pasteur against rabies. BCG was based on a live, attenuated bacillus—*Mycobacterium bovis*—responsible for bovine tuberculosis that conferred a crossover immunity to *M. tuberculosis*, just as Jenner's cowpox gave rise to immunity against variola major.

BCG was given its first mass trial in the 1930s on Native American reservations. Joseph Aronson of the tuberculosis unit of the Bureau of Indian Affairs reported that BCG had an efficacy of 80 percent, although he did so on the basis of statistical data that were never clearly substantiated. Nonetheless, on the basis of this optimistic finding, WHO and UNICEF championed BCG as a solution to the world's tuberculosis problem. Together with the Scandinavian countries, they launched one of the largest public health programs ever devised, in the form of an effort to vaccinate the world's population, known as the International Tuberculosis Campaign (ITC). This effort was the first mass public health campaign of the newly founded WHO, and it served as the inspiration for later international vaccination programs.

Organizationally, the ITC was remarkably successful as a prototype for later international efforts. Between July 1, 1948, and June 30, 1951, it carried out nearly 30 million prevaccination tuberculin tests in twenty-two countries and Palestinian refugee camps, followed by 14 million vaccinations among those whose tuberculin test was negative. Prophylactically, however, the results were disappointing, and the 123 reports produced by the campaign were inconclusive with regard to the central issue of efficacy. A major reason is that the campaign began with no knowledge of the fact that there are multiple strains of *M. tuberculosis* and that the efficacy of BCG is strain-specific, varying from Aronson's famous 80 percent to nil. His results were discovered not to be replicable in other contexts.

Far from moving the world closer to the eradication of TB, BCG has therefore been mired in controversy. Some countries—most notably the United States—refused to participate on the grounds that its efficacy was unproved. American officials also argued that BCG posed the danger of misleading recipients into thinking that they were protected and therefore did not need to take additional precautions. The ITC thus became the most wide-reaching and sustained public health initiative ever made to no measurable effect. But, as its early proponents stated at the outset as one argument in favor of launching the ITC, at least it did no harm.

Far more promising was the opening of a new era in medicine with the discovery of antibiotics, beginning with penicillin, which was discovered in 1928 by Alexander Fleming and put to therapeutic medical uses from 1941. Penicillin had no relevance to TB specifically, but it opened the way to the development of a series of additional "magic bullets" and to the belief that TB could be eradicated globally by a spectacular technological fix. The first of these "wonder drugs" applicable to tuberculosis was the antibiotic streptomycin, which was discovered at Rutgers University by Selman Waksman in 1943. It was used the following year to effect a seemingly miraculous cure in its first trial on a critically ill tuberculosis patient who made a full and rapid recovery. Waksman received the Nobel Prize in 1952 for his work. Streptomycin was then followed by isoniazid in the early 1950s and by rifampicin in 1963.

With these new weapons in its arsenal, the medical profession believed that tuberculosis, the scourge of the previous three centuries, had become a readily treatable and curable disease. In the United States, for instance, the incidence of TB plummeted by 75 percent, from more than eighty thousand cases annually in 1954 to twenty thousand in 1985. The government predicted eradication in the US by 2010 and worldwide by 2025. Fearing obsolescence, the National Association for the Study and Prevention of Tuberculosis changed its name to become the American Lung Association and the *British Journal of Tuberculosis* became the *British Journal of Diseases of the Chest*. Meanwhile, the structures of the war on tuberculosis were systematically dismantled. The sanatoria, emptied of their patient populations, closed one after another, and the dispensaries became superfluous. Research on the disease languished, and funding for public health measures to combat it dried up.

The New TB Emergency

Unfortunately, the decline, both in the United States and abroad, slowed during the 1980s and then began a steep upward climb. The curve charting the incidence of TB in the United States leveled in 1985, when 22,201 new cases of active disease were recorded—the lowest figure ever recorded. The incidence rose slightly in 1986–1987 and then climbed steeply into the 1990s. In 1991, 26,283 new cases were confirmed—an 18 percent increase over 1985. Most alarming was New York, the epicenter of the outbreak, where Margaret Hamburg, the commissioner of health, sounded the tocsin: "New York City is in the midst of a severe tuberculosis epidemic with little or no relief in sight. In 1991 there were close to 3,700 reported cases, representing a 143 percent increase since 1980. With close to 15 percent of all cases nationwide, our case rate is five times the national average."[24]

The conditions promoting the upsurge in the US were multiple: the concurrent epidemic of HIV/AIDS, immigration from countries with a high prevalence of tuberculosis, the spread of drug resistance, and the lack of patient adherence to the standard treatment regimen, especially among people who were homeless, mentally ill, or impoverished. Most important, however, was the decision to dismantle the tuberculosis campaign as a result of the confident assumption that antibiotics would eliminate the disease. In 1993 Congressman Henry Waxman of California also condemned the apathy of the federal government, which failed to respond to the crisis:

> TB is not a mystery or a surprise. . . . Its outbreak has been steadily documented. The needs to respond have been outlined, but actions have been delayed. Little has been done because little money has been provided. If there were a court of public health malpractice, the Federal Government would be found willfully negligent in the case of TB. During the time that little was done, the problem has, of course, compounded. In 1988 the public health service estimated the annual cost of TB control to be $36 million per year. The President's budget that year was just over one-tenth of that.[25]

But the most affected areas globally were Eastern Europe, Southeast Asia, and sub-Saharan Africa. In 1993, far from anticipating eradication, WHO took the unprecedented step of announcing that the TB epidemic was spiraling out of control and proclaimed a world emergency. United Nations figures indicated that in 2014, 9.6 million people fell ill with TB and 1.5 million died, including 140,000 children—all as a result of a disease that had been declared both preventable and treatable. What had happened to destroy the earlier confidence and optimism?

Here a multiplicity of factors was at work. Of these, the most important was the concurrent global pandemic of HIV/AIDS (Chapters 19 and 20). Tuberculosis rapidly established itself as the most important "opportunistic" infection complicating AIDS and as the major cause of death for AIDS patients. As a major immunosuppressive disease, HIV/AIDS activated abortive cases of tuberculosis, transforming them into active disease. At the same time it made sufferers severely vulnerable to new infection and reinfection. The global AIDS epidemic, therefore, is a major determinant of the TB pandemic that followed in its wake. In combination, they are known as HIV/TB.

In some of the world's most resource-poor countries, especially in southern Africa, HIV/AIDS and TB are major afflictions of the general population and major causes of morbidity, mortality, social hardship, and inequality. In the industrial world, on the other hand, the situation is profoundly different. There, HIV/AIDS is not a disease of the general population but rather of marginalized and relatively impoverished groups—racial and ethnic minorities, immigrants, the residents of nursing homes, the prison population, the homeless, intravenous drug users, and people with compromised immune systems from causes other than HIV such as diabetes.

But HIV/AIDS, although it may be the leading factor in the vast contemporary upsurge of tuberculosis, is not alone. Another major cause of the global emergency is drug resistance. Under selective evolutionary pressures, *M. tuberculosis* developed resistance to the "wonder drugs" deployed against it. This was first documented in the 1970s when bacilli became resistant first to one antibiotic and then to all of the first-line medications. This was known as multidrug-resistant tuberculosis (MDR-TB). When antibiotics are overprescribed, or treatment courses are interrupted when symptoms are resolved but a complete cure has not yet been achieved, the evolution of resistance is faster and more powerful. Normally, a cure regimen for tuberculosis lasts six to eight months, but since patients begin to feel better after three weeks, they commonly stop their treatment. On relapse, 52 percent of such noncompliant consumptives have tuberculosis that is drug-resistant. Recent years have seen the emergence of extensively drug-resistant tuberculosis (XDR-TB). In the judgment of a group of physicians in Nigeria, interrupted or improper treatment of TB is worse than no treatment at all, "because the patient whose life is prolonged excretes tubercle bacilli for a longer period and his bacilli may now be drug resistant."[26]

In addition to the general factors of HIV and drug resistance, local conditions also play a factor in the TB emergency. Large-scale displacements of populations due to war, economic disaster, political oppression, and environmental catastrophes lead to unsanitary refugee camps, impoverished diets, and other conditions in which pulmonary diseases thrive. Similarly, the "war on drugs" and the high rates of incarceration that result promote crowded prison cells that favor transmission. The collapse of medical services in Eastern Europe with the fall of communism favored disease by making access to treatment difficult or unavailable.

To confront the TB emergency, WHO and UNICEF announced what they proclaimed as a novel and effective approach that was also thought to be inexpensive and not reliant on new technologies or scientific discoveries.

Announced in 1994 and named DOTS (directly observed treatment, short course), it was conceived—without efficacy trials to determine its usefulness in practice—as an administrative panacea to the TB epidemic. The strategy was premised on the assumption that the problem of noncompliance by TB patients could be overcome by placing patients under surveillance by public health personnel while they took their standardized "short course" of antibiotics throughout the six- to eight-month regimen. As additional components of DOTS, WHO called for political commitment by host nations, a guaranteed drug supply, case detection by sputum smear analysis, and follow-up assessment. More recently, DOTS has also been supplemented by DOTS Plus, a program providing a cocktail of second-line antibiotics for cases of multidrug-resistant TB.

DOTS, of course, was not novel because its essential component—comprehensive surveillance of adherence to the treatment regimen—was, without attribution, a return to the strategy underlying the rest porches and verandahs of the sanatoria. It has remained the strategy of choice, but within a decade of being launched, DOTS had demonstrated that its effect in resource-poor environments was severely limited. In impoverished countries where the impact of TB was most devastating, DOTS could not be fully implemented. Public health facilities and trained personnel were lacking, the population did not have enough education and understanding of the mechanisms of the treatment regimen, medications were often in short supply, and the obstacles people needed to overcome to attend clinics—distance, inadequate transport, lost wages, and ill health itself—were overwhelming. In such settings randomized studies have found little difference in completion rates between DOTS patients and patients who administer their own medications. The new strategy to overcome noncompliance frequently fails where it is most needed.

At the same time DOTS revealed a fundamental strategic flaw. It was intended, like the sanatorium regime on which it was modeled, to deal with TB as if it were a stand-alone epidemic. DOTS in that sense was anachronistic in the midst of a concurrent pandemic of HIV/AIDS; that is, it did not provide a strategy for dealing with the co-epidemic that was its principal driver or with the economic and social conditions that enabled both diseases to flourish. As the twenty-first century reaches its middle decades, the tuberculosis epidemic clearly requires new research, new tools, and new approaches.

CHAPTER 16
The Third Plague Pandemic
Hong Kong and Bombay

The Modern Pandemic

The third pandemic of bubonic plague began in Yunnan province in China, perhaps as early as 1855. After recurring there in successive waves, it spread to Canton, Macao, and Hong Kong in 1894. Following the pattern of the cholera epidemic in Egypt a decade earlier, several countries sent teams of microbiologists to investigate the disaster. Their goals were to determine the pathogen that caused it, unravel its epidemiology, and suggest means to prevent its further diffusion. In June 1894, working independently and competitively, the French and Japanese microbiologists Alexandre Yersin (1863–1943) and Shibasaburo Kitasato (1853–1931) almost simultaneously identified *Yersinia pestis* as the pathogen responsible for bubonic plague. This substantial advancement of science did not, however, lead automatically to a cure, a prophylactic, or a viable public health strategy. From Hong Kong, the bacterium promptly followed the steamship routes east, west, and south to major port cities around the world.

Between 1894 and 1900 the plague traveled east to Kobe and Nagasaki in Japan; across the Pacific to Manila, Honolulu, and San Francisco; and around Cape Horn to Santos, Buenos Aires, Havana, New Orleans, and New York. At the same time it journeyed west and south to Sydney, Bombay (now Mumbai), Cape Town, Madagascar, Alexandria, Naples, Oporto, and Glasgow. In this way the third pandemic became the first authentically global, or "oceanic," pandemic, touching all five continents through their principal seaports. By radically reducing travel times, the transportation revolution of the steamship and the railroad enabled the plague bacillus to reach the Americas for the first time.

This third, or "modern," pandemic followed a path that was radically different from the Plague of Justinian and the Black Death. After disembarking with migratory rats and their fleas, the earlier scourges gave rise to large-scale epidemics in the ports where they put ashore. Then the disease moved irresistibly onward—overland, by river, and by coastal shipping—unleashing a prodigious mortality among all classes, races, and religious groups. Then, having devastated a community, the first two pandemics departed. The oceanic pandemic followed a strikingly different course. Unlike its predecessors, the modern pandemic was markedly unequal in its social profile and limited in its mortality. In the industrial world the disease arrived at major seaports, where it caused only limited outbreaks. Frequently, it did not cease entirely after a season, but continued to smolder. Its characteristic pattern was to produce cases for several years, without bursting into flame and spreading amidst a firestorm of contagion.

An illustrative example in Europe is Portugal's Atlantic port of Oporto, the city most stricken in the industrial world. The disease arrived there in early June 1899 by means of the transshipment of fabrics and grain from India via London, Liverpool, Hamburg, and Rotterdam. Fleas nestling in garments or borne by stowaway rats were the vectors responsible. As a result, the first documented victims in Portugal were among stevedores who offloaded cargo from Bombay. After being declared on June 5, plague persisted for more than a year. Unlike the Black Death, however, pestilence in modern Oporto exclusively attacked the overcrowded and vermin-infested slums of the city center. Furthermore, instead of setting off a major and uncontrollable epidemic, the outbreak was described by surgeon Fairfax Irwin of the US Marine Hospital Service as "singular in its slowness."[1] It struck an average of fewer than ten people a week from August until the end of the year, killing 40 percent of them. In February 1900, the Portuguese government prematurely pronounced the city plague-free, only to experience an ongoing trickle of cases through the summer and fall.

By contrast with the industrial West, the third pandemic produced disasters of a different order of magnitude as it demonstrated in the colonial world of China, Madagascar, Indonesia, and, above all, the British colony of India. These places suffered calamities reminiscent of the Black Death, with devastating mortalities, flight, economic collapse, social tensions, and the application of the draconian antiplague measures devised during the Renaissance. Therefore, this third pandemic produced a lopsided burden of suffering as it spread along international fault lines of poverty, hunger, and deprivation. North America suffered a few hundred deaths from the disease after its onset in 1894, western Europe a few thousand, and South America some twenty thousand. At the opposite end of the spectrum, India, the most afflicted area of the world, endured 95 percent of the total global mortality of the third pandemic before it finally subsided. That country lost 20 million people to *Y. pestis* from a population of 300 million, and the CFR approached 80 percent. Because of this distinctive distribution and its racial connotations, contemporary Western observers referred to the third pandemic as the "Oriental plague" and the "Asiatic pestilence."

This modern plague was also unequal in its impact by economic stratum. Within individual countries and cities, it did not strike indiscriminately but instead exhibited a marked preference for the bodies of the poor. At Manila, Honolulu, and San Francisco, bubonic plague was overwhelmingly an affliction of Chinese "coolies" (manual laborers) and Japanese workers. At Bombay, the epicenter of the disaster, pestilence caused trifling numbers of deaths among Europeans but devastated the slums and their "native" denizens—Hindus, Muslims, and, to a lesser extent, Parsees. Similarly, the most afflicted districts of Bombay were those that were the most notoriously unsanitary and economically depressed—Mandvi, Dhobi Talaq, Kamathipura, and Nagpada. In these neighborhoods plague ran amok, especially among low-caste Hindus. By contrast, the disease largely spared the "other Bombay" of wealth, elegance, and broad boulevards. In this other Bombay a tiny Indian elite—merchants, bankers, industrialists, and professionals—experienced death rates as low as those of resident Europeans.

So pronounced was the differential impact of plague by ethnicity that it reinforced racialized conclusions in medical science. Physicians and health authorities postulated that diseases demonstrate the innate inequality of the races. Flagrantly ignoring the history of plague in Europe during the Plague of Justinian and the Black Death, this argument held that whites enjoy a high degree of genetic immunity to the disease. Those espousing this view construed plague as an affliction of the dark-skinned and the uncivilized and of those unfortunate Europeans who lived in dangerous proximity to natives. Whites were safe unless they pressed their luck. Here was a legitimation for racial segregation.

An important implication was that the impoverished and the ill were responsible for their own medical misfortunes, as the press repeatedly opined. According to an 1894 newspaper article, China, where the modern pandemic originated, was an "unchanging Oriental country" and "the filthiest and nastiest country on the face of the globe."[2] The lives of "Chinamen" were inherently pathological, conducted in "nests" of corruption, vice, and dirt. People holding such beliefs found it logical that Black Death, a disease of medieval Europe, should smite the Orient on the eve of the twentieth century. Plague sprang, they thought, "from a total disregard of all sanitary conditions" and from the refusal of "pig-tails" to accept the wisdom of "opening up that country to Western civilization and enlightenment."[3] In a recurrent opinion, the Orient, having learned nothing during the intervening centuries, was tragically backward. It was therefore condemned to relive the ordeal of the Great Plague of London in 1665–1666 as described by Daniel Defoe. An undeveloped and un-Christian country, China was "paying a big penalty for national uncleanliness and national ill-faith."[4]

This colonial attitude served as the foundation for scientific hubris. The international medical community made two reassuring assertions. The first was that Europe and North America were safely protected by civilization and science. Robert Koch pronounced that plague always retreats before the advance of civilization. The second claim was that, even in the Orient, scientific understanding of the disease would enable colonial authorities to exterminate it—if only the uneducated natives would cooperate. In 1900, the US Surgeon General Walter Wyman triumphantly announced that science had all the knowledge needed to "stamp out" the disease. In his view, which bore little relationship to the course of events that followed, "This disease furnishes a striking illustration of the scientific advance of modern medicine. It was not until 1894 that positive knowledge of its true nature became known. Now its cause, method of propagation, and the means to prevent its spread are matters of scientific certainty."[5]

Hong Kong played a crucial part in the worldwide dissemination of the modern pandemic. As the world's third leading seaport, it was linked by trade and migration to port cities on every continent. When plague first arrived there from the Chinese province of Yunnan in the spring of 1894, there was a clear threat of global propagation. Indeed, outbreaks at places as widespread as Honolulu, Manila, Bombay, and Oporto were directly traceable to the arrival of steamships from Hong Kong. Plague first attacked the colony in the spring of 1894 and subsided in September. Precise numbers of victims were impossible to obtain because of Chinese efforts to conceal cases from the British authorities, so statistics are significant underestimates. Officially, however, in 1894 plague claimed three thousand lives in a population of two hundred thousand, half of whom fled the island. The *British Medical Journal* suggested that the official figures were a fraction of the real numbers, which it estimated at ten thousand.

The victims were almost exclusively Chinese laborers who inhabited the teeming slums of the district of Taipingshan. There, in "health district five," population density reached 960 per acre as opposed to a density of 39 in the prosperous European neighborhood of "health district three" where almost no one among the ten thousand white residents perished. Apart from troops of the garrison and sailors, Europeans lived on the hill leading to "the Peak"—the highest point on the island, eighteen hundred feet above sea level. There, whites inhabited bungalows built to house the economic and racial elite of Hong Kong society. Residents of the hill, noted the press, "enjoy almost entire immunity."[6]

A distinguishing feature of the third pandemic is that it became entrenched in the localities it attacked, returning year after year for decades as a seasonal epidemic. After appearing in 1894, plague spared Hong Kong in 1895; but it returned in 1896 and erupted every February or March thereafter until 1929, although on each occasion it ebbed away reliably in the early autumn. These regular annual outbreaks were highly unequal in their intensity. In some years mortality was slight, whereas in 1912, 1914, and 1922 the colony suffered tragedies that compared with 1894. In aggregate, approximately twenty-four hundred cases were officially reported in the colony, of which more than 90 percent ended fatally. In other words, 10 percent of the residents of the island perished of plague, and in every year the epicenter of the epidemic was the neighborhood of Taipingshan.

The responsibility, according to one daily paper, was that of the coolies, who had "a developed taste for tenement houses" and positively delighted in crowding by the hundreds into "a house built for one ordinary European family. All the accommodation the coolie boarder asks is a shelf commodious enough to smoke opium on."[7] Here in their lairs, the English-language press complained, the Chinese willfully listened only to native practitioners who were "all quacks," coiffing their "absurd decoctions"—"in preference to adopting the common-sense recommendations of European physicians."[8]

Germ Theory, Miasma, and Plague

European physicians and health authorities working in colonial contexts, together with their local colleagues, were frequently misunderstood. It was assumed that they were distant from the latest developments in European science and that their interpretations of plague were a reflection of medical backwardness. In fact, however, medicine in the 1890s was the most cosmopolitan of all professions, and physicians in the farthest reaches of the British Empire carefully followed, and took part in, debates in the field. With regard to bubonic plague in particular, the discovery of the plague bacillus was made in Hong Kong, and both Kitasato and Yersin advised the Board of Sanitation there and discussed health policy extensively with its members. Important public health authorities in the colony such as Drs. James Cantlie and J. A. Lowson were influential figures in the international medical establishment, and the governor, Sir William Robinson, closely followed the medical science of bubonic plague. Their views reflected the latest scientific understanding.

In 1894 the germ theory of disease was therefore the dominant medical philosophy in Hong Kong, just as it was in the metropole and throughout the British Empire. The discovery of *Y. pestis* (originally called *Pasteurella pestis*) was instantly acclaimed. Far from revolutionizing the understanding of plague when they declared it an infectious disease caused by bacteria, Kitasato and Yersin confirmed the insights of Pasteur and Koch. Plague thus joined a cascade of epidemic diseases—including anthrax, tuberculosis, cholera, and typhoid—whose causative microbes had been identified.

Confusingly, however, plague was later determined to be different from the other microbial diseases whose pathogenic bacteria had already been identified. Unlike them, bubonic plague was a vector-borne disease with a far more complex etiology. Simply identifying *Y. pestis* as its cause was a vital but frustratingly restricted advance. Until the Indian Plague Commission in 1908–1909 established the complex role of rats and fleas, the epidemiology of the disease therefore remained a mystery. As the third pandemic struck Hong Kong and began its international travels, critical questions remained. How did the bacillus gain entry to the human body? Why were the poor and the inhabitants of crowded slums so overrepresented among the sufferers? Why did the disease persist in a locality, returning regularly year after year? Where did plague bacteria go during the months between the end of one outbreak and the onset of the next? How, the *British Medical Journal* asked, "in the creed of that phrase which is dear to all those who deal with epidemics," can the disease be "stamped out"?[9]

A preponderance of medical scientists postulated that locality and the hygiene associated with it provided the key to solving these riddles. Their theory held that the soil beneath a stricken city functioned as a giant petri dish that was contaminated by plague bacteria. If the soil was hygienic, as doctors deemed that it was in the civilized West, *Y. pestis* would fail to thrive. Under those sanitary conditions, a flare-up of plague would produce a few scattered cases before being extinguished. On the other hand, in the underdeveloped world, where filth abounded, the land beneath a city was a mix of earth, decaying organic matter, and sewage. The result was to provide plague bacteria with a fertile medium in which to grow luxuriantly.

Kitasato and Yersin announced that they had confirmed this hypothesis during the Hong Kong visitation of 1894. Having separately discovered *Y. pestis* in June, they had been asked by the Board of Sanitation to examine the soil of Taipingshan. Both microbiologists, again working competitively, claimed to have found the pathogen in local soil samples. This second apparent discovery suggested an analogy with the etiology of anthrax. This idea occurred most easily to Kitasato, who had worked in Koch's laboratory from 1885 until 1892. During those years he had investigated anthrax, a disease fundamental to the discoveries of both Koch and Pasteur, and he had published scientific papers on the topic.

Pasteur had famously demonstrated in 1881 at Pouilly-le-Fort that a field contaminated by infected sheep retains anthrax bacteria—later determined to persist as spores. *Bacillus anthracis* can then transmit the disease to healthy, but nonimmunized sheep, even after the passage of years. Following this analogy, Kitasato suggested that when planted in the well-fertilized soil of Hong Kong's slums, the plague bacterium had no need to be imported repeatedly in order to sustain transmission year after year. On the contrary, the claim was that *Y. pestis* established itself permanently in the filthiest microenvironments of the Taipingshan tenements—the soil, floors, drains, and walls. There it lay in wait, ready to germinate. Both Kitasato and Yersin searched for plague spores in order to make the analogy with anthrax fit perfectly, but without success. Nevertheless, they postulated that the mechanisms involved in plague at Hong Kong were similar to those of anthrax at Pouilly-le-Fort. When conditions of temperature, moisture, and nutrients permitted, the plague bacterium persisting in the city flourished and generated disease anew in healthy but impoverished bodies. During the first decade of the third pandemic, this doctrine was termed the "theory of true recrudescence." It offered an explanation of plague's mysterious capacity to recur again and again in a given place over many years.

For proponents of "true recrudescence," transmission occurred in three ways. First, by walking barefoot on manured soil or a filthy floor, humans absorbed plague bacteria through abrasions on their feet. Second, people who slept on the floor provided a portal of entry through their nostrils. And third, it was suggested that bacteria could be carried aloft in a room by dust particles or by rising effluvia. Neither the dust nor the vapors were considered to be poisonous directly. They were lethal because they carried *Y. pestis* into the air, where it was inhaled, first by rodents close to the ground and then by people at a higher level. In a striking linkage of contradictory concepts, plague was held to be a germ that was miasmatically transmitted.

This manner of conceptualizing the etiology of plague had major implications for public health strategies. Having diagnosed the disease prevailing in 1894 as bubonic plague, the Hong Kong Board of Sanitation applied the full rigor of antiplague measures. Declaring the port infected, it claimed emergency powers and introduced quarantine for arriving ships and passengers, dispatched teams of soldiers to locate suspect cases and convey them forcibly to newly activated pesthouses opened in the remote Kennedy Town neighborhood, shut up and disinfected the homes of patients, burned victims' clothing and personal effects, and buried victims' bodies in plague pits and covered them with lime. The implementation of these measures violated the religious beliefs of the population, generated flight, and produced desperate rumors that the colonial government of "foreign devils" was spreading plague as a means to rid itself of the poor while harvesting their body parts for experimentation and allowing "barbarian" soldiers to take women away to satisfy their "sordid intentions."[10] The deployment of troops from the Shropshire Regiment and the appearance of a gunboat in the harbor proved essential to secure compliance and forestall riot.

Such rigorous procedures were judged by the Board of Sanitation to be necessary but not sufficient. The measures could, the board thought, prevent the import and export of *Y. pestis,* and they could reduce its transmission within the city, thereby lowering mortality and morbidity in general and safeguarding resident Europeans in particular. On the other hand, having gained access to the soil and filth of the city, the bacteria lurked beyond the reach of the traditional battery of antiplague measures. Partially contained but not "stamped out," *Y. pestis* survived to unleash fresh epidemics at every favorable opportunity.

History, as understood by the doctrine of "true recrudescence," seemed to offer a solution. The Great Plague of London in 1665–1666 was the last epidemic of bubonic plague to scourge Britain. In retrospect and with an insight derived from the germ theory of disease, public health authorities early in the third pandemic assumed that the triumph over plague in England was the result of the Great Fire. Immediately following the plague, the fire engulfed medieval London in September 1666, and it was thought that the purifying flames thoroughly disinfected the city and its soil, thereby "stamping out" *Y. pestis* and plague, which never recurred. In strict logic, correlation and causation are entirely distinct, but the close temporal association of fire with the permanent end of the plague made this hypothesis irresistibly tempting.

Well informed regarding both the history and the scientific theory then used to explain the history, Governor Robinson drew up a plan based on the experience of seventeenth-century London. His scheme, which he implemented before retiring from his post in 1898, was to wall off tenements in Taipingshan and then torch them so that the accumulated filth of the plague-infested buildings could be cleansed and the soil beneath them permanently disinfected. As Dr. J. Ayres, a Board of Sanitation member, commented early in the outbreak, "The entire neighbourhood should be destroyed, as far as possible, by fire."[11] The *Globe* of Toronto reported that the "Government took possession of the infected quarter, destroyed every building by fire, and filled up all the underground rookeries. Altogether," the paper assessed, "the action of the authorities and of the men forms one of the most splendid pages in the history of British colonial administration."[12]

The housing market in the city facilitated the decision to employ fire because land values were higher on lots without buildings. Thus twentieth-century Taipingshan was to arise on sites cleared and purified by Governor Robinson's flames. The feelings of the Chinese residents can be indicated by the fact that he is the only governor of the colony who has not been remembered by a street named in his honor. Ample reason for such a sentiment is provided by the *Times of India* in 1903, when it asked, "What has become of the thousands of souls rendered homeless when these huge blocks were bitten out of the heart of slumland? The answer lies in the appalling overcrowding now found in the quarters undisturbed. Rents have gone up fifty and seventy-five per cent. Where one family occupied a room, there are now two and three."[13]

Far from being an exotic and parochial medical philosophy, the theory of recrudescence—buttressed by the history of London and the example set by Hong Kong—was widely adopted elsewhere during the early years of the third pandemic. Honolulu, for example, set many "sanitary fires" in the Chinatown area at the end of 1899, only to have them spread by winds and culminate in January in the "Great Chinatown Fire of 1900," which is estimated to have burned thirty-eight acres and destroyed four thousand homes across the neighborhood. By 1903, the *Chicago Daily Tribune* ran the headline "British Give Up Fight on Plague" and commented:

> Fire is a favorite agent of purification employed by the white man in the Orient, one, too, that is used with a ruthless hand, especially during the visitations of cholera and of the plague, and that has served to intensify the hatred with which . . . the foreigner is regarded by the Asiatic. Convinced that the dirty conditions of the native dwellings were responsible for much of the spread of the disease, the torch has been freely applied to them. But it has proved of no use.[14]

The Devastation of Bombay

The principal victim of the modern pandemic was Bombay. A great cosmopolitan metropolis, Bombay was the second largest city in the British Empire, a textile and administrative center, and a major seaport with a famously sheltered deep-water harbor. With a population of eight hundred thousand when plague struck in 1896, the Indian port led all cities in the world by enduring the heaviest burden of sickness and death. Bombay is the essential microcosm in which to examine the conditions that fostered plague, the policies deployed to combat it, and the desperate resistance of the urban poor to the stringent measures that colonial authorities sought to impose.

Although the first case of plague, known as "patient zero," was never identified, *Y. pestis* arrived in 1896 on the bodies of rats that were stowaways from Hong Kong. As a result, the earliest diagnosed cases occurred in the slums of the Mandvi neighborhood immediately adjacent to the waterfront. Here the grain trade that brought countless sacks of wheat and rice into the city played a leading role. As the Indian Plague Commission soon explained:

> All the large granaries in Bombay are situated in Mandvi; . . . these granaries are numerous and always well stocked, and . . . they are kept open only during business hours. . . . As the stored grain is never kept in metallic receptacles, but in gunny bags stacked in tiers from floor to ceiling, all the conditions for successful rat propagation are fully satisfied. Abundance of nutritious

> food supply, suitable hiding and breeding crannies between the bags on the ground tier, perfect safety from intrusion, and full timely warning of the approach of exposure and danger, all these conditions for rat breeding on a stupendous scale are present.[15]

Officially, the epidemic began on September 23, when a case was diagnosed and reported by the Indian doctor A. G. Viegas. As early as July, however, people had noticed die-offs of rodents in the Mandvi warehouses. Then in early August mysterious illnesses broke out among the residents of the apartments located above the granaries. These were blandly diagnosed as "glandular fever," "remittent fever," and "bubonic fever" rather than plague. Such anodyne labels resulted from powerful pressure by merchants, manufacturers, and local authorities who hoped to avoid troubling medical reports that would close European seaports to ships from Bombay, thereby causing significant financial losses. Given the stigma surrounding plague as the product of Oriental barbarism, the municipality made every effort to prevent the lurid press reports that would follow such a diagnosis. From late July, therefore, the first public health strategy adopted in Bombay was that of euphemism and deception—a strategy that lasted until late September when laboratory tests and the sheer number of cases created an official emergency. Plague had broken out on the island that was Bombay, where weeks of silence and inaction had enabled it to establish the first Indian focus of infection (fig. 16.1).

From the autumn of 1896 until the 1920s bubonic plague ravaged India's largest city in a pattern that was distinctive to the third pandemic, as we have seen at Hong Kong. Every year plague erupted anew when cool weather arrived in December and ebbed away with the summer heat of May. The periodicity of the cycle was determined by the temperature and moisture requirements of the "Oriental" rat flea *Xenopsylla cheopis*. The flea was inactive both in the intense heat of the tropical summer and during the moisture of the monsoon season from June to September. So regular were the arrival and departure of plague that the months from December to April were known as the annual "plague season."

During the plague seasons the *Times of India* recorded plague mortalities that peaked, in such fatal years as 1903, at more than two thousand per week in the deadly months of January and February. The figures appeared amidst disclaimers that the statistics were significant underestimates because

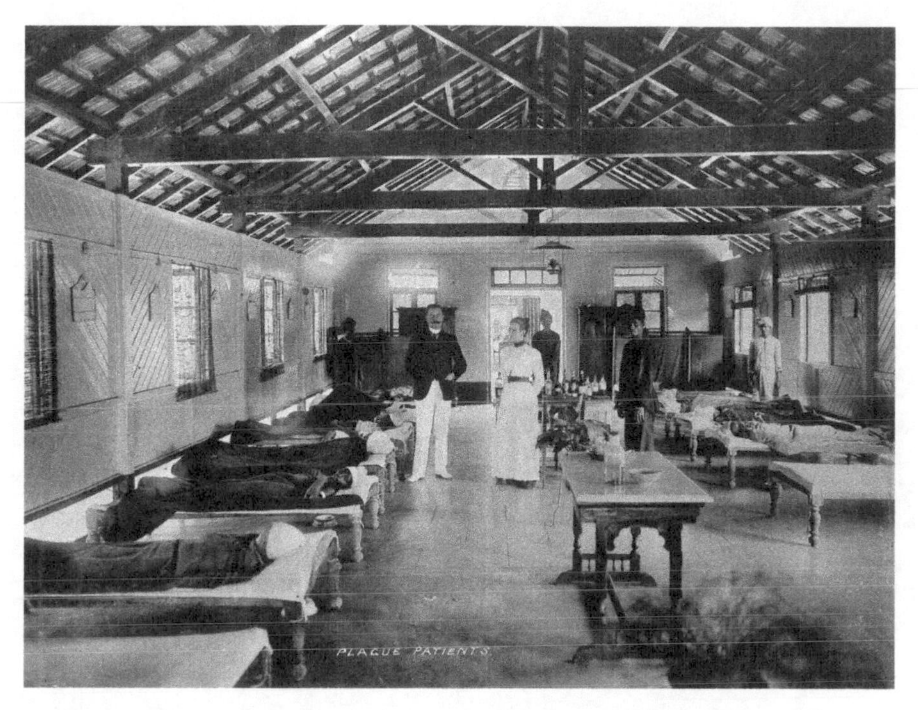

Figure 16.1. Inside a Bombay plague hospital, 1896–1897. (Photograph attributed to Clifton & Co. Wellcome Collection, London. CC BY 4.0)

many victims were never officially discovered. In the words of the eminent physician Alessandro Lustig, then resident in Bombay:

> Exact statistics of disease are not possible in India, where not a few of the inhabitants of the large towns live a nomad life, out of doors, in the streets and squares, without any fixed and stable roof over their heads. Still less can one make an approximation at the number of deaths, for the corpses of many Hindus are thrown into the rivers or sacred ponds in obedience to certain rites, or are burnt in the thickets without the authorities knowing anything of it, even if they should wish to do so.[16]

Bubonic plague, furthermore, spread beyond the island to engulf the whole Bombay Presidency. It also moved farther afield—to Poona (now Pune), Karachi, Calcutta (now Kolkata), and innumerable towns and villages of the north and east of the subcontinent. Linked to Bombay by road and rail for the purposes of trade, labor migration, and pilgrimage, these places experienced per capita mortalities rivaling that of the great seaport.

In late Victorian Bombay, an economic boom had drawn floods of impoverished migrants from the countryside in search of jobs as porters, weavers, spinners, dock workers, sweepers, and construction workers. This throng of immigrants transformed Bombay into one of the world's most congested cities. The population per acre in Calcutta was 208 people, and 221 in certain boroughs of London, but in Bombay it soared to 759. With no provision for waste removal and no building codes, a crush of stables, domestic animals, tanneries, butcher shops, and factories all contributed to ever-accumulating mounds of waste. Open drains and sewers ran in the narrow alleys between buildings where sunlight never penetrated. Dried cow dung was caked, sold, and burned as the poor man's fuel despite the evil stench it emitted. Thus improper drainage, poor ventilation, ubiquitous filth, malnutrition, vermin, and—above all—sheer congestion all played their part in the disaster that soon unfolded in the city's infamous tenements, known as *chawls*. These structures had been hastily built to provide cheap housing for immigrants. The Indian Plague Commission described these dwellings as

> great warren-like buildings with numerous apartments opening off narrow passages, or off each other in series. The apartments are small, dark, and almost without any intentional ventilation. The floors consist of a mixture of earth and cow dung plastered on weekly, the operation being in part a religious ceremonial. . . . Floors of this sort are said to afford a pleasant and cooling surface for the bare feet of the inhabitants. . . . Another important feature is the extraordinary overcrowding of the houses at night, the floors being practically covered with recumbent human bodies.[17]

Most frequently, maintenance of the apartment blocks by absentee landlords was as negligent as their construction was flimsy. Individual units, known as *kholis*, were so tiny and congested that the doors were left permanently open to prevent suffocation.

The British Colonial Antiplague Campaign

After the initial weeks of inertia and denial, September 23 inaugurated a heroic phase of health policy to contain the disease by stringent measures.

The Epidemic Diseases Act of February 1897 had given unlimited authority to a central board of plague commissioners. Knowing from Kitasato and Yersin that bubonic plague was a contagious disease, the commissioners revived the draconian antiplague priorities of the second pandemic—quarantine, the forcible isolation of cases and contacts in pesthouses, and speedy burial. Even the initial rumors that antiplague measures were to be instituted gave rise to terror and elicited a popular response that negated the goal of containing the disease within the city. Men, women, and children took flight, even before the official announcement turned the movement into a stampede. By December two hundred thousand people had departed, and the number continued to soar so that by February 1897 it reached four hundred thousand, or half the population of the city. Inevitably, the fugitives transported *Y. pestis* with them. These escapees dreaded the health measures to be imposed by the British more than plague itself.

Meanwhile, as the port fell silent and all economic activity ceased, the British unveiled the military policies the population so feared. Having divided the city into districts, a newly appointed Vigilance Committee issued a proclamation that caused "much public excitement." It claimed "the right of entry to buildings by authorized officers, for cleansing, for removal of infected articles, for compulsory removal to hospital of any person certified by the health officer to be suffering from the disease, and finally for the isolation of infected houses."[18] With this authority, the committee sent out search parties of health officials accompanied by carters, sepoys (indigenous troops), British constables, and a justice of the peace. They set off early every morning on case-hunting expeditions. Entering homes unannounced, the officials sought all people who were suspected of having plague. Ignoring traditions of caste and religion, they subjected everyone to physical examination. The searchers pursued the buboes and "tokens" of plague, even if the women lived in purdah—the Muslim and Hindu practice of secluding women from public view in their homes.

Having located a case they considered suspect, the searchers marked the house, such as daubing an exterior wall of the premises with a circle and the letters "UHH" (unfit for human habitation) painted over the doorway (fig. 16.2). Then they conveyed the newly found patient by cart or wheeled stretcher to the dreaded Municipal Infectious Diseases Hospital on Arthur Road (fig. 16.3). At the same time the search parties took all contacts sharing the same dwelling into custody and conducted them to hastily established detention centers sardonically termed "health camps." Here the evacuees gathered in quarantine under canvas or in sheds, with no attention to

Figure 16.2. A marked plague house in Bombay, 1896. The open circles represent deaths from plague; the crossed circles, deaths from other causes. Marked houses indicated to officials that sanitary action was needed. (Wellcome Collection, London. CC BY 4.0.)

traditional Indian protocols governing the mixing of people from different castes, religions, ethnicities, and genders. The British divided the ill from their families, imposed Western medical practices, segregated contacts, and violated social and religious conventions.

The painted circles and the designation UHH did not serve the same purpose as the red crosses described by Daniel Defoe in *A Journal of the Plague Year*. London's crosses warned passersby of buildings that were "shut up" and where all movement in or out was forbidden. In Bombay, the markings served the different function of alerting officials to premises requiring further sanitary action. Just as in Hong Kong, the medical theory underlying Bombay's campaign against plague was based on the contemporary orthodoxy regarding the transmission of plague to humans as "true recrudescence." This telluric, or soil-based, theory suggested the work to be performed in premises newly daubed with paint. In villages

Figure 16.3. Wheeled stretcher outside a Bombay plague hospital, 1896–1897. (Photograph by Clifton & Co., Wellcome Collection, London. CC BY 4.0.)

and small towns, where it was possible to provide emergency accommodation for the poor, the sanitarians returned to deport the population en masse to detention camps. As at Hong Kong and Honolulu, they then burned the infected dwellings to the ground. This was public health by eviction and destruction. In terms of modern understandings, such policies were counterproductive because their immediate effect was to send whole colonies of frightened and flea-bearing rodents scurrying to find alternative quarters.

Such a policy, moreover, was impractical in a city as large as Bombay. In the midst of an emergency it was impossible to improvise a program of rehousing on such a scale. For the long term, a new agency—the Bombay City Improvement Trust—was established for wholesale urban renewal. In the words of the medical officer of health, Dr. J. A. Turner, regarding plague, "We must . . . consider it a disease of locality and endeavour to find some means of reducing its vitality by abolishing the places where it thrives and thus gradually exterminating it."[19] In their immediate antiplague measures, the British yielded to necessity and the realities of a great metropolis. Their strategy was to follow *Y. pestis* wherever it was thought to have settled in the polluted urban environment and then to destroy it, but by means short of setting fire to entire buildings, blocks, or neighborhoods.

Following this logic, the Bombay authorities defined the bedding, clothes, and furnishings of victims as plague-bearing fomites, so sanitary teams were instructed to incinerate them. These teams also fumigated and disinfected buildings. Sulfur fumes had a long history as a means to cleanse the air, so they lit sulfur fires in *kholis* thought to be contaminated. For the same purpose the municipality dispatched workers to saturate the walls with disinfectants—mercury perchloride, carbolic acid, and chlorine—in order to destroy the bacteria. They also dug up floors to a depth of six inches and soaked the soil with mercury perchloride to purify it. In the upper stories they removed roofs and cut holes in the walls to admit sunlight and air and to disperse disease-bearing effluvia.

While this sanitizing work was under way indoors, the sanitarians gave their attention to the outdoors as well. Workers lime-washed and steam-flushed walls, swept streets, and flushed gullies and sewers. Finally, wherever the searchers found the dead body of a plague victim, they immediately carted it away for rapid cremation—in defiance, for instance, of the Parsee prohibition of cremation.

It was India's misfortune that the arrival of plague at the end of the nineteenth century coincided with the worst famine in living memory, which the *New York Times* called the "grimmest and most terrible that India has ever experienced."[20] Beginning in 1897 the monsoon failed for three years in succession, causing a drought that destroyed harvests and starved both cattle and humans. This agricultural devastation was enhanced, as in Pharaoh's Egypt of the Bible, by "plagues" of crop-devouring locusts and rats. The scant rainfall and the famine persisted for an entire decade until 1906. By then, 100 million Indians had been affected by hunger; 5 million perished of starvation between 1897 and 1901 alone. Most significantly for the plague, the region that suffered most severely was the vast Bombay Presidency. There the gaunt and malnourished bodies of the population offered a severely compromised resistance to the disease, and plague ran rampant among the Bombay poor, as did smallpox and influenza. Ironically, even shipments of wheat and rice that arrived to alleviate the famine stoked the pestilence, as rats and fleas nestled in the grains.

Resistance and Riot

As the inspectors made their rounds in the distraught and famished city, rumors followed them concerning the nefarious intentions of the British. A popular theory was that the disregard for Hindu, Muslim, and Parsee belief systems was deliberate. The goal, many Indians imagined, was to discredit indigenous religion in order to impose Christianity and strengthen British rule. For the same reason, the theory held, the British were determined to force Indians to accept Western allopathic medical practices, sweeping aside traditional healers once and for all. In addition, a whispering campaign fostered a more sinister paranoia—that the epidemic was not misfortune but homicide. The cause, in other words, was a poison administered by the British to solve the twin problems of overpopulation and poverty. A further theory alleged that Queen Victoria's intention was to sacrifice Indian bodies to placate the god of plague, thereby preserving Britain, not India, from divine wrath. Thirty thousand Indians, it was rumored, were required for this atonement.

Such ideas flourished in a context of deep racial and colonial distrust, and they were propagated by early proponents of Indian nationalism. They also took root because the Indian poor and illiterate possessed no understanding of Pasteur and Yersin; they seldom made contact with the medical profession; and they saw no reason to endure measures imposed by strangers whose motives were suspicious. Hungry and fearful, they had little patience for further intrusions. The anger that resulted was ignited all the more easily since there were authenticated reports of cases in which searchers, police, and soldiers abused the broad power entrusted to them by the plague regulations. There were scattered but explosive accounts of soldiers who sexually molested Indian girls in their charge, of attendants who stole watches from the dying, and of searchers who demanded bribes in order to refrain from daubing paint on a dwelling.

In this politically charged environment, the arrival of the search teams rapidly became a flashpoint for violence. Throughout Bombay people concealed the ill as relatives, friends, and neighbors attempted to protect the sick from trauma and possibly murder. With such little cooperation from the population, the authorities were unable to gauge the extent and location of the epidemic—a factor that complicated the task of taking useful and timely sanitary measures. Furthermore, beyond passive resistance, the citizens of Bombay frequently opposed plague policies more resolutely.

In the city's overcrowded *chawls*, as many as a thousand people might occupy a single building. In such cramped quarters, news and rumor traveled rapidly, and it rarely took long for crowds to gather and to drive searchers away under a barrage of abuse. Already in October 1896 a thousand outraged Hindus attacked the hospital on Arthur Road. There was also a general strike of the city's dock and transport workers and a series of strikes by weavers and spinners. Then, in the spring of 1897, a tremor of fear shook the expatriate British community. Plague Commissioner W. C. Rand and three other officials were shot in retaliation for the draconian policies they had implemented. India, the *New York Tribune* warned, was "like a volcano."[21]

Perhaps the most dramatic events of the pandemic were the riots that caused the *Times of India* and other English-language observers to fret that the British Raj was on the verge of a broad "Indian awakening" or perhaps even a general mutiny. The most famous revolt took place on March 9, 1898, in the slums of north Bombay bisected by Ripon Road. There, eighty thousand Muslim textile workers inhabited *chawls* overshadowed by the tall, dark chimneys of the mills. The mood of the weavers and spinners was already enflamed by industrial disputes over piecework and the practice of enforcing labor discipline by paying wages a month or even two months in arrears. Hungry, indebted to moneylenders who tided them over the gap between workday and payday, frightened by the disease, and distrustful of authority, the mill hands did not welcome visiting officials.

A search party consisting of two police officers and two assistants under the direction of an English medical student visited a tenement on Ripon Road in the early hours of March 9. There they found a feverish young woman in a delirious state. Her general appearance raised the suspicion of plague, but her father steadfastly refused to allow his daughter to be examined by an unknown Englishman. Nor did he relent when the searchers summoned the assistance, first, of a Muslim *hakim* (or healer) and then of a female English nurse. By that time, a noisy and defiant throng had gathered, and the search party withdrew. As it retreated, the crowd of several hundred people followed the fugitives and hurled insults and stones at their backs.

The search party escaped, but the crowd continued to grow while its members armed themselves with bamboo cudgels and rocks. Groups broke off to attack Europeans whenever they appeared, to set fire to ambulance carts, and to kill two British soldiers who happened on the scene of the fray. In the meantime, British reinforcements arrived—police, soldiers, and sepoys—to confront the demonstrators. A magistrate ordered the rioters to disperse, but when the crowd responded with hisses and more stones, the soldiers opened fire. The rioters fled in confusion, leaving behind twelve dead or mortally wounded where they fell.

The *Times of India* reported that uproar spread through the "native quarters" of Bombay with the usual "oriental rapidity."[22] Hindus, Muslims, and Parsees, so often divided among themselves, made common cause against the British. Crowds continued to attack Europeans, especially solitary police en route to disturbances. Rioters also made their way to Arthur Road, where they laid siege to the plague hospital. At first they attempted to liberate the patients, symbolically reversing social roles by burning down the lazaretto as the British had done with their homes. By the end of the day there were more than one hundred casualties.

Similar confrontations occurred at Poona, Calcutta, and Karachi, as well as in a host of towns and villages—wherever authorities tried to implement military-style antiplague measures. The British responded with a combination of repression in the political field but concessions in the medical arena. Politically, the sedition act was invoked to raid and close down newspapers, to deport agitators without trial to the remote Andaman Islands penal colony, and to initiate prosecutions for treason. The goal was to suppress rebellion, and the press suggested that the frequency of police raids suddenly rivaled the repression in tsarist Russia.

To balance such rigor, Lord Sandhurst, governor of Bombay, announced a more conciliatory policy in the campaign against bubonic plague with the goal of removing the sources of such widespread ill will. Compulsion, it was decided, was suitable only for limited outbreaks in small towns, not for a raging epidemic in a major metropolis where large crowds threatened to overwhelm security forces. In an all-out confrontation, the balance of force would be unfavorable for the British, whose total military force was just 230,000 troops in the whole of India.

The decision to abandon compulsion was rendered all the easier when the Indian Plague Commission released its report. It concluded that the impact of medieval rigor was nil with regard to containing the disease and negative from the standpoint of public order. Far from being stamped out, plague returned every cool season and extended its geographical reach. Furthermore, the implementation of rigorous antiplague policies was ruinously expensive. Rejecting stringency as dangerous, ineffective, and costly, Sandhurst turned in 1898 to voluntary compliance.

New Directions in the Antiplague Campaign

After Sandhurst's announcement, all forms of coercive public health—search parties, compulsory physical exams, the imposition of allopathic therapeutics, and forcible isolation on Arthur Road—ceased. The forbidding lazaretto remained, but under new regulations, and small neighborhood hospitals began to appear for each indigenous religious group. These establishments were open to friends and relatives, provided treatment by traditional healers, and respected norms of caste and gender. Before the final departure of the modern plague, thirty-one of these institutions opened in Bombay. To entice the population to comply, the British instituted a discretional relief fund that provided compensation for wages lost through quarantine, replacement costs for personal effects destroyed by disinfectors, funeral expenses for patients who had sought treatment, and relief for their families.

Sera and Vaccines

Enticement rather than force dramatically lowered social tensions and brought the period of near-revolt to a close. It proved ineffective, however, in reducing plague morbidity and mortality, which persisted in the new century after voluntarism had become the rule. Hope to end the epidemic rested instead on new tools emerging from the laboratory and the rapidly developing discipline of immunology. The first was antiplague serum. Yersin and Lustig recognized that horses experimentally infected with *Y. pestis* did not develop bubonic plague. They successfully used this observation to produce sera consisting of the blood plasma of the immunized horses. The Yersin and Lustig sera were first administered as heroic therapy for desperately ill patients and then as a measure of prophylaxis, achieving limited but positive results in both uses. In a trial of 480 plague patients, Lustig's serum achieved a recovery rate of 39.6 percent versus a rate of 20.2 percent for an identical number of controls. Emboldened by this trial at a time when no more successful treatment was available, the Arthur Road hospital made extensive use of both sera.

More important locally was a vaccine developed in Bombay itself. The Bombay Presidency had decided early in the pandemic to fund plague research by the eminent scientist Waldemar Mordecai Haffkine (1860–1930), a Ukrainian disciple of Jenner and Pasteur. By the time of the Indian emergency, he had established an international reputation through his part in producing a cholera vaccine. Working at the Government Plague Research Laboratory in Bombay, Haffkine in 1897 developed a vaccine of killed plague bacteria that passed efficacy and safety trials, first on rats and then on human volunteers. In both trials the vaccine provided only a partial and temporary immunity to the disease, but it compared favorably with an armamentarium that was otherwise empty. In 1898 the British plague commissioners therefore decided on a campaign of mass vaccination with Haffkine's experimental product. They were emboldened by the Indian Plague Commission, which authoritatively pronounced that all strategies and tools employed before Haffkine's breakthrough were useless. The new vaccine alone provided a glimmer of optimism.

Unfortunately, the vaccination campaign failed to achieve the desired results. One reason was technical. Haffkine's vaccine required multiple doses, all administered by medically trained personnel, who were in short supply. Assuring delivery to the vast population of the Bombay Presidency was a daunting logistical challenge. Furthermore, the campaign provoked resistance more than compliance. In part, popular opposition resulted from the fact that the vaccine provided only partial immunity, and everyone soon knew of people who had been vaccinated but still contracted plague. Furthermore, the vaccine produced unpleasant side effects—vertigo, headache, enlargement of the lymph nodes, and painful swelling at the vaccination site—that incapacitated recipients for days. Such complications reinforced suspicion that the British were dispensing poison.

Anxiety hardened into certainty after a major accident. Pressed to supply the quantities of vaccine required, the Plague Research Laboratory compromised on safety. The result was the Mulkowal disaster of October 1903, when the laboratory contaminated a batch of vaccine with tetanus and then compounded the error by omitting the failsafe precaution of adding carbolic acid to the fluid. As a result, nineteen people who were inoculated with the tainted vaccine died excruciating deaths—not from plague but from tetanus. Thus, in the medley of successive emergency policies, sera and vaccine failed no less than the resort to medieval antiplague measures.

"No Rats, No Plague"

The final phase of antiplague strategy began with the belated vindication of Paul-Louis Simond's 1898 theory of the rat-flea nexus in the transmission of plague—a theory the Indian Plague Commission had originally greeted with skepticism (see Chapter 4). Reexamining both the biology and the epidemiology of the disease, however, the commission reversed its position. Instead of rejecting Simond's findings, it now confirmed them through extensive research, thereby establishing the rat-flea mode of transmission as the new orthodoxy. The result, after fourteen years of unsuccessful antiplague activity, was yet another plan to control the disease. Policy, now focused by Simond's discoveries, carefully targeted the killing of rats—the so-called gray peril—as the new antiplague priority. The campaign's slogan was "No Rats, No Plague."

In order to implement this approach in Bombay, Dr. J. A. Turner launched an extermination campaign in 1909. The rodent hunt began with circles drawn on a map of the city, each circle representing the area that a crew could treat during a three-day visit. Teams of hunters were dispatched to their assigned neighborhoods bearing brooms, traps, barrels of disinfectant, and flushing engines. The rat hunters devoted the morning of the first day to the streets, where they swept out gullies, cleared drains, and removed waste in order to deprive the rodents of food. In the afternoon every team set hundreds of traps baited with fish and cubes of bread dipped in powdered sugar and flour, and then laced with arsenic or strychnine, and ground glass. The baited traps were placed wherever inspectors identified a rat runway or nest.

The hunters returned the following day to collect the rats, both dead and alive, and to convey them in tin boxes to the laboratory, where their bodies, each bearing a tag indicating its place of origin, were inspected for buboes and then incinerated. On the third day, the hunters returned to the circled areas with brooms and flushing engines to sweep and disinfect indoors. If the lab had found buboes on a rat's body, the team returned to the site on an additional day to carry out further disinfection. Then the rat-catchers moved to a new neighborhood. Their goal was to drive surviving rats from their nests and deter them from returning. The campaigners' resolve was strengthened by the fact that similar anti-rat strategies had been successfully adopted in England, South Africa, Australia, the Philippines, and Japan.

In 1910, the second year of intensive rat hunting in Bombay, the teams collected half a million animals for examination and destruction. The municipal lab estimated—perhaps optimistically—that a quarter of the total rat population had been culled and the disease had receded. That year bubonic plague claimed just five thousand victims in the city—as compared with ten thousand to fifteen thousand in a "normal" year and twenty thousand in the terrible year of 1903. For the first time, by 1910 people began to wonder whether the plague was dying out.

Unfortunately, rat reduction did not provide the high road to plague eradication. Both fertile and intelligent, the rodents rebounded from the assault and even learned to avoid the bait, rendering the traps and their poison an ever-diminishing asset. A further serious complication was that the Hindu population, believing in the sanctity of animal life, stubbornly resisted killing rats and freed them from traps. This obstacle was especially serious because so many Indians suspected the British of homicidal intentions. In their minds, British traps—not Indian rats—disseminated plague. In addition, the hunts were counterproductive in ways that Turner had not anticipated. Driving rats from their nests propagated the disease more widely, if less intensively. Thus bubonic plague lingered. Although its rate of transmission slowed, it returned stubbornly every winter to claim thousands of victims. Finally, the numbers of annual victims declined during the 1920s, and then—after nearly half a century—subsided entirely around 1940.

Global Lessons and Priorities

But after the recession of the epidemic, a new anxiety arose: Could plague be reintroduced to the city? The social and economic conditions that had originally made Bombay so susceptible remained. Poverty, crowding, slums, and rodents all persisted. Mercifully, however, the plague did not return. Apart from a smaller population of rodents, the reasons reflect both policy and biology. In terms of policy, one of the global lessons mastered during the third pandemic was the role of rats and steamships in transmitting the disease around the world. After Simond's discoveries regarding the rat-flea nexus, an international effort began to rat-proof ships in order to prevent rodents from traveling.

The first step, taken in 1903, was to confirm the role of rodents in the global diffusion of *Y. pestis*. Beginning at Savannah, Georgia, and Tampa, Florida, rat infestation services undertook a demonstration by fumigating all incoming steamers with sulfur. By driving the rodents from their nests with the acrid fumes, the experiment vividly demonstrated the extraordinary size of the rat population plying the global trade routes. By catching and examining escaping animals, investigators also proved conclusively that large numbers of the seaborne rats carried bubonic plague. Hundreds of ships in fifty-five countries were officially documented as plague-infested.

Since keeping plague at bay was a major economic and public health priority for every nation, rapid international agreement was reached about practical measures. Shipping companies were required to deploy scientific rat-proofing and extermination services. Chlorine gas and hydrogen cyanide were used to destroy the unwelcome passengers. Ships were then rat-proofed more permanently by sealing empty spaces in bulkheads and ceilings where rats had once established runways and harborage. Shipyards also began to construct steamers according to specifications that would deny access and space to rodents. As a result, by the 1920s the world's oceans had become impassable barriers to rodents, not highways for them.

Two biological factors also worked silently to bring the long Indian tragedy to an end and to prevent a recurrence. One was the violent ongoing progress of *Rattus norvegicus* (the brown rat) as it ousted *Rattus rattus* (the black rat) from its niches across the subcontinent. The result was to make India far less vulnerable to plague. Since the brown rat was aggressive but retiring, it created a new distance between humans and rodents, and the fleas those rodents bore. This factor was especially influential because some Indians had always treated the black rat more as a pet than as vermin. Thus a violent war between rodent species had major implications for human health.

A second biological factor is unconfirmed but potentially consequential. This is the speculative possibility that rodent immunity may have played a part in diminishing Bombay's exposure to plague. The persistence of bubonic plague as an endemic disease of rodents subjected rats in the Bombay Presidency to severe evolutionary pressure over many generations. A plausible result may have been the development of herd immunity and resistance to *Y. pestis* among rats. If the antisocial brown rat was also partially protected, then the rate of transmission of plague among humans would have fallen sharply. Possibly as a result, the plague sputtered and departed, and to date has found it difficult to return.

A final question remains. Why did the third pandemic follow an epidemiological pattern so different from the first and second? Since there are no comprehensive studies of the issue, any answer is necessarily provisional. It is known, however, that there are various strains of *Y. pestis* and that the strain responsible for the modern pandemic was overwhelmingly dependent on rats and fleas for its transmission. Rarely was it propagated by droplets or person-to-person transmission in the manner of the more virulent pneumonic plague that was so prominent during the two earlier pandemics. This difference caused the third pandemic to disseminate less rapidly. To the extent that the modern strain was also less virulent, *Y. pestis* could have established itself endemically among rodent populations that it culled but did not extinguish. This quality would have permitted the establishment in Hong Kong and Bombay of permanent reservoirs of infection. Plague then would have persisted, only to reemerge at intervals under favorable conditions of temperature and humidity. It is also possible that still unknown variables affected the rat flea and its efficiency as a vector. A combination of all such factors could plausibly have produced a pestilence that, like the third pandemic, was slow moving, recurrent, and less lethal in aggregate than the all-destroying Plague of Justinian and the Black Death.

CHAPTER 17

Malaria and Sardinia

Uses and Abuses of History

Malaria is one of the oldest of human afflictions, and some scholars suggest that in aggregate, it may also be the disease that has caused the largest total burden of illness in human history. Two reasons for this are that, unlike plague, smallpox, and cholera, it has collected its grim annual tribute continually since the earliest days of our species. In addition, its geographical reach is enormous.

Today it continues to be one of the most serious of human afflictions, despite the fact that it is both preventable and treatable. In 2017 the cautious statistics of WHO indicated that half of the world's population—3.2 billion people—were at risk of contracting the disease, which in 2017 sickened 219 million people and killed 435,000 in 106 countries. Most severely affected was sub-Saharan Africa, which experienced 92 percent of the malaria morbidity and 93 percent of the mortality, especially in the four countries of Nigeria, Mozambique, Democratic Republic of the Congo, and Uganda. It is estimated that a child dies of the disease nearly every minute, making malaria—together with HIV/AIDS and tuberculosis—a leading public health emergency.

The burden is undoubtedly heavier than statistics for just death and disease indicate. Malaria is a major complication during pregnancy, leading to high rates of miscarriage, of maternal death through hemorrhage and severe anemia, and of all the sequelae that follow from severely low birth weight. Since malaria can be transmitted from mother to fetus, it can also lead to the birth of infants who are already infected.

Furthermore, malaria is a major immunosuppressive disease, which is to say, its victims are highly vulnerable to other diseases, particularly respiratory infections such as tuberculosis, influenza, and pneumonia. In areas of the tropical world where it is hyperendemic and transmission continues throughout the year, the at-risk population can be infected, re-infected, or superinfected every year. Those who survive possess a partial immunity, but it comes at a high and enduring cost because repeated bouts of malaria often lead to severe neurological deficit and cognitive impairment. The results are ineradicable poverty, increased illiteracy, compromised economic growth, a stunted development of civil society, and political instability. The direct costs alone of malaria in terms of treatment, illness, and premature death are estimated at $12 billion annually, with indirect costs that are difficult to quantity but are thought to be several times that figure. Malaria thus makes a major contribution to the global inequalities between North and South, and of the dependence of the Third World. Ronald Ross, the British Nobel Prize–winner who discovered the mosquito theory of malaria transmission, argued that malaria enslaves those it does not kill.

Malaria Parasites and Their Life Cycle

Human malaria is not a single disease but five different diseases caused by five different species of parasite known as "plasmodia." Of the five, two are overwhelmingly the most important afflictions in terms of prevalence, morbidity, mortality, and impact on human history: *Plasmodium falciparum* and *Plasmodium vivax*. (The others are *Plasmodium ovale, Plasmodium malariae,* and *Plasmodium knowlesi*.) These parasites do not exist free in the environment, but complete their complex life cycle in the bodies of human beings or of certain species of the mosquito genus *Anopheles*.

A classic vector-borne disease, malaria is transmitted by infected female *Anopheles* mosquitoes that, like flying syringes, inoculate plasmodia directly into the bloodstream by stabbing through the skin with their proboscis, thereby initiating the disease process. Instead of remaining in the circulating blood, where they are vulnerable to the body's immune system, the newly inoculated plasmodia migrate rapidly to the safety of the tissues of the liver, where they escape detection. There they replicate in the liver cells until, having achieved a critical number, they burst the host cells and return to the open bloodstream to initiate the next phase of their life history. An important difference between *P. vivax* and *P. falciparum* is that the former continues to nestle in the liver, where it can initiate a new infection months or years later. In that sense, even a person who has apparently made a full recovery from vivax malaria is susceptible to relapses later in life without being reinfected.

Having entered the bloodstream, the plasmodia of all species are similar in that they attack red blood cells, or erythrocytes, and invade them. The plasmodia then replicate asexually in ameba-like fashion inside the erythrocytes, bursting and destroying them in such a synchronized manner that the plasmodia are once again discharged into the bloodstream simultaneously throughout the body. There they repeat the process of attack, replication, and return to the circulating blood at fixed intervals of either forty-eight or seventy-two hours, depending on the species.

The incubation period varies in length but normally lasts from nine to fourteen days for *P. falciparum* and from twelve to eighteen days for *P. vivax*. Disease begins when, through repeated replication in geometric progression, the plasmodia achieve the critical mass necessary to activate the immune system. This ends the incubation period and causes the first symptoms—paroxysms of high fever and chills that are not constant but intermittent and give malaria one of its names: intermittent fever. One of the defining features of the disease is that patients suffer recurring bouts of fever every forty-eight or seventy-two hours, with the onset of fever marking each return of the parasites to the bloodstream. For both *P. vivax* and *P. falciparum* the interval is forty-eight hours, so the fever is called "tertian," and—historically—the two diseases were known, respectively, as "benign tertian fever" and "malignant tertian fever."

After many rounds of asexual replication and return to the bloodstream, the parasite reaches a new phase of its life cycle by producing male and female cells known as "gametes." These are capable of sexual reproduction, but not in the human body. To complete their life cycle, plasmodia are dependent on returning to the body of the mosquito. This is achieved when a female *Anopheles* mosquito takes a blood meal and in the process sucks up gametes from the blood vessel it attacks. Gametes then mate in the body of the mosquito and produce offspring that, after a further stage in the gut of the mosquito, migrate to the proboscis of the insect. Known as "sporozoites," these cells are ready to be inoculated into a human bloodstream to renew the full cycle of asexual reproduction in humans and sexual reproduction in the mosquito.

Symptoms

The classic symptoms of malaria infection are intermittent fever accompanied by high fever, chills, profuse sweating, and headache. Vomiting, severe diarrhea, and delirium are common. Particularly in falciparum malaria, the most virulent of malarial diseases, the density achieved by plasmodia in the blood is often overwhelming: 40 percent of the body's erythrocytes can be infected. Patients with such intense parasitemia became anemic and debilitated. In the case of *P. falciparum* a further effect on the red corpuscles is to cause them to become sticky, to adhere to the walls of blood vessels, and to agglutinate in the capillaries and venules of internal organs. There they cause occlusion and hemorrhage and can lead to an array of symptoms that often mimic other severe infections, making malaria one of the most mercurial of all diseases in its symptomatology. The outcomes are also swiftly lethal if the brain, lungs, and/or gastrointestinal tract are affected and if the patient is a child or a pregnant woman. Malaria is one of the most severe complications of pregnancy, as pregnant women invariably suffer miscarriage and frequently hemorrhage to death. In severe cases, death can also result from acute respiratory distress, hypoglycemia leading to coma, and profound anemia.

In less severe falciparum malaria and in most cases of vivax and malariae malaria, the disease is self-limiting. Instead of indiscriminately attacking all red corpuscles, *P. vivax* and *P. malariae* demonstrate a marked preference for young and aging erythrocytes. The effect is a much lower level of infection. The immune system of white cells, circulating and fixed, successfully contains and eventually eliminates the parasites from the circulation, although relapses—characteristic of *P. vivax*—are typical. But malaria yields a very imperfect and costly immunity so that reinfection and superinfection, that is, illness with more than one species of plasmodium in the bloodstream at the same time, are frequent in highly malarial areas. In such cases the bouts of fever, caused first by one species and then by another, are often daily, resulting in a disease that was historically termed "quotidian fever."

Even in so-called benign cases of the disease, chronic disabilities often result. Sufferers experience painful enlargement of the spleen (splenomegaly), emaciation, fatigability, anemia, and mental impairment that can lead at last to the state of cachexia—total apathy and indifference. Such a patient moves with difficulty, has a lifeless stare, and is troubled by nervous disorders. The Italian naturalist writer Giovanni Verga in his short story "Malaria" described such an unfortunate:

> The malaria doesn't finish everybody. Sometimes there's one who will live to be a hundred, like Cirino the simpleton, who had neither king nor kingdom, nor wit nor wish, nor father nor mother, nor house to sleep in, nor bread to eat. . . . He neither took . . . medicines, nor did he catch the fever. A hundred times they had found him stretched out across the road, as if he was dead, and picked him up; but at last the malaria had left him, because it could do no more with him. After it had eaten up his brain and the calves of his legs, and had got into his belly till it was swollen like a water-bag, it had left him as happy as an Easter Day, singing in the sun better than a cricket.[1]

Clearly, people suffering from such sequelae are inattentive at school, unproductive at work, and do not participate in civil society. In these ways the disease undermines the economy of countries and regions where it is endemic, and it contributes substantially to illiteracy and poverty.

Besides being a classic immunosuppressive disease, "intermittent fever" also lowers the resistance of its victims to chronic occupational diseases of the lungs, such as pneumoconiosis (black lung disease) among coalminers and silicosis among glass workers.

Transmission

A range of different *Anopheles* mosquitoes are efficient vectors of human malaria. They occupy different ecological niches and vary in their feeding habits. Some breed in freshwater, others in brackish; some are attracted to warm-blooded animals other than humans, others are exclusively "anthropophagic" and feast exclusively on people. Some mosquitoes bite indoors, others outdoors; some bite only at night, others by day. In all cases, however, the male of the species lives on the nectar of fruit, while the female depends on a blood meal to secure the protein that enables her eggs to mature.

Anopheles mosquitoes deposit their eggs on the surface of water. This water can be as extensive as a swamp, but the most efficient vectors do not require such large expanses and delight in ponds, puddles left in riverbeds and along riverbanks, even stagnant rainwater in the footprint of an ox or a mule. Historically, this dependence on expanses of stagnant water has meant that malaria is primarily a disease of rural areas although the most dreaded of all vectors—*Anopheles gambiae*—has evolved in manners that enable it to breed in urban and peri-urban locations.

Having hatched, the mosquito eggs yield larvae that in turn develop into pupae and then adults. Mosquitoes are weak fliers and rarely travel farther than about two miles from where they hatch. Rising into the air, they orient themselves toward human habitations by means of the carbon dioxide that emanates from them and is detected by the sensors on their antennae. Approaching closer, their sensors also detect prey by the odors given off from sweat and by warm body temperature. Finally, as they prepare to land, their eyes respond to movement and to expanses of exposed flesh. If the victims already have malaria, it is advantageous for the mosquito because they are likely to be still and less likely to disturb her in her quest. If there are gametes in their blood, then the mosquito in turn is infected although the presence of plasmodia in her gut does the insect no apparent harm.

Having feasted on blood, *Anopheles* mosquitoes are not normally satisfied. After a pause to rest and digest, they bite repeatedly, with the double result that they efficiently transmit malaria and are prepared to find an expanse of water suitable for the establishment of the next generation of insects.

A series of factors intervene, of course, to determine the intensity with which disease is transmitted. Plasmodia can be spread and mosquitoes can reproduce only within a narrow temperature range; cold weather in particular prevents both. Tropical climates enable year-round transmission and are therefore ideal; temperate climates usually confine malaria to an annual malaria season during warm weather and make the hold of the disease on the population more precarious. The existence of standing water is also critical, and drought or human intervention to drain or remove water collections slows or halts the diffusion of disease. On the other hand, environmental degradation, through deforestation, for example, causes flooding and therefore a vast multiplication of breeding sites. Climate change toward warmer temperatures further extends the breeding and transmission season.

At the same time housing conditions are decisive in favoring malaria. Habitations that are porous and open to flying insects, and congestion that accommodates many bodies densely packed into a single space—these circumstances promote malaria. Calamities such as war and natural disaster are also ideal as they bring crowds of people together under canvas or even outdoors in refugee camps.

The physical condition of the mosquitoes' prey is another major variable that determines the outcome of the disease. Poverty-stricken people are malnourished and have compromised immune systems, which increases the likelihood that an infectious bite will result in active disease. Lack of appropriate clothing, screens, or bednetting also facilitates malaria transmission by making it easier for mosquitoes to feast, and to do so repeatedly. In highly malarious places, people are bitten multiple times every day, and the proportion of bites that are infective is high.

For all of these reasons, malaria is a complex disease that can be variously and accurately described as an occupational disease, an environmental disease, a disease of poverty, and a disease of warfare and human disaster.

The Global Importance of Sardinia

In 1944–1945, as the Second World War ended, Italy experienced a major epidemic of malaria driven by the devastating consequences of warfare. Malaria, a disease transmitted by certain species of *Anopheles* mosquito, had been the leading public health problem in Italy since its unification (completed in 1871) and was even known as the "Italian national disease." Since 1900, however, it had been the target of a multifaceted campaign that had radically reduced the burden of disease. By the eve of World War II the ultimate goal of the campaign—the eradication of the disease—seemed attainable. Unfortunately, the antimalarial campaign itself was a casualty of the conflict, which monopolized personnel and resources. The postwar epidemic struck the island of Sardinia with special severity. In a population of 794,000, some 78,173 people contracted malaria in 1944. Most of the victims, moreover, suffered from falciparum malaria, the most virulent form.

This emergency furnished an occasion for the scientists of the Rockefeller Foundation in the United States to do health work in Europe. The foundation had long been involved in a global effort to control malaria. As early as 1925 it had adopted Italy as a base for its research, even establishing a field station at Porto Torres in Sardinia. As World War II ended and Sardinia experienced a major epidemic, the foundation resolved to undertake the experiment of eradicating the disease from the island while also providing an arresting display of US technology. Ignoring the indigenous prewar campaign, the Rockefeller scientists persuaded the US and Italian governments to implement their own alternative strategy. The idea was to follow the victorious war against the Axis powers with a new military campaign to annihilate a species of insect.

The target was the mosquito *Anopheles labranchiae*, the vector responsible for transmitting malaria in the region. The Rockefeller Foundation malariologist Fred Soper had succeeded in eradicating *An. gambiae* from

Brazil and Egypt. Never before, however, had an attempt been made to exterminate an indigenous anopheline. This experiment, known as the Sardinian Project, was planned in 1945 and launched in 1946 under the auspices of a specially created agency known as ERLAAS (Ente Regionale per la Lotta Anti-Anofelica in Sardegna [Regional Agency for the Anti-Anopheline Battle in Sardinia]). The effort continued until 1951, and Sardinia was declared malaria-free in 1952—for the first time since the Phoenicians had arrived on the island in 1200 B.C. Small numbers of *An. labranchiae* survived, but the transmission of disease ended.

Ironically, the Sardinian Project also triumphed over the indigenous traditions of Italian medical science. For the first half of the twentieth century, malariology had been dominated by the so-called Italian School led by such figures as Angelo Celli, Giovanni Battista Grassi, and Ettore Marchiafava. Having unraveled the etiology, epidemiology, and pathology of the disease, these scientists established a national antimalarial campaign relying on a multipronged program combining health education, chemotherapy, environmental sanitation, social uplift, and agricultural improvement.

The US intervention in Sardinia marked the rejection of this approach through reliance instead on the pesticide DDT (dichlorodiphenyltrichloroethane) as its single weapon of choice. This strategy embodied what W. L. Hackett in the 1930s termed the "American thesis" that malaria could be vanquished by employing technology alone to eradicate mosquitoes, thereby halting the transmission of the disease without the need to address complex social and economic issues. With the discovery of DDT, it was tempting to suspect that the intricate mysteries of malaria could be reduced to "an entomological rather than a social problem"—one to be resolved by this single weapon, the "atomic bomb of the insect world."[2]

The unrivaled insecticidal properties of DDT were discovered in 1939 by the chemist Paul Hermann Müller. As World War II ended, it was used experimentally in Italy at Castel Volturno and subsequently in the Pontine Marshes and the Tiber Delta. Success in these early trials prompted enthusiasm for the larger deployment of the chemical in Sardinia, whose geographical isolation and moderate size appealed to experimenters. The remoteness and economic underdevelopment of the island were also attractive for an experiment whose ultimate purpose was to devise an antimalarial methodology for the Third World. The experience of Sardinia thus acquired international significance by providing a model for the global antimalarial campaign under WHO auspices between 1955 and 1969.

Victory in Sardinia marked the ascendance of the "American School" of malariology with its stress on technology over the once dominant Italian School, whose approach had stressed the social, educational, and environmental determinants of disease. The triumphant narrative of this American "victory" in Sardinia dominated the official history of the five-year spraying campaign, *The Sardinian Project: An Experiment in the Eradication of an Indigenous Malaria Vector,* which was written by its director, John Logan, in 1953. Logan described in detail this pioneering attempt to destroy malaria by annihilating mosquitoes—an approach that he predicted would forever "push forward the frontiers of public health practice."[3]

Logan's enthusiasm bore fruit under the inspiration of the Rockefeller malariologists. The foremost foundation authorities were Soper, who had eradicated mosquitoes in Brazil and Egypt, and Paul Russell. Echoing Logan's account of the power of DDT, Russell proclaimed "man's mastery of malaria" in his 1955 book of the same name and urged the Eighth World Health Assembly at Mexico City to launch the "DDT era" of malariology. Explicitly invoking the Sardinian experience, Marcolino Gomes Candau, director-general of WHO, called on the assembly to establish a global eradication campaign using DDT and a one-size-fits-all methodology. Directing the WHO effort that resulted, Emilio Pampana outlined in his 1963 book *A Textbook of Malaria Eradication* a four-step strategy of preparation, attack, consolidation, and maintenance.

The WHO campaign, unlike the Sardinian Project, ultimately foundered, and it was abandoned as a failure in 1969. By then, however, malariology had ushered in a period of unparalleled hubris that dominated the field of infectious disease. Between 1945 and the recognition of emerging diseases in the 1990s, infectious diseases were thought to be easily eradicable. Led by malariology, the prevailing reasoning held that technology could be counted on to develop weapons of such power as to eliminate communicable diseases one by one. The experiment in Sardinia thus helped to foster the expectation that the global conquest of all infections would be swift and painless.

Public health policy needs to be informed by history. A policy that either ignores the past or draws misguided lessons from it can all too easily result in serious mistakes and a colossal waste of resources. WHO demonstrated this danger when it misread the Sardinian history in its global eradication program. This chapter examines the Sardinian Project and the boundless confidence in DDT it inspired. The tendency to celebrate the Sardinian display of the stand-alone power of DDT has long obscured a more complex history. The result has been to nourish belief in "magic bullets" and vector control as the sole antimalarial methodologies. But success in Sardinia was based on a more varied set of factors than Logan, Soper, Russell, and WHO have suggested. The pesticide DDT was a potent tool, but it was only one instrument in a multifaceted approach.

Malaria and Sardinia as Synonyms

Malaria had afflicted Sardinia since antiquity, but—as for Italy as a whole—the late nineteenth century marked a vast upsurge in its ravages. As the leading historian of malaria on the island writes, "The extraordinary tightness of the vise in which malaria held Sardinia by the end of the century was a recent phenomenon that emerged . . . after national unification."[4] The very symbols of modernity—national unification, railroads, and demographic growth—combined to create the environmental catastrophe of deforestation, which in turn had dire consequences for public health. With a rugged terrain dominated by hills and mountains punctuated by countless torrents and river valleys, and with an agriculture characterized by small peasant farming, Sardinia had a hydrological system that was immensely vulnerable to the destruction of its forests.

Population growth, the enclosure of common land, and the pressure of an onerous and regressive tax system impelled peasants to clear ever higher slopes to plant wheat on virgin soil that was fertile for a period until it succumbed to erosion. At the same time the growth of railroads and a national market provided both a market for lumber and the means to remove it. The result was a general invasion of the hills with axes, fire, mattocks, and plows. Herds of goats and sheep frequently completed the decimation of forests of beech, pine, chestnut, and oak, which had performed multiple functions in regulating the flow of water. The canopy broke the force of falling rain and reduced its volume by providing broad leaf surfaces for evaporation. At the same time roots and undergrowth anchored the topsoil and protected the underlying limestone from erosion by wind and rain. Drenched by downpours in the absence of this cover, the progressively denuded slopes generated torrents that swept away soil and rock, set off landslides, and silted up riverbeds downstream. Fed in this manner by rushes of water, soil, and detritus, rivers and streams repeatedly overflowed, creating stagnant ponds in valleys and along the coast. The report from the health council at Oristano commented on this phenomenon in the 1880s: "Sardinia, which was once rich as a result of its ancient and flourishing forests, is now becoming a desert steppe through the vandalism of greedy speculators. Through love of lucre they are transforming into charcoal an immense number of plants that represent the patient legacy of centuries."[5]

Since agriculture was extensive and provided few countervailing opportunities to anchor the soil or control drainage, the result was the creation of countless expanses of stagnant or slow-moving water that were ideal for breeding mosquitoes. The Sardinian Project recorded more than a million such breeding places at all altitudes, but above all in the valleys and narrow coastal plains where the soil remained fertile and which peasants and farm laborers cultivated at the simultaneous height of the agricultural and malarial seasons. The degraded environment of the island was ideal for its primary vector—*An. labranchiae.* This mosquito could breed in mountain pools up to altitudes of three thousand feet, on the edges of streams and rivers, and in both fresh and brackish water along the coastline. By day it rested; by night, it feasted on human blood.

Anopheles labranchiae also took advantage of opportunities that developed underground with the growth of the mining industry that exploited deposits of lead, zinc, iron, silver, copper, antimony, and manganese. Mining had powerful incentives: the legislation of the Liberal political regime that ended the crown's monopoly over the subsoil and opened it to exploration and extraction, the demand for raw materials for industry on the mainland, and the transport revolution that made continental markets accessible. Mining therefore expanded exponentially during the closing decades of the nineteenth century. Output, as measured in tons, quintupled between 1860 and 1900, and the workforce tripled from five thousand to fifteen thousand. It was the miners' misfortune that mosquitoes bred in damp mine shafts and that poverty, a deficient diet, and substandard housing were powerful risk factors for the disease. By 1900 malaria had become the chief health problem of the mining population. At the mining center of Montevarchi, for instance, malaria dominated all other causes of admission to the company infirmary, and 70 percent of miners reported in 1902 that they had suffered from the disease in the previous year.

Along with the immediate circumstances that followed deforestation and the development of mining, Sardinia offered a distressing illustration of the close relationship between malaria and human misfortune. The population of the island consisted principally of impoverished peasants who performed outdoor manual labor at the height of the malaria season in the valleys and coastal plains. The peasants lived in housing that was porous and open to flying insects; their immune systems were compromised by malnutrition;

they were inadequately protected by clothing; and their pervasive illiteracy and sanitary ignorance led to their inability to defend themselves through knowledge.

The symbiotic relationship between malaria and poverty was well known to health officials. The eminent physician Giuseppe Zagari, for example, stressed that all Sardinians bore on their bodies the stigmata of chronic malaria—emaciation and splenomegaly, a painful swelling of the spleen. But the poor, whom he identified as those who lived on a diet of beans, cornmeal, and snails, suffered far more than their better-fed neighbors. They also experienced cachexia—the extreme neurological deficit caused by malaria that renders patients indifferent to their surroundings and unable to work, learn, or take part in civil society.

The free market economic policies pursued by the newly united kingdom of Italy promoted inequality, fostered indigence, and therefore directly undermined the health of Sardinians. The integration of the region into a national and then international market through unification and modern means of transportation had severe economic consequences. The undercapitalized agriculture of southern Italy was unable to compete with the modern farms of northern Italy and the Midwest in the United States. Prices for grain products plummeted, unemployment soared, and malnutrition prevailed. This agricultural crisis struck hardest between 1880 and 1895, greatly widening the inequality between a relatively advantaged north and an economically deprived south.

The grievances of southern Italians in these circumstances led to the emergence of the "Southern Question" in Italian history as the expression of the southern sense of injustice. For many *meridionalisti*, or spokesmen for the south, malaria became a symbol of this sectional issue as the disease became the leading public health problem of the region. Sardinia in particular earned an unhappy reputation as the most malarious Italian region. Indeed, popular sayings held that every Sardinian had a swollen belly and that the words "Sardinia" and "malaria" were synonymous.

The First Campaign against Malaria: Before DDT

Malaria was Italy's leading public health problem, causing one hundred thousand deaths per year at the close of the nineteenth century. Not coincidentally, malariology as a discipline emerged as the pride of Italian medical science. It was Italian scientists who played the leading role in unraveling the secrets of the disease, and the Italian School of malariology led by

Camillo Golgi, Angelo Celli, and Giovanni Battista Grassi led the field internationally. Above all, the demonstration of the mosquito theory of transmission by Grassi and Celli in 1898, combined with the sudden availability of quinine on the international market, suggested the idea of a campaign to eradicate the disease.

Quinine is a natural antimalarial present in the bark of the cinchona tree native to the Andes. It had long been known to indigenous populations, but its properties were discovered by Europeans only in the seventeenth century. The bark was then taken to Europe by the Jesuits and began its long career as the source of quinine, which destroys plasmodia when they are circulating freely in the bloodstream. Its twin uses, therefore, as was widely recognized, were as a prophylactic administered before exposure and as a remedy after the onset of fever. The problem until the late nineteenth century was that the supply of South American bark was limited, and the cinchona tree is extremely fastidious and resistant to transplanting to other areas of the world. But the successful plantation of the "fever tree" both in Java and, to a lesser extent, in India transformed the possibilities for its use on a mass scale and enabled the Italian campaign from the turn of the twentieth century to rely heavily on the first successful antimalarial.

Other spurs to action along with faith in quinine were also at work: a new understanding of the economic, human, and social costs of the disease; and the pressure of enlightened landlords, mine owners, and railroad entrepreneurs whose businesses were compromised by an ill and unproductive workforce.

Thus encouraged to act, the Italian parliament passed a series of measures between 1900 and 1907 that launched the world's first national campaign to eradicate malaria. Deploying the "magic bullet" of quinine, the campaign proceeded resolutely until 1962, when victory was declared and Italy was pronounced malaria-free. At the outset, the strategy was single-pronged—to use quinine to medicate the entire population at risk from malaria for several malaria seasons. Since quinine kills plasmodia in the bloodstream, the idea was to distribute tablets both prophylactically and therapeutically with the goal of breaking the chain of transmission. Prophylactically administered, quinine would prevent people from contracting the disease after being bitten by infected mosquitoes. At the same time, quinine used therapeutically would "sterilize" the blood of infected patients, thereby protecting mosquitoes when they took their blood meals. Thus defended against plasmodia by a protective chemical barrier, neither mosquitoes nor humans would be infected, and transmission would cease. The most optimistic members of the Italian School, including Grassi, expected their method to yield victory—which Grassi defined as the *finis malariae*—within a few years.

Simple in theory, the quinine strategy proved ineffective in practice as a stand-alone antimalarial methodology. It was impossible to administer medication to a remote and isolated population without making medical care readily accessible. In order to "quininize" the entire population, the campaign therefore created a new institution—the rural health station—that became the sheet anchor of the initiative. The Sardinian provinces were dotted with health stations staffed by physicians, medical students, and nurses from the mainland who provided care and quinine.

The quinine regimen, however, is lengthy and complicated, and it requires faithful compliance. Furthermore, quinine frequently produces unpleasant side effects, including vertigo, nausea, tinnitus, rash, mental confusion, blurred vision, dyspnea, and headache. All were disincentives for continuing the regimen. Thus the common practice of patients was to take the medication until their fever subsided, when they discarded it. Compliance, in any case, could not be expected from an illiterate population that had no understanding of the disease, coupled with the necessity for disciplined observance. The campaigners rapidly learned this lesson and urged that education was no less important as an antimalarial than quinine. Thus, the rural health station was soon flanked by a second institution: the rural school, which tackled the gigantic problem of illiteracy among both children and adults with a special stress on promoting "health consciousness" and knowledge of the fundamentals of malaria. Malaria, it was said, could be defeated only by an alliance of *medico* and *maestro* (doctor and teacher).

The campaign therefore adjusted course, shifting from its original focus on a single magic bullet to address the "social determinants" of disease. The first step was to establish a context for administering quinine through access to care at health stations and to education in rural schools. Soon the campaign created other institutions such as malaria sanatoria and summer camps for children to remove people from the zones of greatest risk and to provide them with a sound diet, clothing, and instruction in malaria transmission. In time, the "war on malaria" further expanded its mission to target working conditions, wages, housing, land drainage, and environmental sanitation, including the control and removal of impounded water on farmsteads.

The passage from the Liberal regime to the fascist dictatorship after Mussolini's seizure of power in 1922 marked, in theory, a transformation of antimalarial strategy. Viewing the persistence of the disease as proof of the

Liberal regime's failure, fascism made the defeat of malaria a test of its legitimacy. To fascists, Liberal Italy had been weak and indecisive, like all parliamentary democracies, and totalitarianism alone possessed the resolute will to carry the project to victory. Referring to pre-fascist Italy as "petty-minded and flabby," the medical authority Achille Sclavo assured his audience in 1925 that fascism alone would fulfill the nation's promises to Sardinia.[6] The new regime, he claimed, replaced the weakness of democracy with the all-conquering will of Il Duce. Mussolini would rid Italy of malaria, that shameful symbol of backwardness and an enduring threat to weaken the "race" in its quest for greatness.

Furthermore, Mussolini promised Sardinia a special place in his vision of a new Roman Empire. As the only Italian region that could plausibly be described as underpopulated, a medically redeemed Sardinia could serve as an integral part of demographic and territorial expansion. Italy's second largest island could absorb settlers produced by the regime's pronatalist policies on the mainland. Pronatalism, territorial expansion, and malaria eradication were all facets of a single project.

Despite the bombast of fascist rhetoric, however, the substance of antimalarial policy changed far less than the language used to justify it. In theory, the fascist approach to malaria was to attack it through an integrated program of land reclamation, settlement, and intensification of agriculture known as *bonifica integrale*. In practice, this approach required such a gigantic commitment of planning and resources that it was never seriously implemented on a national level. There were, however, two areas of Italy where conditions of extensive marshland and low population density most clearly fit the strategy of *bonifica integrale*. These were the Pontine Marshes and Sardinia, which were intended to showcase Mussolini's program. The Pontine Marshes were the great success of the integrated fascist approach; Sardinia, where vested interests, expense, and bureaucratic inertia triumphed, was its great failure.

By the eve of the Second World War, the Pontine Marshes had been successfully drained, the land settled, and malaria largely controlled. Sardinia, by contrast, boasted only oases of drainage and intensive cultivation. The island as a whole was still characterized by environmental degradation, an unregulated hydrology, extensive farming, and pervasive malaria. Advances were made, but slowly and in line with a recognizably traditional methodology rather than the "totalitarian" solution promised by Il Duce.

In practice, therefore, the antimalarial campaign of the fascist era on Sardinia marked a continued application of the tools and patient work that had been pioneered during the Liberal era—quinine distribution, rural health stations, rural schools, and localized environmental sanitation. Other methodologies were taken up and applied selectively—antilarval attacks by means of predatory Gambusia fish and spraying with the insecticide Paris green as well as what was termed "bonification"—a full program of land reclamation, settlement, and cultivation. Progress continued, but incrementally rather than in a manner consistent with the loudly proclaimed "fascist revolution."

Unfortunately, statistics on the progress of the Italian campaign against malaria are notoriously unreliable, and nowhere more so than in Sardinia. Distance, poor means of communication, severely limited access to medical care, a shortage of laboratory facilities, and the protean nature of the symptomatology of malaria that enables it to mimic other diseases and therefore to confound physical diagnosis—all of these factors precluded an accurate statistical profile of the disease, especially with regard to numbers of cases rather than deaths. Data for mortality were always more reliable than those for morbidity.

With these caveats about accuracy, the official figures for average annual mortality from "malarial fever and swamp cachexia" per one hundred thousand inhabitants by region clearly establish three critical points. The first is that the campaign against malaria achieved considerable, and ongoing, success throughout the nation. The second is that despite major gains, malaria in Sardinia remained a significant and endemic problem of public health at the end of the period. Finally, although Sardinia progressed in absolute terms, it lagged relatively. And third, Sardinia consolidated it's "sad primacy" as the most afflicted region of Italy.

A further feature of malaria in both Italy as a whole and Sardinia in particular is that the steep decline in mortality does not indicate an equally steep decline in morbidity and a concomitant slowing of transmission. This was a consequence of the campaign's reliance on quinine. The availability of the antiplasmodial meant that cases could be treated, even if incompletely, and death prevented even though transmission continued. Deaths from malaria therefore declined sharply while cases persisted at a much higher level although no reliable statistics on morbidity exist.

The Crisis after World War II

The advance toward the control of malaria was not an unbroken, linear progression. On the contrary, the severity of the annual summer epidemic varied according to the vicissitudes of the weather. Years of abundant rainfall and high summer temperatures were the delight of mosquitoes. The most severe and lasting setbacks, however, were the results of war. Both world wars reversed the gains of decades of patient effort and left dramatic malaria epidemics in their wake.

The reasons behind the public health calamity that followed the wars are multiple and interlocking. Warfare effectively brought the antimalarial campaign to a halt as medical personnel were conscripted and as the collapse of the international supply of quinine deprived the movement of its chief tool.

A second cascade of public health consequences seriously affected the two mainstays of the Sardinian economy—agriculture and mining. With regard to agriculture, the world wars systematically diverted resources from the countryside to industry and the military. Draft animals and machinery were requisitioned; men were conscripted; fertilizers, parts, fuel, and equipment became unavailable; investment ceased; land drainage was halted; and nonimmune women and children replaced absent men in the fields. (Women and children lacked immunity because they were far less exposed to the disease as a result of not performing agricultural work and living in hill towns high above the malarial plains and valleys.) But draft animals were themselves coefficients of health: their disappearance deprived mosquitoes of blood meals apart from humans, so they feasted on them instead. Furthermore, with shortages of every kind, production declined sharply while the infrastructure of irrigation and drainage systems was neglected. In addition, the collapse of the transportation system of boats, trains, motor vehicles, and horse-drawn carts disrupted distribution networks so that food shortages developed even when supply was adequate. Prices rose remorselessly, and the problems worsened dramatically under the influence of hoarding and black marketeering.

Throughout the mining districts the impact of the conflict was equally devastating. With supplies, equipment, and transport all unavailable, it became impossible to maintain the shafts, which flooded and were closed. Workers were dismissed en masse, and the industry was paralyzed. With no work, the living conditions of miners were no better than those of slaves.

Military action of course contributed to the epidemic crisis. The mobilization of large numbers of young men, and their deployment under crowded and unsanitary conditions in malarial zones both in Italy and in the Balkans, exposed them to infection. The consequences of Mussolini's military adventures devastated health. Defeat, occupation, the disintegration of the army after Italy changed sides in the war on September 8, 1943, and the transformation of the peninsula into a war zone further decimated agriculture, industry, and the transport system; produced large numbers of displaced and homeless people; and subjected the occupied nation to the German policy of viewing Italy as a place to be pillaged for raw materials, industrial plant, and labor. A series of Allied bombing raids on Sardinia also caused extensive destruction and unleashed floods of refugees. These were all factors that destroyed resistance to disease.

The situation confronting ERLAAS at the outset of its project in 1945 was dramatic, with the agricultural sector further crippled by the worst drought ever recorded on the island. Unhappily, 1946 was also exceptionally dry. Yields per acre for every crop were a fraction of prewar levels, and vast areas were left unplanted as the soil was too hard and dry for plowing or planting. Fodder crops failed as well, so farm animals were slaughtered or faced starvation while the oxen that survived were too lean to drag plows. Forest fires blazed across the arid landscape, destroying orchards, vineyards, and crops. There were almost biblical plagues of locusts and grasshoppers. Inflation was an inevitable result as consumers were caught in a scissors crisis between wages for a manual laborer that increased ninefold over prewar levels and prices that increased twentyfold.

The island faced hunger, and even state rations raised the daily intake of the average Sardinian to only nine hundred calories of the twenty-six hundred considered necessary for health. A typical family during the crisis years 1945–1946 spent 90 percent of its income on food. Faced with empty shops and no money, people dressed in rags and went barefoot. Such declines in the standard of living found clear expression in a crisis of public health. Although malaria and tuberculosis were by far the most serious concerns, authorities also faced serious outbreaks of other diseases—trachoma, syphilis, and gastroenteritis among adults and scabies, vitamin deficiencies, and pertussis among children.

A crime wave also swept over Sardinia. War veterans, ex-partisans, and prisoners of war returned to the island to face unemployment and starvation. They joined unemployed farmworkers and escaped convicts to form marauding bands armed with hand grenades, rifles, and submachine guns diverted from the war effort. These gangs carried out robbery, kidnapping, extortion, and murder while the understaffed police forces struggled to reestablish authority.

The Second Campaign against Malaria: ERLAAS and DDT

American policies toward Sardinia evolved as the extent of the social, economic, and medical calamity emerged. The first task was to prevent a public health disaster from several epidemic diseases that threatened to cause immense suffering, to unravel the economy, and to complicate the job of restoring stable government. Typhus, typhoid fever, and tuberculosis all caused concern and sparked intervention, but the overwhelming threat was that of malaria. ERLAAS confronted the greatest public health problem of the postwar era.

ERLAAS needs to be understood in its political as well as its public health context. As the decision to deploy DDT was being made, the Cold War was beginning, and the campaign against *An. labranchiae* reflected the logic of the conflict between the West and communism. The Rockefeller Foundation, which directed ERLAAS, had genuine humanitarian concern to promote health. But at the same time it was fully aware that medicine, science, and public health were not simply means to relieve suffering, but also instruments of "soft power" to promote US hegemony. Defeating malaria by means of DDT first in Sardinia and then on a global scale would provide a stunning display of the might of US science and technology.

Enlightened self-interest also indicated that a postwar global market economy required healthy consumers and producers and that an inexpensive and rapid means of achieving health would pay lasting dividends. The prospect was all the more appealing since US firms such as DuPont and Monsanto would play a prominent role in supplying DDT. Furthermore, the "American solution" to the problem of malaria would obviate the need to attend to recalcitrant problems of poverty and environmental degradation. Raising those issues smacked of social medicine and socialism.

ERLAAS strategy, devised in 1945 and first implemented in 1946–1947, relied on a military approach to controlling malaria. The organization employed martial terms and metaphors in its internal documents, followed military principles of hierarchy, and made use of army equipment in its work. This "second Normandy landing" and the quasi-military occupation that followed took no notice of the history of the island, the living conditions of the population, the Sardinian economy, or even the prior record of the antimalarial campaign that had begun in 1900. Instead, the planners divided Sardinia into a hierarchical series of administrative units—regions, divisions, sections, districts, and sectors. The sector, defined as the area that a single spraying team could treat with DDT in a week, was the most important operational unit.

For each of the 5,299 sectors, the mappers employed scouts to conduct a survey of all structures where *An. labranchiae* could rest—houses, mineshafts, public buildings, churches, stores, barns, bridges, pigsties, stables, sheds, chicken houses, and the ancient towers known as *nuraghi* that are a feature of the Sardinian landscape. Caves and grottoes were also designated for treatment. Having thus defined and identified the task, ERLAAS recruited teams of sprayers from the local population, who had detailed knowledge of the geography of the area. One team was employed per sector and was equipped with hand and shoulder sprayers and canisters of 5 percent DDT solution suspended in oil. They were instructed to cover the walls and ceilings of each structure with a film of DDT at the rate of two grams per square meter. This application would kill insects on contact and also provide a lethal residual effect that would destroy mosquitoes for months until the subsequent round of spraying.

This original strategy was based on the unprecedented killing power of DDT and on the assumption that *An. labranchiae* was endophilic, that is, spent most of its time nestling indoors before and between blood meals. The goals were twofold. The first was immediate—to destroy all actively feeding mosquitoes before they could deposit their eggs. The second was to poison all resting places where anophelines overwintered in the malarial off-season between November and February so that few would emerge in the spring to breed and reestablish the cycle of transmission. A further refinement was that at the completion of each week of spraying, scouts returned to verify that the spray application had been carried out to specification and to find and count mosquitoes. The directors of the project, gathering in a "war room" in Cagliari, collected the information, projected it onto maps of the island, and planned the succeeding campaign. They also awarded prizes for successful spraying and scouting, punished slackers, and fostered a spirit of competition between the teams.

Entomological evidence soon revealed, however, that the habits of *An. labranchiae* were not so domestic as the designers of the project had assumed. Rather than sheltering solely indoors, the Sardinian mosquito was largely sylvatic, resting out-of-doors and entering dwellings chiefly for the purpose of feeding after dark. As a result, the idea of eradicating the species by means of residual interior spraying was problematic. From 1947, therefore, ERLAAS reduced indoor spraying to secondary importance and stressed the alternative strategy of treating outdoor breeding places in the spring and summer.

This spraying followed the upward curve of cases that first appeared in May, climbed steeply in June and July, and then peaked in August before fading away in November.

The teams followed a technique of climatic zoning according to altitude. The squads began their work at sea level in March when anophelines first bred along the coast, and the teams then worked their way progressively into higher elevations as warm weather advanced. The administrative units of the project and the reliance on teams of sprayers and scouts persisted as the work moved outdoors, but the emphasis shifted from adult mosquitoes to larvae, and from structures to mountain pools, puddles in riverbeds, the edges of streams, marshes, lakes, irrigation ditches, and wells.

Outdoor spraying was far more labor-intensive than residual indoor treatment. The spray had to be renewed weekly throughout the breeding season. Furthermore, there were more than a million breeding sites, and many were hidden beneath thick brambles. Spraying therefore often had to be preceded by draining marshes, clearing streams, dynamiting, and working with riverbeds to increase their flow. Outdoor spraying thus required an array of equipment not needed for indoor applications—brush hooks, axes, shovels, picks, scythes, boats, rafts, tractors, ditchers, mowers, pumps, and explosives. In addition, some expanses of water were so extensive that they could not be effectively treated by hand and so were sprayed aerially instead. Pilots from the Italian Air Force flew converted bombers with fuel drums filled with spray instead of weaponry under their wings.

Outdoor spraying also posed the problem of popular resistance. ERLAAS records make no mention of resistance on the scale of the opposition that greeted quinine distribution during the Liberal era. Schools and health stations had broadcast the necessity of antimalarial measures; DDT created its own popularity by ridding homes of flies and bed bugs; and enlistment in the army of sprayers provided a strong financial inducement. Nevertheless, certain sectors of the population actively impeded the work of eradication. Shepherds and fishermen, fearing that DDT would kill the animals on which they depended, hid water collections from the work teams and occasionally opened fire with their rifles. Bandits attacked ERLAAS payroll deliveries, and communists denounced the entire program as a display of US imperialism. Faced with such obstacles and with an arduous task stretching across thousands of square miles of rugged terrain, ERLAAS soon found itself the largest employer in Sardinia, with as many as thirty thousand workers, of whom twenty-four thousand were engaged in the heavy manual labor of draining and clearing rather than spraying itself.

Thus configured, the Sardinian Project reached its peak in 1947–1948 with full-scale spraying under way throughout the island, indoors and out. The years 1948–1949 marked a turning point. Scouts reported large swaths of territory exhibiting both "larval negativity" outdoors and "adult negativity" indoors, so the emphasis shifted in 1949 from all-out spraying to mopping up sectors where scouts had found indications of "positivity." By 1951 mosquitoes of all species had been severely decimated, but here and there *An. labranchiae* persisted. Since the initial objective of the project had been to determine the feasibility of eradicating an indigenous mosquito species, the campaign was technically a failure. From the standpoint of public health, however, the chain of transmission had been broken, Sardinia was malaria-free, and the Sardinian Project ceased. In the words of the Rockefeller Foundation, "At the end of the effort it was still possible to find an occasional *Anopheles labranchiae* mosquito—and so one could not say that the campaign for eradication had been successful. . . . But while the answer to the eradication question was negative, the results of the experiment in terms of public health advancement were rewardingly affirmative."[7] The numbers of officially reported cases traced a dramatic downward trajectory, from 75,447 in 1946, to 15,121 in 1948, to just 9 in 1951. John Logan, noting that the project had attracted "world-wide interest," predicted confidently as early as 1948 that it "will point the way to an eradication technique applicable on a large scale, heretofore impossible."[8]

Additional Factors in Eradication

There is consensus that DDT was a potent factor in eradicating malaria from Sardinia, but the conclusion that it acted as a stand-alone instrument is misleading. The postwar spraying took place in the midst of overlapping interventions unrelated to DDT that are not mentioned in Logan's official account or subsequent histories. One factor all too readily overlooked is that the establishment of ERLAAS transformed the Sardinian labor market. Once the organization had shifted to outdoor spraying, it employed tens of thousands of workers at wages higher than government rates. ERLAAS thereby significantly contributed to combating poverty and unemployment—two of the principal social preconditions that rendered Sardinians vulnerable to malaria. The total of US $11 million that ERLAAS spent on the island also stimulated the regional economy. In that sense the DDT program silently introduced a second important variable into the experiment, and the trustees of the Rockefeller Foundation even noted that project expenditures marked the "economic rehabilitation" of the island.

Furthermore, even Paul Russell, the proponent of the "DDT era of malariology," noted that the spraying introduced a further variable into the experiment. In his view, the Sardinia Project initiated a progressive upward spiral in which mosquito reduction and agricultural improvement mutually reinforced one another. As early as 1949, he wrote that the spraying program had produced "collateral benefits to Sardinia." Peasants were able to "open up new lands for agriculture" and to "go ahead with bonifica projects formerly impossible because of malaria."[9] Such environmental sanitation and agricultural improvement were significant antimalarials in their own right.

ERLAAS and Rockefeller Foundation records do not acknowledge the long history of patient antimalarial work over the half century before the Sardinia Project. The impression they convey is that US intervention began de novo. There is thus an in-built institutional bias in the documentation toward proving that the "American" technological approach to malaria eradication by DDT alone was effective. In fact, however, ERLAAS owed much of its success to the fact that the ground had been prepared in advance. A major example is the ease with which the sprayers carried out their work. They encountered scattered opposition from bandits, shepherds, and communists; but in general, official documents make it clear that the overwhelming majority of Sardinians welcomed the ERLAAS workers onto their fields and into their homes. This enthusiastic reception marks a striking contrast to the widespread antagonism encountered early in the century when physicians and public health officials began to distribute quinine capsules. At that time the antimalarial campaigners faced the popular suspicion that the medicine was a poison and part of a nefarious plot by the state to solve the problem of poverty by settling scores with the poor. The health of Sardinians was so compromised and fevers so familiar that sufferers were often unaware of the specific importance of malaria, and they were therefore unwilling to ingest government-issued tablets. Just as in the time of plague before, the countryside teemed with rumors of poisoners and of a diabolical plot.

Thus one of the greatest initial difficulties of the first campaign was stubborn resistance by those most in need of its benefits. Peasants, farmworkers, miners, and shepherds refused to attend the newly opened clinics. They barricaded themselves in their homes and turned away visiting physicians and nurses. Alternatively, they accepted the suspect medicine they were offered but then hoarded it for resale or barter for cigarettes, spat the capsules on the ground after the intruders had departed, or fed the offending tablets to their pigs. A portion of the subsidized Italian quinine also made its way to the black market, through which it was reexported to meet the demand of malarial areas of North Africa, where Italian state quinine commanded a high price because of its guaranteed purity. Sometimes parents swallowed their own medicine but refused to administer it to their children. Most commonly of all, patients who were seriously ill took quinine long enough to suppress their fever, after which they ignored the remainder of the regimen. Thus public health officials estimated in 1909 that the majority of the quinine distributed was not consumed.

This suspicion and ignorance seriously complicated the work of the campaign, and a major effort was devoted to finding ways to overcome it. These included education, sermons advocating quinine, the previously mentioned health stations, and public health displays in which village notables demonstrated their confidence by publicly swallowing the medication. In this way the tireless campaign between 1900 and the establishment of ERLAAS in 1945 promoted a public health consciousness that was the backdrop to the enthusiasm that greeted the Rockefeller sprayers. Quinine, in a sense, had prepared Sardinians to understand the necessity for DDT.

Even the ability of the Sardinian Project to deploy scouts and sprayers effectively depended on the prior education of the population. ERLAAS relied on the fact that its workers already understood the rudiments of malaria transmission and the importance of their assignment, and the Rockefeller Foundation knew that eradication had to be very well-organized such that it could run like an efficient machine. Such a machine could run smoothly only if the island possessed abundant personnel already educated in the scientific essentials of the disease.

It is suggestive that ERLAAS built on the legacy of decades by carrying out an educational mission of its own. It offered weekly classes covering Grassi's "doctrine" of mosquito transmission for its staff; it provided schools with a syllabus on which to base lessons for all Sardinian children; it broadcast programs on malaria and the mission to eradicate it; and it printed leaflets, posters, and bulletins for distribution across the region. ERLAAS also deployed one of the oldest weapons in the public health arsenal by ringing the island with a sanitary cordon. In order to prevent *Anopheles* mosquitoes from being reintroduced to the island after the campaign, health authorities boarded and sprayed all arriving boats and airplanes.

Thus the performance of the Sardinian Project is misconstrued if one ignores the long history of public health work that smoothed the way for the postwar campaign. In addition, the history of ERLAAS is distorted by confining attention to the records and archives pertaining to the DDT experiment. Those materials, like John Logan's official account, deal with the project as if it had proceeded in isolation, ignoring the initiatives undertaken by authorities other than ERLAAS. Those initiatives had purposes that were ostensibly nonmedical but that nevertheless had a major impact on the vulnerability of the population to malaria. The Sardinian Project is therefore best understood in a broader context of efforts to confront the social and economic as well as the medical dimensions of the crisis.

The most important intervention operating simultaneously with the Sardinian Project was the United Nations Relief and Rehabilitation Administration (UNRRA), which functioned in Italy with US funding from 1945 to 1947 and was succeeded by the Marshall Plan. Like the Sardinian Project, UNRRA and Marshall aid combined humanitarian intentions and Cold War priorities. In terms of the Sardinian Project, UNRRA was the decisive program because it was more tightly linked with the battle against malaria, while the impact of the Marshall Plan was felt after the decisive battles against *An. labranchiae* had already been won. UNRRA, in other words, directly helped the project to achieve victory, and the Marshall Plan consolidated the triumph.

UNRRA's objectives in Italy were both long-term and short-term. "Relief" was the immediate aim, and it entailed combating the twin evils of "disorder" and "disease." "Disorder" signified strikes, demonstrations, riots, advances in the strength of the Italian leftist parties, and any degree of economic hardship that would drive workers and peasants into supporting left-wing trade unions and political parties. The task was at once a humanitarian "international responsibility" and a form of "world insurance" against revolution.

The first priorities in combating disorder were to alleviate hunger and to contain inflation, and the thinking of American planners was that both goals could be achieved by the massive importation of the necessities needed to keep Italian families healthy. The means was the "three ships a day policy" by which three ships bearing goods from the United States docked daily at Italian ports and unloaded urgently needed supplies, with each ship representing 550 Italian railroad cars of material. These supplies were transported to each of the eight regions into which Italy had been divided.

After the supplies arrived in specific communes, they were distributed to needy people by committees consisting of local notables—the mayor, parish priests, doctors, schoolteachers, business leaders, and other authorities—whose standing and legitimacy would also be enhanced by the largesse. The supplies made available to hungry Sardinians included flour, powdered milk, lard, vegetables, semolina, sugar, and canned fish. Clearly, food supplies and the control of inflation had important implications not just for political stability and the relief of suffering, but also for the enhancement of the population's resistance to disease. Improving the diet and increasing the purchasing power of impoverished Sardinians was a powerful coadjutant to the efforts of the Sardinian Project, coming to the island, it was said, "like oil to a lamp that was dying out."[10]

In addition to dealing indirectly with disease, UNRRA furnished assistance that was immediately linked to the antimalarial effort. This assistance included hospital and medical supplies, and in particular the synthetic antimalarial quinacrine, which largely replaced quinine in the immediate postwar armamentarium. Quinine was reserved for intravenous administration for the most serious comatose cases. Along with vector control by DDT, the traditional Italian effort to attack plasmodia in the bloodstream recommenced. At the same time UNRRA funded the establishment of sanatoria for malaria patients and summer camps for children at the seaside and in the mountains. There they could be shielded from biting insects, furnished a sound diet, and educated in the basics of malaria prevention. Pregnant and nursing women were also offered special rations, clothing, and shoes. Concurrent efforts were launched to repair damaged housing and to provide accommodation for refugees and the displaced. Large numbers of the most vulnerable sectors of the population were thus at least partially protected from infection.

In addition to "relief," the UNRRA mission stressed "rehabilitation," which meant repairing the shattered Italian economy. The goal was to restore industry and agriculture to prewar levels of production. Postwar American planners were convinced that the totalitarian regimes of interwar Europe and the Second World War had their roots in the Great Depression and autarchic economic policies. To prevent the spread of communism and renewed warfare, the United States therefore offered massive support to restore Italian production and to reintegrate the nation into a global free market economy.

In Sardinian terms, this intervention primarily affected agriculture. UNRRA provided seeds, fertilizer, fuel, mechanical equipment, and assistance to repair damaged and neglected irrigation and drainage networks. Crops were also dusted with insecticide to destroy locusts. The direct result was to restore output. Indirectly, however, these interventions attacked malaria by intensifying agriculture in ways that controlled water and eliminated mosquito breeding sites.

UNRRA also contributed directly to antimalarial campaigning in Sardinia in two ways. The first was that proceeds from the sale of American goods provided free to the Italian government were used to establish a Lire Fund to underwrite public health. In Sardinia the Lire Fund was the financial mainstay of ERLAAS. In this way the Sardinian Project is clearly intelligible only in a context that includes the broader relief activities of UNRRA, which financed it. But UNRRA funding also restored the infrastructure of the prewar campaign, allowing provincial antimalarial committees, health stations, and schools to resume work and to distribute quinacrine. In this sense the Sardinian Project did not work in isolation but alongside the traditional efforts of the prewar period.

Conclusion

The successful eradication of malaria in Sardinia has significance as a hopeful example of victory over a deadly and incapacitating disease that continues to hold much of the world, and especially sub-Saharan Africa, in its thrall. Today, approximately 1 million people die annually of malaria, which is both treatable and preventable. It remains the most significant tropical disease and, synergistically with HIV/AIDS and tuberculosis, perhaps the world's most serious infectious disease.

It is therefore important to draw accurate conclusions from the Sardinian experience—to provide, that is, a historically valid account. WHO and the international community initially drew an inaccurate lesson, reaching the logically fallacious conclusion that because eradication followed DDT spraying, the eradication was caused by the spraying.

But the reality of the Sardinian Project is more nuanced. It does of course demonstrate the importance of technological tools, as the role of both quinacrine and DDT amply confirms. The control of malaria requires ongoing scientific research and the practical use of its results. On the other hand, the antimalarial campaigners who worked at the health stations in Sardinia and throughout rural Italy learned even before the First World War that relying on a single scientific weapon was misguided. Malaria, they reported, is the infectious disease that most closely reflects the totality of the relations of human beings with their environment and with one another.

They argued that it is at once a disease of poverty, environmental degradation, poor nutrition, inadequate housing, illiteracy, neglect, population displacement, and improper agricultural cultivation.

Free quinine—the first magic bullet—was distributed in a context that included improved housing, wages, and literacy; adequate nutrition; and the moral commitment of the nation. These factors were antimalarials of no less importance than the quinine itself. Malaria receded with advancing social justice as well as a powerful technological tool. But even with the most powerful technologies, the question remains: What is the appropriate context for their deployment?

Angelo Celli, one of the founders of the antimalarial campaign, replied to this question with a motto that has great contemporary relevance: "Do one thing, but do not omit others."[11] As Celli suggested, the lesson of Sardinia may be that an effective antimalarial program should involve a multifaceted awakening. It requires partnership, the moral conscience of affluent interests, education to teach populations to protect their own health, access to care, affordable treatment, environmental sanitation, and the tools provided by basic scientific research. In addition, Celli's motto suggests another lesson of Sardinia—the need for malaria control to be based on a long-term commitment rather than a "quick fix." Eradication occurred on the island only after half a century of campaigning. Finally, Sardinian success illustrates the importance of international assistance. Sardinia realized the final goal of eradication with the financial and technical support of the United States and an explicit recognition that disease is an international problem in which the world community has a vital stake. Malaria, like all epidemic diseases, is a crisis not of nations, but of humanity.

Despite such somber lessons regarding the complexity of eradication, the Sardinian Project presents a hopeful example of the role that eradication can play in repaying the effort by unleashing the resources of a region. The postwar development of Sardinia presupposed the condition that malaria no longer continued to stunt productivity, limit education, consume resources, and enforce poverty. Sardinia today is an illustration of the social, economic, and cultural possibilities that become available when malaria is vanquished.

CHAPTER 18
Polio and the Problem of Eradication

The movement to rid the world of poliomyelitis, or polio, is a direct outgrowth of what may be termed the "age of eradicationism," which began at the end of the Second World War. During these euphoric decades, there was a consensus that the decisive battle in the long contest between humans and microbes had been joined and that the moment had arrived to achieve final victory. Not surprisingly, this vision took root most deeply in the United States, where victory in war led to unbridled confidence in the power of science and technology. Secretary of State George Marshall declared in 1948 that the world now had the means to eradicate infectious diseases from the earth. Similarly, in 1969 Surgeon General William H. Stewart announced that the time had come to "close the book" on infectious diseases.

Efforts against polio reflected the new spirit. Many leading figures fighting the disease were veterans of other campaigns, and they brought the eradicationist perspective with them. Fred Soper and Alexander Langmuir were pioneering figures in malariology; D. A. Henderson had served as medical director of the global smallpox campaign; and Aidan Cockburn had theorized the end of all infectious diseases.

In addition, the movement received enormous stimulus from the example and encouragement of President Franklin Delano Roosevelt, who in his late thirties developed a paralytic illness diagnosed at the time as poliomyelitis. Roosevelt galvanized the campaign against polio through the two philanthropic organizations he was instrumental in establishing—the Georgia Warm Springs Foundation and, above all, the National Foundation for

Infantile Paralysis. This foundation, which was known as and eventually renamed the March of Dimes, became the major instrument for raising funds for research, patient care, and publicity regarding polio. As David Oshinsky writes, "The National Foundation became the gold standard for private charities, the largest voluntary health organization of all time," an organization that was able "to redefine the role—and the methods—of private philanthropy in the United States."[1]

Polio, the Disease

Poliomyelitis is a highly infectious viral disease caused by three strains of poliovirus. Distinguishing the strains is important because a patient who recovers from polio has a robust and lifelong immunity to the strain that caused the illness, but no crossover immunity to the other two. In addition, poliovirus 1 is more virulent than the other strains and causes about 85 percent of the paralytic and fatal cases.

The virus is transmitted primarily through the oral-fecal route when a person ingests contaminated food and water or touches objects that are contaminated and then puts unwashed hands into the mouth. It can also be disseminated directly from person to person through exposure to the phlegm or mucous of an infected individual who coughs or sneezes. The virus multiplies in the tissues of the throat and the mucosa of the lower gastrointestinal tract for an incubation period that lasts from one to three weeks. In most cases, the infection is asymptomatic and the patient is unaware of its presence. It is critical to the epidemiology of polio, however, that such people nevertheless shed the virus and function as asymptomatic carriers.

In approximately a quarter of infected people, the infection progresses to symptoms and illness of varying severity. Eventually the virus drains from the intestines into the lymph system and then into the bloodstream, which transports it throughout the body and enables it to attack nearly all organs. In most cases, however, the infection causes a mild, flu-like illness—fever, headache, fatigue that is not alleviated by rest, sore throat, nausea, and abdominal pain—that typically persists for up to two weeks. In the transmission of the disease, this "abortive" or "nonparalytic" form of polio plays an important part because such patients, unaware of the severity of their illness, are highly contagious and pose a serious danger to those around them.

For perhaps one person of every two hundred suffering from abortive polio, the disease progresses to the far more severe paralytic form. Onset is marked by what is called paresthesia, which consists of a feeling of "pins and needles" in the limbs. This signals the involvement of the central nervous system—the spinal cord and the brain. Most common is "spinal polio," in which the virus invades the motor neurons of the spinal cord. Indeed, the word "poliomyelitis"—coined in the nineteenth century—is derived from the Greek words *polios* (pale), *myelos* (marrow), and *itis* (inflammation). The affliction was therefore originally conceptualized as a disease of the white marrow of the spine.

The precise mechanisms by which spinal involvement occurs are not yet clearly understood, but once the poliovirus reaches the spine, it invades and destroys the motor neurons responsible for the movement of muscles throughout the body. Consequently, deprived of electrical input by nerve stimulation, the muscles lose function and rapidly atrophy, causing paralysis of one or more of the limbs and frequently of the chest and diaphragm muscles responsible for breathing. Often the result is death. The onset of such paralysis is sudden, and its severity is variable. Paralysis can be partial and transient, but it can frequently be total and permanent. The limbs of one or both sides of the body become loose and floppy, and the affected limbs themselves become deformed, as well as the ankles and feet. This is the origin of the syndrome known as "acute flaccid paralysis."

Less frequent than spinal polio is the involvement of the brain. "Bulbar polio," an affliction of the brainstem, affects nerve centers that control muscles responsible for sight, swallowing, breathing, and the movement of the tongue. This leads to distressing and often fatal consequences, such as the buildup of mucous in the air passages, leading to asphyxiation. Bulbar polio also produces abnormal reflexes, severe headache, spasms, mental impairment, and the inability to concentrate.

Patients who survive paralytic polio and recover are not, however, finished with their ordeal. Frequently a lifetime of physical impairment, disability, and deformity follows. Furthermore, at an interval varying from fifteen to forty years after the initial infection, roughly a third of the victims suffer post-polio syndrome—a condition that is progressive in its severity. Beginning with muscle and joint weakness, fatigue, and intolerance of cold temperatures, it can lead to muscle atrophy, joint deterioration, breathing and swallowing difficulties, skeletal deformities, and mental symptoms such as mood swings, depression, and loss of memory.

Modern Polio

The Experience of the Industrial World

Historically, poliomyelitis was also known also as "infantile paralysis" and seen as an exotic disease of early childhood. Although it could both kill and maim, its sufferers were mercifully few. Between 1890 and the First World War, however, a dramatic transformation occurred in Europe and North America. During those years polio began a new career as a virulent epidemic that struck older children, adolescents, and young adults. Passing violently through the industrial world during the warm summer months, polio began to rival tuberculosis as a cause of concern. It was increasingly referred to as a "plague" and "the crippler." People also began to call it the "new polio" and "modern polio."

The first polio epidemic ever recorded struck Sweden in 1881, followed by Vermont in 1894, Scandinavia again in 1905, New York in 1907, and Vienna in 1908. But it was in 1916 that modern polio came of age when it devastated New York and a large swath of the northeastern United States. After 1916, substantial epidemics of poliomyelitis became a recurring feature of summer in the industrial West. In the United States polio reached a peak of severity from 1949 to 1954 (Table 1).

The fear associated with modern polio reflected a variety of factors—its sudden emergence as a major infectious disease, the absence of a cure, and the burden of illness and death that it exacted. Equally disturbing was its capacity to paralyze and disfigure as well as to kill. A representative reaction, published in 1935 in *Ladies' Home Journal*, stressed that polio caused "mutilation worse than death" and therefore terrified parents more than any other epidemic.[2]

In the intervals between epidemics it was impossible to forget polio, because of the many young bodies tormented by withered limbs, supported by calipers and metal braces, confined to wheelchairs, or imprisoned in iron lungs. Jonas Salk (1914–1995), who was first to develop an effective vaccine against polio, argued, "We know only too well the uniqueness of paralytic poliomyelitis, among all of the infectious diseases; the unusual combination that exists here is the terror and tragedy inflicted to a degree that is out of all proportion to the frequency of attack."[3] For the first time society was forced to confront the largely neglected issue of disability.

Furthermore, the terror associated with modern polio was enhanced by the mystery that enveloped it. When the disease struck New York in 1916, even physicians were confounded by its major features. There was little sci-

Table 1. Cases of paralytic poliomyelitis reported in the United States, 1915–1954

Year	*No. of cases*	*Cases (per 100,000 population)*
1915	1,639	3.1
1916	27,363	41.4
1917	4,174	5.0
1918	2,543	2.9
1919	1,967	2.3
1920	2,338	2.4
1921	6,301	6.1
1922	2,255	2.0
1923	3,589	2.9
1924	5,262	4.6
1925	6,104	5.2
1926	2,750	2.2
1927	10,533	8.8
1928	5,169	4.2
1929	2,882	2.3
1930	9,220	7.5
1931	15,872	12.8
1932	3,820	3.0
1933	5,043	4.0
1934	7,510	5.9
1935	10,839	8.5
1936	4,523	3.5
1937	9,514	7.4
1938	1,705	1.3
1939	7,343	5.6
1940	9,804	7.4
1941	9,086	6.8
1942	4,167	3.1
1943	12,450	9.3
1944	19,029	14.3
1945	13,624	10.3
1946	25,698	18.4
1947	10,827	7.6
1948	27,726	19.0
1949	42,033	28.3
1950	33,300	22.0
1951	28,386	18.5
1952	57,879	37.2
1953	35,592	22.5
1954	38,741	24.0

Source: National Foundation for Infantile Paralysis, *Poliomyelitis 1957: Annual Statistical Review* (N.p., 1957).

entific or medical understanding of the disease and its course. Physicians could recommend no cure, no palliative therapy, no prophylactic measures of protection, no strategy of rehabilitation for those it left disabled. Its mode of transmission, its portal of entry, its pathology—all were unsolved puzzles.

A particularly confounding aspect of polio in the West was the social profile of its victims. There appeared to be no correlation linking sanitation, social class, and housing standards with infection. "Poliomyelitis is not the penalty of poverty," the *Ladies' Home Journal* pointed out. "Once the terror stalks, mere wealth cannot buy immunity. The well-fed babies of the boulevards are no safer than gamins from the gutter from the mysterious universality of the crippling midget, once it's on the rampage. . . . There's hardly a human plague . . . that is as mysterious as this subvisible torturer and maimer of our children."[4]

Thus, unlike many other major epidemics, polio was not a "social disease" from which the wealthy could escape. On the contrary, it seemed perversely to be the exact reverse of a disease of poverty—as it had a predilection for affluent neighborhoods, suburbs, and rural areas. In the United States, polio attacked white children in preference to ethnic minorities, thereby enhancing its salience. A poll conducted in 1948 revealed that Americans feared only nuclear war more than polio.

Significantly, the ravages of poliomyelitis reached new peaks simultaneously in Russia and the United States. This shared burden of suffering and fear formed the background to the extraordinary collaboration across Cold War barriers between Albert Sabin and Dorothy Horstmann on one hand and their Soviet colleagues Mikhail Chumakov and A. A. Smorodintsev on the other. Together they promoted the first mass oral vaccine trials. In 1959 Chumakov oversaw the administration of strains of attenuated poliovirus from Sabin's laboratory to more than a million Soviet citizens. Facing the threat of a new epidemic the following summer, Chumakov persuaded the Russian Academy of Sciences and its political masters that the vaccine would not only forestall the anticipated fury of the epidemic, but could also eradicate the disease. While the Cold War deepened and threatened to ignite into armed conflict, US and Soviet scientists carried out a large-scale program to exchange information and collaborate peacefully in combating a common microbial threat.

A ubiquitous disease transmitted by the oral-fecal route, polio was transformed by hygienic advances in the industrial world. Protected by sanitary defenses, children avoided poliovirus during infancy and early childhood and thereby failed to build up an acquired immunity. As a result, susceptible people in the population accumulated, providing the basis for periodic large-scale epidemics that struck at later ages. It was this mechanism that explained what the Yale epidemiologist John Paul called "modern polio," which could no longer be described as "infantile paralysis."

Polio in the Third World

While modern polio increased the pace of its ravages in the West, proof emerged of the heavy tribute of suffering it caused in the resource-poor tropical world. Here there was considerable surprise. The old dogma prevailing in the mid-twentieth century held that poliomyelitis was an affliction of modernity and sanitation and that it was therefore a minor public health problem in the developing world. But during World War II British and US troops—contrary to all expectations—fell victim to polio in Third World postings such as Egypt and the Philippines (Table 2). This fact revealed how much polio, rather than being a "minor" problem, saturated the developing world.

This finding was confirmed by evidence from serological examinations, rectal swabs, and the direct laboratory analysis of the microbial bounty of Third World sewers. Even more disturbingly, "lameness" surveys conducted during the 1970s and 1980s in such places as India disclosed that the incidence of flaccid paralysis there was at least equal to the burden in the West. Indeed, Albert Sabin (1906–1993) argued in 1983 that the incidence of paralytic poliomyelitis in the tropics was significantly higher than in the developed world at its acme during the prevaccination era.

Spreading through direct contact and excrement, polio in fact thrived in urban tropical environments. But despite the fact that multitudes fell victim to flaccid paralysis or even death every year in developing nations, their suffering—the "old polio" of authentic infantile paralysis—was unreported

Table 2. Confirmed cases of paralytic poliomyelitis in the US Army, 1942–1945

Year	*No. of cases*
1942	48
1943	248
1944	350
1945	680

Source: R. Prentiss, Letter to Albert Sabin, October 17, 1949, Albert B. Sabin Archives, Series 3, Box 23, Item 294, Cincinnati, Ohio.

and invisible. It was unreported because the impoverished children of the tropical world had limited access to care, because physicians were not trained to diagnose a disease that medical orthodoxy held did not exist in this environment, and because the illness of the poor was simply unremarkable.

In the West, children afflicted with polio stood out with their braces, calipers, crutches, wheelchairs, and iron lungs, especially since the National Foundation devoted enormous resources to publicizing their ordeal. In the developing world, infants who were stricken with paralysis died or grew up crippled, but unnoticed, to join the ranks of beggars on the streets of New Delhi, Cairo, and Jakarta. Their plight was unseen in part because poor people fatalistically expected that their children would suffer and die in disproportionate numbers, and in part because medical orthodoxy steadfastly denied their condition and averted its gaze. As Sabin reported, however, the sudden revelation of the burden that polio imposed on Third World nations roused governments and health officials from what he described as their apathy regarding polio and predisposed them to welcome the idea of global eradication. Polio suddenly became a matter of urgency both in the West and in the developing world. There was a new awareness of the "hidden universality" of infection and an appreciation that eradication could save the world community the $1.5 billion that it spent annually on immunization and treatment.

New Scientific Understandings: From Hope to Despondency

Along with a sense of urgency in both the industrialized and the developing worlds, the postwar decades provided new grounds for optimism that polio was a realistic target for eradication. From the turn of the twentieth century, the disease attracted an upsurge of medical and public health interest due to the onset of major epidemics and the hundreds of thousands of paralytics they left in their aftermath. But the "new polio" remained a mystery even as it grew in ferocity. Even though poliovirus had been discovered in 1908 by Karl Landsteiner, the natural history of the disease was still not understood in 1948. The exact mode of transmission and in particular the portal of entry of the virus into the body were unknown, as was the issue of whether the poliovirus existed in a single or in multiple serotypes and strains.

Similarly, the immune mechanisms by which the body defends itself against polio's onslaught were still unknown. Indeed, the lessons derived from experimental poliomyelitis as pioneered by Simon Flexner, who induced the disease in monkeys, misled the medical profession for decades. Here the critical mistake was the assumption that the pathology of experimental poliomyelitis in monkeys was the same as that of the natural disease in humans. Three mistaken conclusions resulted: that the disease overwhelmingly attacked the nervous system, that its portal of entry was the nose rather than the alimentary tract, and that the virus made its way from the nose to the spinal cord and the base of the brain via the nervous system. From the standpoint of vaccine development, these assumptions led to a cul-de-sac because their corollary was that, since the virus spread via the nervous system rather than the bloodstream, there was no opportunity for antibodies to mount a defense. Vaccination, in this understanding, was impossible.

The year 1948 marked a decisive breakthrough, when John Enders, Thomas Weller, and Frederick Robbins discovered that poliovirus could be cultured in vitro in nonnervous human tissue. This advance, which earned them a shared Nobel Prize, revived hope for a vaccine. It did so by drastically reducing the cost of research on poliovirus, by leading from the idea of the nose as the portal of entry to the alimentary tract, by uncovering a phase in which antibodies could attack the virus in the open bloodstream, and by enabling researchers to discover not only the three poliovirus serotypes, but also various strains of each. With these discoveries, the ground was prepared for a vaccine, and in fact not one but two were rapidly developed. The first was Jonas Salk's formalin-inactivated polio vaccine (IPV), which was tested in 1954, officially pronounced safe and effective in April 1955, and administered en masse to children in the United States immediately thereafter. The other was Albert Sabin's live oral polio vaccine (OPV), tested extensively during 1959–1960 and approved in 1962.

Confidence that the new vaccines could achieve global eradication swept aside rational calculation. Belief in the total defeat of poliomyelitis became an article of faith, both for the scientific community and for the general public. Indeed, the strength of the belief that polio would soon be eradicated reflected the intensity of the "icy fear" the disease had instilled. Consequently, news that trials of the Salk vaccine proved it "safe and effective" produced a wave of euphoria and adulation of Jonas Salk. He had demonstrated his hypothesis that "the presence of neutralizing antibody in the circulating blood is an effective barrier in reducing the likelihood of paralysis by the poliomyelitis virus."[5]

The day April 12, 1955, was unique in the history of public health. On no other occasion has a nation awaited the results of a medical trial with such rapt attention. On that day the Poliomyelitis Vaccine Evaluation Center at Ann Arbor, Michigan, which had been charged with assessing the Salk vaccine, released its findings. Thomas Francis, director of the center, delivered the report. In his televised words, Francis found that the Salk vaccine was "safe, effective, and potent." Its rate of success was 80 to 90 percent, and no statistically significant side effects had been found.

Newspapers announced this scientific verdict under banner headlines such as "Dread Disease Believed Dead," "World Cheered by Salk Success," "The Conquest of Poliomyelitis," "Churches Plan Prayers of Joy over Vaccine," "Dr. Salk a Potential President of the U.S.," and "Victory over Polio." The normally cautious *New York Herald Tribune* concluded that polio was finished and that the common cold, heart ailments, and cancer would all be "next."[6] The Eisenhower Administration broadcast this triumph of American science worldwide through Voice of America. Even the stock market applauded as share prices surged, led by those of pharmaceutical companies.

Sabin was equally confident about the imminent demise of polio. As Salk's IPV was being tested, Sabin wrote emphatically that he shared the common goal—"to completely eradicate" poliomyelitis. But Sabin believed that only his own live vaccine could accomplish that goal. His lesson from the trial in 1960 was that those who received the OPV shed attenuated poliovirus to such an extent as to saturate the environment of a community, thereby raising its herd immunity and protecting even the unvaccinated—"for free." As he explained, "It is naturally occurring virulent viruses that one is attempting to eliminate by a rapid production of so many resistant intestinal tracts that they will have no place in which to multiply."[7]

The issue that divided virologists was the question of which vaccine—"Salk's shot" or "Sabin's dose"—was more effective. Skepticism soon emerged about the capacity of Salk's injections to eradicate the disease. Although IPV could stimulate the production of antibodies, it was a killed virus and did not produce mucosal immunity in the intestine as Sabine's live vaccine did. Furthermore, IPV posed a serious practical objection. "Salk's shot" required qualified vaccinators, making the vaccine labor intensive and too expensive for worldwide use. In the United States, for example, during the 1960s a Salk vaccine injection cost twenty-five to thirty dollars, whereas "Sabin's dose," which could be swallowed on a sugar cube, cost only three to five dollars. Complicating the picture still more, the Salk vaccine required booster shots.

By 1960 Alexander Langmuir (1910–1993), CDC chief of epidemiology, concluded that Salk's IPV had led the nation into a prophylactic dead end despite unprecedented efforts and positive results. The scale of the effort was undeniable. Since 1955, 93 million Americans had been vaccinated with inactivated virus. Surgeon General Leroy Burney even declared that the achievement was "monumental" and had no parallel in the history of medi-

Table 3. Poliomyelitis cases reported in the United States, 1955–1961

Year	*No. of cases*	*Cases (per 100,000 population)*
1955	28,985	17.6
1956	15,140	9.1
1957	5,894	3.5
1958	5, 787	3.3
1959	8,425	4.8
1960	3,190	1.8
1961	1,327	0.7

Source: National Foundation for Infantile Paralysis, *Statistical Review: Poliomyelitis, Congenital Defects, and Arthritis* (N.p., June 1962).

cine. Furthermore, the positive impact of Salk's vaccine was notable, as the incidence of poliomyelitis in the United States plummeted dramatically (Table 3). In addition, Salk's killed poliovirus could not revert to virulence, so it could be safely administered to immunocompromised patients and to their household contacts.

On the other hand, the downward trajectory of the incidence of poliomyelitis in the United States stalled unexpectedly and even turned upward in 1959, leading to press reports that polio was "striking back" and that a new and more virulent strain was circulating. It turned out that the mass administration of Salk's IPV altered the nature of the annual epidemic. Under pressure from the killed vaccine, polio retreated to those reservoirs of susceptible people beyond the reach of the vaccinators. Instead of remaining the "new polio" of the comfortable and the clean, the disease became instead an affliction of the poor, of ethnic minorities, and of religious groups that opposed vaccination (such as the Dutch Reformed Church).

No mechanism was in place to reach those underserved or recalcitrant pockets of the population. The *New York Times* reported that the incidence of polio among African Americans in inner cities and among Native Americans on reservations was four to six times the national average. Instead of disappearing, polio had retreated to the "domestic Third World" of impoverished America. Thus, Langmuir, who had been a leading eradicationist, became a skeptic in 1960. In his considered view, eradication had been an unrealistic goal:

> Five and one-half years ago, when the Salk vaccine first became available, some epidemiologists, including the present speaker,

> were hopeful that the disease could rapidly be eradicated. Since that time, its incidence has declined due in large measure to the use of this vaccine. But polio seems far from being eradicated. The dreamed-for goal has not been achieved. In fact many students of the problem question that eradication of poliomyelitis infection with inactivated vaccine is a scientifically tenable concept.

The end result, Langmuir lamented, was that "an overestimate of potential vaccine acceptance was made. Large parts of the population remain unimmunized. . . . Large 'islands' of poorly vaccinated population groups exist—in our city slums, in isolated and ethnically distinct communities, and in many rural areas." Therefore, he said, since "killed-virus vaccine so far has not reached substantial non-immune segments of the U.S. population," there was a need for "some new approach."[8]

The Cutter Incident

The dream of eradication was further frustrated by the disastrous "Cutter incident" that occurred during the first campaign using Salk's vaccine in 1955. The Cutter Laboratories of Berkeley, California, were one of six major pharmaceutical companies (in addition to Allied Laboratories, Eli Lilly, Merck, Parke-Davis, and American Home Products) contracted to produce polio vaccine for the campaign. After just two weeks of elation, as people jostled to be among the first to be injected, the nation received distressing news. On April 27 the public health director of Illinois, Dr. Roland Cross, issued a statement indicating that the vaccine produced by Cutter Laboratories "may not be safe" and cautioned physicians not to use the Cutter product until further notice. The Public Health Service responded by placing an embargo on all vaccine produced by Cutter. Then on May 8 Surgeon General Leonard A. Scheele halted the entire vaccination program while the government conducted an investigation.

The CDC discovered, according to an account in the *New England Journal of Medicine*, that

> two production pools made by Cutter Laboratories (accounting for 120,000 doses) contained live poliovirus. Among the children who had received vaccine from these pools, abortive polio (characterized by headache, stiff neck, fever, and muscle weakness)

> developed in 40,000; 51 were permanently paralyzed; and 5 died. Cutter's vaccine also started a polio epidemic: 113 people in the children's families or communities were paralyzed, and 5 died. It was one of the worst pharmaceutical disasters in U.S. history.[9]

Although the CDC report on the incident concluded that the "exact causes" of the contamination had eluded investigators, the authors regarded three factors as most harmful. The first was the negligence of the Public Health Service, which had not drawn up detailed regulations for vaccine production. It had instead allowed the companies to determine for themselves what safeguards were appropriate. In the caustic comment of Senator Wayne Morse, "The federal government inspects meat in the slaughterhouses more carefully than it has inspected the polio vaccine."[10]

Second, in this laissez-faire environment, Cutter had come under intense pressure to produce a vast quantity of vaccine in haste. Eager to meet the demand, the company had compromised on safety. Specifically, it did not inactivate the virus adequately, which contaminated six lots of vaccine—the ones from which the infected recipients had obtained their shots. Third, the pharmaceutical giant had neglected the failsafe procedure of testing for the presence of live virus in the final product.

Thus the "incident" unfolded in the press as a tale combining inadequate federal oversight, corporate greed, and human tragedy. The alarm was deepened by revelations of prior incidents of corporate malfeasance. Cutter Laboratories had been accused of product safety violations in 1949 and of price fixing and fraud in negotiating the vaccine contract in 1955. These tawdry revelations created enormous public distrust. Indeed, in 1956—the year after the scandal—one family in five across the country feared vaccination more than the disease and refused IPV. Furthermore, the rippling effects of the Cutter tragedy involving scores of children who were sickened, crippled, or killed were not confined to 1955 and 1956; doubt persisted for years. As a result, American children continued to be exposed to infection, and the recession of polio was delayed.

The Push toward Global Eradication

After the surge of discouragement, two new developments—one medical and the other operational—created a new burst of eradicationist optimism. The scientific breakthrough was the series of successful large-scale trials of Sabin's competing oral vaccine. The first limited trials from 1954 to 1957 were conducted on volunteers at the Federal Correctional Center at Chillicothe, Ohio. Then between 1958 and 1960 mass trials took place both domestically at Cincinnati and Rochester, New York, and abroad—in the USSR, Hungary, Czechoslovakia, Singapore, and Mexico. These trials demonstrated that "Sabin's dose" was not only safe and effective, but also dramatically easier to administer than Salk's injection.

Operationally, a novel campaign strategy provided a framework within which Sabin's vaccine could be maximally effective. This was the practice of community-wide vaccination days, known in the United States as Sabin Oral Sundays (SOS) and globally as national immunization days (NIDs). Immunization days provided the formerly missing mechanism needed to reach Langmuir's recalcitrant islands of unvaccinated people. Instead of limiting immunization to children who visited private physicians, the immunization days took polio vaccine to the people.

Cuba pioneered this strategy in 1962 where the grassroots organizations of Fidel Castro's revolution—the "committees for the defense of the revolution"—carried out house-to-house surveys to locate every child on the island. After this census-taking effort, the committees returned to the home of every child who had been identified. On this return visit, vaccinators—recruited by the committees and trained for just half an hour—administered candies or sugar cubes permeated with live attenuated poliovirus. The goal was to immunize every susceptible child in Cuba.

Castro's invention of vaccination days, coupled with the use of Sabin's doses, achieved rapid success in interrupting transmission and making Cuba the first country to eliminate polio. This success raised the question of whether the technique could be applied in a noncommunist country. The United States provided an affirmative demonstration. Sabin Oral Sundays began in Maricopa and Pima counties in Arizona, in which Tucson and Phoenix are located. These counties furnished a model for the nation. In Tucson, leadership was provided not by a "committee for the defense of the revolution" but by the Maricopa County Medical Association and the local pediatric society. Their members volunteered their services alongside local pharmacists, county health department officials, nurses, Boy Scouts, schoolteachers, clergy, the local press, and homeowners. The method also departed from Cuban precedent in that children were taken to fixed vaccination centers set up in schools instead of having peripatetic vaccinators travel from house to house.

The result was described by the press as an event of a kind never seen before as more than six hundred thousand people, or more than 75 percent of all the city's residents—took part in a quasi-festival occasion on each designated Sunday in January and February of 1962. The citizens of Tucson, for example, after being greeted by Boy Scouts and PTA volunteers, stood in line to pay twenty-five cents for their cubes of sugar. Especially if they were children, they also rejoiced that Sabin's doses eliminated the need for a needle. For those unable to pay, the sugar cubes were free because the principle of the day was that no one would be turned away. Sabin himself argued that it was immoral to sell medicine as if it were a commodity like fabric to be purchased at so much per yard. Such Sabin Oral Sundays were soon observed throughout the nation under the aegis of local medical societies.

Polio eradication efforts during the 1960s received powerful encouragement from the progress made against smallpox with the apparently identical weapon of mass vaccination. In the mid-1960s, the temptation to make a last push against that disease had become irresistible, and technological innovations made the objective of eradication more feasible. By 1959 WHO embarked on a global eradication attempt against smallpox, and as we have seen, the program achieved success in 1977 when the world's last case of variola major occurred.

This first intentional eradication of a human disease fueled hopes of a similarly successful campaign against polio. Three factors were deemed to be of critical importance: (1) like smallpox, polio had no reservoir of infection apart from human beings; (2) there was an effective and easy-to-administer vaccine to interrupt transmission; and (3) modern diagnostic tools were available to detect infection. In 1997 these preconditions for successful eradication were even officially codified. In that year Berlin hosted a workshop—the Dahlem Workshop on the Eradication of Infectious Diseases—to establish the criteria for deciding when an infectious disease is a good candidate for an attempt at eradication rather than a more modest goal such as control or elimination.

Good news from the United States heightened eradicationist euphoria. Sabin's vaccine and Sabin Oral Sundays made good on their promise to immunize even the unvaccinated Americans whose numbers had so dismayed Langmuir. The US campaign therefore achieved the objective of halting transmission by the end of the 1960s—the very decade that had begun in pessimism when the progress made by Salk's vaccine faltered.

Against this background, the push to eradicate polio globally began soon after the examples of American eradication of polio and of global victory over smallpox. Three preliminary steps were taken between 1984 and

1988. The first occurred in March 1984 at the initiative of Jonas Salk and former secretary of defense Robert McNamara. At their urging, a group of sponsors—UNICEF, WHO, the World Bank, and the UN Development Programme—established the Task Force for Child Survival. Its director called for a campaign against poliomyelitis as central to any mission to save children.

The second step resulted from a suggestion by Albert Sabin. He pressed Rotary International—the Gates Foundation of the era—to establish a consultative committee to consider sponsoring worldwide eradication. Appointed in 1984, the committee took up Sabin's challenge by calling for the global vaccination of children. Under the name "PolioPlus," the committee chose 2005 as its target for polio eradication. Rotary International's commitment was decisive. An organization of community notables in the realms of business, finance, and the professions, Rotary had multiple resources to deploy—money, an international infrastructure of thirty-two thousand clubs and 1.2 million members, access to governments and ministers of health, and a philanthropic mission of service.

The third and final step took place in March 1988 at Talloires in France. There the newly created Task Force for Child Survival convened a meeting whose final resolution—the Declaration of Talloires—was decisive. The declaration explicitly called for the eradication of polio by the year 2000. This recommendation was then taken to the Forty-First World Health Assembly two months later, where 166 member states adopted the Talloires objective as WHO's operational goal. They agreed to launch what became the most ambitious campaign in the history of public health—the Global Polio Eradication Initiative (GPEI).

From the outset in 1988, GPEI adopted a strategy that joined Sabin's vaccine as its tool of choice with Cuba's operational device of community-wide immunization days. The goal was to provide the most susceptible populations with at least two oral doses. Immunization days, targeting children from birth to six years of age, took place every year in two rounds of one or two days each, spread four to six weeks apart. Where coverage was deemed inadequate, the national days could be supplemented by subnational or regional immunization drives. These supplementary days particularly targeted inaccessible or high-risk sectors of the population. They provided publicity in local dialects, involved community leaders in the planning and conduct of operations, and deployed teams of vaccinators to make house-to-house visits along the lines of the Cuban precedent. A vital component of the effort was intense community participation, including the mobilization of churches, women's groups, NGOs, and prominent local figures.

In addition, the strategy incorporated a lesson derived from the smallpox experience. This was the importance of laboratory surveillance to detect ongoing transmission. To make surveillance possible in resource-poor settings, WHO established a Global Polio Laboratory Network that consisted of 145 laboratories charged with analyzing rectal swabs from all suspected cases of paralytic polio. The goal was to enable the campaign to monitor its progress so that it could organize house-to-house "mop-up" campaigns wherever transmission persisted.

Taken together, these measures constituted WHO's "four-pronged methodology"—routine immunization, supplemental mass immunization, poliovirus surveillance, and rapid outbreak response. As in the earliest trials of Sabin's vaccine, the goal was to replace virulent wild poliovirus with attenuated strains that were live and infective but not virulent. In the attainment of this goal, the method was to immunize even the unvaccinated population. The nonvirulent virus shed by people who had received vaccine would saturate the community, inducing mild infection and therefore immunity but avoiding paralysis.

To carry out its mission, GPEI established a vast operation funded not only by Rotary International, but also by national governments and UNICEF. The campaign employed Sabin vaccine strains produced by international laboratories and delivered to communities via cold chain technology. It reached remote populations through technical and logistical support from WHO and the CDC; it distributed publicity furnished by the Geneva offices of WHO and a variety of international agencies, national governments, and community leaders; and it deployed an army of vaccinators. NIDs were particularly labor intensive, necessitating the participation of more than 10 million people. In India, for instance, the NIDs vaccinated as many as 90 million children in a single day. In immunizing such unprecedented numbers, the campaign took full advantage of the simplicity of Sabin's doses. Instead of requiring skilled personnel and syringes, the oral vaccine only required that vaccinators possess the single skill of counting to two—the two drops dripped into a person's mouth.

Between launch and 2003, the progress made by the war against polio was rapid and astonishing, even if the goal of eradication by 2000 was impossible. In 1988, WHO estimated that there were 350,000 cases worldwide of paralytic polio—known as acute flaccid paralysis—and that the disease was endemic in 125 countries. Already in its inaugural year GPEI succeeded in halting transmission in Europe. Similar triumphs took place in the Americas by 1991 and in the Pacific region by 1997. In 2001, WHO announced a record low of 483 cases worldwide in just four endemic countries—Afghanistan, Pakistan, India, and Nigeria. In 2002, there was an increase to 1,918 cases, but a compensating advance occurred when the organization reported that type 2 poliovirus was extinct. The world seemed poised for final victory.

Setbacks, 2003–2009

Between 2003 and 2009, however, the campaign experienced a series of significant setbacks, generating pessimism and raising the question of whether the eradication of smallpox had been a misleading special case rather than a valid precedent. There was a suggestive analogy between the microcosm of the United States in 1960 described by Langmuir and the global macrocosm in 2003. In the United States, poliomyelitis had retreated to its strongholds of ethnic minority groups in the inner cities and areas of rural poverty. Similarly, by 2003 the disease, readily vanquished in the economically developed world, remained endemically entrenched in some of the most unsanitary, impoverished, and insecure places on the planet—war-torn and remote areas of Afghanistan; the Muslim-dominated states of northern Nigeria; northern Sind in Pakistan; and Bihar and Uttar Pradesh in India.

The northern states of Nigeria, especially Kano, were striking cases. There, because of political and religious complications, the campaign was entirely halted for thirteen months beginning in 2003. As a result, the disease flared up, causing more than five thousand cases of paralysis in the three years ending in 2006. From Nigeria polio spread widely, causing paralysis in eighteen countries, primarily in West and Central Africa, that WHO had previously pronounced polio-free. The fragility of earlier progress was evident.

The problems that emerged between 2003 and 2009 were complex and multifaceted. In Nigeria key components were religious and political conflict. In Kano, Muslim leaders, suspicious of Western intentions after the launching of war in Afghanistan and Iraq, preached that Sabin's doses were not a public health measure at all. The vaccine, they warned, formed part of a sinister plot to sterilize Muslim children with a reproductive poison.

Furthermore, Western motives appeared mysterious. Why was the international community so intent on ridding Nigeria of polio when other local needs were more pressing? Nigerians questioned the plausibility of a campaign against polio when they were far more concerned with safe water, poverty, and diseases that were more prevalent, such as malaria, tuberculosis, and HIV/AIDS. And why was the campaign distributing the so-called polio vaccine free of charge?

In addition, the emir of Kano attributed malevolent intentions to Christian nations because many had chosen to use Salk's inactivated shots as an endgame strategy for their own populations while insisting that Nigerian Muslims receive Sabin's live virus drops, which presented some risk of reverting to virulence. Indeed, scattered cases of "vaccine-derived polio" occurred in Nigeria, highlighting the danger. Finally, religious and political opposition to the vaccination campaign was inflamed by more local concerns. The Nigerian Ministry of Health represented a federal government dominated by the Christian southern states, which were relatively privileged. The ministry therefore became a proxy for the sectional, political, and religious discontents of the impoverished northern states.

The result was a boycott of the antipolio campaign that began in 2003 and ended in 2004 only after a series of reassurances. Muslim-operated laboratories in India analyzed the polio vaccine supplied to Kano by the West and found it to be harmless; pharmaceutical laboratories directed by Muslims in Indonesia agreed to supply all further shipments of Sabin's vaccine to northern Nigeria; and prominent Islamic leaders around the world pressed their coreligionists in Nigeria to support the international antipolio effort.

By 2004, however, considerable and lasting harm had been done to the campaign. It was not clear how soon and to what extent Muslim populations would change their minds by allowing vaccinators into their homes, and many health officials doubted that the spread of the virus beyond Nigerian borders was reversible. Indeed, there was a major upsurge in paralytic cases of polio, which rose from an all-time low of 56 in 2001 to 355 in 2003, 831 in 2005, and 1,143 in 2006. A World Health Assembly progress report on Nigeria in early 2009 was far from encouraging. It noted that there were major gaps in coverage, with the result that 60 percent of children had not been fully vaccinated. Furthermore, all three poliovirus serotypes were still circulating in northern Nigeria, and viruses originating in Nigeria had spread to Benin, Burkina Faso, Chad, the Côte d'Ivoire, Ghana, Mali, Niger, and

Togo. Fortunately, the reversal of the gains achieved against polio was temporary, and GPEI redoubled its efforts. Tangible and encouraging advances were indicated by the downward general trajectory in the global number of cases between 2006 and 2016:

Year	*No. of cases*
2006	2,233
2007	1,527
2008	1,903
2009	1,947
2010	1,377
2011	758
2012	319
2013	505
2014	458
2015	114
2016	46

Nevertheless, troubling obstacles arose in India in Uttar Pradesh and Bihar, the epicenters of polio endemicity on the subcontinent, where accessibility, security, and the fears of a minority Muslim population became issues. More disturbing were biological problems that posed the question of whether the concept of eradication was scientifically valid in a tropical context. This was the problem of "interference" that Albert Sabin had posed half a century before. In the aftermath of the discovery in the 1950s of Coxsackie and Echo viruses, it became clear that the three serotypes of poliovirus form part of a much larger family of enteroviruses that populate the human alimentary tract. In unsanitary tropical conditions, Sabin worried, the intestinal flora of enteroviruses could be so dense and varied that the poliovirus contained in the vaccine would be unable to establish an infection. It would be "interfered" with, and immunization would therefore be impossible.

In India Sabin's theoretical concern proved to be a practical problem when it was discovered that some children who had been vaccinated as many as ten times still had not developed any protective immunity. Interference also gave rise to worries about the possible consequences if children were excessively dosed with poliovirus. Multiple vaccinations had not been envisaged by the campaign at the outset, but in India some children were vaccinated as many as twenty-five times during the first five years of their lives because of the unprecedented combinations of routine vaccination, national and regional immunization days, and mop-up operations.

Other problems also emerged in the turbulent years from 2003 to 2009. One was that, as live viruses, the attenuated Sabin poliovirus strains always possessed the potential to revert by mutation to virulence and neurotropism, leading to outbreaks of "vaccine-associated paralytic polio." This possibility—that the vaccine itself would unleash epidemic disease—is not simply theoretical. Outbreaks of vaccine-associated polio occurred in the Philippines (2001), Madagascar (2002), China (2004), and Indonesia (2005). Reversing progress against the disease, such outbreaks occasioned the quip that it is impossible to eradicate polio without OPV, but it is also impossible to eradicate polio with it. Vaccine-associated polio also implies that the campaign has no logical endpoint because it will always be necessary to immunize populations against epidemics that can be initiated by the vaccine itself. The difficulty is all the more intractable in the case of people with immunodeficiency disorders as they can continue to shed the vaccine virus for as long as ten years. If the campaign is allowed to lapse, therefore, the accumulation of nonvaccinated susceptible people can result in devastating epidemics among the immunologically naive.

Of the incidents of vaccine-associated polio, the most important example, as of mid-2018, is ongoing in the Democratic Republic of the Congo (DRC). There, in the wake of campaigns administering Sabin's OPV, type 2 virus has reverted to virulence and has caused thirty cases of paralysis in three widely separated provinces. Worryingly, genetic analysis has revealed that the outbreaks in the three provinces are caused by separate strains of the virus that have been circulating silently for several years.

The DRC outbreaks raise four principal considerations. The first is that the presence of separate strains suggests that vaccine-associated type 2 virus has been circulating widely, but undetected, across great swaths of the country. In addition, the fact that one of the outbreaks happened near the border with Uganda raises the immediate threat of the international spread of the disease. Third, the return of polio to the DRC is inherently problematic because problems of security and poor transportation in the country complicate the tasks of surveillance, case tracking, and vaccine administration. Finally but ironically, the only means to contain the upsurge is to administer yet more type 2 monovalent OPV, creating the danger of further outbreaks through reversion to virulence.

Therefore, in the assessment of Michel Zaffran, the director of the global antipolio campaign, the upsurges of the disease in the DRC constitute "absolutely" the most serious crisis facing the eradication effort—far more serious than the endemic transmission of "wild" polio virus in Afghanistan, Pakistan, and Nigeria. The DRC outbreaks, in his view, place the global campaign in jeopardy either of a lengthy delay or even of final defeat.[11]

Even apparent victory both in the DRC and in the remaining endemic countries will leave prolonged uncertainty. WHO policy for declaring eradication would be to rely on the one single indication of the end of transmission—the absence of acute flaccid paralysis. Polio, however, is notorious for its silent circulation because the vast majority of infections are either entirely asymptomatic or produce mild flulike symptoms. In this regard polio is radically different from smallpox, which was eradicable in part because of the florid symptoms that facilitated tracing cases. In polio, fewer than 1 percent of infections progress to central nervous system involvement and paralysis. The absence of paralysis is thus a poor indicator of the end of transmission, even after several years. It is all too easy to declare a country polio-free prematurely, as happened in Albania. There, an indigenous case of polio appeared almost ten years after the last case of acute paralysis, and even a single case of paralytic polio normally indicates at least one hundred additional asymptomatic cases.

Further complicating matters, cases of flaccid paralysis can be induced by a variety of noninfectious conditions, such as Guillain-Barré syndrome, traumatic neuritis, acute transverse myelitis, and neoplasms. Furthermore, as Sabin noted, clinical conditions identical to paralytic polio can result from infections caused by seventeen other enteroviruses. In other words, one of the complexities of the attempt to eradicate polio is that poliovirus can circulate for extended periods without producing flaccid paralysis, while flaccid paralysis can occur in the absence of poliovirus.

Clearly, the history of GPEI demonstrates that the uncritical optimism generated by the comparison with smallpox was misplaced. Smallpox was exceptional in its vulnerability to attack. It had no animal reservoirs; it had no range of serotypes to pose the problem of crossover immunity; its symptoms were florid and readily detectable; it gave rise in survivors to a robust and lasting immunity; and the vaccine used to attack it was effective after a single dose while posing no threat of unleashing vaccine-associated epidemics.

Polio is a far more formidable adversary, and it reminds us that the original eradicationist perspective was illusory. Final victories over disease are to be celebrated as exceptions rather than as the expected steps toward a germ-free Eden. The conquest of polio, so near, but so elusive, would make it just the second infectious human disease ever to be eradicated. Whatever the final outcome of the campaign, which currently hangs in the balance, the difficulties that have arisen clearly demonstrate that the eradication of an epidemic disease requires adequate tools, extensive funding, careful organization, sustained effort, and good fortune.

CHAPTER 19
HIV/AIDS
An Introduction and the Case of South Africa

Origins of AIDS

The AIDS pandemic began when a mutation occurred in an affliction of monkeys and apes known as simian immunodeficiency virus. This mutation enabled the virus to become an example of a zoonosis—a disease that crosses the species barrier from animal to human. We cannot know when this momentous event first happened, but a current hypothesis is that there were isolated cases as early as the 1930s that were never identified. No later than the early 1950s, stable transmission from human to human was occurring, launching the emerging new human immunodeficiency virus (HIV) on its modern career.

This crossing of the species barrier by the immunodeficiency virus came about in two distinct African settings, and it resulted in two different biotypes of HIV. In Central Africa, in the border area where Burundi, Rwanda, and the Democratic Republic of the Congo converge, HIV took its place as an ongoing affliction of humans. This was HIV-1, which is the more virulent biotype and the one that is preponderantly responsible for the modern HIV/AIDS pandemic. At roughly the same time in West Africa, HIV-2 emerged as a human affliction that is slower acting and less contagious.

How the species barrier was crossed from monkey or ape to human is a matter of conjecture. One popular theory is propounded by Edward Hooper in his 1999 book *The River: A Journey to the Source of HIV and*

AIDS. Hooper's idea is that the widespread testing of an oral polio vaccine in Africa in 1958 provided the opportunity. He targets ethically questionable biomedical research carried out in Third World countries and criticizes the unseemly haste of researchers to win a Nobel Prize for enabling AIDS to develop. In reaching his conclusion, Hooper conducted extensive epidemiological research modeled on John Snow's early work with cholera, but his evidence is purely circumstantial. An alternative popular theory is that HIV was transferred to humans through the consumption of infected bush meat, probably of chimpanzees in Central Africa and of monkeys in West Africa.

HIV and the Body

There is an exotic debate over whether viruses should be classified as living organisms. Certainly, HIV is extremely simple in its structure, consisting of two strands of genetic material in the form of ribonucleic acid (RNA) containing just ten genes encased in a membrane with two glycoproteins on its surface. By contrast, a bacterium contains five thousand to ten thousand genes. HIV is incapable of independent motion, metabolism, growth, or reproduction except through the process of invading host cells and transforming them into the means of producing viruses. In this sense, HIV is entirely parasitic, capable of carrying out the essential processes of life only by converting living cells to its purposes.

Having once gained access to the human bloodstream, the glycoprotein Gp21 on the surface of the microorganism enables it to target and then invade certain cells, in particular the white blood cells known as CD4 cells (or T-helper cells). These cells regulate the body's immune system by activating it when they detect invading microbes. Inside a host CD4 cell invaded by HIV, an enzyme known as reverse transcriptase converts the viral RNA into deoxyribonucleic acid (DNA). This process of reverse transcription—discovered separately in the 1970s by Howard Martin Temin and David Baltimore, who shared the 1975 Nobel Prize in Physiology or Medicine with another recipient—is the reason for calling HIV a "retrovirus." Until 1970 a central tenet of evolutionary biology held that DNA creates RNA. The discovery of the reverse process—the creation of DNA from an RNA template—made it possible to understand the biology of HIV/AIDS.

Reverse transcription is critically important, first because of its impact on the infected body. The newly converted DNA is encoded into the genome of the host cell. This enables the invader to take over the machinery of the cell, converting it into a means of generating new microorganisms and destroying the cell itself. CD4 cells, the regulators of the immune system, become "viral factories" generating HIV virions that return to the bloodstream to invade further CD4 cells and repeat the cycle. This destruction of CD4 cells disables the network of signaling molecules that mobilize the body's defenses. It is therefore the foundation of immunosuppression in AIDS and the key to the many opportunistic infections it gives rise to. In addition, beyond the body of the individual sufferer, the process of reverse transcription is central to the epidemiology of HIV/AIDS because it is highly prone to error. Thus it causes mutations with great frequency, creating varieties, or "clades," of HIV and underlying the development of drug resistance.

Invasion of the bloodstream by HIV is followed by an incubation period of six to eight weeks that leads to a "primary infection" that lasts about a month. Often, the primary infection is asymptomatic and unknown to the patient, who nevertheless undergoes seroconversion, meaning that HIV antibodies reach detectable levels and the patient becomes HIV-positive. Many patients, however, experience symptoms similar to influenza or mononucleosis—fever, aches and pains, fatigue, swollen lymph nodes, headache, sore throat, diarrhea, and sometimes a rash. Then, twelve weeks from the original infection, the symptoms disappear.

Even after the resolution of symptoms, however, the disease progresses silently for a latent period that lasts for years. The life cycle of HIV involves the continual invasion of CD4 cells, the replication of HIV virions, the destruction of the host lymphocytes, and the return of virions to the bloodstream to renew the process. The virus progressively reduces the number of CD4 cells while increasing the "viral load," or number of viruses per cubic milliliter of blood. Each day the virus destroys approximately 5 percent of the CD4 cells in the bloodstream while they are replaced at a slower rate. In time, therefore, the balance between virus and T-cells is decisively and irreversibly tipped in favor of the virus.

Meanwhile, latency plays a critical role in the epidemiology of the disease. Classified as a lentivirus, HIV is especially slow-acting in its course in the body. This extended latency means that infectious people, unaware of their condition, can remain apparently healthy and sexually active for years.

As a disease, HIV/AIDS passes through four stages as recognized by WHO, with each stage determined by the CD4 count—the number of CD4 cells per cubic millimeter of blood. (Other health authorities have adopted a three-stage approach to HIV/AIDS.) A count in the range of 800—1,200 is considered normal. After the primary infection, the first two stages mark the period of latency. As the CD4 count declines, the body becomes increasingly susceptible to opportunistic infections. It is these infections, rather than the underlying condition of HIV positivity, that give rise to the characteristic presenting symptoms that normally appear in Stages 3 and 4, initiating "active AIDS." A spectrum of possibilities varies from case to case.

Stage 1. CD4 count less than 1,000 but higher than 500. In this latent phase there are usually no presenting symptoms, although enlarged lymph nodes are not uncommon.

Stage 2. CD4 count less than 500 but higher than 350. At this level of immunosuppression, clinical signs vary. Normally they are not sufficiently pronounced to support an accurate physical diagnosis. The duration of the period of latency and the onset of severe opportunistic infections is determined not only by the progress of HIV, but also by the general health of the patient. Sound nutrition, regular exercise, and an avoidance of drugs, alcohol, and smoking forestall the onset of active AIDS. In Stage 2, however, patients frequently experience unexplained weight loss, fungal nail infections, sore throat, cough, recurrent oral ulcerations, and respiratory tract infections such as bronchitis and sinusitis.

Stage 3. CD4 count between 200 and 350. Advanced immunosuppression signals active AIDS. Common signs and symptoms include severe weight loss, intermittent or constant fever with night sweats, bilateral patches on the borders of the tongue, loosening of the teeth, chronic diarrhea, oral candidiasis, gingivitis, pulmonary tuberculosis, and bacterial infections, including pneumonia and meningitis.

Stage 4. CD4 count less than 200. Severe clinical signs include dry cough, progressive shortness of breath, chest pain, dysphagia, retinitis, headache, pulmonary tuberculosis, pneumocystis carinii, Kaposi's sarcoma, toxoplasmosis, cognitive or motor dysfunction, and meningitis. Worldwide, the principal complication of HIV/AIDS is pulmonary tuberculosis, which is the immediate cause of death in the majority of patients. In untreated cases of Stage 4 AIDS, death normally ensues within three years.

Transmission

Sexual Transmission

HIV is infectious in all four stages, and the virus is present in all bodily fluids. It is minimal in sweat, tears, and saliva; significant in semen and vaginal fluids; and maximal in blood. Historically, HIV/AIDS has been predominantly a sexually transmitted disease. Studies suggest that more than three-quarters of cases to date have originated through sexual intercourse.

Initially, the understanding of HIV/AIDS involved an extrapolation from the course of the epidemic in the United States, where the alarm was first raised and where homosexual men were overrepresented among the victims in the early decades of the epidemic. Homosexual men were indeed at risk through a variety of factors, including anal intercourse, with its high probability of abrasions that provide perfect portals for the virus to enter directly into the bloodstream. They were also susceptible to infection through mass migration to urban centers such as San Francisco and New York where they developed a culture that promoted the spread of sexually transmitted diseases. Unfortunately, the transmission of preexisting herpes, syphilis, and chancroid facilitated the progress of HIV via lesions that were ready portals of entry. The major risk factor, however, was not homosexuality, but having many sexual partners. In Africa, the epicenter of the current pandemic, sexual behaviors involving multiple partners account for two-thirds of all new infections.

Globally, heterosexual intercourse has been the dominant mode of transmission, with women more susceptible to infection than men. The reasons for the disproportionate susceptibility of women are biological, cultural, and socioeconomic. Biologically, women are more vulnerable then men because semen infected with HIV remains for a prolonged period in the vagina. Women may also be subjected to violent or abusive sex, which is likely to cause abrasions, as in the case of homosexual men. Additionally, previously existing venereal diseases in women create ulcers that provide HIV with easy means to bypass the body's first line of defense—the skin.

In many societies women begin to have sex at significantly earlier ages than men. They also have less access to education—a major risk factor in all sexually transmitted diseases. Globally, women who have not completed primary school are twice as likely to contract AIDS as those with greater educational attainment. Women's inequality also plays a critical part because it frequently creates situations in which women are unable to demand safe sex practices with their partners or reject unwelcome sexual advances. On average, their sexual partners are older, stronger, more educated, and more likely to control women's financial means. Furthermore, the greater likelihood that women will be impoverished drives the commercial sex industry, where women are disproportionately represented and therefore disproportionately at risk from HIV/AIDS.

Prevailing norms of masculinity in some cultures also have an adverse impact on the health of women. Often, for example, it is a badge of male honor to have multiple sexual partners, and using condoms and being sexually rebuffed are dishonoring. Such ideas expose women to enhanced risk of contracting HIV/AIDS by denying them the right to equality in determining the rules of sexual engagement.

Vertical Transmission

Although sex is by far the chief factor in diffusing HIV/AIDS, other pathways are significant, even if it is impossible to quantify their role. Transmission from mother to fetus or infant occurs through a variety of ways. One is transplacental transmission, and another is transmission during the birth process itself. Finally, there is the possibility of infants becoming infected through breast milk. In all of these circumstances, the level of risk can be significantly reduced through the use of some HIV drugs, such as Nevirapine.

Transmission via Blood

Blood provides a significant mode of transmission for HIV in multiple ways. The profusion of invasive medical procedures that require blood transfusion multiplies opportunities for transfusing contaminated blood, especially when unsafe practices are permitted, such as the recourse to paid blood donors or in settings where blood banking is not carefully regulated. In such conditions people, particularly hemophiliacs, are exposed to danger.

Contaminated blood also transmits disease among intravenous drug users who share needles. The risk is amplified when societies experience severe disruption, as in Eastern Europe after the fall of communism. In the climate of demoralization and hopelessness that ensued, heroin use surged, and HIV claimed its victims. Elsewhere, high rates of incarceration can exacerbate the problem given the numbers people in prison who engage in risky behaviors. Alcohol abuse, which can negate sound decision-making and safe sexual practices, also increases the spread of the disease. As regards the development of needle-exchange programs, moralism divorced from fact-based epidemiology compounds the problem by creating faith-based obstacles that stand in the way of a demonstrably effective way of preventing infection.

Needles have also led to a trickle of HIV infections among health-care workers in hospitals, dental offices, and health clinics who are accidentally pricked through haste, overwork, and improper disposal of syringes. In resource-poor settings, the absence of adequate sterilization and safety training magnifies this problem.

Treatment and Prophylaxis

There is no specific remedy for HIV/AIDS. Treatment is based on the development of antiretroviral therapy since the discovery in 1987 of the first antiretroviral, known as AZT (azidothymidine, or zidovudine). Antiretroviral therapy does not cure, but it does radically reduce the viral load, thereby extending life by slowing the destruction of the immune system, transforming HIV/AIDS into a chronic affliction.

Antiretroviral therapy also functions as a prophylactic strategy because its ability to reduce viral loads makes HIV-positive patients much less infectious. The odds of vertical transmission from mother to fetus and from mother to infant during the birth process, for instance, are therefore greatly reduced. Similarly, the risk of sexual transmission declines precipitately. In this way antiretroviral therapy blurs the distinction between prevention and cure, which are tightly interlocked.

After the breakthrough with AZT, six classes of retrovirals—with several choices of drugs in each class—have been developed, enlarging the range of tools available to clinicians. Each class targets a specific stage in the life cycle of the virus. Clinically, the availability of multiple classes of treatment options enables practitioners to adapt drug choices and multiclass combinations to a variety of variables, such as side effects, resistance to a particular drug, pregnancy, and the presence of comorbidities and complications.

Unfortunately, the benefits of antiretroviral therapy are qualified. In the first place, all retrovirals currently available are toxic. Their adverse side effects range from rashes, diarrhea, anemia, and fatigability to osteoporosis and damage to the liver, kidney, or pancreas. In addition, antiretrovirals impose on patients the necessity of a lifetime's adherence to a complicated and expensive pharmaceutical regimen. Severe problems of compliance arise with patients who are homeless, cognitively impaired, addicted to alcohol or drugs, or uninvolved with the health-care system because of poverty, illiteracy, or recent immigrant status.

In impoverished countries, making antiretrovirals available has frequently confronted insoluble financial barriers. These medications therefore raise issues of inequality, priorities in the use of resources, and the ethics of market principles in matters of health. Furthermore, antiretroviral treatment prolongs the life and therefore the sexual activity of HIV-positive patients, who continue to be infective, although at a lower level, for a much longer time. Prophylactic gains are partially offset by prolonged opportunity for transmission.

Antiretroviral treatment is further complicated by the problem of drug resistance, which developed rapidly. In order to overcome resistance, treatment protocols involve combined regimens of three different medications. Pharmaceutical laboratories find themselves in an arms race with the virus. Will it be possible to develop ever newer medications as resistance emerges to each? Or is a future in which antiretroviral treatment is no longer effective a legitimate fear?

An additional pharmacological strategy for confronting HIV/AIDS is preexposure prophylaxis (PrEP). This strategy is appropriate for HIV-negative people who have sexual partners who are HIV-positive. It consists of taking a daily pill that contains two medications that prevent the virus from establishing an infection, and it can be employed in conjunction with male condom use. The CDC regards preexposure prophylaxis as 90 percent effective when properly deployed, but it is suitable only for defined segments of the population, and it too poses serious problems of compliance.

In the meantime, research continues for a vaccine while the sheet anchor of present prophylactic efforts is behavior modification. This includes needle-exchange programs; "safe" sexual practices, such as the use of condoms; people becoming sexually active at older ages; and the empowerment of women.

The Pandemic in South Africa

After crossing the species barrier by the middle of the twentieth century, HIV/AIDS followed two distinct epidemic patterns—one in Africa, where it originated, and the other in the industrial world, to which it spread. In Africa, AIDS became a disease of the general population transmitted primarily through heterosexual intercourse. In the industrial world, it developed instead as a "concentrated" epidemic that claimed its victims among populations that were socially or economically marginalized such as gay men, intravenous drug users, and ethnic minorities. We examine the case of South Africa because of its centrality in the contemporary pandemic. South Africa presently has the highest number of people living with HIV in the world; 7 million South Africans in a population of 48 million are HIV-positive, for a prevalence rate of 12.9 percent—a figure that rises to 18 percent if one excludes children.

In addition, South Africa is a uniquely important case because it is the only major example of a severely afflicted country with a strong industrial base and democratic governance. Of all the countries in sub-Saharan Africa, it has the greatest resources to confront the pandemic. Other African countries therefore look to it for leadership in finding solutions to confront the crisis, and, as the *New York Times* put it, South Africa "is the natural leader in the fight against a disease that has devastated Africa."[1] For this reason the XIII International AIDS Conference convened at Durban, South Africa, in 2000, the first time the conference met in a developing country.

HIV/AIDS spread rapidly in South Africa after its appearance by the middle of the twentieth century. In the early decades its presence was unnoticed. The political tensions that accompanied decolonization, apartheid, and the Cold War; the absence of a health-care system for the black majority; the lack of public health surveillance; and the high prevalence of other diseases diverted attention from the unknown invader. The first case of HIV/AIDS to be diagnosed in South Africa, and the first official death, dated from 1983. Already by 1980, however, sub-Saharan Africa was rapidly becoming the storm center of the new pandemic, with forty-one thousand infections—as compared with eighteen thousand in North America and only one thousand each in Europe and Latin America.

In South Africa, as in the United States, the earliest recognized cases were among homosexuals, hemophiliacs, and intravenous drug users. By 1989, however, the transmission of HIV among heterosexuals in South Africa overtook the "concentrated" disease. The numbers of gay men attending clinics leveled off while patient numbers in the general population rose exponentially, first in urban centers and then in rural areas as well. Furthermore, women outnumbered men. From that time, HIV/AIDS followed a path that diverged sharply from that in the United States.

The Legacy of Colonialism and Apartheid

One of the factors that made South Africa highly vulnerable to HIV/AIDS was the legacy of colonialism and apartheid. Having established transmission initially among urban homosexuals, the virus spread along the geography of racial difference, ravaging the general population of black Africans. In 2005, prevalence rates at antenatal clinics revealed that whereas only 0.6 percent of whites and 1.9 percent of Indians were HIV-positive, 13.3 percent of blacks were. The extent of the racial patterning of HIV/AIDS is captured by the sardonic title of Susan Hunter's 2003 book *Black Death: AIDS in Africa.*

A salient aspect of apartheid was its impact on black African families. The white supremacist National Party that held power from 1948 espoused the aim of "separate development" for each racial group. This goal was described by British prime minister Alec Douglas-Home in a 1971 press conference as the "principle of unhindered self-determination along parallel lines"—as enshrined in two legal embodiments: the Group Areas Act of 1950 and the Group Areas Amendment Act of 1966. These acts sanctioned social engineering to be achieved by dispossessing nonwhites of all immovable property and of all rights of residence except in designated areas. The remainder of the country was reserved for whites, including the centers of South Africa's major cities. This geographical separation by race was buttressed by apartheid in employment, or "job reservation." Enacted by the Industrial Conciliation Act of 1956, the system stipulated that jobs be assigned by racial group and that skilled, well-paid jobs be "reserved" for whites.

To establish a sharp line of racial demarcation in both residence and employment, the ruling party resorted to compulsory relocation. This measure, which ultimately affected 3.5 million people, began by razing buildings occupied by Africans, Indians, and coloreds. Military units then forcibly removed the nonwhites to one of two designated areas. The first consisted of "locations," or townships, adjoining major cities, such as Soweto near Johannesburg, Umlazi near Durban, and Tembisa near Kempton Park. Unprepared for the sudden influx of people, these black townships were densely crowded shantytowns without water, sewerage, electricity, transport, and sanitary housing.

The alternative destination for the dispossessed consisted of small, fragmented, and resource-poor "group areas" designated as "homelands," or Bantustans. In aggregate, the homelands covered 14 percent of the land mass of South Africa—an area intended to accommodate 75 percent of the population. In the homelands the relocated people were systematically denied political participation and employment opportunities. Taking the logic of exclusion to its extreme, the regime divided the homelands along lines of African ethnicity. Zulus, Xhosas, Sothos, Tswanas, and Swazis were all consigned to separate areas.

Once resettled and denied freedom of movement by internal passports, black Africans were deprived of citizenship except in their designated homelands—that is, they were reduced to the status of foreigners in their own country. Ultimately, National Party rhetoric proclaimed, the Bantustans would evolve into "independent" states. In reality, however, de facto South African domination of every aspect of their affairs was guaranteed by their miniature size, the absence of defensible borders, and their lack of economic resources.

Understandably, the regime imposed upon them by the Nationalists was illegitimate in the eyes of black South Africans. It was no accident that the Bantu Affairs Department, which issued passports and directed the policy of "influx control," was one of the first targets in the riots that marked the twilight years of apartheid rule; that the buildings housing the black administrators of the homelands were known derisively as "Uncle Tom's cabins"; and that perennial targets of crowd violence were the beer halls that were a state monopoly whose proceeds financed homeland administration.

The vision motivating the various complex institutional structures of apartheid was clearly expressed by the National Party leader P. W. Botha (1916–2006) when he argued:

> I . . . believe that there can be no permanent home or any permanence, for even a section of the Bantu in the area of White South Africa. In other words, White South Africa must gradually free itself from the possibility of a Black proletariat gaining a hold over its economic future, because otherwise this urbanized Black proletariat will, by means of intermingling with the Coloureds, achieve a position of power over the whole of South Africa.[2]

In order to survive in their group areas, African families were compelled to separate for long periods as men sought employment as migrant workers in the now all-white cities, on farms, and in mines. Black women, children, and elderly or disabled males were relegated to the Bantustans, which served as dumping grounds for those whose presence was surplus to the needs of the white areas. The condition of the black men who worked in South Africa's mines exemplified the working of the new system. With no alternative employment available, miners were compelled to serve as guest workers in a foreign country. There they lived eleven months a year in company compounds, returning "home" just once.

Mass resettlement caused a rapid deterioration of living conditions in the homelands. Population density increased between 1955 and 1969 from sixty to one hundred people per square mile, so that ever larger numbers of rural families became landless in a homeland economy that provided no employment. The young were thus forced to migrate to townships, mines, or white-owned farms. The scale of the problem is made clear by the fact that in the early 1970s, eighty-five thousand Africans entered the labor market every year, and of these forty thousand were from the homelands.

Such a system promoted a series of behavioral patterns that facilitated the transmission of sexually transmitted diseases. Both the men who resided in compounds or slums and the women who were confined to their homelands or commuted from "locations" to perform domestic service in white cities frequently had recourse to multiple concurrent sexual relationships. Often, these relationships were transactional or openly commercial.

The residential regime contemplated by the Group Areas Act was coercive and inimical to public health. Over time, moreover, it developed further pathological ramifications. One was that young African males, separated from the influence of families and older role models, developed norms of masculinity in all-male compounds, urban gangs, and prisons. The result was a culture of "masculine" sexuality that promoted a sense of entitlement, of sexual conquest as a mark of status, and of violence as an attribute of manliness. This tendency was reinforced by the broader context of a repressive and racist regime that socialized the population in a culture of violence.

South Africa, both during apartheid and in its aftermath, thus became a country with the world's highest rate of rape per capita. Estimates suggest that 1.7 million rapes occur per year and that sexual violence has been "normalized." More than 50 percent of the cases brought to South African courts in 2015 were for sexual assault, and the majority of these involved gang rape. The titles of Pumla Dineo Gqola's 2015 book *Rape: A South African Nightmare* and the 1999 film *Cape of Rape* (referring to Cape Town) tell the story. The *South African Medical Journal* issued a call to action to confront the emergency of rape in the country. The point here is that such violence against women actively promotes the spread of HIV. In addition, the demoralization and foreshortened opportunities that accompanied apartheid also fostered excessive use of alcohol, which fueled such behavior.

A further aspect of apartheid was that the long periods men spent on their own in townships and compounds promoted sexual relationships of men with other men. These relationships took place, however, in a social atmosphere that was aggressively "masculine" and treated gay relationships with stigma and sometimes with physical assault. As a result, homosexual partnerships were concealed, with the consequence that cases of disease were not brought to medical attention, or at least not at an early stage. Homosexuality was a risk factor in the transmission of HIV, but one submerged in an epidemic spread primarily through heterosexual sex.

For women's health the impact of apartheid was especially adverse. Relegated to homelands and townships, African women were consigned to poverty, low educational attainment, and poor access to medical care. They therefore lacked the information they needed to adopt safe sexual practices. Furthermore, women were placed at a severe disadvantage in negotiating sexual relationships. It is not surprising that more women than men suffer from HIV/AIDS in South Africa and that women contract the disease above all during their teenage years—five years earlier than their age cohort of males. The South African government reported in 2014 that girls between the ages of fifteen and nineteen were seven times more likely to be HIV-positive than boys, with comparative prevalence rates of 5.6 percent and 0.7 percent, respectively.

Urbanization and Poverty

One of the hallmarks of the HIV/AIDS pandemic is the prominent role that major urban centers have played in its epidemiology. Sexually transmitted diseases have always flourished in cities, which promote networks of sociability, large numbers of young people, and a culture of escapism, often fortified with drugs and alcohol. Thus it is understandable that the epidemic in South Africa spread in part because of the economic development of the country and the flows of migrant labor that followed in its wake.

Here apartheid had played a dual role with regard to urban growth. On the one hand, the economically unsustainable homelands were designed to ensure a steady flow of labor to the towns. On the other hand, the restrictions on population movement—known as "influx control" in the bureaucratic jargon that cloaked repression—controlled the flow, confining it to those men and women whose labor was needed for low-paying jobs in industry and domestic service.

The collapse of white hegemony in 1994 transformed the situation by introducing freedom of movement. Ironically, however, sudden mass mobility in conditions of limited employment led to consequences that were conducive to the spread of disease. The immediate result was like the breaking of a dam, releasing a staggering flood of people to urban centers just as the HIV/AIDS epidemic was gathering momentum. Deputy Minister of Co-operative Governance and Traditional Affairs Andries Nel commented in 2015:

> South Africa is urbanising rapidly. The United Nations estimates that 71.3% of South Africans will live in urban areas by 2030. . . . South Africa's urban population is growing larger and younger. Two-thirds of South African youth live in urban areas. . . . Urban areas are dynamically linked to rural areas—flows of people, natural and economic resources. Urban and rural areas are becoming increasingly integrated, as a result of better transport, communications and migration.[3]

HIV followed the lines of communication and the free population movement from rural areas to the cities, accelerating just as Thabo Mbeki took office and chose to ignore the conclusions of medical science regarding its public health implications.

But HIV thrived in South Africa's townships not only because of the legacy of apartheid, the inequality of women, rapid urbanization, and modern transportation networks. In addition, mass poverty promoted disease by undermining the diet of millions and lowering their immunity to disease, despite the fact that South Africa is classified internationally as a middle-income country. Maldistribution and unemployment rather than an absolute lack of resources lay at the root of the problem. Inequality also took the form of foreshortened educational opportunities, thereby depriving large swaths of the population of the knowledge they needed to protect themselves. In the immediate aftermath of apartheid's downfall, surveys in 1995 indicated poverty levels—based on a monthly income of 352 Rand—of 61 percent for black Africans, 38 percent for coloreds, 5 percent for Indians, and 1 percent for whites. Observers termed such rates of poverty for blacks "shocking," especially when measured against poverty rates for other middle-income countries, such as Chile, Mexico, and Indonesia, where it was 15 percent, or Jamaica, Malaysia, and Tunisia, where it was 5 percent.[4]

Democratic government on the principle of "one man, one vote" has not on its own alleviated the problem of poverty. At the time of its election to office in 1994, the African National Congress Party (ANC) announced a radical plan for eliminating poverty and economic injustice—the Reconstruction and Development Programme. This program proclaimed a "Better Life for All" to be achieved by large-scale government spending on social welfare and infrastructure. Under pressure from the international business community and local elites, however, the ANC did an economic about-face, abandoning the program and closing its offices. In its stead Nelson Mandela's government implemented an entirely different plan termed Growth, Employment and Redistribution, which stressed fiscal conservatism, the control of inflation, and austerity in accord with market-driven neoliberal priorities. Redistributing wealth and reducing poverty were no longer on the agenda of the ruling party.

As a result, the level of poverty, measured in 2011 by Statistics South Africa, remains extreme, as defined by the inability of a household to provide "the minimum amount of money you need to survive." Poverty, in other words, was defined by the lowest standard—that of maintaining life. This measure is the "Food Poverty Line"—the capacity to purchase a daily energy intake of twenty-one hundred calories. By that standard, 21.7 percent of the population, or approximately 12 million people, are unable to achieve the nutritional requirements needed to sustain health.[5]

Inevitably, 23 percent of children younger than six were found to be stunted in their growth as a result of undernourishment, with the percentages spiking in rural areas. In terms of perceptions the 2011 survey revealed that more than half of South African blacks believed that they lacked adequate food to meet their daily needs. The conclusion was that the "racialized inequities of apartheid" had given away to "market inequities" that were also severe.[6] Over the whole of the period since the onset of HIV as an epidemic in the 1980s, South Africa, governed first by apartheid and then by the market, has provided the material prerequisite—malnutrition—for the transmission of disease by undermining people's immune systems. Furthermore, poverty also condemned South African victims of HIV/AIDS to a much more rapid progression from HIV positivity to active AIDS, because they were unable to afford antiretroviral treatment. In a vicious spiral, HIV/AIDS and poverty are both cause and effect of the other—what some have called the "poverty-AIDS cycle."

The South African State: From Indifference to Denial

As the disease spread—initially among urban gay men—the fact that the political leadership of the country had other priorities played a fatal role. It was South Africa's misfortune that the onset of HIV/AIDS occurred during the 1980s under a National Party government led by P. W. Botha. Botha was a Cold Warrior, a committed white supremacist, and a law-and-order man.

He was haunted by the specter of communism—Russian, Cuban, and Chinese—that he saw advancing from without in alliance with a fifth column of alienated Africans, whom he termed a "black proletariat."

To forestall this dual threat, Botha pressed for closer ties with NATO and for vastly increased expenditure on the South African military. His goal was to ensure the ability of his country's military to crush subversion at home and throughout southern Africa. He was convinced that Russia had launched a "total onslaught" against the South African regime and that an apocalyptic showdown was imminent. At stake was white hegemony, which the USSR was determined to overthrow as part of its "grand design" for world domination.

Confronted with this challenge, the Afrikaner leader responded with a superficial willingness to reform apartheid. On secondary matters of "petty apartheid" that irked Africans without significantly buttressing white supremacy, he was willing to make concessions in order to lower social tensions. He stunned his followers by such unprecedented steps as a willingness to have dialogues with black leaders, to abolish the segregation of public facilities, and to decriminalize interracial sex.

However, Botha was intransigent in defending the white monopoly of political and economic control. As social tensions escalated amidst strikes and demonstrations, he declared a state of emergency that lasted for most of the 1980s with its corollaries of bans, arrests, and violent repression. The pillars of apartheid, including the Group Areas Act and the white monopoly on political representation, were defended with force.

Thus faced with a life-and-death struggle for political survival, the National Party regime gave no attention to public health. HIV was unimportant to fundamentalists like Botha. He regarded the disease as a manifestation of God's wrath that afflicted only two groups: "deviants" such as homosexuals, drug addicts, and sex workers; and blacks, who were now defined as foreigners. There was no urgency, therefore, to educate the public, to teach safe sex practices, to discover new modes of treatment, and to care for the victims.

Far from implementing measures to contain HIV, Botha seized on the opportunity to condemn homosexuals for the crime of sodomy. In 1985 he asked the President's Council to make recommendations on enforcement of the Immorality Act that criminalized homosexual practices. Ignoring the issue of disease, the council reiterated its condemnation of homosexuality and urged widening the ban to include gay women. Meanwhile, when dealing with black Africans, Botha felt no sense of responsibility toward "foreigners." Following the logic of apartheid, his solution to HIV among black miners, therefore, was to repatriate them to their homelands if they fell ill and could no longer work. The priority was to protect a healthy white nation besieged from without by foreign migrants and from within by immoral degenerates. Throughout the 1980s the state turned a deaf ear to warnings about HIV/AIDS. Only in its final years under President F. W. de Klerk, who sought to buttress a collapsing regime with reforms, did the National Party consider a change of course. By then it was too late, and the old order was swept aside entirely in 1994.

In principle the opposition ANC, led by Nelson Mandela (1918–2013), had a strong commitment to developing a strategic national plan against the epidemic. In 1990, while operating as a banned organization, the ANC attended an international conference in Mozambique and took part in drafting the "Maputo Statement." This called for strong public health policy to confront HIV/AIDS, which had already infected sixty thousand people in South Africa. With the numbers doubling every eight months, the statement condemned the response of the state as "grossly inadequate" and as demonstrating a refusal to apply the lessons drawn by the experience of community-based organizations.

Boldly declaring that HIV/AIDS was a social disease, the conference called for a combination of initiatives by the state and local communities. The state was summoned to correct the social conditions that promoted the transmission of HIV—poverty, migrant labor, population relocation, homelessness, forced removal, unemployment, lack of education, and poor housing. In addition, the conference called upon the state to foster public awareness, establish access to nondiscriminatory and nonstigmatizing health care, repeal repressive legislation against homosexuals and sex workers, and promote the universal distribution of condoms.

At the same time, the Maputo Statement summoned the institutions of civil society to action, calling on community-based organizations representing workers, youth, and women as well as faith-based institutions to promote behavioral changes that would encourage safe and consensual sex among their members. Finally, the Maputo signatories pressed for the establishment of a National Task Force to surveil the epidemic and make further recommendations from the perspective of health as a human right.

Strong in theory, the ANC's commitment to a vigorous anti-AIDS policy withered in practice when the movement came to power. South Africa's first racially inclusive democratic elections in 1994 brought the liberation movement and Mandela to power. As he took office, the prevalence rate of

HIV in the general population reached 1 percent—the internationally recognized threshold for a "generalized and severe epidemic." The ANC had made a strong anti-AIDS campaign one of its central policy planks, but once in office it did not make public health central to its program. As late as 1997 Mandela had not given a single speech dealing with AIDS, and even when he first broached the topic, he did so during a trip abroad. Edwin Cameron, a justice on the South African Constitutional Court and a victim of the disease, recalled his disappointment ten years after Mandela's election. In his view, Mandela's thinking on AIDS was overtaken by other priorities and he failed to act on AIDS because

> he had a set of pressing priorities, which took precedence over AIDS. He had the question of military and political stability against a diminishing but still powerful racist white minority. . . .
>
> He had the problems of economic policy. He was melding economic policy within a government that consisted of an alliance between the Communist party, the ANC, and the Congress of South African Trade Unions.
>
> He had the question of reconciliation between moderate whites and moderate blacks, who had been kept apart for 300 years of history. He had, vitally, the question of international relations. South Africa had been the polecat of the world for 30 or more years, and Mandela was reintroducing South Africa. . . .
>
> Now, I'm going to say something that is going to be harsher than most people have said about Mandela: There is no doubt that he was flattered and seduced by the thrill, the enticement, the allure of the international adulation. I remember when the Spice Girls came to visit South Africa. I remember thinking acridly . . . he has spent more time with the Spice Girls than he has on AIDS![7]

Mandela's chosen successor, Thabo Mbeki, who assumed office in 1999, took a far more retrograde stance. An ideologue rather than a believer in science, Mbeki aimed to promote an "African Renaissance." In its name, South Africa would eschew "colonial" medical science in favor of indigenous healing practices. With regard to HIV/AIDS, Mbeki found inspiration in the views of the AIDS denier Peter Duesberg of the University of California at Berkeley, who rejected the consensus opinion of the international scientific community regarding the disease. Duesberg's appeal to Mbeki was that he espoused a conspiracy theory regarding Western medical science. Biomedicine, Duesberg believed, was dominated by an international cabal that excluded all independent and unorthodox opinion. Although Duesberg accepted the existence of AIDS, he believed that HIV was not its cause. AIDS was not a viral disease at all, he argued, but a disorder of the immune system rooted in malnutrition and drug abuse. "AIDS is not contagious," he wrote, "and HIV is just a passenger virus."[8] Accordingly, Duesberg claimed, both the prophylactics and the therapeutics of Western medicine were useless and possibly lethal. Since HIV did not cause the disease, safe sex and condoms were worthless prophylactically, while antiretrovirals were not treatments but poisons. Trained in medicine only by searches on the internet, Mbeki espoused Duesberg's pseudo-science as the official policy of his government—with tragic consequences.

A first result was that South Africa vacated its position as "natural leader" of the war against HIV/AIDS in the Third World. Instead of providing hope and pioneering a strategy to contain the scourge, the country took a position that made it a pariah in the international scientific community. The world's leading scientists threatened to boycott the long-awaited International Aids Conference at Durban. In the end, they attended, but five thousand drafted the "Durban Declaration" that condemned HIV denialism as unscientific and certain to cause countless deaths.

More important was the impact on the people of South Africa. In 2000, 6 million South Africans—or one in eight people—were HIV-positive, and seventeen hundred additional people were infected daily. At that time of crisis Mbeki's government, far from providing leadership and support for the campaign against AIDS, argued that the disease was a secondary problem. It was, the South African president claimed in a speech at Fort Hare University in 2001, a myth propagated by Eurocentric racists who wished to portray Africans as "germ-carriers, human beings of a lower order that cannot subject its passions to reason."[9] Playing the "race card" for all it was worth, in an April 2000 five-page letter to President Bill Clinton and others, he accused Western leaders of a "campaign of intellectual intimidation and terrorism" akin to "racist apartheid tyranny."[10]

In defiance of all the evidence, Mbeki claimed that he did not know of a single person who had died of AIDS, and he denounced coroners who issued death certificates mentioning AIDS as a cause of death. In line with this position, South Africa refused to distribute antiretroviral medications and argued that sex education was irrelevant to prevention. In addition, public hospitals and clinics were forced to turn away AIDS patients because the government withheld funds. Health ministry officials who demurred were accused of "disloyalty" and dismissed.

At the moral center of the controversy was the harrowing issue of pediatric AIDS. As the twenty-first century began in South Africa, fifty thousand babies were born annually with the disease through vertical transmission from mother to fetus or from mother to infant during birth. By radically lowering the mother's viral load, antiretrovirals halved the rate of vertical transmission, which would save twenty-five thousand newborns every year. Mbeki's government, however, refused to distribute antiretroviral medication to pregnant women who were HIV-positive.

This position made South Africa the only country in the world that refused to distribute such medication to pregnant women, not as a matter of availability but as an issue of principle. Reactions were deeply felt. Dr. Malegapuru Makgoba, director of the South African Medical Research Council at the time, accused his president of "genocide"; and Kenneth Kaunda, the ex-president of Zambia, said that Mbeki was ignoring the equivalent of "a soft nuclear bomb" dropped on his own people.[11] Mark Weinberg, president of the International AIDS Society, was equally resolute: "Those who contend that HIV does not cause AIDS are criminally irresponsible and should be jailed for the menace they pose to public health. . . . People have died as a consequence of the Peter Duesbergs of this world."[12]

The tide began to turn against Mbeki in 2005. After Mandela's only surviving son died from AIDS, he broke with Mbeki over the issue of public health, saying, "Let us give publicity to HIV/AIDS and not hide it."[13] The ninety-three-year-old Mandela then devoted the remaining years of his life and his unrivaled authority to campaigning actively against the disease and to rallying the ANC to the cause.

Meanwhile, community activists organized to press the ANC to develop a strategic plan to combat the epidemic. Particularly important were the National AIDS Convention of South Africa and the Treatment Action Campaign. The result was the unprecedented spectacle of a major political confrontation within the ANC centered on debate over a medical issue—the efficacy of antiretroviral treatment.

Pressure also arose from the predations of the disease itself. In the gloomy assessment of the Joint United Nations Programme on HIV/AIDS (UNAIDS) in 2008, "The estimated 5.7 million South Africans living with HIV in 2007 make this the largest HIV epidemic in the world. . . . In South Africa, total deaths (from all causes) increased by 87% between 1997 and

2007. During this period, death rates more than tripled for women aged 20–39 and more than doubled for males aged 30–44."[14]

In the terrible year 2006, the epidemic reached its peak as 345,185 South Africans died of AIDS-related causes, accounting for nearly half (49.2 percent) of the mortality from all causes, and life expectancy fell to 52.3 years for men and 54.7 for women, as compared with 68.2 for men and women combined in 1998. Retrospective assessments suggest that as many as half a million people in South Africa died as a result of the implementation of denialist policies.

Finally in 2008, after decades first of passivity and then of denial, a new government under Kgalema Motlanthe reversed South Africa's position on HIV/AIDS. Barbara Hogan, the minister of health, stated clearly, "The era of denialism is over completely in South Africa."[15] When Durban again hosted the International AIDS Conference in 2016, the picture of HIV/AIDS in South Africa had changed radically. By that time the country had launched an all-out campaign against the disease; it had instituted the world's largest treatment program, with 3.4 million people receiving antiretroviral drugs; and it had established an extensive sex education campaign.

Figures for 2017 indicate, however, that HIV/AIDS continued its ravages, as 7.06 million South Africans, or 12.6 percent of the population, were HIV-positive. On the other hand, the epidemic was clearly receding rather than expanding. In that year 126,755 people in a population of 56.5 million died of AIDS-related causes, which accounted for 25.03 percent of total mortality. Life expectancy had risen to 61.2 years for men and 66.7 years for women. New HIV infections declined from 500,000 in 2005 to 380,000 in 2010 and 270,000 in 2016; and HIV incidence per 1,000 population declined from 11.78 in 2010 to 8.37 in 2010 and 5.58 in 2016. A well-funded and multipronged effort is yielding results. The question now is whether the commitment will be sustained as the epidemic has been checked but not defeated.

CHAPTER 20

HIV/AIDS

The Experience of the United States

The AIDS epidemic in South Africa constitutes the extreme instance of a "generalized" epidemic. Now we turn to the epidemic in the United States, which provides the major example of a "concentrated" epidemic. It began in the 1980s among marginalized and high-risk groups—white gay males, intravenous drug users, and hemophiliacs. This example is especially important in part because the disease was first identified in the United States, and many of the mechanisms of its etiology, epidemiology, symptomatology, and treatment were unraveled there.

Origins in the United States

Officially, the AIDS epidemic in the United States began in 1981, which was the year when the disease was first recognized and given a name. Silently, however, HIV was almost certainly present from 1976 and probably from the 1960s. At that stage it claimed the lives of people whose deaths were attributed to other causes.

The outbreak of HIV in North America derived from the networks of transmission that had begun in central and western Africa. As a result of globalization, medical events on the two continents were tightly linked. In *The Communist Manifesto* (1849) Karl Marx wrote prophetically of the "need of a constantly expanding market" to "nestle everywhere, settle everywhere, establish connextions everywhere." This meant that "new wants [required] for their satisfaction the products of distant lands and climes. In place of the old local and national seclusion and self-sufficiency, we have intercourse in every direction, universal inter-dependence of nations."[1] Since Marx's pamphlet was structured as a Gothic horror tale, he imagined that the new global world would produce unintended consequences, some of which would prove uncontrollable. Thus his metaphor to describe the advanced industrial world, "a society that has conjured up such gigantic means of production," was that of a sorcerer "who is no longer able to control the powers of the nether world whom he has called up by his spells."[2] The airplane and the cruise ship completed the process he had envisaged.

A link between Central Africa and the Americas was forged at the time of the Belgian Congo's independence in 1960. Thousands of Haitian professionals took jobs in the Congo, and in time many of them repatriated, some carrying the virus of a newly emerging disease in their bloodstreams. Haiti in turn had extensive links with the United States. One was provided by thousands of refugees who sought political asylum from the brutal dictatorship of François "Papa Doc" Duvalier and his infamous paramilitary Tonton Macoutes. Duvalier seized power in 1957 and was succeeded by his equally oppressive son Jean-Claude "Baby Doc" Duvalier. Annually over the three decades of Duvalier rule, seven thousand Haitians arrived as permanent immigrants to the United States and twenty thousand more as holders of temporary visas. In addition, significant but unknowable numbers of desperate "boat people" landed on the shores of Florida. At the same time, waves of American holiday-makers visited Port au Prince, which enjoyed great notoriety as a major destination for "sexual tourism." All of these movements of people—between the Congo and Haiti, and between Haiti and the United States—were perfect pathways for a deadly but still unknown sexually transmitted disease.

Popular mythology identified a "patient zero"—the French Canadian flight attendant Gaëtan Dugas—who was vilified for a time as the person responsible for the onset of the North American AIDS epidemic. Dugas attracted attention because of his lifestyle, which was defiantly flamboyant. Crisscrossing the continent by air, he boasted of having hundreds of sexual partners a year. When challenged by health officials for knowingly presenting a serious risk to others, he infamously declared that is was none of their "goddam business," claiming it was his right to do what he wanted with his body. Dugas undoubtedly made some small contribution to the epidemic, but his role has been dramatically exaggerated. He played a tiny part in an unfolding catastrophe.

First Recognized Cases

The conventional starting date of the AIDS epidemic in the United States is June 5, 1981, when the CDC published a troubling notice in its *Morbidity and Mortality Weekly Report* (*MMWR*). What the CDC announced was a cluster of opportunistic infections that normally occurred only in rare cases of immunosuppression—pneumocystis pneumonia and Kaposi's sarcoma. The fact that both occurred in a network of five young gay men in Los Angeles was striking, and it was soon followed by news of similar clusters in New York City and San Francisco. By July there were forty cases of Kaposi's sarcoma among the gay communities of those two cities, and by the end of the year 121 men had died of the new disease.

HIV had been present in Africa since at least the 1950s, and almost certainly by the 1970s in the United States. But the *MMWR* in 1981 provided the first official acknowledgment of its presence and its devastating potential for public health. The epidemiological pattern presented by the *MMWR* instantly enabled some public health officials, such as the CDC epidemiologist Don Francis, to understand that an immunosuppressive virus was responsible for the clusters of pneumocystis pneumonia and Kaposi's sarcoma and to fear that a public health disaster was already under way. Francis at the time was working on the development of a hepatitis vaccine, and he had a long-term research interest in retroviruses. On reading the *MMWR* piece, he instantly recognized the implication: a still unknown retrovirus was responsible for the immunosuppression that predisposed people to rare cancers and opportunistic infections. Indeed, just the year before—in 1980—Dr. Robert Gallo of the National Cancer Institute had shown that a retrovirus he dubbed human T-cell lymphotropic virus (HTLV) caused a type of leukemia common in Japan. HTLV was contagious, and it had a frighteningly long incubation period. Immediately Francis called for research to isolate the pathogen responsible.

Meanwhile, members of the gay community had also followed events and understood their meaning. Michael Callen and Richard Berkowitz, who lived in New York and had already fallen ill, published a pamphlet in 1982, *How to Have Sex in an Epidemic: One Approach*. This booklet promoted condom use and is perhaps the earliest known call for safe sex.

Biomedical Technology

Ironically, another factor responsible for the spread of HIV/AIDS in the United States was biomedicine itself—through its hypodermic needles, blood banks, and invasive surgical techniques. One of the first cases retrospectively diagnosed as AIDS was the Danish surgeon Grethe Rask, who had gone to work in the Congo in 1964. She worked for years in a rural hospital that lacked surgical gloves, so she did surgery bare-handed. Rask fell ill in 1976, was repatriated under emergency conditions, and died in 1977 of pneumocystis pneumonia. According to friends, she had no possible route of infection other than performing surgery, as she was celibate and spent her whole life working.

Another iatrogenic pathway, that is, through medical treatment or procedures, for the transmission of HIV was unregulated blood banks. Hemophiliacs were one of the first groups reported to have the new disease because they needed the blood-clotting protein Factor VIII to prevent bleeding. At the time, Factor VIII was concentrated from pooled serum from many units of blood, some of which were donated commercially and not subject to screening. By 1984, 50 percent of hemophiliacs in the country were HIV-positive.

And finally, of course, the tools of modern medicine do not stay locked safely in hospitals and clinics. Syringes made their way onto the streets for the use of intravenous drug users, who soon became a high-risk group for HIV.

Early Testing and Naming

The alarm sounded by Don Francis fell mostly on deaf ears, though a few scientists did pay attention. Gallo at the National Cancer Institute in particular was convinced that Francis was right, and he dedicated his lab to the hunt for the new pathogen. Similarly, Luc Montagnier of the Pasteur Institute in France and Jay Levy in San Francisco set out to isolate a virus from the mysteriously immunosuppressed patients.

The first breakthroughs occurred in near-record time in two separate laboratories. Gallo and Montagnier independently and almost simultaneously announced in 1984 that they had identified the virus responsible, and the next year they applied for patents for an enzyme-linked immunosorbent assay (ELISA) that tested for HIV. The result was another unedifying scientific and national rivalry between Gallo and Montagnier of the kind we already know from the enmities dividing Pasteur and Koch, and Ross and

Grassi. A settlement reached in 1987 named them codiscoverers and divided the royalties from the blood test. The Nobel Prize, however, went exclusively to Montagnier in 2008.

The ELISA developed by Gallo and Montagnier was the first diagnostic test to determine HIV by detecting the presence of antibodies, and it remains the most common means to screen for HIV infection. Its development was a landmark event because it enabled doctors and health officials to screen high-risk populations to determine who was infected. As a result, they possessed a tool for containing the disaster by identifying carriers and their contacts and thereby interrupting transmission. The assay also made it possible to make blood banks safe by screening donors and therefore preventing transmission to hemophiliacs and recipients of blood transfusions.

A different test—the CD4 cell count—permits physicians to follow the progress of the disease. As we saw in the last chapter, HIV targets CD4 cells in the blood. Researchers discovered that the progress of the disease could be followed by counting CD4 cells and tracking their destruction. When CD4 counts are less than 200 per cubic millimeter, the patient is considered to be immunodeficient and unable to fight off opportunistic infections.

In 1982 the new disease was named gay-related immune deficiency (GRID), and it was derisively called the "gay plague." Both terms were clearly inaccurate in view of the epidemiological pattern of the disease in Africa, where it had become prevalent in the general population and was primarily transmitted heterosexually. But even in the United States, health authorities already knew that approximately half of those affected were not gay. Since the disease in North America affected hemophiliacs, heroin users, Haitian immigrants, and homosexuals, it gave rise to the alternative term of "the 4H disease." Then in 1984 the causative pathogen was renamed HIV, the human immunodeficiency virus.

Stigma

Two other features of the North American epidemic were vitally important although less obvious. The first of these was the presence of stigma. Here it is important to remember the climate of bigotry and oppression that characterized the mid-twentieth century. Internationally, Nazi Germany was the extreme case, as homosexuals there were forced to wear pink triangles and were sent to concentration camps for destruction along with Jews, communists, disabled people, and gypsies. In Britain the poignant case of Alan Turing reminds us of the toxicity of widely prevailing homophobic attitudes.

Turing was the mathematical genius who helped crack the codes of the Nazi Enigma machine during World War II and was thereby responsible for saving the lives of untold numbers of Allied troops. Instead of being honored by his country, he was arrested, tried, and convicted under Britain's antihomosexual "public indecency" laws in 1952. In 1954 he committed suicide.

Cold War America was of course swept by a wave of rabid anticommunism that provided fertile terrain for witch hunts by Senator Joseph McCarthy (1908–1957), FBI director J. Edgar Hoover (1895–1972), and "Red squads" in municipal police departments. But a "lavender panic" targeting homosexuals paralleled the "Red scare." Indeed, McCarthy and Hoover regarded communists and homosexuals as intertwined threats to US security. In the cosmology of the American political right, homosexuals were akin to communists—both were secretive, untrustworthy, eager to make converts, and open to blackmail. Furthermore, the two threats were often embodied in the same person. Leading figures, or former leaders, in the communist movement, including the later repentant Whittaker Chambers, had homosexual relationships. At the same time the founder of the principal gay rights group of the 1950s and 1960s—the Mattachine Society—was the communist Harry Hay. Further fueling such anxieties, the reports on male and female sexual behavior by Alfred Kinsey (1894–1956) revealed the unexpected prevalence of homosexual behavior throughout American society, while Red baiters warned that homosexuals had thoroughly infiltrated the civil service and that they owed allegiance to a shadowy "Homosexual International" that operated in league with the Communist International.

Motivated by such fears, vice squads joined Red squads in major cities to entrap homosexual men and arrest them, applying laws that made sodomy illegal in every state. The usual punishment was not jail but public humiliation and the loss of employment. In a similar fashion, the federal government purged homosexual men and women from its bureaucracy, and gay people were not allowed to immigrate to the United States. Openly gay people were also subjected to violent attacks from the public.

In this threatening environment gay people moved to places where they believed they would be accepted or at least tolerated, particularly in the anonymity of cities. Gay communities sprang up in particular in New York, Washington, DC, and San Francisco. Author Randy Shilts wrote that "the promise of freedom had fueled the greatest exodus of immigrants to San Francisco since the gold rush. Between 1969 and 1973, at least 9,000 gay men moved to San Francisco followed by 20,000 between 1974 and 1978. By 1980 about 5,000 homosexual men were moving to the Golden Gate every year. This immigration now made for a city in which two in five adult males were openly gay."[3]

In these centers gay communities were socially active and politically cohesive. The gay immigrants were thrilled to come out of the closet and form and attend gay churches, bars, bathhouses, community centers, medical clinics, and choirs. In 1977 Harvey Milk won election to San Francisco's Board of Supervisors—the first openly gay person to be elected to public office in California. In a backlash of hatred, however, he was assassinated in City Hall by his fellow supervisor Dan White.

Transmission

One aspect of urban gay culture, however, provided a perfect avenue for sexually transmitted diseases. Gay men had long been accustomed to the covert and anonymous expression of their sexuality, and for many, bathhouses provided enormous new sexual freedom. Sexual promiscuity also increased the opportunity for one to contract sexually transmitted diseases, including hepatitis B, giardia, gonorrhea, syphilis, and now HIV. Indeed, preexisting sexually transmitted diseases enormously increased the probability of transmitting HIV during sex, because the lesions associated with them breach the body's outermost defenses and allow HIV entry into the bloodstream of the infected person's partner.

A second, specifically American feature of the transmission of HIV was put forward by Randy Shilts in his 1987 book *And the Band Played On: Politics, People, and the AIDS Epidemic*. Shilts recognized that in order to explain the epidemiology of the epidemic in the United States, one needs to regard HIV as present in the US population well before its identification in 1981. He argues that the bicentennial celebrations of July 1976 offered the disease an important opportunity. Tall ships came to New York from all over the world, and the city witnessed a frenzy of partying. Later public health studies demonstrated that the first US babies with congenital HIV/AIDS were born nine months later.

In summation, the preconditions for the emergence of the HIV/AIDS epidemic in the 1980s included globalization, invasive modern medical technology, and the effects of homophobia. An additional factor that propelled the epidemic was a prolonged period during which the political leadership of the country refused to confront the gathering public health emergency, just as was the case in South Africa.

"The Wrath of an Angry God" and AIDS Education

The early identification of HIV as a "gay plague" among homosexuals helped to define it not as a disease but as a sin. Many conservative Protestant evangelicals and Catholics played a leading role in this regard, associating gays with treason, believing that homosexuality was a mental illness, and supporting laws in every state that made homosexual practices illegal. For those enthralled by the fears of the Cold War, the terrifying vision of a secret fraternity of proselytizing gays who plotted to betray their country to the Soviets was very real. In their view, gays stood poised to overthrow the nation in league with communists.

Against this background, the onset of a "gay plague" revived the oldest of all interpretations of epidemic diseases—that they are the "wages of sin" meted out by a wrathful God. Recalling biblical strictures about the evil of "sodomites," some conservative religious leaders took the lead in propounding this view. Jerry Falwell, founder of the Moral Majority, gained instant notoriety by famously declaring that AIDS was God's punishment not just for homosexuals, but for a society that tolerated homosexuality—a view that Billy Graham and televangelist Pat Robertson echoed.

Popular extremist religious works dealing with the disease in the 1980s similarly rejected both science and compassion in order to blame the victims of disease and define their suffering as divine punishment. As Anthony Petro explains in his 2015 book *After the Wrath of God: AIDS, Sexuality, and American Religion,* such Christians went beyond the biomedical understanding of HIV as a viral affliction to confront their personal conviction that it was instead a moral scourge. A corollary of this view was that public health measures to confront the epidemic could be palliatives only. The real means to defeat the disease were to implement the moral preferences of the writers—abstinence until marriage and faithfulness within a heterosexual, monogamous relationship thereafter.

There was an alternative Christian view, however. In cities where the disease first claimed large numbers of victims—San Francisco, Los Angeles, Chicago, and New York—clergy members were confronted with the sufferers and with the clergy's Christian duty to minister to "sinners." This was a view of the disease that called for the exercise of charity and compassion. Protestant clergyman William Sloane Coffin was a well-known advocate of this position. Such analyses were scattered, however, and they were outweighed by the deafening silence of mainstream Christianity during the early years of the HIV/AIDS pandemic and the drumbeat of the Christian

Right, which asserted a claim to define AIDS and the religious and ethical response to it.

With HIV/AIDS understood by so many as a moral disease, it is not surprising that during the crucial early 1980s when AIDS gained a foothold in the United States, the Republican leadership under President Ronald Reagan was unenthusiastic about taking robust public health measures against the HIV emergency. Thus, like P. W. Botha in South Africa, Reagan was preoccupied with winning the Cold War and protecting Americans from the Soviet "Evil Empire." A disease that, in Reagan's view, affected only marginal and despised groups could make little claim to his attention.

Furthermore, the reasoning that "sinful behavior" caused the epidemic led logically to the conclusion that the proper remedy was behavioral rather than medical. The onus was perceived to be on the "sodomites" to end the disease by returning to righteous American values. The Reagan Administration held that a moral stand was more important—and more effective—than scientific public health, which would not attack the problem at its roots.

The conflict between the biomedical and moral interpretations of AIDS appeared most clearly in what is known as the 1987 Helms Amendment. After seeing a comic book that showed two men having safe sex, Senator Jesse Helms said on the Senate floor: "This subject matter is so obscene, so revolting, it is difficult for me to stand here and talk about it. . . . I may throw up. This senator is not a goody-goody two-shoes. I've lived a long time . . . but every Christian ethic cries out for me to do something. I call a spade a spade, a perverted human being a perverted human being."[4] Helms then sponsored the amendment banning the use of federal funds in support of HIV/AIDS prevention and education on the grounds that teaching "safe sex" and condom use meant promoting homosexual activities in violation of antisodomy laws and moral values. The result was to prevent the federal government from taking action to defend public health.

The official onset of the American HIV/AIDS epidemic in 1981 coincided with Reagan's assumption of office, and six years of unbroken silence regarding the deadly disease followed. At a time that required strong leadership to contain a public health crisis, Reagan chose instead to ignore the insistent warnings of the CDC and gay rights groups concerning the need to prevent further suffering and death. Far from acting vigorously to combat AIDS, he slashed federal budgets. It was not until May 31, 1987—after 20,849 Americans had died of AIDS and the disease had spread to all fifty states, Puerto Rico, the District of Columbia, and the Virgin Islands—that he first publicly mentioned AIDS and, then only reluctantly and under intense

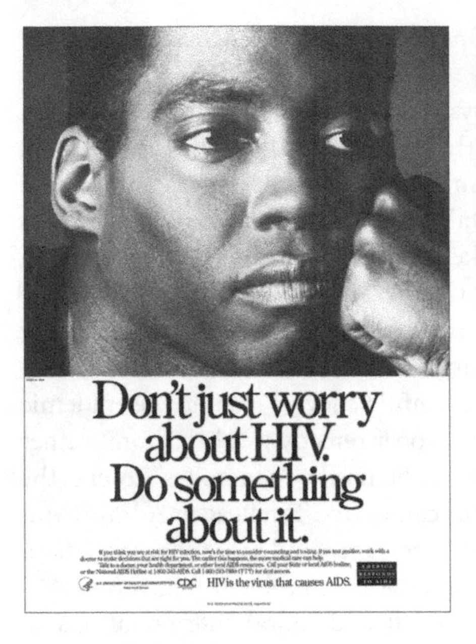

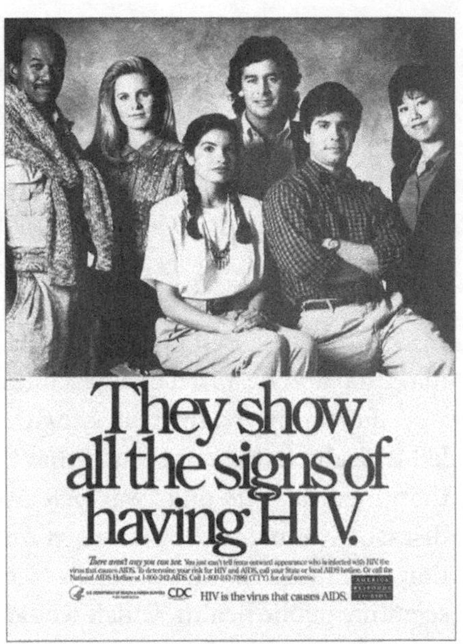

pressure, asked his Surgeon General, Dr. C. Everett Koop, to prepare a report on the disease.

In a manner that the president had not anticipated, Koop prepared an analysis of the epidemic that was thorough, explicit, and nonjudgmental. The brochure, *Understanding AIDS* (1988), contributed significantly to the public comprehension of the crisis, especially since Koop decided to send the brochure to all 107 million households in the United States—the largest public health mailing ever undertaken in the country. Rumor has it that Reagan himself did not find out about the mailing until he received his own copy in the post.

Unfortunately, Koop was subsequently gagged by the administration. He had urged the CDC and the Department of Health and Human Services to produce educational material discussing HIV transmission in an explicit manner. He advocated naming the practices, such as anal intercourse, that heightened the risk of infection and those, such as condom use, that diminished it. But Koop's nemesis and bureaucratic superior, Secretary of Education William J. Bennett, ideologized this vital issue of public health. In Bennett's view, a blunt and frank policy was unconscionable. He argued that for the federal government to produce materials explicitly mentioning sodomy and condom use outside of wedlock was to legitimize them as mor-

Figure 20.1. The America Responds to AIDS campaign from 1987 to 1996 included posters in its efforts to get out the message that everyone could be at risk from HIV/AIDS, not just gay white men or intravenous drug users, and how important it was to face the epidemic and to act on education and prevention. Published by the Centers for Disease Control. (US National Library of Medicine.)

ally acceptable. In the estimation of Congressman Barney Frank (who himself came out in 1987 as a gay man), Bennett "and some others" seemed to be inculcating values—"in fact, Mr. Bennett's values"—and believed that education which suggests behaviors "that Mr. Bennett doesn't approve of . . . would be wrong."[5]

As a result, the federal government produced informational material that was sanitized and intentionally vague, misleading, and unhelpful. As Congress learned in hearings in 1987, generic warnings about "intimate sexual activity" and the "exchange of bodily fluids" entirely failed to "differentiate between the potential riskiness of anal and vaginal intercourse and the apparent safety of, say, mutual masturbation."[6] "Dirty" words and "dirty" practices could not be named.

In the early years of the epidemic, the nature of the public education campaign was especially important (fig. 20.1). As the National Academy of Sciences noted in the spring of 1987, there was no vaccination to prevent the disease, nor effective medications to treat sufferers. The only tools available to stem the epidemic were education and the changes in behavior that would, it was hoped, result. Unfortunately, as the academy explained to Congress, "The present level of AIDS-related education is woefully inadequate. It must be vastly expanded and diversified." It was imperative to "fill the void in the federal efforts."[7] As a representative of the National Academy of Sciences stressed,

> We can create an environment in which it is all right to talk about sex, and . . . to talk about behaviors that will prevent the transmission of HIV. That's the kind of leadership that I think the Federal Government can give. . . .
>
> We cannot proscribe certain activities, and we cannot write off certain people. We must bring the education to all people who are at high risk, recognizing that there are differences in sexual behavior and sexual orientation. We must be helpful to the people . . . we're serving, not condemning of certain actions.[8]

The failure of political leadership with regard to the gathering epidemic was not confined to the national level or to the Republican Party. In New York City, the gay activist Larry Kramer, author of the play *The Normal Heart* and founder of the activist AIDS Coalition to Unleash Power (ACT UP), repeatedly castigated the Democratic administration of Mayor Edward Koch for its refusal to organize public health campaigns in the city. In Kramer's view, the motivation for Koch's inaction was different from that of Reagan, Falwell, and Helms. Kramer regarded Mayor Koch as a victim of homophobia—that is, Kramer viewed Koch as a closeted homosexual who was paralyzed by self-loathing and by a morbid fear of being "outed" if he rallied to the defense of the gay population. A progressive on many issues, Koch chose silence and inaction with regard to HIV/AIDS. It was tragic that he did so at the very center of the epidemic.

In the view of the physician David Fraser, president of Swarthmore College and member of the Institute of Medicine, San Francisco was the only city where adequate measures were undertaken to educate the public and where appropriate behavioral changes—involving condom use and reduction in the number of sexual partners—had followed. Credit for these advances, however, was not due to the federal government. The list of those engaged in the intense local educational campaign included instead the city Board of Supervisors under Mayor Dianne Feinstein; the legislature of the state of California, which subsidized the city's efforts; NGOs such as the Red Cross; and the city's highly organized gay community.

In the absence of political leadership from the top, two poignant media events belatedly triggered a national discussion. The first occurred on July 25, 1985, when the publicist for movie star Rock Hudson announced that

Hudson was suffering from AIDS. This created a shock wave because of Hudson's standing as one of Hollywood's best-known romantic male lead actors. Just over two months later, in early October, Hudson died at age fifty-nine—the first celebrity to perish of AIDS in full public knowledge. His death caused the media and the public to undertake a painful reassessment of the meaning of AIDS and the nature of the stigma attached to it. Reagan himself was Hudson's friend, and he later admitted that the actor's plight caused him to rethink his moral condemnation of homosexuality.

Then in 1991, Earvin "Magic" Johnson, a darling of the American sporting public and one of the best basketball players of all time, announced that he was HIV-positive. Johnson played point guard for the Los Angeles Lakers in the National Basketball Association (NBA) and was twelve times an NBA all-star and three times the NBA's most valuable player. He had also played as a member of the "Dream Team" that won an Olympic gold medal for the US in 1992, and he was elected to the Basketball Hall of Fame in 2002. His nationally televised press conference on November 7, 1991, followed by a series of public appearances, stunned the American press and the public.

In 2004 ESPN called Johnson's declaration one of the seven most influential media events of the previous quarter century. What he said was extraordinary in terms of popular HIV awareness because he told the nation that he was exclusively heterosexual and had never been an intravenous drug user. He also shattered conventional stereotypes by being an African American at a time when the public perceived AIDS to be a disease of white gay males. He repeatedly said, "It can happen to anybody, even me." Crucially, he followed his announcement by taking on the role of spokesman for an alternative view of the disease.

A Complex Epidemic

While the federal government and the city of New York chose inaction and silence during the critical early years of HIV, the epidemiological pattern of the disease altered profoundly. In the early 1980s HIV spread as a "concentrated" epidemic among marginalized social groups at high risk. By the middle of the decade, however, the concentrated epidemic was first paralleled and then overtaken by a second epidemic pattern—that of a disease transmitted heterosexually among the general population of African Americans. Thus as the 1980s unfolded, the United States experienced a "complex" epidemic exhibiting two different patterns: a concentrated epidemic among high-risk social groups and a general epidemic among the population of ethnic minorities—above all African Americans but also Latinos and Native Americans.

By 1993 the gathering disaster among African Americans was clearly documented: approximately 360,000 AIDS cases had been reported cumulatively since the start of the epidemic, and of these, 32 percent were African Americans, who represented just 12 percent of the population. The rate of AIDS was fifteen times higher among African American females than among white females, and five times higher among black males than white. Moreover, as the epidemic gathered momentum, the disproportionate representation of the black population among the victims grew increasingly apparent. In 2002, 42,000 Americans were diagnosed with HIV, and of them 21,000, or 50 percent, were African American. Given African American demographic features, the epidemic among the general population of blacks was geographically specific: it occurred in cities in the Northeast, the South, and the West Coast.

By 2003, the prevalence rate of HIV among the overall African American population reached 5 percent, when 1 percent was the standard indication for a "generalized severe epidemic." HIV among African Americans had attained levels comparable to those of sub-Saharan Africa.

What accounted for this discrepancy between blacks and whites? Public health officials had a ready answer. In the words of the associate director of the District of Columbia office of the CDC, "Although we all know that race and ethnicity are not risk factors themselves for HIV transmission, they are markers for underlying social, economic, and cultural factors that affect health."[9] What were those factors?

Poverty

The CDC reported in 2003: "Studies have found a direct relationship between higher AIDS incidence and lower income. A variety of socioeconomic problems associated with poverty directly or indirectly increase HIV risks, including limited access to quality health care and HIV prevention education."[10] Economic disadvantage was a major feature of the African American population, and the CDC reported that one in four blacks lived in poverty. This feature of African American life was a major driver of HIV transmission not simply because malnourishment in general lowers resistance, but also because it increases the specific risks for sexually transmitted diseases. The CDC concluded in a 2009 report that "HIV prevalence rates in urban poverty areas were inversely related to annual household income: the lower the income, the higher the HIV prevalence rate. This inverse relationship between HIV prevalence and socioeconomic status (SES) was observed for all SES metrics examined (education, annual household income, poverty level, employment, and homeless status)."[11]

A further effect of poverty was less immediately visible. An important feature of the gay white communities in New York and San Francisco was that they were educated, affluent, and highly organized. As a result, organizations representing them successfully promoted medical awareness and sex education, thereby slowing transmission. By contrast, the African American community was relatively impoverished, poorly educated, and unorganized. As the US Congress investigated the HIV epidemic, representatives of the African American community repeatedly noted that their churches, members of Congress, and authority figures kept silent about HIV/AIDS.

Finally, poverty did not just increase the risk of contracting HIV; it was also correlated with rapid progression from HIV-positive status to active AIDS. Among those infected individuals who progressed to active AIDS within three years, African Americans of low socioeconomic standing were overrepresented.

Family Disintegration

An important and lasting legacy of slavery was its impact on the African American family, which was often broken apart and its members sold. Historically, therefore, the tendency was for the black family to be headed by a female, and the generations since emancipation provided little in the way of countervailing forces. On the contrary, migration, unemployment, and incarceration provided additional fractures and absences and made the position of black males insecure. By the time of the onset of HIV, fully one-third of African American families with children were maintained by the mother alone.

Here the clearest influence is that of incarceration. A decisive factor was the "war on drugs," which introduced a rigid policy of criminalization and mandatory prison sentences for drug possession and use. Largely on the strength of this "get tough on crime" policy, the prison population of the United States more than quadrupled between 1980 and 2008, from half a million to 2.3 million, and the US became the world's foremost jailor, with 25 percent of the world's prisoners despite just 5 percent of the world's population.

In this enormous lockup, African Americans were disproportionately affected. Representing 12 percent of the U.S. population, African Americans accounted for 48 percent of arrests for drug offenses. Incarcerated at a rate six times that of whites, they constituted 1 million of the total of 2.3 million prisoners in the country. According to the NAACP, "If current trends continue, one in three African-American males born today can expect to spend time in prison during his lifetime."[12] The war on drugs greatly magnified the disparities among ethnic and racial groups because African Americans are imprisoned at ten times the rate of whites for drug offenses, and they receive much longer sentences.

In terms of the HIV/AIDS epidemic, such high—even unique—levels of imprisonment had a major impact. Most obviously, the incarceration of such large numbers of young men led to the spread of multiple concurrent sexual relationships, which are a major risk factor for all involved in the sexual network. Concurrent partnerships also promoted HIV indirectly by encouraging the transmission of other sexually transmitted disease (STDs) such as chancre, gonorrhea, herpes, and syphilis, which were already overrepresented among African Americans and which can cause abrasions that facilitate the transmission of HIV. In 2003 the CDC observed:

> African Americans . . . have the highest STD rates in the nation. Compared to whites, African Americans are 24 times more likely to have gonorrhea and 8 times more likely to have syphilis. . . . The presence of certain STDs can increase the chances of contracting HIV by three to five-fold. Similarly, because co-infection with HIV and another STD can cause increased HIV shedding, a person who is co-infected has a greater chance of spreading HIV to others.[13]

Furthermore, once imprisoned, African American men were more likely to engage in homosexual acts and to share needles and other sharp objects for injecting drugs or tattooing. Sex in prison was also never safe sex because, taking the religious high ground that homosexual acts are immoral, correctional facilities did not distribute condoms. Finally, in a vicious downward spiral, imprisonment enhanced the unraveling of family relationships and exacerbated the preexisting problems of poverty and unemployment that were already rampant in the black community.

Finally, the last stage in the disintegration of a family—homelessness—is a problem above all of inner-city African Americans. The homeless have high rates of HIV infection because their condition correlates with high-risk behaviors of many kinds, such as exchanging sexual favors for shelter, drugs, or food; lack of sex education; limited access to health care; malnutrition; and a tendency to self-medicate with drugs and alcohol and then to engage in risky sexual practices.

Cultural Factors

Culture plays a major role in the vulnerability of the African American population to HIV. The legacy of a long history of oppression and societal neglect made it difficult for blacks to believe the messages of the CDC and the Department of Health and Human Services that federal, state, and local officials were suddenly interested in their well-being. On the contrary, the results of a 1999 survey indicated that half of the black population regarded it as either "likely" or "possible" that the government welcomed the AIDS epidemic as a means of ridding itself of a troublesome surplus population. The lives of black welfare recipients and drug users were expendable. This semiconspiratorial view seemed all the more plausible in the wake of the Tuskegee Syphilis Study of 1932–1972, which involved the systematic abuse of the trust of impoverished black syphilis patients.

Furthermore, the educational and governmental messages were often couched in terms that made them incomprehensible to and deeply counterproductive for the black community. Raquel Whiting, a twenty-two-year-old African American woman who worked as a policy analyst for the National Pediatric HIV Resources Center, carefully elaborated on this point in her testimony before Congress in 1993. On the basis of her experience working with young African Americans, she had learned that they did not protect themselves during sex and that one of the reasons was that the educational materials had a misleading message. Posters, magazine articles, brochures, television ads—all conveyed the prevailing social construction of HIV/AIDS that it was a disease of educated, middle-class, white homosexuals and of intravenous drug users. In Whiting's words, even though HIV/AIDS had become predominantly an affliction of the black population, "The media and larger society continue to portray images of gay white men as the face of HIV. People of color regardless of their sexual orientation are absent from the picture."[14] Therefore, black young people concluded that they were not at risk.

A further problem, Whiting reported, was that the educational campaign against AIDS relied on scare tactics with slogans such as "Get High,

Get Stupid, Get AIDS." In inner-city African American neighborhoods where gang violence, drugs, and shootings were integral parts of daily life, such slogans carried no conviction and elicited mirth rather than behavioral change.

Finally, in her analysis, the anti-AIDS campaign failed the black community in general and black young people who were most at risk in particular because the message concentrated on schools. For a community with high numbers of dropouts and absentees, such an approach often failed to reach those who most needed the message. As she stated starkly, "The prevention message is not reaching this group." How badly the message misfired became apparent in Whiting's experience with female gang members in inner-city Philadelphia. There, it was an accepted practice to sleep with male gang leaders known to be HIV-positive to demonstrate that the strongest and fittest young women would not catch the disease, giving them respect among their peers. Until the message, the messenger, and the location changed, Whiting said, "Young people of African-American descent will not protect themselves."[15]

Conclusion

As it assessed the state of HIV/AIDS in its June 15, 2018, issue, the journal *Science* summarized the epidemic under the revealing title "Far from Over." As the disease became overwhelmingly a disease of African Americans and Latinos, it also followed a different geographical pattern, with its heaviest burden falling on the South and the District of Columbia. Most worrying is Florida, which suffers, according to *Science*, from a "startlingly high HIV infection rate" and where Miami is the "epicenter of HIV/AIDS in the United States with the highest new infection rate per capita of any U.S. city: 47 per 100,000 . . . , more than twice as many as San Francisco, New York City, or Los Angeles."[16]

Miami, Fort Lauderdale, Jacksonville, Orlando, and the state's other urban centers present AIDS campaigners with a compendium of all the factors driving the contemporary US epidemic: a large and constant influx of immigrants; a sizable African American population with little access to medical care and therefore little hope of successful antiretroviral treatment; a booming industry of sexual tourism; significant numbers of homeless people; pervasive inequality and a large urban underclass; widespread stigma that prevents the groups most at risk from seeking diagnosis and therefore knowing their HIV status; a state legislature that suffers "donor fatigue" regarding HIV/AIDS and prioritizes other health concerns; a severe problem of heroin addiction; and a Bible Belt culture that, in the words of the CDC, "suffers from homophobia and transphobia, racism, and general discomfort with public discussion of sexuality."[17] Under the administration of Donald Trump, the federal government has also declined to provide leadership, to devise strategies to confront the problem, and to fund existing programs that tackle HIV/AIDS.

CHAPTER 21

Emerging and Reemerging Diseases

An Age of Hubris

In the long contest between humans and microbes, the years from the mid-twentieth century until 1992 marked a distinctive era. During those euphoric decades, there was a consensus that the decisive battle had been joined and that the moment was at hand to announce the final victory against disease. Almost as if introducing the new period, the US secretary of state George Marshall declared in 1948 that the world now had the means to eradicate infectious diseases from the earth. Marshall's view was by no means exceptional. For some, in the early postwar years, the triumphant vision applied primarily to a single disease. The heady goal arose first of all within the field of malariology, where the Rockefeller Foundation scientists Fred Soper and Paul Russell thought that they had discovered in DDT a weapon of such unparalleled power that it would enable the world to eliminate the ancient scourge forever. With premature confidence in 1955, Russell published *Man's Mastery of Malaria*, in which he envisaged a global spraying campaign that would free humankind from the disease—cheaply, rapidly, and without great difficulty. Rallying to Russell's optimism, WHO adopted a global campaign of malaria eradication with DDT as its weapon of choice. The director of the campaign, Emilio Pampana, elaborated a one-size-fits-all program of eradication through four textbook steps—"preparation, attack, consolidation, and maintenance." Russell's followers Alberto Missiroli, the director of the postwar campaign in Italy, and George Macdonald, the founder of quantitative epidemiology, reasoned that so signal a victory over mosquitoes could be readily expanded to include the elimination of all other vector-borne tropical diseases, ushering in what Missiroli called a "contagion-free Eden," where medicine would make humans not only healthy but happy.

Malariologists, who dominated the international public health community, launched the idea of the final conquest of infectious diseases, but it rapidly developed into the prevailing orthodoxy. Chest specialists firmly believed that the combination of two technological innovations—the bacillus Calmette-Guérin (BCG) vaccine and such "wonder drugs" as streptomycin and isoniazid—would enable them to eradicate tuberculosis, and they even set target dates of 2010 for the United States and 2025 for the planet. E. Harold Hinman, chief malariologist to the Tennessee Valley Authority and member of the WHO Expert Committee on Malaria, extrapolated from the conquest of malaria to victory over all contagion in his influential 1966 work *World Eradication of Infectious Diseases*.

Aidan Cockburn, a distinguished epidemiologist at Johns Hopkins and WHO advisor, gave expression to this new creed in his revealingly titled work *The Evolution and Eradication of Infectious Diseases* (1963). As Cockburn noted, "'Eradication' of infectious disease as a concept in public health has been advanced only within the past two decades, yet it is replacing 'control' as an objective."[1] Although not a single disease had yet been destroyed by the time he wrote his book in the early 1960s, he believed that the objective of eradication was "entirely practical," not just for individual illnesses but for the whole category of communicable diseases. Indeed, he argued, "it seems reasonable to anticipate that within some measurable time, such as 100 years, all the major infections will have disappeared."[2] By that time, he wrote, only "their memories in textbooks, and some specimens in museums" should remain. "With science progressing so rapidly," he explained, "such an end-point is almost inevitable, the main matter of interest at the moment is how and when the necessary actions should be taken."[3]

Cockburn's timetable of total eradication by 2060 was, in fact, too slow for some. Just a decade later, in 1973, the Australian virologist and Nobel laureate Frank Macfarlane Burnet went so far as to proclaim, together with his colleague David White, that "at least in the affluent West," the grand objective had already been reached. "One of the immemorial hazards of human existence has gone," he reported, because there is a "virtual absence of serious infectious disease today."[4] WHO also saw the entire planet as ready to enter the new era by the end of the century. Meeting at Alma-Ata (now

Almaty, Kazakhstan) in 1978, the World Health Assembly adopted the goal of "Health for All, 2000."

What led to such overweening confidence in the power of science, technology, and civilization to vanquish communicable disease? One factor was historical. In the industrialized West, rates of mortality and morbidity from infectious diseases began to plummet during the second half of the nineteenth century, in large part as a result of "social uplift"—dramatic improvements in wages, housing, diet, and education. At the same time, developed nations erected solid fortifications of sanitation and public health. These included, as we have seen, sewers, drains, sand filtration, and chlorination of water as defenses against cholera and typhoid; sanitary cordons, quarantine, and isolation against bubonic plague; vaccination against smallpox; and the first effective "magic bullet"—quinine—against malaria. Meanwhile, improvements in food handling—such as pasteurization, retort canning, and the sanitation of seafood beds—yielded major advances against bovine tuberculosis, botulism, and a variety of foodborne maladies.

Already by the early twentieth century, therefore, many of the most feared epidemic diseases of the past were in headlong retreat for reasons that were initially more empirical than the result of the application of science. Science, however, soon added new and powerful weapons. Louis Pasteur and Robert Koch had established the biomedical model of disease that promoted unprecedented understanding and yielded a cascade of scientific discoveries and new subspecialties (microbiology, immunology, parasitology, and tropical medicine). Meanwhile, the dawn of the antibiotic era, with penicillin and streptomycin, provided treatments for syphilis, staph infections, and tuberculosis. Vaccine development also dramatically lowered the incidence of smallpox, pertussis, diphtheria, tetanus, rubella, measles, mumps, and polio. And DDT seemed poised to abolish malaria and other insect-borne pathogens. So by the 1950s scientific discoveries had provided effective tools against many of the most prevalent infectious diseases. Extrapolating from such dramatic developments, many concluded that it was reasonable to expect that communicable diseases could be eliminated one at a time until the vanishing point was reached. Indeed, the worldwide campaign against smallpox provided just such an example when WHO announced in 1979 that the disease had become the first ever to be eradicated by intentional human action.

Those who asserted the doctrine of the conquest of infection viewed the microbial world as largely static or only slowly evolving. For that reason there was little concern that the victory over existing infections would be challenged by the appearance of new diseases for which humanity was unprepared and immunologically naive. Falling victim to historical amnesia, they ignored the fact that the past five hundred years in the West had been marked by the recurrent appearance of catastrophic new afflictions—for instance, bubonic plague in 1347, syphilis in the 1490s, cholera in 1830, and Spanish influenza in 1918–1919.

Burnet was typical. He was a founding figure in evolutionary medicine who acknowledged, in theory, the possibility that new diseases could arise as a result of mutation. But in practice he believed that such appearances are so infrequent as to require little concern. "There may be," he wrote, "some wholly unexpected emergence of a new and dangerous infectious disease, but nothing of the sort has marked the last fifty years."[5] The notion of "microbial fixity"—that the diseases we have are the ones that we will face—even underpinned the International Health Regulations adopted worldwide in 1969, which specified that the three great epidemic killers of the nineteenth century—plague, yellow fever, and cholera—were the only diseases requiring "notification." Notification is the legal requirement that, when diagnosed, a given disease be reported to national and international public health requirements. Having framed notification in terms of a short list of three known diseases, the regulations gave no thought to what action would be required if an unknown but deadly and transmissible new microbe should appear.

If belief in the stability of the microbial world was one of the articles of faith underpinning the eradicationists' vision, a second misplaced evolutionary idea also played a role. This was the doctrine that nature is fundamentally benign because, over time, the pressure of natural selection drives all communicable diseases toward a decline in virulence. The principle was that excessively lethal infectious diseases would prevent their own transmission by prematurely destroying their hosts. The long-term tendency, the proponents of victory asserted, is toward commensalism and equilibrium. New epidemic diseases are virulent almost by accident as a temporary maladaptation, and they therefore evolve toward mildness, ultimately becoming readily treatable diseases of childhood. Examples were the evolution of smallpox from variola major to variola minor, the transformation of syphilis from the fulminant "great pox" of the sixteenth century into the slow-acting disease of today, and the transformation of classic cholera into the far milder El Tor serotype.

Similarly, the doctrine held a priori that in the family of four diseases of human malaria, the most virulent, or falciparum malaria, was an evolutionary newcomer relative to the less lethal vivax, ovale, and malariae diseases, which were held to have evolved toward commensalism. Against this background, the standard textbook of internal medicine in the eradicationist era—the seventh edition of *Harrison's Principles of Internal Medicine* of 1974—claimed that a feature of infectious diseases is that they "as a class are more easily prevented and more easily cured than any other major group of diseases."[6]

The most fully elaborated and most cited theory of the new era was the "epidemiologic transition" or "health transition" concept represented by Abdel Omran, professor of epidemiology at Johns Hopkins University. In a series of influential writings between 1971 and 1983, Omran and his colleagues analyzed the encounter of human societies with disease during the modern period. According to them and their journal—the *Health Transition Review*—humanity has passed through three eras of modernity in health and disease. Although Omran is ambiguous about the precise chronology of the first era—the "age of pestilence and famine"—it is clear that it lasted until the eighteenth century in the West and was marked by Malthusian positive checks on demography—epidemics, famines, and wars.

There followed the "age of receding pandemics," which extended from the mid-eighteenth century until the early twentieth in the developed West and until later in non-Western countries. During this period mortality from infectious diseases declined progressively. Here tuberculosis seemed to be leading the way.

Finally, after World War I in the West and after World War II in the rest of the world, humanity entered the "age of degenerative and man-made diseases." Whereas in the earlier stages of disease evolution, social and economic conditions played the dominant role in determining health and the risk of infection, in the final phase medical technology and science took lead. Under their influence, mortality and morbidity from infectious diseases were progressively displaced by alternative causes of death. These included degenerative diseases such as cardiovascular disease, cancer, diabetes, and metabolic disorders; human-made diseases such as occupational and environmental illnesses; and accidents. As US Surgeon General Julius B. Richmond noted in 1979, adopting the perspective of "health transition" theory, infectious diseases were simply the "predecessors" of the degenerative diseases that succeeded and replaced them in a process conceived as simple and unidirectional.[7]

Memory of the power of public health and science provided impetus to the overconfidence of the transitionists, but forgetfulness also contributed. The idea—expressed by Surgeon General William H. Stewart in 1969—that the time had come to "close the book on infectious diseases" was profoundly Eurocentric. Even as medical experts in Europe and North America proclaimed victory, infectious diseases remained the leading cause of death worldwide, especially in the poorest and most vulnerable countries of Africa, Asia, and Latin America. Tuberculosis was a prominent reminder. Sanatoria were closing their doors in the developed North, but TB continued its ravages in the South; and it continued to claim victims among the marginalized of the North—the homeless, prisoners, intravenous drug users, immigrants, and racial minorities. As Paul Farmer has argued in his 2001 book *Infections and Inequalities: The Modern Plagues*, tuberculosis was not disappearing at all: the illusion persisted only because the bodies it affected were either distant or hidden from sight. Indeed, conservative WHO estimates suggested that in 2014 there were approximately as many people ill with tuberculosis as at any time in human history. WHO also reported that in 2016, 10.4 million people fell ill with tuberculosis and that 1.7 million died of it, making TB the ninth leading cause of death worldwide and the top cause of death from infectious diseases, ahead of HIV/AIDS.

Raising the Alarm

Already by the early 1990s the eradicationist position had become untenable. Rather than witnessing the rapid fulfillment of the prediction that science and technology would eliminate infectious diseases as a class from the world, the industrial West discovered that it remained vulnerable, and to a degree that had so recently seemed unimaginable. The decisive event, of course, was the arrival of HIV/AIDS, which was first recognized as a new disease entity in 1981. By the end of the 1980s it was clear that HIV/AIDS embodied everything that the eradicationists had considered unthinkable: it was a new infection for which there was no cure; it reached the industrial world as well as developing countries; it unleashed in its train a series of exotic additional opportunistic infections; and it had the potential to become the worst pandemic in history, as measured not only by mortality and suffering, but also by its profound social, economic, and security consequences.

From the front lines of the battle against AIDS, a series of voices sounded the alarm in the 1980s about the severity of the new threat. Most famous of all was the case of the US Surgeon General C. Everett Koop. As we have seen, in 1988 he had the brochure *Understanding AIDS* mailed to every household in the nation. Working in sub-Saharan Africa, Peter Piot, who later directed UNAIDS, warned in 1983 that AIDS in Africa was not a "gay plague" but an epidemic of the general population that was also transmitted by heterosexual intercourse and affected women more readily than men.

These warnings of the 1980s, however, were confined to HIV/AIDS: they did not directly confront the larger issue of eradicationism or announce a new era in medicine and public health. That task fell instead to the National Academy of Science's Institute of Medicine (IOM) and its landmark publications on emerging diseases that began in 1992 with *Emerging Infections: Microbial Threats to Health in the United States.* Once raised by the IOM, the cry of alarm was taken up widely and almost immediately—by the CDC, which devised its own response to the crisis in 1994 and founded a new journal, *Emerging Infectious Diseases,* devoted to the issue; by the National Science and Technology Council in 1995; and by thirty-six of the world's leading medical journals, which agreed to take the unprecedented step of each devoting a theme issue to emerging diseases in January 1996, which they proclaimed "Emergent Diseases Month." In the same year President Bill Clinton issued a fact sheet titled "Addressing the Threat of Emerging Infectious Diseases" in which he declared such diseases as "one of the most significant health and security challenges facing the global community."[8] Hearings on emerging infections also took place in the US Congress before the Senate Committee on Labor and Human Resources, during which committee chair Senator Nancy Kassebaum noted: "New strategies for the future begin with increasing the awareness that we must re-arm the Nation and the world to vanquish enemies that we thought we had already conquered. These battles, as we have learned from the 15-year experience with AIDS, will not be easy, inexpensive, nor quickly resolved."[9] Finally, to attract attention at the international level, WHO, which had designated April 7 of each year as World Health Day, declared that the theme for 1997 was "Emerging Infectious Diseases—Global Alert, Global Response," with the lesson that in a "global village" no nation is immune.

In addition to the voices of scientists, elected officials, and the public health community, the press gave extensive coverage to the new and unexpected danger, especially when the lesson was driven home by three events of the 1990s that captured attention worldwide. First was the onset of a large-scale epidemic of Asiatic cholera in South and Central America, beginning in Peru in 1991 and rapidly spreading across the continent until four hundred thousand cases and four thousand deaths were reported in sixteen countries (see Chapter 13). Since the Americas had been free of the disease for a century, the arrival of the unwelcome visitor reminded the world of the fragility of advances in public health. Because cholera is transmitted through the contamination of food and water by fecal matter, it is a "misery thermometer"—an infallible indicator of societal neglect and substandard living conditions. Therefore, its outbreak in the West in the late twentieth century caused shock and a sudden awareness of vulnerability. Indeed, the *New York Times* informed its readers of the "Dickensian slums of Latin America," where the residents of Lima and other cities drew their drinking water directly from the "sewage-choked River Rimac" and similarly polluted sources.[10]

The second news-making event in the matter of epidemic diseases was the outbreak of plague in the Indian states of Gujarat and Maharashtra in September and October 1994. The final toll was limited—seven hundred cases and fifty-six deaths were reported—but the news that plague had broken out in both bubonic and pneumonic forms unleashed an almost biblical exodus of hundreds of thousands of people from the industrial city of Surat. It cost India an estimated $1.8 billion in lost trade and tourism, and it sent waves of panic around the world. The disproportionate fear, the *New York Times* explained, was due to the fact that "plague" was an emotive word with uniquely resonant memories of the Black Death, which killed a quarter of the population of Europe, followed by half a millennium of disasters in its train. India's plague, the paper continued, "is a vivid reminder that old disease, once thought to have been conquered, can strike unexpectedly anytime, anywhere."[11]

The third major epidemic shock of the 1990s was an outbreak of Ebola hemorrhagic fever (later called Ebola virus disease, or just Ebola) in the city of Kikwit, Zaire (now the Democratic Republic of the Congo), in 1995. This outbreak inspired terror not by unleashing a major epidemic—it infected only 318 people between January and July—but by dramatically exposing the lack of preparedness of the international community to confront a potentially global health emergency; by awakening primordial Western fears of the jungle and untamed nature; and by feeding on racial anxieties about "darkest" Africa. As a result, the Kikwit outbreak generated what the *Journal of Infectious Diseases* called "extraordinary" and "unprecedented" press coverage that amounted at times to the commercial "exploitation" of human misery and a "national obsession."[12] Descending onto the banks of the Kwilu River, the world's tabloids stressed in vivid hyperbole that Ebola had sprung directly from the jungles of Africa as a result of the encounter of charcoal burners and bush hunters with monkeys and that it now threatened the West. The Sydney, Australia, *Daily Telegraph,* for instance, ran the revealing headline "Out of the Jungle a Monster Comes." Even judicious investigators, however, were disturbed to discover that Ebola had eluded public health attention for twelve weeks between the death of the index case on January 6 and the notification of the international community on April 10, despite the fact that the virus had left clusters of severely ill and dying patients in its train. With such a porous surveillance network in place, Ebola might spread unnoticed more than three hundred miles from Kikwit to Kinshasa and then throughout the world by means of the Zairian capital's intercontinental airport. There it could be loaded onboard as a "Ticking, Airborne Time Bomb," as a headline in the *New York Daily News* put it.

Most of all, however, the Kikwit outbreak commanded attention because Ebola is highly virulent and its course in the human body, not unlike bubonic plague, is excruciating, dehumanizing, and dramatic. Commenting on the scenes that he had observed in Zaire, the author Richard Preston stoked anxiety by lively exaggeration. After explaining on television that the CFR was 90 percent and that there was no known remedy or prophylactic, he continued:

> The victims suffer what amounts to a full-blown biological meltdown. . . . When you die of Ebola, there's this enormous production of blood, and that can often be accompanied by thrashing or epileptic seizures. And at the end you go into catastrophic shock and then you die with blood pouring out of any or all of the orifices of the body. And in Africa where this outbreak is going on now, medical facilities are not all that great. I've had reliable reports that doctors . . . were literally struggling up to their elbows in blood—in blood and black vomit and in bloody diarrhea that looks like tomato soup, and they know they're going to die.[13]

In combination with the announcement by scientists that the world was highly exposed to new pandemics of just such infections, these events in Latin America, India, and Zaire generated sensationalist headlines, such as "Killers on the Loose," "Bug War," "Doomsday Virus Fear," "Heat from the Hot Zone," and "Revenge of the Microbes." Apocalyptic images were invoked of civilization perched on the slopes of an erupting volcano, of the West besieged by invisible hordes, and of nature exacting its revenge for human presumption. As Forrest Sawyer reported on ABC news in February 1995, the Western world had thought that civilization was safe from such invisible killers, but now the West had discovered that civilization was "as vulnerable as ever" to parasites, bacteria, and viruses, specifically the Ebola virus.

In addition, there was an outpouring of movies and books devoted to pandemic disasters, such as Wolfgang Petersen's thriller *Outbreak* (1995), Lars von Trier's horror film *Epidemic* (1987), and Steven Soderbergh's later movie *Contagion* (2011); Richard Preston's bestseller *The Hot Zone* (1994); Laurie Garrett's nonfiction *The Coming Plague: Newly Emerging Diseases in a World Out of Balance* (1994); and surgeon William T. Close's 1995 report *Ebola: A Documentary Novel of Its First Explosion in Zaire by a Doctor Who Was There*. In the words of David Satcher, director of the CDC, the result was the "CNN effect"—the perception by the public that it was at immediate risk even when the actual danger was small.

A More Dangerous Era

In this climate of anxiety, Joshua Lederberg, winner of the Nobel Prize in Physiology or Medicine, coined the term "emerging and reemerging diseases" to mark a new era. "Emerging infectious diseases," he wrote, "are diseases of infectious origin whose incidence in humans has increased within the past two decades or threatens to increase in the near future."[14] Emerging diseases, such as AIDS and Ebola, were previously unknown to have afflicted humans; reemerging diseases, such as cholera and plague, were familiar scourges whose incidence was rising or whose geographical range was expanding.

Lederberg's purpose in devising a new category of diseases was to give notice that the age of eradicationist euphoria was over. Instead of receding to a vanishing point, he declared, communicable diseases "remain the major cause of death worldwide and will not be conquered during our lifetimes. . . . We can also be confident that new diseases will emerge, although it is impossible to predict their individual emergence in time and place."[15] Indeed, the contest between humans and microbes was a Darwinian struggle with the advantage tilted in favor of microbes. The stark message of the IOM was that, far from being secure, the United States and the West were at greater risk from epidemic diseases than at any time in history.

An important reason for this new vulnerability was the legacy of eradicationism itself. The belief that the time had come to close the books on infectious diseases had led to a pervasive climate that critics labeled variously as "complacency," "optimism," "overconfidence," and "arrogance." The conviction that victory was imminent had led the industrial world to premature and unilateral disarmament. Assured by a consensus of the leading medical authorities for fifty years that the danger was past, federal and state governments in the United States dismantled their public health programs dealing with communicable diseases and slashed their spending; investment by private industry on the development of new vaccines and classes of antibiotics dried up; the training of health-care workers failed to keep abreast of new knowledge; vaccine development and manufacturing were concentrated in a few laboratories; and the discipline of infectious diseases no longer attracted its share of research funds and the best minds. At the nadir in 1992, the US federal government allocated only $74 million for infectious disease surveillance as public health officials prioritized other vital concerns, such as chronic diseases, tobacco use, geriatrics, and environmental degradation. For these reasons, informed assessments of American preparedness to face the unexpected challenges of emerging contagious diseases were disheartening. In the sobering words of the CDC in 1994, for example,

> The public health infrastructure of this country is poorly prepared for the emerging disease problems of a rapidly changing world. Current systems that monitor infectious diseases domestically and internationally are inadequate to confront the present and future challenges of emerging infections. Many foodborne and waterborne disease outbreaks go unrecognized or are detected late; the magnitude of the problem of antimicrobial drug resistance is unknown; and global surveillance is fragmentary.[16]

Similarly, but more bluntly, Michael Osterholm, the Minnesota state epidemiologist, informed Congress in 1996: "I am here to bring you the sobering and unfortunate news that our ability to detect and monitor infectious disease threats to health in this country is in serious jeopardy. . . . For twelve of the States or territories, there is no one who is responsible for food or water-borne disease surveillance. You could sink the *Titanic* in their back yard and they would not know they had water."[17]

Lederberg and other theorists of emerging and reemerging diseases developed a critique of eradicationist hubris that went deeper than a mere protest against a decline in vigilance. They argued that, unnoticed by the eradicationists, society since World War II had changed in ways that actively promoted epidemic diseases. One of the leading features most commonly cited was the impact of globalization in the form of the rapid mass movement of goods and populations. As William McNeill noted in *Plagues and Peoples* (1976), the migration of people throughout history has been a dynamic factor in the balance between microbes and humans. Humans are permanently engaged in a struggle in which the social and ecological conditions that they create exert powerful evolutionary pressure on microparasites. By mixing gene pools and by providing access to populations of nonimmunes, often in conditions under which the microbes thrive, globalization gives microorganisms a powerful advantage.

In the closing decades of the twentieth century, the speed and scale of globalization amounted to a quantum leap as the number of passengers boarding airplanes alone surpassed 2 billion a year. Elective air travel, however, was only part of a far larger phenomenon. In addition there are countless involuntary immigrants and displaced persons in flight from warfare, famine, and religious, ethnic, or political persecution. For Lederberg and the IOM, these rapid mass movements decisively tilted the advantage in favor of microbes, "defining us as a very different species from what we were 100 years ago. We are enabled by a different set of technologies. But despite many potential defenses—vaccines, antibiotics, diagnostic tools—we are intrinsically more vulnerable than before, at least in terms of pandemic and communicable diseases."[18]

After globalization, the second factor most frequently underlined is demographic growth, especially since this growth so often occurs in circumstances that are the delight of microorganisms and the insects that transmit them. In the postwar era, populations have soared above all in the poorest and most vulnerable regions of the world and in cities that lack the infrastructure to accommodate the influx. The global urban population is currently soaring at four times the rate of the rural population, creating sprawling and underserved megacities with more than 10 million inhabitants. By 2017 there were forty-seven such conurbations, such as Mumbai in India, Lagos and Cairo in Africa, and Karachi in Pakistan. Their hallmarks include teeming urban and periurban slums lacking sanitary, educational, or other provisions.

Such places—where millions of people live without sewers, drains, secure supplies of drinking water, or appropriate waste management—are readymade for the transmission of disease. Historically, as we have seen, nineteenth-century diseases flourished in the conditions created by chaotic urbanization in Europe and North America. In the final decades of the twentieth century and the opening decades of the twenty-first, a much larger global process of urbanization is reproducing similarly anomalous sanitary conditions. These prevail in the shantytowns of cities like Lima, Mexico City, Rio de Janeiro, and Mumbai.

The Lessons of Dengue Fever and Cholera

Such urban poverty furnished the social determinants for the global pandemic of dengue fever that began in 1950 and has continued unabated as 2.5 billion people are at risk every year and 50 million to 100 million people are infected. Dengue is the ideal type of an emerging disease. An arbovirus transmitted primarily by the highly urban, day-biting, and domestic *Aedes aegypti* mosquito, dengue thrives in crowded tropical and semitropical slums wherever there is standing water. The mosquito breeds abundantly in gutters, uncovered cisterns, unmounted tires, stagnant puddles, and plastic containers, and it takes full advantage of societal neglect and the absence or cessation of vector control programs.

Particularly important for the theorists of emerging diseases was the manner in which dengue demonstrated the hollowness of the reassuring dogma that infectious diseases evolve inexorably toward commensalism and diminished virulence. Dengue virus is a complex of four closely related serotypes that have been known to infect humans since the eighteenth century. Until 1950, however, dengue infections in any geographical area were caused by a single serotype indigenous to the region. "Classical" dengue arising from one serotype alone leads to a painful illness marked by fever, rash, headache behind the eyes, vomiting, diarrhea, prostration, and joint pains so severe that the infection earned its nickname "break-bone fever." But classical dengue is a self-limiting disease that is followed by lifelong immunity.

Global mobility, however, has allowed all four serotypes to spread indiscriminately around the world, not only expanding the areas at risk from the disease, but also producing epidemics characterized by the interaction of more than one serotype. Since there is no crossover immunity from one serotype to another, individuals who have recovered from classical dengue (the result of infection by one of the four dengue serotypes) can be subsequently infected by one or more of the other three serotypes. Through mechanisms that are still imperfectly understood, the disease is far more virulent in those patients who suffer such successive reinfection from a different serotype. Instead of becoming milder, therefore, dengue is now a far more serious and ever growing threat that often gives rise even to sudden epidemics of the severe disease known as dengue hemorrhagic fever (DHF) and its lethal complication of dengue shock syndrome (DSS).

In the Americas, the first epidemic combining more than one serotype of dengue fever broke out in 1983 in Cuba. That outbreak resulted in 344,000 cases overall, of which 24,000 were diagnosed as DHF and 10,000 as DSS. Moreover, since the dengue vectors *Aedes aegypti* and *Aedes albopictus* are present in the United States, scientists at the National Institute of Allergy and Infectious Diseases, including its director Anthony S. Fauci, have predicted that dengue and DHF will include the continental United States in its ongoing global expansion.

Dengue therefore demonstrates the following important evolutionary lessons: (1) infectious diseases that do not depend on the mobility of their host for transmission (because they are borne by vector, water, or food) are under no selective pressure to become less virulent; (2) overpopulated and unplanned urban or periurban slums provide ideal habitats for microbes and their arthropod vectors; and (3) modern transportation and the movements of tourists, migrants, refugees, and pilgrims facilitate the process by which microbes and vectors gain access to these ecological niches.

Recent research on Asiatic cholera, as we have seen in Chapter 13, adds a further dimension to the debate over microbial virulence. At first glance, the evolution from "classical" *Vibrio cholerae* with its high CFR to *V. cholerae* El Tor O1 seems to support the optimistic conclusion of an inherent evolutionary tendency toward commensalism. Classical *V. cholerae* causes a high CFR and violent symptoms, while *V. El Tor* produces a low CFR and relatively moderate symptoms. On the other hand, the subsequent history and evolution of *V. El Tor* demonstrates the possibility of mutations that revert to virulence, giving rise to more severe and violent epidemics than the optimistic view would anticipate. Atypically violent outbreaks in Haiti, Pakistan, and Bangladesh in 2010 are examples. The reason, it is suspected, is that *V. El Tor* is adapted to persistence in natural environmental reservoirs, where it survives and replicates for long periods. As a result, the microbe is not dependent for its survival and propagation on human-to-human transmission, which occurs only episodically when particularly favorable climatic and societal conditions promote its spillover to human populations. For a pathogen so adapted to nonhuman and even abiotic environments, the evolutionary pressure for a decline in virulence in human populations is no longer determinative. Future cholera outbreaks may therefore hold unpleasant surprises.

Nosocomial Infections and Microbial Resistance

Paradoxically, the very successes of modern medical science also prepares the way for emerging infections. By prolonging life, medicine gives rise to ever-growing populations of elderly people with compromised immune systems. As part of this process, medical interventions also yield increasing numbers of immunocompromised cohorts even at earlier ages—diabetics, cancer and transplant patients undergoing chemotherapy, and AIDS patients whose disease has been transformed into a chronic illness by antiretroviral treatment. Furthermore, such nonimmunes are concentrated in settings where the transmission of microbes from body to body is amplified, such as hospitals, facilities for the elderly, and prisons. The proliferation of invasive procedures has also increased portals through which microbes can enter the body. Previously rare or unknown nosocomial infections (those acquired in hospitals) emerge in these conditions and have become a major problem of public health as well as an ever-growing economic burden. Of these infections, the so-called superbug *Staphylococcus aureus*—the leading cause of nosocomial pneumonia, of surgical site infections, and of nosocomial bloodstream infections—is the most important and widespread. A recent study notes that, in the United States by 2008, "Each year approximately two million hospitalizations result in nosocomial infections. In a study of critically ill patients in a large teaching hospital, illness attributable to nosocomial bacteria increased intensive care unit stay by 8 days, hospital stay by 14 days, and the death rate by 35%. An earlier study found that postoperative wound infections increased hospital stay an average of 7.4 days."[19]

A further threatening by-product of the advance of medical science is the emergence of ever-increasing antimicrobial resistance. Already in his 1945 Nobel Prize acceptance speech, Alexander Fleming issued a prophetic warning. Penicillin, he advised, needed to be administered with care because the bacteria susceptible to it were likely to develop resistance. The selective pressure of so powerful a medicine would make it inevitable.

Echoing Fleming's warning, theorists of emerging diseases argue that antibiotics are a "nonrenewable resource" whose duration of efficacy is biologically limited. By the late twentieth century, this prediction was reaching fulfillment. At the same time as the discovery of new classes of antimicrobials had slowed to a trickle, the pharmaceutical marketplace staunched the flow by inhibiting research on medications that are likely to yield low profit margins. Competition, regulations mandating large and expensive clinical trials, and the low tolerance for risk of regulatory agencies all compound the problem.

But while antimicrobial development stagnates, microorganisms have evolved extensive resistance. As a result, the world stands poised to enter a post-antibiotic era. Some of the most troubling examples of the emergence of resistant microbial strains are plasmodia that are resistant to all synthetic antimalarials, *S. aureus* that is resistant both to penicillin and to methicillin (MRSA), and strains of *Mycobacterium tuberculosis* that are resistant to first-line medications (MDR-TB) and to second-line medications (XDR-TB). Antimicrobial resistance threatens to produce a global crisis, and many scientists anticipate the appearance of strains of HIV, tuberculosis, *S. aureus*, and malaria that are not susceptible to any available therapy.

In part, the emergence of antimicrobial resistance is a simple result of Darwinian evolution. Tens of thousands of viruses and three hundred thousand species of bacteria are already known to be capable of infecting human beings, and many of them replicate and evolve billions of times in the course of a single human generation. In this context, evolutionary pressures work to the long-term disadvantage of human beings. But profligate human action dramatically hastens the process. Farmers spray crops with pesticides and fruit trees with antibiotics; they add subtherapeutic doses of antibiotics to animal feed to prevent disease, promote growth, and increase the productivity of chickens, pigs, and feedlot cattle. Indeed, half the world output of antimicrobials by tonnage is used in agriculture.

At the same time, the popular confidence that microorganisms will succumb to a chemical barrage has led to a profusion of antimicrobials in domestic settings where they serve no sustainable purpose. Physicians—pressured to give priority in clinical settings to the immediate risk of individual patients over the long-term interest of the species, and to meet patients' expectations—often adopt prescribing fashions by which antibiotics are administered even for nonbacterial conditions for which they are unnecessary or useless. The classic case in this regard is the pediatric treatment of middle ear infection, for which the overwhelming majority of practitioners in the 1990s prescribed antibiotics even though two-thirds of the children derived no benefit from them. Widespread possibilities of self-medication in countries with few regulations and opportunities for unregulated access to medication created by the internet amplify the difficulties. In the case of diseases such as malaria and tuberculosis that require a long and complicated therapeutic regimen, some patients interrupt their treatment after the alleviation of their symptoms instead of persevering until the underlying condition is cured. Here the problem is not the overuse but the underuse of antibiotics.

A further issue is the overly rigid conceptualization of disease by eradicationists, who drew too sharp a distinction between chronic and contagious diseases. Infectious diseases, it became clear in the 1990s, are a more expansive category than scientists previously realized because many diseases long considered noninfectious in fact have infectious origins. In demonstrating these causal connections, the decisive work was that of the Australian Nobel laureates Barry J. Marshall and Robin Warren with regard to peptic ulcers in the 1980s.

Peptic ulcers are a significant cause of suffering, expense, and even death as one American in ten develops one during the course of his or her lifetime. More than 1 million people are hospitalized by ulcers every year, and six thousand die. But until the work of Marshall and Warren, the chronic etiology of peptic ulcer was universally but misleadingly accepted as scientific truth. In Marshall's words: "I realized that the medical understanding of ulcer disease was akin to a religion. No amount of logical reasoning could budge what people knew in their hearts to be true. Ulcers were caused by stress, bad diet, smoking, alcohol and susceptible genes. A bacterial cause was preposterous."[20] Marshall and Warren demonstrated, in part by means of an auto-experiment, that the bacterium *Helicobacter pylori* was the infectious cause of the disease and that antibiotics rather than diet, lifestyle change, and surgery were the appropriate therapy. This was nothing short of a medical watershed.

This insight led to the realization that many other nonacute diseases, such as certain forms of cancer, chronic liver disease, and neurological disorders, are due to infections. Human papillomavirus, for instance, is thought to give rise to cervical cancer, hepatitis B and C viruses to chronic liver disease, *Campylobacter jejuni* to Guillain-Barré syndrome, and certain strains of *E. coli* to renal disease. There are indications as well that infections serve as an important trigger to atherosclerosis and arthritis, and there is a growing recognition that epidemics and the fear that accompanies them leave psychological sequelae in their wake, including posttraumatic stress. This understanding of these processes is what some have termed a new awareness of the "infectiousness of noninfectious diseases."

Finally, and most emphatically, the concept of emerging and reemerging diseases was intended to raise awareness of the most important threat of all—that the spectrum of diseases that humans confront is broadening with unprecedented rapidity. The number of previously unknown conditions that have emerged to afflict humanity since 1970 exceeds forty, with a new disease discovered on average more than once a year. The list includes HIV, Hantavirus, Lassa fever, Marburg fever, Legionnaires' disease, hepatitis C, Lyme disease, Rift Valley fever, Ebola, Nipah virus, West Nile virus, SARS, bovine spongiform encephalopathy, avian flu, Chikungunya virus, norovirus, Zika, and group A streptococcus—the so-called flesh-eating bacterium. Skeptics argue that the impression that diseases are emerging at an accelerating rate is misleading. Instead, they suggest, it is largely an artifact of heightened surveillance and improved diagnostic techniques. WHO has countered that not only have diseases emerged at record rapidity, as one would expect from the transformed social and economic conditions of the postwar world, but that they gave rise between 2002 and 2007 to a record eleven hundred worldwide epidemic "events." A careful examination of the question, published in *Nature* in 2008, involved the study of 335 emerging infectious disease (EID) events between 1940 and 2004, controlling for reporting effort through more efficient diagnostic methods and more thorough surveillance. The study concluded: "The incidence of EID events has increased since 1940, reaching a maximum in the 1980s. . . . Controlling for reporting effort, the number of EID events still shows a highly significant relationship with time. This provides the first analytical support for previous suggestions that the threat of EIDs to global health is increasing."[21]

There are no rational grounds, the public health community concluded, to expect that as diseases emerge in the future, none of them will be as virulent and transmissible as HIV or the Spanish influenza of 1918–1919. Thus, discussion shifted dramatically from the question of *whether* new diseases would emerge and old ones resurge to the issue of how the international community would face them *when they did*. In the stark words of the US Department of Defense, "Historians in the next millennium may find that the 20th century's greatest fallacy was the belief that infectious diseases were nearing elimination. The resultant complacency has actually increased the threat."[22]

CHAPTER 22
Dress Rehearsals for the Twenty-First Century
SARS and Ebola

Rearmament

A major aspect of the official response to the challenge of emerging and re-emerging diseases is that microbes are regarded as threats to the security of states and to the international order. For the first time, not only public health authorities but also intelligence agencies and conservative think tanks have classified infectious diseases as a "nontraditional threat" to national and global security. A turning point came in 2000, when the CIA's national intelligence estimate for that year was devoted to the danger of epidemic disease, now viewed as a major security challenge.

In the first section of the report, "Alternative Scenarios," the CIA outlined three possible scenarios for the course of infectious diseases over the next twenty years: (1) the optimistic contemplation of steady progress in combating communicable disease; (2) the forecast of a stalemate, with no decisive gains either by microbes or by humans; and (3) the consideration of the most pessimistic prospect of deterioration in the position of humans, especially if the world population continues to expand and if megacities continue to grow with their attendant problems of overcrowding, poor sanitation, and unprotected drinking water. Unfortunately, the CIA regarded the optimistic first case as extremely unlikely.

Against this background, the following sections of the report, "Impact" and "Implications," outlined a series of likely economic, social, and political results that would occur in the new age of increasing disease burdens. In the most afflicted regions of the world, such as sub-Saharan Africa, the report anticipated "economic decay, social fragmentation, and political destabilization." The international consequences of these developments would be growing struggles to control increasingly scarce resources, accompanied by crime, displacement, and the degradation of familial ties. Disease, therefore, would heighten international tensions. Since the consequences of increasing burdens of communicable diseases in the developing world are certain to impede economic development, the report also predicted that democracy would be imperiled, that civil conflicts and emergencies would multiply, and that the tensions between North and South would deepen.[1]

Three years later, motivated by the CIA's report, the influential RAND Corporation turned to the intersection of disease and security when it attempted to provide "a more comprehensive analysis than has been done to date, encompassing both disease and security."[2] *The Global Threat of New and Reemerging Infectious Diseases* envisaged even more somber probabilities than the CIA in the new global environment. The intelligence report had two leading themes. The first was that in the postwar era, the importance of direct military threats to security has declined sharply. The second was that there has been a corresponding rise in the impact of "nontraditional challenges," of which diseases are the major component. The era of emerging and reemerging diseases marked the opening of a period in which infectious diseases would profoundly affect the ability of states to function and preserve social order.

The starting point of a plan to confront emerging diseases envisaged by the CDC, the National Institute of Allergy and Infectious Diseases, and the White House was the Institute of Medicine's description of the Darwinian struggle under way between humans and microbes. In the IOM's analysis, microbes possess formidable advantages: they outnumber human beings a billionfold, they enjoy enormous mutability, and they replicate a billion times more quickly than humans. In terms of evolutionary adaptation, microbes are genetically favored to win the contest. As Joshua Lederberg observed, "Pitted against microbial genes, we have mainly our wits."[3] Taking the IOM's analysis as its starting point, the US response to the new challenge is best seen as an attempt to organize and deploy human wit, backed by newly found financial resources, to counter the microbial challenge.

A 1996 White House "Fact Sheet" on the threat of emerging infectious diseases declared in clear alarm, "The national and international system of infectious disease surveillance, prevention, and response is inadequate to protect the health of U.S. citizens." To remedy the situation, the White House established six policy goals:

1. Strengthen the domestic infectious disease surveillance and response system, both at the Federal, State, and local levels and at ports of entry into the United States, in cooperation with the private sector and with public health and medical communities.
2. Establish a global infectious disease surveillance and response system, based on regional hubs and linked by modern communications.
3. Strengthen research activities to improve diagnostics, treatment, and prevention, and to improve the understanding of the biology of infectious disease agents.
4. Ensure the availability of the drugs, vaccines, and diagnostic tests needed to combat infectious diseases and infectious disease emergencies through public and private sector cooperation.
5. Expand missions and establish the authority of relevant United States Government agencies to contribute to a worldwide infectious disease surveillance, prevention, and response network.
6. Promote public awareness of emerging infectious diseases through cooperation with nongovernmental organizations and the private sector.[4]

In pursuit of goals 2, 3, and 4, the National Institute of Allergy and Infectious Diseases established a research agenda to develop new weapons to combat epidemic diseases, giving rise to an explosion in knowledge. In the decade after 1995 its budget quadrupled from $50 million in 1994 to more than $1 billion in 2005, while publications on infectious diseases burgeoned. Indeed, the agency director, Anthony S. Fauci, claimed in 2008 that HIV/AIDS in particular had become the most extensively studied disease in human history. The work of the federal agency, moreover, has been complemented by the activity of private organizations, most notably the Bill and Melinda Gates Foundation, and by university and pharmaceutical industry laboratories.

At the same time that the IOM stressed basic research, the CDC developed a defensive strategy against emerging pathogens in compliance with goal 1 of the White House directive. In two seminal works published in 1994 and 1998 it outlined its objectives in four principal areas: surveillance, applied research, prevention and control, and the enhancement of the infrastructure and trained personnel needed for diagnostic laboratories at the federal, state, local, and international levels. In addition, the CDC strengthened its links with the international public health community and with other surveillance agencies, such as the Food and Drug Administration and the Department of Defense; it enhanced its capacity to respond to outbreaks of disease; it launched the journal *Emerging Infectious Diseases* as a forum to pool information on communicable diseases; and it sponsored a series of major international conferences on the topic of emerging and reemerging diseases.

Following discussion with the National Intelligence Council, President George W. Bush took the further step of establishing the two most extensively funded initiatives ever launched to combat a single disease. The first, founded in 2003 and renewed as recently as 2018, was the President's Emergency Plan for AIDS Relief (PEPFAR), overseen by the State Department Office of the Global AIDS Coordinator. Concentrating on twelve sub-Saharan African countries plus Vietnam, Haiti, and Guyana, PEPFAR has provided billions of dollars for prophylactic measures and antiretroviral treatment as well as such actions as building health infrastructure, supporting AIDS orphans, and training medical personnel. Under the supervision of the AIDS coordinator, PEPFAR oversees the involvement of an array of federal agencies—the US Agency for International Development; the Departments of Health and Human Services, State, Defense, Commerce, and Labor; and the Peace Corps.

The second major program targeting a specific epidemic disease was the President's Malaria Initiative (PMI), which President Bush established under the directorship of Admiral Timothy Ziemer in 2005. Like PEPFAR, PMI combines humanitarianism with enlightened self-interest. Its purpose is to counter the ravages of malaria in sub-Saharan Africa with ample funding in order to develop health infrastructure and to support the deployment of such antimalarial tools as artemisinin-based combination therapy, insecticide-treated mosquito nets, health education, and vector control.

At the global level, WHO also took major steps to strengthen international preparedness for the ongoing siege by microbial pathogens. A first was the creation in 1996 of the disease-specific UNAIDS, whose functions are to raise awareness, to mobilize resources, and to monitor pandemic.

Funding levels in the fight against the disease increased from $300 million in 1996 to nearly $9 billion a decade later. A further step was that, like the United States, the United Nations announced that it regarded infectious diseases as threats to international security. In acknowledgment of this new development, the UN Security Council took the unprecedented step in June 2001 of devoting a Special Session to the HIV/AIDS crisis. The session adopted the "Declaration of Commitment on HIV/AIDS: Global Crisis—Global Action," which called the global epidemic a "global emergency and one of the most formidable challenges to human life and dignity."[5] Five years later, in June 2006, the UN General Assembly reaffirmed its commitment to the campaign and adopted the "2006 Political Declaration on HIV/AIDS," whose chief goal was the establishment of national campaigns to improve access to care and treatment.

A third step was the establishment in 2005 of a new set of International Health Regulations (IHR 2005) to replace the outdated IHR 1969. Whereas the old framework required notification only in the event of plague, yellow fever, and cholera, the new rules required notification for any public health emergency of international concern, including unknown pathogens and emerging infections. The 2005 regulations specified the nature of the events that should trigger international concern and committed all of the 193 WHO member states to improve their capacity for surveillance and response. In addition, recognizing that microbes do not acknowledge political frontiers, IHR 2005 called for effective responses wherever necessary to contain an outbreak on the basis of real-time epidemiological evidence instead of concentrating on taking defensive measures at international borders.

Finally, WHO organized a rapid response capacity. This was the Global Outbreak Alert and Response Network (GOARN), which was established in 2000 with the goal of ensuring that even the most resource-poor countries would have access to the experts and means needed to respond to an epidemic emergency. To that end, GOARN pooled the resources of sixty countries and organized five hundred experts in the field. In addition, it stockpiles vaccines and drugs and supervises their distribution during epidemic events.

Severe Acute Respiratory Syndrome—SARS

In practice, the first test of the new structures was the SARS pandemic of 2002–2003, which was the first major emerging disease threat of the twenty-first century. After appearing in the Chinese province of Guangdong in November 2002, SARS erupted as an international health threat in

March 2003, when WHO received notification and declared a global travel alert. Between March and the declaration on July 5 that the disease had been contained, SARS affected 8,098 people, caused 774 deaths, brought international travel to a halt in entire regions, and cost $60 billion in gross expenditure and business losses to Asian countries alone.

As retrospective studies have demonstrated, SARS presented many of the features that most severely expose the vulnerabilities of the global system. It is a respiratory disease capable of spreading from person to person without a vector; it has an asymptomatic incubation period of more than a week; it generates symptoms that closely resemble those of other diseases; it takes a heavy toll on caregivers and hospital staff; it spreads easily and silently by air travel; and it has a CFR of 10 percent. Moreover, at the time it appeared, its causative pathogen (SARS-associated coronavirus) was unknown, and there was neither a diagnostic test nor a specific treatment. For all of these reasons, it dramatically confirmed the IOM's 1992 prediction that all countries were more vulnerable than ever to emerging infectious diseases. SARS demonstrated no predilection for any region of the world and was no respecter of prosperity, education, technology, or access to health care. Indeed, after its outbreak in China, SARS spread by airplane primarily to affluent cities such as Singapore, Hong Kong, and Toronto, where it struck relatively prosperous travelers and their contacts and hospital workers, patients, and their visitors rather than the poor and the marginalized. More than half of the recognized cases occurred in well-equipped and technologically advanced hospital settings such as the Prince of Wales Hospital in Hong Kong, the Scarborough Hospital in Toronto, and the Tan Tock Seng Hospital in Singapore.

In terms of response to the crisis, the SARS outbreak vindicated the reforms taken on both the national and international levels. After the debacle of Chinese obfuscation at the start of the epidemic, national governments cooperated fully with IHR 1969. The world's most equipped laboratories and foremost epidemiologists, working in real-time collaboration via the internet, succeeded, with unprecedented speed, in identifying the SARS coronavirus in just two weeks. At the same time the newly created GOARN, together with such national partners as the Public Health Agency of Canada, the CDC, and WHO's Global Influenza Surveillance Network, took rapid action to issue global alerts, monitor the progress of the disease, and supervise containment strategies before it could establish itself endemically. Ironically, given the high-tech quality of the diagnostic and monitoring effort, the containment policies were based on traditional methods dating from the public health strategies against bubonic plague of the seventeenth century and the foundation of epidemiology as a discipline in the nineteenth century—case tracking, isolation, quarantine, the cancellation of mass gatherings, the surveillance of travelers, recommendations to increase personal hygiene, and barrier protection by means of masks, gowns, gloves, and eye protection. Although SARS affected twenty-nine countries and five continents, the containment operation successfully limited the outbreak primarily to hospital settings, with only sporadic community involvement. By July 5, 2003, WHO could announce that the pandemic was over.

Although the global sanitary defenses withstood the challenge of SARS, serious doubts surfaced. The Chinese policy of concealment between November 2002 and March 2003 had placed international health in jeopardy and revealed that even a single weak link in the response network could undermine the international emergency response system. For four months following the November outbreak of SARS in the province of Guangdong and its subsequent spread to Beijing, the Chinese authorities followed a policy of concealment and obfuscation. Given the extent of links connecting the People's Republic with Hong Kong and Taiwan—via trade, investment, family ties, and tourism—it was impossible to prevent all awareness of a new disease from reaching the outside world. A complete news blackout was especially difficult in the age of the internet and social media, as well as at the time of the lunar new year in February, which is the peak time for travel within China. On the other hand, the Chinese party-state was determined to project an imagine of mastery and control, both to its own people and to the outside world. The Communist Party also feared that too much information would expose the reality in the country of impoverished living standards, an inadequate health-care system, and a lack of preparedness for public health emergencies.

For those reasons, the regime led by Premier Wen Jiabao persisted, until March 2003, in resisting all international pressure for accurate and timely information. The Chinese instead minimized the extent of the crisis, manipulated figures for public consumption, banned all unfavorable news from appearing in the press, and denied WHO teams access to affected areas of the country. Only in March, when WHO assumed the role of whistleblower and released the partial information it possessed, did China lurch in a new direction. On April 17 the Politburo reversed course by promising timely notice of SARS cases, granting WHO access to Guangdong and Beijing, and setting up a SARS task force under Vice Premier Wu Yi. At the same time the official *People's Daily* admitted that the country had been poorly prepared, and the director of China's CDC apologized.

If the regime thus became more open, it did not become less heavy-handed and authoritarian. Indeed, the party imposed quasi-military measures of quarantine and isolation backed by severe sanctions that extended to capital punishment and rewards for informants willing to denounce violators to the authorities.

Chinese policies clearly reveal, therefore, that a major factor in the containment of SARS was serendipity. The world was fortunate that the disease is spread by droplets and therefore requires extended contact for transmission, unlike classic airborne diseases such as influenza and smallpox. It was relatively easy to contain because it is not readily communicable from person to person—except in the still poorly understood case of so-called super-shedders. Most SARS sufferers infect few if any secondary contacts. An epidemiological assumption that has rarely been challenged is that all patients are more or less equally infectious. In the case of SARS, however, an outsize role in the epidemic was played by a small group of patients who disproportionately affected large numbers of contacts and were therefore called super-shedders. Dependence for its transmission on a small proportion of those affected was a significant limitation in the propagation of SARS.

As poorly transmissible as it was, however, SARS exposed the absence of "surge capacity" in the hospitals and health-care systems of the prosperous and well-resourced countries it affected. The events of 2003 thereby raised the specter of what might have happened had SARS been pandemic influenza, and if it had traveled to resource-poor nations at the outset instead of mercifully visiting cities with well-equipped and well-staffed modern hospitals and public health-care systems. Furthermore, SARS arrived in peacetime rather than in the midst of the devastation and dislocations of war or a natural disaster. In that respect it did not repeat the challenge of the "Spanish Lady" of 1918–1919—the influenza pandemic that spread extensively with the movement of troops during World War I. SARS also appeared as a respiratory disease in Southeast Asia, where the WHO sentinel system had been created to monitor and respond to exactly such an emergency. The physician Paul Caulford, who fought the SARS epidemic on the front lines at Scarborough Hospital in Toronto, raised these matters. In December 2003, after the emergency had passed, he reflected:

> SARS must change us, the way we treat our planet, and how we deliver health care, forever. Will we be ready when it returns? SARS brought one of the finest publicly-funded health systems in the world to its knees in a matter of weeks. It has unnerved

> me to contemplate what the disease might do to a community without our resources and technologies. Without substantive changes to the way we manage the delivery of health care, both locally and on a worldwide scale, we risk the otherwise preventable annihilation of millions of people, either by this virus, or the next.[6]

At the end of the victory over SARS, the nagging question therefore remained: Even after the impressive efforts at rearmament against epidemic disease since 1992, how prepared is the international community for upcoming emerging diseases? Have we been "forever changed"?

The Challenge of Ebola

In December 2013 the small child Emile Ouamouno, who lived in a village in the forests of southeast Guinea, died of Ebola. His home was located in the Mano River Basin where the borders of three West African countries intersect—Guinea, Liberia, and Sierra Leone. When it was revealed months later, the location of Emile's death confounded the international public health community. Although there had been a long series of small outbreaks of the disease since 1976, all of them had occurred in Central Africa and especially the Democratic Republic of the Congo (or the Congo). The Congo even provided the name "Ebola," after the river that flows through the territory where the 1976 outbreak had occurred.

A highly virulent infectious disease whose initial appearance caused panic around the world had seemingly settled into a reassuring pattern. In all previous upsurges in the Congo and Uganda, Ebola, in the prose of the international press, always appeared suddenly "from the jungle" but then disappeared each time as rapidly as it arrived. The morbidity and mortality figures for all known outbreaks before 2013 taken together totaled 2,427 and 1,597, respectively. The largest single outbreak was that of Uganda from October 2000 to January 2001, which gave rise to 425 cases and 226 deaths.

International and local surveillance systems, focused on Central Africa, were therefore caught off guard. Unexpectedly, Ebola spread from the forested areas of Guinea throughout the Mano River Basin. By March 2014 it became a major international epidemic transmitted person-to-person in the overcrowded capitals and urban centers of three countries. There were even brief flare-ups in neighboring states such as Mali, Nigeria, and Senegal, where small clusters of victims contracted the disease. Of these, the largest numbered twenty people in Senegal. Threatening to escape all control, the epidemic lasted for two years—until WHO finally declared it over in December 2016. By then, even on the basis of conservative official statistics that seriously understate the extent of the disaster, the epidemic had caused 28,652 cases and 11,325 deaths (40 percent).

This public health emergency radically transformed medical and epidemiological understandings of Ebola. It also tested the emergency response systems that were intended to cope with emerging infectious diseases after the inadequacies exposed by SARS. In Toronto, Dr. Caulford, as we have seen, had commented in 2003 that the world needed to be "forever changed." His hope was that no outbreak would ever again find the international public health system so unprepared. To be sure, resolutions had been passed and reforms promised. Sadly, however, West Africa from 2013 to 2016 demonstrated that, in matters of sickness and health, short-sightedness and cost-cutting had prevailed. On the ground in Guinea, Liberia, and Sierra Leone, the policies actually implemented were eerily reminiscent of the hasty improvisations of the Black Death.

Symptoms

Ebola virus is a member of the family of filiviruses (*Filoviridae*), and the disease it causes was originally classed as a hemorrhagic fever because bleeding was seen as a chief symptom and leading cause of death. The experience of large-scale case management, however, led to a change in terminology when "Ebola hemorrhagic fever" was rechristened "Ebola virus disease." It was discovered that in the course of this disease, bleeding is frequently entirely absent. In those cases where it does occur, it is seldom profuse but is limited to bleeding of the gums and nose, bleeding at injection sites, and blood in vomit or diarrhea. Such bleeding, furthermore, does not correlate with a negative prognosis as it rarely progresses to massive hemorrhaging, except among pregnant women, who are especially vulnerable.

After a variable incubation period of two to twenty-one days, the "dry stage" of the disease begins. The symptoms of this stage mark the onset of both the illness and the infectious period, but they are nonspecific and misleadingly similar to those of seasonal influenza—fever, headache, muscle pain, fatigue, and sore throat. After several days, the critical manifestation of Ebola begins. This is the "wet stage," characterized by unstoppable loss of fluid from vomiting, by diarrhea, and, in many cases, by the loss of blood from bodily orifices. Patients are also tormented by pain in the chest and abdomen, violent hiccoughs, and conjunctivitis. In the majority of cases, the loss of fluid leads to death from multiple causes—dehydration, kidney failure, respiratory distress and asphyxia, severe cardiac arrhythmias, and heart failure. It is the loss of virus-bearing fluids that also makes the patient highly infectious during the wet stage and then just after death. In all cases, the prognosis is unfavorable because the CFR of Ebola ranges from 60 percent to 90 percent, depending on the virus strain and the availability of supportive therapy and nursing care. At the end of the seventh day the majority of sufferers lapse into coma, followed by death. A minority, however, begin slowly to recover step by step, with progressively less pain, less loss of fluid, and more energy.

But even survivors face an extended ordeal. Convalescence is protracted, and the disease often leads to post-Ebola syndrome—an array of disabling symptoms including joint pain, headaches, memory loss, hearing deficits, tinnitus, depression, and posttraumatic stress accompanied by violent dreams and visions. Most common, however, is the ocular condition of uveitis that leads to blurry vision, light sensitivity, or permanent blindness. Since the virus is sequestered in the body for months after recovery, survivors also continue to be infectious through certain bodily fluids—breast milk, semen, vaginal secretions, tears, and spinal fluid. For these reasons, many recovered Ebola patients suffer not only physical pain, but also societal stigma on the part of frightened communities. Many lose their jobs and are shunned by friends and families and abandoned by partners.

In 2014–2016 this fear of Ebola was enhanced by the fact that no effective preventative or curative means were known. In well-resourced hospitals the standard of care involved what are termed "advanced life support therapies"—mechanical ventilation, hemodialysis, and intravenous rehydration, supplemented by medication to calm the purging and to relieve pain. In West Africa, experimental treatments were also adopted in desperation, all with disappointing results. These therapies included (1) the antiviral ZMapp and the antiretroviral lamivudine, which proponents hoped would prevent Ebola virus from replicating; (2) statins, such as Lipitor, which it was hoped would "calm" the immune system after infection; (3) the antimalarial drug amodiaquine, which potentially worked by mechanisms that were not understood; and (4) transfusions of the blood of convalescent patients in the hope that the donors' antibodies would enhance the immune response of recipients. Unfortunately, these strategies failed to save or prolong life, and supplies in any case were not adequate for the purpose of mass treatment.

Spillover to Humans

Ebola is now known to be a zoonotic disease whose natural hosts are fruit-eating bats (family Pteropodidae) in which the virus replicates easily but without giving rise to disease in the animal. Spillover into human populations from the reservoir among bats, however, is a rare event determined by the nature of both forest land use and human interactions with the environment. In principle, spillover can occur due to contaminated bush meat when forest dwellers hunt, butcher, and eat bats or other previously infected animals. A handful of instances of such transmission has been documented. In West Africa, however, an extensive and semicolonial mythology appeared in the press concerning the allegedly bizarre practices of African natives, their strange practices in the jungle, and their dietary predilection for roasted bat wings and bat stew. Even the health ministries of the affected countries took up this vision of the onset of the epidemic. In the early months they devoted energy and resources to a campaign to persuade villagers to alter their cuisine in the name of public health. In testimony to the US Congress on the medical emergency, the anthropologist and physician Paul Farmer, who had served as an emergency responder during the crisis, was emphatic on this point when he declared, "We should be very clear that the rapid spread of Ebola is not due to 15,000 episodes of bush meat eating frenzy."[7]

In fact, the spillover to which Emile Ouamouno fell victim had a far more complex gestation. In order to understand it, one needs to dispel another legend that shrouds the events in West Africa. In press accounts, two of the most frequent descriptors applied to the forested region that was "Ground Zero" are "remote" and "inaccessible." The implication is that the area was almost a virgin forest cut off from urban centers and the larger world beyond. In that description, the transmission of Ebola to the capital cities of Conakry in Guinea, Freetown in Sierra Leone, and Monrovia in Liberia is intelligible in terms of the tribalistic movements of the Kissi ethnic group. Since their forest homeland straddles all three countries that were infected, it was said that Ebola traveled with Kissi villagers as they "made kinship visits," or, in everyday parlance, dropped by to see relatives.

In reality, the forestlands of the three countries are not "remote" in any meaningful sense at all. On the contrary, the countries had been deeply integrated into world markets from the closing decades of the twentieth century through thick and overlapping networks of trade, investment, mining, logging, and agrobusiness. Not by accident, the three countries afflicted by Ebola were subject to a frenetic pace of deforestation and land clearance to meet international demand for resources from the woods. Clearest and most illustrative is the example of the palm oil industry, which has been the most dynamic sector in world agriculture since the 1990s, when its production tripled, with the forests of West and Central Africa as an important center. A 2016 book described this upsurge of palm oil, alongside that of the soybean, as "the world's most recent agricultural revolution."[8]

The oil palm is native to West Africa, and its scientific name—*Elaeis guineensis*—even specifies its origins in Guinea. There, forest dwellers had long used it to produce the medicinal remedies employed by traditional healers; to cut fronds for thatching and fencing; to harvest the edible and prized palm heart; and to yield a range of culinary ingredients. What was new in the late twentieth century was the project to cut down the forests by clear cutting to establish a monoculture of large oil palm plantations. Capital for this project was provided by the World Bank, the African Development Bank, and the International Monetary Fund, together with their "partners"—the local governments of the three countries involved. Since the land was held by small subsistence-producing peasants, one of the great "external economies" that made the venture profitable was that the West African states undertook to dispossess the villagers, whose holdings were based on custom rather than legally recognized titles. This large-scale enclosure, which some sources describe as a "land grab," provided cheap and extensive acreage to plantation owners, while the army enforced the entrepreneurs' claims against rural resistance to mass dispossession. Driven off the land and into a choice between migration and employment at low plantation wages, peasants mounted a stubborn opposition to the onrush of "development."

For local governments, the oil palm was highly attractive. As a cash crop, palm oil targeted export markets, thereby alleviating foreign debt and providing foreign exchange. Companies involved also generated substantial profits, and they were willing to acknowledge officials who brokered deals and promoted their interests with generous rewards. Furthermore, in brochures, mission statements, and reports, the industry represented itself as environmentally sensitive in its promotion of an indigenous tree, as economically progressive in creating jobs, and as "modern" in its technology and management practices. Every concern found an answer in one or another company document. Plantations would yield jobs, infrastructure, vocational training, and education. In the gushing language of promoters of the industry, palm oil is nothing less than "liquid gold" that helps develop nations.

Actual implementation of the monocropping scheme was carried out by large companies such as the giant Guinean palm oil and rubber company founded in 1987 known as SOGUIPAH (Société Guinéenne de Palmier à huile et d'Hévéa). This firm was partly state-owned and was based in Conakry. Palm oil was appealing to agrobusiness because it met a gamut of industrial and consumer uses. Oil from the palm kernel is a constituent element of biodiesel fuel, and it is used to produce cosmetics, soap, candles, detergent, and lubricants. Oil from the fruit is edible, and it is eagerly sought by the food industry to make margarine, ice cream, cookies, pizza, and an array of processed foods. Home cooks also use it abundantly as a cooking oil. It is estimated that half of the items for sale in a modern supermarket contain palm oil as an important component. Finally, the leftover kernel cake provides high-protein feed for livestock.

Favorable political conditions in West Africa, where agrobusiness found enticing subsidies in terms of capital, labor, land, and sympathetic access to government, attracted planters such as SOGUIPAH. Equally decisive was the fact that tropical forests such as those available in the Mano River Basin provide optimal environmental conditions for the oil palm to thrive. *Elaeis guineensis* grows most rapidly and produces the greatest abundance of fruit in the conditions of temperature, humidity, wind, and soil prevailing in the tropical rainforest. The combination of these circumstances placed the palm oil industry on a collision course with the primary forests of the region.

Palm oil companies comprehensively transformed the landscapes they encountered, and in ways that were not conducive to the health of the population or the environment. They began by destroying the existing primary forest by fire and bulldozer. Having cleared the land, they then established a monoculture of oil palm cultivated in large plantations. There is a burgeoning literature on the negative social, economic, and environmental impact of the new palm oil monoculture, and a vociferous opposition to it has been mounted by such "green" NGOs as the World Rainforest Movement, the Union of Concerned Scientists, and Greenpeace. They point to such negative features as the loss of biodiversity, the contribution of deforestation to the greenhouse effect and global warming, population displacement, the low wages and harsh working conditions of plantation workers, the unfavorable long-term position of nations that develop on the basis of producing raw materials in the global market, and the inability of perennial crops like palm oil to respond to market fluctuations.

In addition, the emergence of Ebola demonstrates that deforestation also has direct implications for health and disease. The areas where Ebola outbreaks have occurred since 1976 map perfectly onto the geography of deforestation in Central and West Africa. The link between Ebola and deforestation is the fact that the fragmentation of African forests disrupts the habitat of fruit bats. Before the arrival of agrobusiness, the bats normally roosted high in the forest canopy, far from human activities. In the wake of clear-cutting, however, these "flying foxes," as they are known locally, forage ever closer to human settlements and grow dependent on household gardens with their scattered trees and crops. As more than three-quarters of the Mano River Basin primary forests have been destroyed since 1990, the bats have been brought progressively into much closer and more frequent contact with villagers. In the words of a 2009 report, "Three nations have deforested more than 75 percent of their land, forcing the inexorable meet-up between Ebola-carrying bats and people."[9]

This transformation allowed Ebola to "spill over" from bats to humans in West Africa in the wake of deforestation. And it proved fatal to Emile Ouamouno, a toddler who played in the hollow of a fruit tree adjacent to his home just as a child in the developed world might climb a nearby apple tree. It was his tragic misfortune that the tree stood on the edge of his village of Méliandou, which was no longer located in a forest area but was "surrounded by a landscape strongly reshaped by plantations."[10] As a consequence of this change from forest to palm oil plantation, the hollow of the tree in which Emile chose to play—just fifty yards from his home—harbored thousands of roosting fruit bats, whose droppings were almost certainly the source of his infection.

Furthermore, high-resolution satellite data have made it possible to correlate the index cases of all known outbreaks of Ebola virus disease since 2004 with changes in patterns of land use that have occurred during the same period. The results indicate that Emile's bad luck was part of a larger tendency at work in Central and West Africa. In the twelve outbreaks of Ebola known to have occurred between 2004 and 2016, the index cases have been consistently traced to the edges of forest fragmentation and deforestation that happened during the previous two years. Eight of the twelve confirmed index cases were traced to "fragmentation hot spots." In addition, of the three apparent exceptions, one occurred very close to an area of high fragmentation, and a second was associated with hunting and poaching in the forest. Only one of the twelve index cases was a genuine outlier.

A further consequence of forest fragmentation also became apparent. In comparison with bat populations in the canopies of primary forest, the fruit-eating species that are the reservoirs of Ebola virus are overrepresented in fragmentation areas because the insect-eating species that do not transmit the disease are not drawn to the new habitat. This tendency may be the result of the destruction of insect habitat by deforestation. Therefore, not only does forest fragmentation bring humans into more frequent contact with bats, but it also ensures that such contact involves precisely those species that carry the disease. In the summary of a 2017 report:

> Our results indicate that *Ebolavirus* spillover events from wildlife reservoirs to humans preferentially occur in areas that are relatively populated and forested, yet where deforestation is reshaping the forest boundaries by increasing forest fragmentation. . . . High degrees of forest fragmentation and their increase over time can be good indicators of enhanced opportunities for human contact with wildlife . . . and, possibly, also improved habitat for some reservoir species.[11]

Human-to-Human Transmission

Ebola is highly infectious from person to person, but only through direct contact between a healthy person and an infected person's bodily fluids. The environments that patients occupy during their illness teem with viruses, including surfaces they have touched, bedclothes and linen, the insides of vehicles, and personal effects. The fact that patients are initially infective when their condition resembles flu rather than a more serious affliction increases the opportunities for the epidemic to spread, as the patient is still likely to underestimate the danger and to remain mobile rather than taking to bed. Sexual transmission and transmission from mother to child via breastfeeding are also possible for months after someone has recovered. Given these modes of transmission, the epidemic of 2013–2016 occurred most frequently via certain nodal points. Three were most significant: homes, burial grounds, and hospitals.

Patients in their homes posed a lethal danger to family, friends, and all who cared for them or entered their contaminated sickrooms. Ebola therefore initiated its onward spread not at a distance but through tightly linked networks of family members and caregivers who shared intimate domestic spaces with the sufferers. Thus Emile Ouamouno's death was soon followed by the demise of his mother, his three-year-old sister, his grandmother, a village nurse, and a midwife. The mourners at the grandmother's funeral and her caretakers then contracted Ebola.

Similar considerations made funerals and burials a second major site of transmission throughout the 2013–2016 epidemic (fig. 22.1). At no time does a victim of Ebola shed more virus particles than immediately after death. At that very time, however, local customs draw relatives and members of the community into the highly contaminated sickroom. Tradition and religion among the Kissi people of Guinea call for a series of funeral rituals that are highly dangerous in the midst of a spreading Ebola outbreak.

Figure 22.1. Gravedigger Saidu Tarawally in March 2015, Bombali Cemetery in Sierra Leone, located at the epicenter of the Ebola epidemic. Undertakers and gravediggers were at particular risk of developing the disease, as the viral load is highest just after death. (Photograph by Daniel Stowell, MPH, CDC Public Health Images Library.)

When a member of a community dies, the body is kept at home for several days in order to allow mourners time to visit and pay their last respects by touching or kissing the head of the deceased. Family members then ceremonially wash the body and wrap it in a winding sheet, and the community gathers to accompany its departed member to the grave.

Failure to observe these practices, it is thought, prevents the dead person's soul from proceeding in peace to the afterlife. Instead, the tormented spirit remains to haunt the living. Even after knowledge of the mechanisms of Ebola became more widespread, the disease was sometimes less feared than possible retribution from the dead and remorse over perceived injury to a deceased friend, neighbor, or relative. Indeed, the difficulty of establishing safe burial practices was one of the principal preoccupations of the effort to contain the disease. As Dr. Hilde de Clerck, an early responder, commented: "Often, convincing one member of the family is simply not enough. To control the chain of disease transmission it seems we have to earn the trust of nearly every individual in an affected family. This is a mammoth task, which is why greater involvement from the religious and political authorities in raising awareness about the disease is crucial."[12] For this reason, anthropologists and linguists were important as consultants to the public health movement.

The third major site for Ebola transmission was the treatment center or hospital. No job was more dangerous in West Africa during the epidemic than serving as a health-care worker—orderly, nurse, or physician. Those who staffed the front lines paid a heavy tribute to Ebola, certainly in death and disease, since 20 percent of the Ebola victims in the three countries of the Mano River Basin are estimated to have been health-care personnel. But caregivers also suffered severely from fear, overwork, and demoralization. There were many reasons for this, including the sheer infectivity and lethality of the disease. All direct contact with patients was hazardous.

But the specific conditions of the health-care systems in West Africa greatly magnified the inherent dangers involved. Guinea, Liberia, and Sierra Leone were among the world's poorest nations. The United Nations Human Development Report of 2016 estimated that Liberia ranked 177 out of 188 member nations in its "Human Development Index," which is a measure of overall economic well-being as assessed by a number of matrices. In terms of income alone, the per capita annual income of Guinean residents was US $1,058. Liberia, whose citizens earned US $683 per capita annually, was ranked 177; and Sierra Leone, with per capita earnings of $1,529, was ranked 179. The percentages of the population estimated to be living in "severe multidimensional poverty" were 49.8, 35.4, and 43.9, respectively.[13]

Poverty of such depths profoundly affected the ability of West Africa to build health-care infrastructures. This problem was compounded by the fact that the states in the region had different priorities. At the Abuja, Nigeria, summit of African health ministers in 2001, a widely applauded resolution pledged all participating countries to move rapidly toward a spending target of 15 percent of gross national product on health. But in 2014, Sierra Leone, Guinea, and Liberia lagged far behind that goal, with expenditures of only 1.9 percent, 2.7 percent, and 3.2 percent, respectively. Education, welfare, housing, and transport suffered from comparable neglect, and the only relatively well-funded institution was the army of each country. Far from moving toward the Abuja targets, Guinea and Sierra Leone actually reduced their budgets for the health ministry in the years following the conference, even though the expansion of palm oil and other industries brought economic growth and created pockets of wealth amidst the prevailing poverty. The story of Ebola is not only a narrative of poverty, but also one of the poor distribution of resources and the suspect quality of prevailing moral priorities.

As Ebola erupted, preparedness in all three countries was nonexistent. For example, there were virtually no health-care workers. West Africa had the world's fewest trained physicians, nurses, and midwives per capita. Liberia had 0.1 doctor per 10,000 citizens, and comparable figures for Sierra Leone and Guinea were 0.2 and 1.0, respectively, as contrasted, for example, with 31.9 doctors per 10,000 in France and 24.5 in the United States. A large Canadian hospital had more physicians than all of Liberia, where the severe shortage had been exacerbated by civil war at the turn of the twenty-first century. War drove a majority of the limited number of Liberian doctors away from the country; thus, as Ebola began, more Liberian physicians lived in the United States than in Liberia, where 218 doctors and 5,234 nurses remained to serve a population of 4.3 million. These caregivers, moreover, were concentrated in Monrovia, leaving the remainder of the country entirely without health-care provision except the arts of traditional healers.

Hospitals fared equally badly. They seldom possessed isolation wards, electricity, or running water and had no diagnostic facilities, protective equipment for staff, or training in response to a public health emergency. Already overcrowded, they also lacked surge capacity in the event of an emergency. In such conditions, morale was low, contributing during the emergency to large-scale desertion of hospitals by medical staff members who were frightened, poorly paid, and overworked. They were also overwhelmed by a sense of despair at their inability to help sufferers and by the atmosphere of distrust by the general population, who regarded hospitals as places to die. Thus, when the virus appeared, one *New York Times* article described the health-care system in the three affected countries as "invisible."[14]

In such an environment, it was inevitable that many of the health-care workers who remained at their posts would contract the disease. Local medical staff had none of the tools, facilities, equipment, or training to protect themselves, and their high rate of contracting Ebola themselves further undermined a health-care system that was already crumbling.

All of the risk factors for Ebola would not have generated a major epidemic, however, if the disease had remained confined to the forest region. What transformed the epidemiological history of Ebola was the fact that West African forests are intimately linked to urban centers. Having established a focus in Méliandou among a stricken child and his family, the Ebola virus was well-positioned to spread throughout Guinea and the other two countries that abutted the prefecture where the eighteen-month-old Emile fell ill. By 2013 palm oil had established extensive connectivity between the forested area and the outside world. Labor migration on the part of dispossessed peasants and plantation workers; travel by company officials, government bureaucrats, and troops from Conakry; the increasing pace of the movement of goods by river upstream and downstream; and the opening of networks of dirt roads—all of these factors made the movement of people, goods, and equipment across borders and into cities a constant feature that bound the whole of the Mano River Basin into the web of globalization.

Furthermore, the interconnectedness of the forests with urban West Africa was established by multiple industries, not palm oil alone. During the decades preceding the outbreak of 2013, multiple businesses had invaded the forests of Guinea, Liberia, and Sierra Leone. Logging companies and rubber planters sought land, and mining firms were drawn by deposits of diamonds, gold, bauxite, and iron. Construction companies, accommodating the burgeoning demand for lumber that resulted from large-scale migration and urban growth, took their share of the trees. All of these forces set people, goods, and trade in motion, both within the forested prefectures and between the forests and the outside world. In an exception to the prevailing tale of a distant jungle inhabited by bush-meat hunters, the *Irish Times* noted that the explosion of Ebola revealed exactly the opposite. "Indeed," one reporter commented, "only a tiny proportion of the wider Upper Guinea rainforest belt remains unexploited after a deforestation process that has accelerated considerably in recent decades. This has caused significant disturbance to bat populations, creating the preconditions it appears for an outbreak."[15]

Arriving as a volunteer on the front lines of the disease in the forest region, the Irish virologist Christopher Logue noted that the woods were not the unspoiled sylvan idyll he had expected. On the contrary, the landscape bore all the signs of bustling activity and commerce. The area, he wrote, "is a vast patchwork of bright green vegetation with terra-cotta-coloured dirt paths, that we later discovered were the roads, weaving in and around these forested areas, linking small villages to each other and to the network of estuaries and rivers."[16] Above all, mines were a school for migratory behavior. Probing ever deeper into the forests, they set in motion an immense movement of young men hungry for work and willing to travel.

For twelve weeks after the death of Emile Ouamouno in December 2013, Ebola circulated silently in the forested region, unseen by a health system that wasn't there. Health officials in the capitals noticed an uptick in mortality, but they attributed it to gastroenteritis and cholera, which are endemic to the region. Thus misdiagnosed, Ebola reached Conakry, Monrovia, and Freetown unopposed. Retrospective investigation determined that the disease reached Conakry, a city of 2 million people that is located about 250 miles from Méliandou, on February 1, 2014. The route the virus took—via an infected member of Emile's extended family who traveled to the capital—illustrates the connections between forest and city. Thereafter, Ebola erupted in slums that, like the hospitals within them, lacked sanitation, adequate space, and facilities of every kind. If transformation of the forest environment allowed Ebola to break out, the degraded and overcrowded conditions of the built environment of West Africa's cities enabled it to spread exponentially.

The Early Emergency Response

In March 2014 the response to Ebola began as the first diagnosed cases in Conakry attracted concern—but not on the official level. It was the private-sector charities Samaritan's Purse and, above all, Médecins Sans Frontières (MSF, or Doctors without Borders) that intervened in March. With laboratory confirmation that the "mysterious cases" reported by the ministry of health in Guinea were actually Ebola, MSF responded immediately. On March 25 the Paris-based NGO deployed sixty health-care workers immediately and dispatched tons of medical equipment and supplies to support them.

By the end of the epidemic MSF had strained its resources to a breaking point. As 2014 began, the agency had prioritized the need to confront humanitarian crises in Sudan, Syria, and the Central African Republic. Suddenly the international charity was confronted with the need to control a public health emergency on an unprecedented scale. MSF rapidly devoted itself to four principal tasks in West Africa: (1) to open and equip a network of Ebola treatment centers, (2) to staff them with volunteer medical personnel from abroad, (3) to treat victims of the disease, and (4) to contain the spread of the epidemic while it sounded the alarm in order to secure the intervention of WHO and various governments. Overwhelmed by patient numbers, MSF by midsummer took another unprecedented step. Regarding the absence of hospitals in West Africa as an "emergency within the emergency," the philanthropic organization built and equipped large-scale treatment centers where volunteers operated reception and triage units, diagnostic labs, isolation wards, and recovery rooms—all functioning within wooden sheds and tents enclosed within perimeter fencing.

MSF doctors immediately recognized the threat that Ebola posed to spiral out of control. Having reached the capital cities of three West African countries, this highly virulent and untreatable disease presented an immediate danger. Already overrunning the originally afflicted nations, it was poised to travel via the region's international airports to other countries both in Africa and abroad. MSF had never imagined confronting the emergency alone and knew that the unfolding disaster was beyond its resources and experience. The organization had been formed in 1971 to deliver, as it says on its website, "emergency aid to people affected by armed conflict, epidemics, healthcare exclusion and natural or man-made disasters." Its mission was to act as a first responder in humanitarian crises while galvanizing local governments, WHO, and First World states to assume major responsibility.

The situation in West Africa, however, was different. As the title of MSF's report on its first year dealing with Ebola in the Mano River Basin announced, the organization was "pushed to the limit and beyond." By the end of its mission, it had treated more than five thousand Ebola patients, or a quarter of the total reported by WHO. Its institutional problem was that it rapidly found itself immersed in the crisis but without a viable "exit strategy" because the international response to its cry of alarm was lamentably tardy, half-hearted, and disorganized.

In theory, WHO had responsibility for leading the campaign to contain and eliminate Ebola. In practice, however, it proved unequal to the task.

On March 31, 2014—three months after the outbreak began—MSF declared that the crisis in West Africa was an "unprecedented emergency" demanding an immediate and coordinated international effort. Far from acting, WHO engaged in a war of words with the messenger bearing bad news it preferred not to hear. Gregory Hartl, the Geneva-based agency's spokesman, minimized the escalating calamity. Safely seated at a desk in Switzerland, Hartl contradicted the MSF assessment and the view of all experts, announcing: "The fortunate thing with Ebola is that it's quite difficult to transmit. You have to touch someone. Fortunately for the greater population, the risks are quite small."[17] Furthermore, in defiance of all the evidence, WHO reported in late May that Ebola had not reached the cities of Sierra Leone and that there was no reason to station international health workers in the country.

The background to Hartl's extraordinary comments, and the inaction that followed, was that WHO was far from learning the lessons of the SARS crisis. Adopting instead the perspective of the industrial world, it decided no longer to prioritize infectious diseases. Accordingly, it drastically cut its budget for surveillance and response and dismissed senior and experienced experts in the field. The organization therefore lacked the competence, staff, and will to confront Ebola. Furthermore, WHO was paralyzed by a bureaucratic turf war between its headquarters in Geneva and the Regional Office for Africa in Brazzaville, Congo. Like the Swiss head office, the African regional office had more than halved its budget for epidemic disease preparedness from US $26 million in fiscal 2010–2011 to US $11 million in fiscal 2014–2015. When officials in Brazzaville, now without seasoned experts to advise them, chose to believe that the epidemic in West Africa was no more serious than prior outbreaks in the Congo and would soon burn itself out spontaneously, the Geneva office deferred to their views without investigation.

As people continued to die by the thousands in three countries and the epidemic spread, the standoff continued. In June 2014 MSF announced that Ebola was "out of control," with more than sixty infectious hot spots, and it again denounced what it had already criticized in March as a "global coalition of inaction."[18] In reply, WHO organized a conference of West African health ministers at Accra, the capital of Ghana, where it offered bland reassurance. Without evidence, the hapless UN spokesman opined, "This is not a unique situation—we have faced it many times—so I'm quite confident that we can handle this."[19] MSF officials were aghast. Dr. Brice de la Vigne, the MSF operations director, was appalled that "the response of the international community is almost zero."[20] Even voices outside MSF now began to be raised. The *New York Times,* for example, was scathing in its comments on WHO "leadership" as the disease worsened during July and August. The agency, it noted, "has snoozed on the sidelines for months"; its response was "shamefully slow"; and its African regional office was "ineffective, politicized, and poorly managed by staff members who are often incompetent."[21]

Governments both in the afflicted area and abroad were no more forthcoming. Locally, authorities in the three besieged countries heeded the reassurances from WHO and the supposed lessons of Ebola in the Congo, but they did so for reasons that had far more to do with economics than with either science or concern for the health of their populations. Their great fear was that the outbreak of a dreaded disease in West Africa would cause investors to rethink their commitment to the development plans that were under way; that it would cause international airlines to cancel flights to the region, crippling tourism; that lucrative kickbacks from mining companies and agrobusiness would dry up; and that the disease would taint the infected countries with the stigma of backwardness and tribal practices. Prevarication and concealment were therefore the strategies of choice.

In this spirit, President Alpha Condé of Guinea opted to paint a rose-tinted picture of events in order to avoid alarming mining and palm oil companies: his government reported only a fraction of the known and suspected cases. Instead, it gave priority to a campaign to encourage villagers to change their culinary habits by banning the sale and consumption of bush meat. He thereby lowered the public profile of the emergency, but he also eliminated the standard public health strategy of tracing contacts. Furthermore, Condé made no concerted effort to recruit personnel to staff hospitals and treatment centers. On the contrary, during the early months of the epidemic, he adopted the alternative strategy, it was said, of applying quarantine not to the virus, but to journalists who reported its predations. His police censored and warned reporters who depicted medical matters truthfully. Condé was as resolutely upbeat and optimistic as WHO. In a visit to Geneva at the end of April, when MSF had proclaimed a global emergency, Condé explained his nonchalant view to the press. "The situation," he declared, "is well in hand. And we touch wood that there won't be any more cases."[22]

Farther afield, medical authorities and political leaders in the developed world adopted a laissez-faire policy. The European Union, Russia, and China, for example, folded their hands while somber statistics accumulated and physicians on the front lines appealed for help. And everywhere, politicians looked to the United States because it was the sole remaining superpower and it possessed in abundance the resources that were needed in the

Mano River Basin. Furthermore, the Atlanta-based CDC set the international standard for all organizations intended to carry out medical surveillance and emergency epidemic response.

Inaction on the part of the United States was due to a form of public health isolationism. What America wanted to know was whether the epidemic in West Africa posed a direct threat of crossing the Atlantic and causing death, not in Monrovia or Conakry, but in New York, Houston, and Los Angeles. Until July 2014 the consensus regarding the answer to that question was a resounding "no!" Screening at West African airports, reinforced by disease surveillance in the US; the reassuring abundance of physicians and nurses; and robust American sanitary infrastructures gave the nation a sense of invulnerability. In the *New York Times,* David Quammen spoke for those who felt safely ensconced behind the bulwarks of American modernity, science, and civilization, penning an article complacently titled "Ebola Is Not the Next Pandemic." Quammen acknowledged that Ebola was a terrible and excruciating disease for its victims, but he was confident that it had no relevance to the United States. It was, frankly, a rare disease caused by the "grim and local misery" endured by a small number of Africans who "are obliged by scarcity of options to eat bats, apes and other wild creatures, found dead or captured alive."[23]

A sea change in attitudes occurred in July 2014, shattering Americans' confident sense of distance from African maladies, after two US medical volunteers, Kent Brantly and Nancy Writebol, contracted Ebola. They were the first foreigners to develop the disease and were evacuated to Emory University Hospital in Atlanta, where they received the "advanced supportive therapy" available only at technologically equipped medical centers. They also received fast-tracked and still experimental antiviral medication and symptomatic treatment to lower fever, reduce pain, and slow the vomiting and diarrhea. Both survived, but their plight attracted endless international attention. Brantly and Writebol literally brought the disease home to the United States and revealed that white people, too, were vulnerable to the deadly disease.

The illness of Brantly and Writebol was a politically transformative experience, as fear spread across the United States with the realization that the country could be in danger from Ebola. As Dr. Joanne Liu, the international president of MSF, said: "The fact that we had some foreigners infected, that drew a lot of attention. All of a sudden, people said, 'On my God, it's knocking at my doorstep.' All of a sudden, people are paying attention."[24] Opinion polls conducted in mid-August demonstrated a sea change in attitudes. By that date 39 percent of Americans polled were convinced that a large outbreak would occur in the United States, and 25 percent that either they or a family member would contract the disease. The message was reinforced throughout the summer and fall. In September and October, eight additional heath-care workers who had volunteered in West Africa contracted the disease. Then every American's worst fears materialized in September 2014. A traveler not involved in patient care, the Liberian Thomas Eric Duncan, flew to Dallas from West Africa. There he fell ill, was misdiagnosed as having sinusitis, and discharged from Texas Health Presbyterian Hospital. He was readmitted and died on October 8, but only after infecting two nurses who treated him (they both recovered). In addition, thirty European medical volunteers were infected and transported to Spain, the United Kingdom, France, Germany, and Italy for care.

CDC director Thomas Frieden also contributed to the tide of opinion favoring intervention. At the end of August he made a fact-finding trip to Liberia to assess the situation. His account of what he found was devastating. The situation, he said, was an absolute emergency, and only massive and prompt outside help could prevent disaster.

Foreign Assistance

A full 180-degree turnaround occurred in August 2014—eight months after the onset of Ebola. On the first of the month, Margaret Chan, WHO's director-general at the time, held a meeting with the presidents of Guinea, Liberia, and Sierra Leone. She informed them that the disease was moving more quickly than the efforts to contain it and warned of "catastrophic" consequences. Then, for the third time ever, WHO declared a Public Health Emergency of International Concern (PHEIC), its highest level of alarm and its call to action. After a further six long weeks of preparation, the agency set up the UN Mission for Ebola Emergency Response, which was tasked with coordinating policy and managing the campaign to defeat the disease. Officially at least, WHO had taken charge.

Other nations followed suit. In West Africa, the three presidents of the countries directly involved experienced a change of heart and declared states of emergency. They also appealed for assistance from abroad. President Ellen Johnson Sirleaf of Liberia directly implored President Barack Obama for help. The US president was stunned by a wave of criticism in Congress and in the press, which described his administration as "rudderless" and "inadequate."[25] In early September the *Washington Post* denounced a US response that it deemed "feeble" and "irresponsible." The United States had a moral imperative to act, the paper said, because no one else had the resources and organization to mount an effective and immediate campaign.[26]

Meanwhile, President Obama was convinced that the epidemic constituted a risk to the United States, first as a medical threat that could—and did—reach American shores. In addition, he realized that it could also precipitate the political failure of three West African states, and perhaps eventually their neighbors as well. Serious diplomatic, medical, and security complications therefore loomed. Ebola was no longer simply a humanitarian crisis in distant lands but a matter of national security. As the first week of September ended, Obama declared the epidemic a national security crisis and directed the Department of Defense to deploy three thousand US soldiers to Liberia in engineering and logistical support roles—a mission termed Operation United Assistance. The US Army would secure the shipment of medical supplies to the epicenter of the outbreak, and it would construct and equip large treatment facilities.

In the meantime, the CDC started training courses in Alabama to prepare volunteer medical personal in the use of personal protective equipment (PPE), which consisted of gloves, goggles, face shields, rubber gowns, biohazard overalls, and rubber boots—the modern-day equivalent of the plague costumes that physicians wore to confront the Black Death (fig. 22.2). Intensive instruction in the use of these suits was needed because wearing the outfit was difficult. Karen Wong, a medical volunteer and CDC official, compared working with protective equipment to scuba diving, which requires careful predive planning. Donning and doffing the equipment correctly while following a precise sequence was essential to preventing exposure to the virus. Furthermore, while imprisoned within the suit, caregivers were overwhelmed by heat and knew that dehydration, prostration, and oxygen hunger were immediate dangers. It was safe to wear the modern plague costume for only fifteen minutes at a time. Another challenge was that the equipment muffled sound, so it was complicated to communicate with patients and co-workers, and to avoid colliding with them and equipment. The CDC courses also provided instruction in the diagnosis and treatment of Ebola patients.

Meanwhile the World Bank, the International Monetary Fund, and UNICEF designated money to support the relief effort. More quietly than the Department of Defense, the CDC intervened directly in West Africa, initiating its largest effort ever in combating an epidemic. It mobilized and deployed teams of responders, set up diagnostic and surveillance facilities,

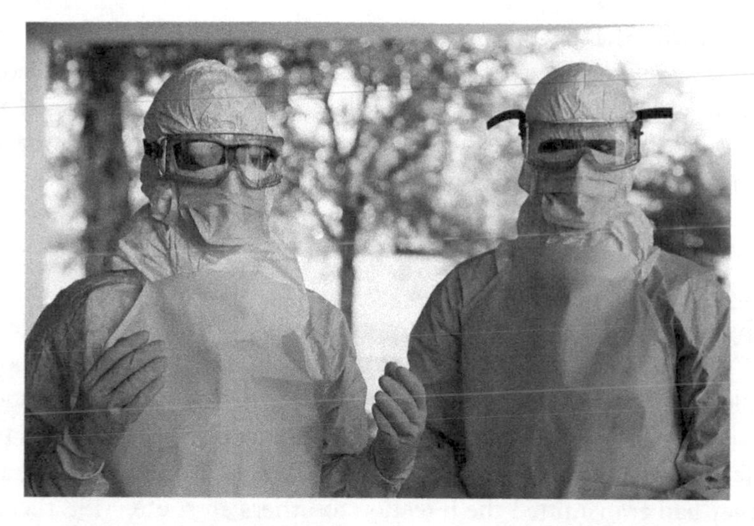

Figure 22.2. The CDC trained physicians, nurses, and other providers in the use of the personal protective equipment needed to interact with Ebola patients safely. (Photograph by Nahid Bhadelia, M.D., CDC Public Health Images Library.)

provided epidemiologists to collect and analyze statistical data, and opened instructional courses on Ebola for medical and public health personnel. It also furnished logistical support and instituted exit screening at West African airports.

Other nations made parallel interventions. France deployed response teams in Guinea, as the United Kingdom did in Sierra Leone. Canada donated supplies and medical personnel, while Cuba, Ethiopia, and China sent teams of doctors and nurses. Strikingly, the aid of Western countries followed the geography of colonial connections and present-day national interests. Thus the United States intervened to assist Liberia, which had been founded by former US slaves; the United Kingdom aided its ex-colony Sierra Leone; and France did the same in its former possession of Guinea. Cuba stood apart as a resource-poor nation that both intervened beside the major powers and sent its aid—4,651 trained doctors and nurses—without regard to national borders. By early 2015, a total of 176 organizations participated in the international effort.

Apart from the tardiness with which foreign assistance arrived, the principal criticism made regarding it was that it was given in accord with priorities determined in Washington, London, Geneva, and Paris rather than in harmony with the evolving experience on the ground in West Africa. MSF, the agency with the longest history of confronting Ebola, felt that US intervention was top-down and took little account of what MSF had learned and what its needs were at various stages of the epidemic. In the spring of 2014, for instance, MSF was primarily concerned with the inadequacy of the West African hospital system and its collapse under the strain. Indeed, at that time the agency even undertook the construction of temporary hospital complexes of its own, including the Kailahun Ebola treatment center near the epicenter of the epidemic in Liberia.

By the summer and autumn, however, MSF priorities had changed. By then it was alarmed above all by the proliferation of epidemic hot spots at multiple locations. The agency had also learned that, as a result of distance, distrust, and poor communications, patients frequently never reached treatment facilities at all, or arrived only when the disease was far advanced and after they had transmitted the infection to others en route. MSF therefore felt the need not for large hospital-like centers such as Kailahun, but rather for a multiplicity of rapid-response medical teams linked electronically with advanced diagnostic laboratories. Such teams, it argued, provided the best means of snuffing out disease foci as they erupted and before further transmission could occur. It therefore protested that US intervention was clumsy, unresponsive to the ever-changing situation in the field, and therefore largely obsolete even as it got under way in October. It was a medical version of the tendency of generals to fight the battles of the previous war. A quip made at the time was that the US had taken a knife to a gunfight.

The Anti-Ebola Campaign

Inevitably, West African governments shaped the nature of the anti-Ebola campaign as it mustered a major effort in the fall, following broadly similar trajectories in all three countries. A principal determinant of the response that unfolded was the quality and availability of the instruments at hand. One resource that the three nations deployed was communications. The governments of Guinea, Liberia, and Sierra Leone all had access to the airwaves, the press, billboards, leaflets, loudspeakers, and megaphones that could deliver messages in marketplaces where people gathered and along city streets. In the spring, the message delivered by these means was heavy-handed, bombastic, and attuned to First World perceptions rather than those rooted in West African reality.

Thus the early medical propaganda aimed to persuade the population that Ebola was real, present, and dangerous. Unfortunately, this message primarily delivered fear. Early placards proclaimed, "Ebola spreads quickly, and it kills!," thereby spreading a frenzy of Ebola phobia that was unhelpful. Terror fostered counterproductive behaviors such as the avoidance of treatment centers. It also promoted stigma toward patients, recovered sufferers, and health-care workers. State messages also stressed the need for giving up the now infamous dietary recourse to bush meat. Since such official notices provided so little genuinely helpful information, residents devised their own measures of safety—avoiding handshakes, wearing gloves whenever possible, and carrying small bottles of bleach with them.

The state also depended on the army, which many observers viewed as the most reliable tool available, rather than the health-care system, to deal with the crisis. Not surprisingly, therefore, the campaign at the outset was thoroughly militarized. Many of the coercive means adopted echoed early modern Europe's effort to defend itself against bubonic plague, such as extraordinary executive powers, sanitary cordons, quarantines, curfews, and lockdowns. Compulsory treatment facilities surrounded by troops even closely resembled lazarettos. Daniel Defoe would have found the response familiar.

Interestingly, presidents Condé, Johnson Sirleaf, and Ernest Bai Koroma implemented these policies against the advice of their own health ministries and the consensus opinion of Ebola experts. Health ministers and Ebola-trained physicians deployed exactly the same arguments that plague and cholera doctors had made in their time, reasoning that coercion would cause the epidemic to spread (1) by severing communication between the population and the state as people resorted to concealment to protect family members, (2) by causing people to take flight, (3) by triggering civil disorder and riot, and (4) by breaking the trust between communities and the medical profession. For besieged chiefs of state, however, the exceptional nature of the threat seemed to justify energetic countermeasures that gave the appearance that they were in control. There was also understandable doubt about what else to do, and generals assured presidents that they had the means to bring the situation under control. In the words of a *New York Times* reporter in Sierra Leone, "The government here is left to pursue the only means at its disposal: coercion."[27]

Liberia's Johnson Sirleaf led the way in deploying force soon after Margaret Chan's terrifying warning in early August and after WHO's PHEIC declaration. This bewildering acronym was all the more concerning because so few knew what the letters signified, except that they announced an emergency. Taking her cue, Johnson Sirleaf proclaimed a full state of emergency and deployed the police and army in force and combat gear. She then curtailed civil liberties, closed schools, banned assemblages of people, instituted a three-day work week, curbed press freedoms, and announced that Liberia's land borders were sealed. This was the launching of her "action plan"—as two journalists called it, "the toughest measures yet imposed by a West African government to halt the worst Ebola outbreak on record."[28]

Famously, Liberia then quarantined strategic urban communities, in particular the slum neighborhood in Monrovia named West Point because of its location on a sandy peninsula. There, Ebola virus was widely circulating among its seventy thousand residents, who lived in congested plywood shacks with corrugated metal roofs. These structures lacked all sanitary facilities such as running water and flushing toilets, and they faced onto unpaved streets covered in all manner of refuse.

Coincidentally, West Point was also a stronghold of opposition to Johnson Sirleaf's political party. Deploying a military cordon there on August 20, sending coast guard cutters to patrol the waters, and cutting the area off from the outside world generated fear and resistance. For some, it bore the hallmarks of a settling of political scores; for all, it entailed shortages, steep rises in the cost of necessities, and hunger. The cordon seemed especially harsh since the presidential office announced that it would remain in force for ninety days. A boiling point was reached when the quarantine was followed by the opening of a "holding camp," in effect, a lazaretto, for patients transported there from the rest of Monrovia. The perception was that West Point had been chosen for sacrifice. Corruption further heightened tensions because numerous residents were able to cross the line of soldiers by bribes or as a result of cronyism.

Such conditions remind us of similar circumstances in the history of plague and cholera, when military measures of public health gave rise to violence, as they did again in India during the late nineteenth and early twentieth centuries. They also created large-scale upheavals when implemented against cholera wherever they were attempted, from Moscow to Naples. So it is hardly surprising that violence broke out in West Point in the summer of 2014 when frightened and hungry residents were trapped behind military lines.

Especially dangerous flashpoints were the distribution areas for emergency food supplies trucked into West Point by the military. Crowds gathered in the heat, and emotions boiled as people jostled one another and pressed forward to gain access to foodstuffs in case supplies ran out. All the while they also worried about the invisible and poorly understood dangers of touching one another. Suddenly shockwaves overcame the throng as they discovered that the price of rice had mysteriously tripled, from $0.30 to $0.90 per bag, or the supply of prized necessities had in fact run short. Then, members of the crowd began to pelt the soldiers with rocks and bottles, and pistol shots rang out. Running battles ensued as young men turned in fury against soldiers they saw as their tormentors, bombarding them with every missile they could locate. Residents also stormed the holding camp, liberating patients, destroying equipment, and distributing contaminated mattresses, linen, and instruments that they seized in their frenzy. Ebola thus found new means of spreading. The security forces reasserted control with baton charges, tear gas, and sudden bursts of rifle fire that left people wounded and bleeding on the ground.

Tensions were not limited to West Point and greater Monrovia. Concerned by proliferating epidemic hot spots, as MSF was, the government decided to root out and sequester suspiciously ill patients who had not been reported to the authorities. To implement this plan, Johnson Sirleaf declared a nationwide lockdown and dusk-to-dawn curfew beginning September 19. To enforce the action plan, long convoys of troops fanned out across the country, setting up roadblocks and checkpoints, where they stopped everyone, monitored their temperatures, and took into custody anyone with a temperature higher than 98.6°F. Armed platoons patrolled the streets to detain anyone found out of doors in violation of the lockdown. Then seven thousand teams of health officials and community workers, sardonically dubbed "health sensitizers," set out on their mission. Accompanied by the police, they performed house-to-house searches to hunt down unreported patients in their hiding places. Meanwhile, the military stationed guards at treatment centers to prevent patients and people taken into compulsory surveillance from escaping.

In the countryside the resistance was smaller in scale and attracted less press coverage than riots in the slums of Monrovia, but it was no less tenacious. The international media often chose to portray opposition as the backward-looking resistance of an illiterate populace to modern medicine and science and its atavistic attachment to ancient rituals and tribal customs. The arrival of armed soldiers sent from the capital, however, built on the tensions that surrounded land enclosure. Outsiders, particularly when armed, were deeply suspect, given a long history of unhappy encounters with officialdom and a searing memory of the civil conflicts of the recent past. As in West Point, there were especially sensitive flashpoints in rural areas. The most important was burial. New state decrees included provisions that the dead be unceremoniously disinfected, packed into double body bags, and hastily buried—normally in unmarked graves—by officially appointed gravediggers wearing protective equipment. This new regulation prevented family members and friends from honoring loved ones, and it negated religious observance. The discovery of a body by a search team thus furnished ample potential for physical confrontations, just as a similar decree had led to clashes in plague-stricken Bombay in 1897–1898.

This tense atmosphere was inflamed by multiple conspiracy theories. One Canadian reporter wrote that people "tell me stories about witchcraft, Ebola witch guns, crazy nurses injecting neighbours with Ebola and government conspiracies."[29] *Untori*, or plague spreaders, were said to be at work, as in the days of the Black Death described by Alessandro Manzoni. Some regarded health-care workers as cannibals or harvesters of body parts for the black market in human organs. The state, rumor also held, had embarked on a secret plot to eliminate the poor. Ebola perhaps was not a disease but a mysterious and lethal chemical. Alternatively, the ongoing land grab was deemed to have found ingenious new methods. Perhaps whites were orchestrating a plan to kill African blacks, or mine owners had discovered a deep seam of ore nearby and wanted to clear the surrounding area.

With this background, resistance in many forms flared up, not as the pitched battles of West Point, but as the guerrilla actions of small rural communities. Villagers erected barriers to prevent army vehicles from advancing, and they fired upon all who approached. Elsewhere, armed with machetes, terrified peasants raided treatment centers to snatch away their relatives, killing or wounding the staff and all who opposed them. In other places, afraid of quarantine more than of the spirits of the dead, people brought out bodies and left them in the streets so that they could not be traced as the deceased patients' contacts. In a number of villages, residents attacked burial teams, forced them to drop their body bags, and chased them away. Everywhere people avoided seeking medical treatment and hid whatever ailments they had in order not to be taken into custody.

In two respects, the preponderance of evidence suggests that popular resistance prevailed. First, it is clear that the state did not emerge from its "action plan" better informed about the true extent of Ebola than it had been before. Second, the action plan did not run its course of ninety days but was set aside in October because it was recognized to be ineffective and counterproductive. Coercion threatened to complicate the task of governance, and it was of no visible use in containing the epidemic. In September and October, the graphs of mortality and morbidity spiked as Ebola reached its height instead of falling.

Even more persuasively, perhaps, after October, coercion lost its raison d'être. The belated but massive international effort to support, and substitute for, local health systems that had collapsed arrived just in time. In August and September the consensus among agencies assessing Ebola in West Africa was that the disease had reached a tipping point where it was poised to escalate out of all control and to create a far larger international pandemic. Already overcrowded treatment centers found themselves forced to turn away infected patients. Joanne Liu commented: "It is impossible to keep up with the sheer number of infected people pouring into facilities. In Sierra Leone, infectious bodies are rotting in the streets. Rather than building new care centers in Liberia, we are forced to build crematoria."[30]

The sudden intervention of major outside powers with trained medical personnel, diagnostic facilities, protective equipment, and an array of well-supplied and well-staffed treatment centers transformed the situation. In October it was possible to abandon coercion and turn instead to strategies based on science—rapid diagnosis, contact tracing, and isolation. In addition, it proved important to persuade communities of the wisdom of having special teams take over burial functions to provide what was termed "safe and dignified burials." Wearing the personal protective equipment, they disinfected and bagged dead bodies. MSF had practiced these functions from the outset, but its resources were unequal to the scale of the emergency.

Results were quickly forthcoming. By November 2015 it became apparent that the international effort was beginning to break chains of transmission. The incidence of new cases trended downward for the first time, and mortality figures declined with them. This downward trend continued without interruption. In the spring of 2015 the campaign was largely one of extinguishing the remaining foci rather than seeking any longer to contain an expanding epidemic. By May Liberia considered itself the first West African country to be free of Ebola. Unfortunately, the announcement was premature as several clusters of cases flared up subsequently, and only at the end of the year was the disease genuinely eliminated. On January 14, 2016, Liberia declared victory. The other two countries followed in its wake. Sierra Leone declared victory on March 7 and Guinea in June. A milestone was passed on March 29 when WHO lifted its PHEIC. Then in December 2016 the United Nations declared the epidemic officially over.

Effects of the Epidemic

From 1976 to 2014, Ebola has appeared in various areas of West and Central Africa (fig. 22.3). But the impact of the 2014 epidemic on West Africa was particularly immense on multiple fronts, most obviously in terms of the burden of death and suffering it imposed on the three nations it ravaged. Patients who survived are often still suffering from the enduring effects of post-Ebola syndrome, and many thousands lost husbands, wives, parents, and other family members. But the indirect medical costs are perhaps far greater, as the epidemic destroyed the already inadequate health-care systems of all three countries.

When Ebola arrived in the Mano River Basin, it forced the closing of the few hospitals and clinics that existed; it decimated the tiny numbers of trained health-care personnel the region possessed; and it entirely monopolized the time and energy of all health-care personnel. As a result, all medical services other than those directed toward Ebola were suspended. Vaccinations were not administered to children, and mathematical models have suggested that as many as sixteen thousand children may have died as a result. At the same time trauma victims injured in road traffic or industrial accidents were turned away; pregnant women received no care before, during, or after birthing; and the campaigns set up to confront other infectious diseases—especially malaria, tuberculosis, and HIV/AIDS—were halted. Those great infections, already prevalent in West Africa, experienced a fearful upsurge during the two years of Ebola. Since agriculture was decimated by the dislocations imposed by both the disease and the coercive measures to combat it, and since real wages tumbled in three of the world's poorest nations, hunger and malnutrition resulted, severely compromising immune systems and preventing normal childhood development. These costs cannot be measured with accuracy, but health officials concur that they are several times higher than the direct costs of fighting Ebola. Maternal deaths alone, resulting from the lack of medical care, are thought to be severalfold higher than the deaths that Ebola caused directly.

Clearly, however, not all costs are medical. The economic effects of the epidemic were also profound. Economists estimate that the direct cost of containing and extinguishing the epidemic of 2013–2016 was approximately US $4.3 billion. But that figure does not measure important secondary effects. Some sectors of the economy were devastated, with tourism the clearest and most evident example. During the epidemic many airlines, including British, Emirates, and Kenya Airways, canceled their flights for a period, and

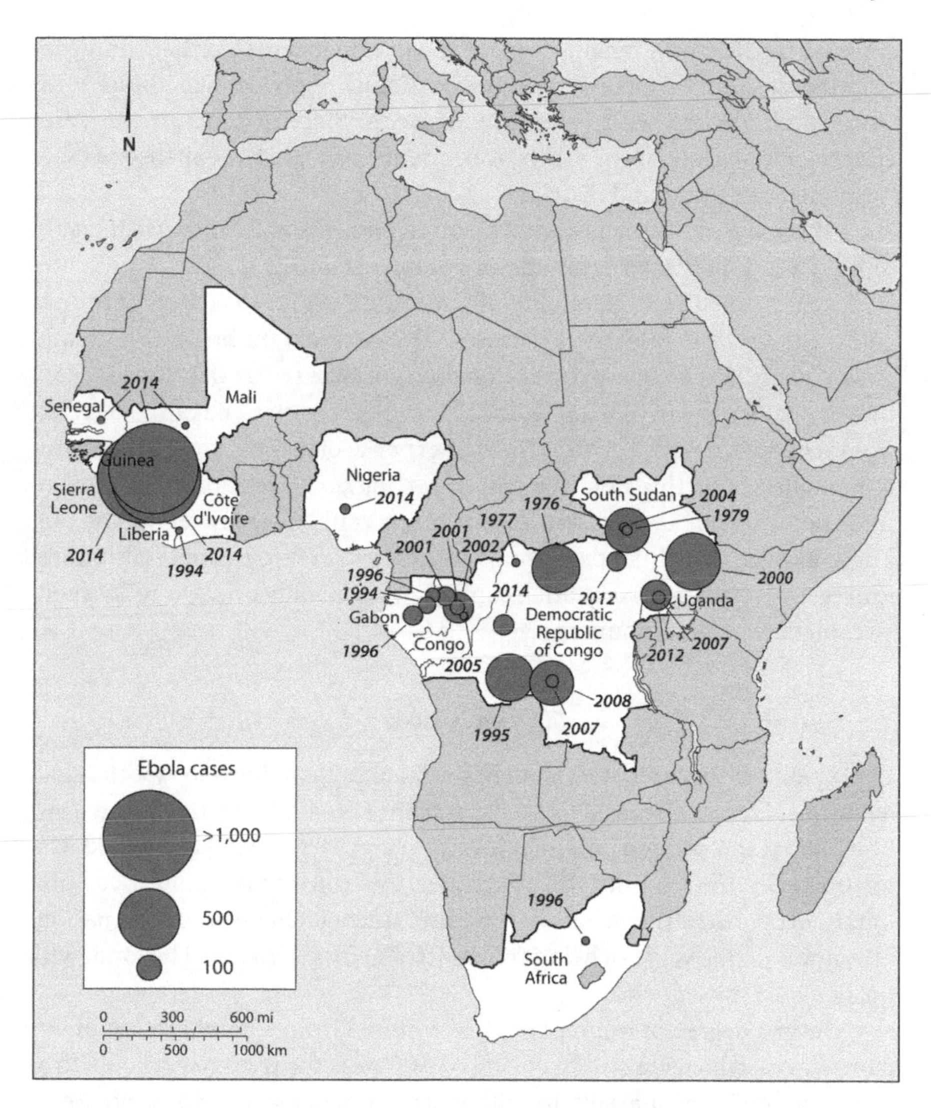

Figure 22.3. Ebola in Africa, 1976–2014. (Map adapted by Bill Nelson.)

travelers, advised by their governments and by common sense, stayed away. Similarly, investment dried up, with significant consequences for employment, growth, and foreign exchange. Businesses closed to protect their employees, and retailers lost their customers. Agriculture was so devastated that production levels were halved in 2015 and unemployment surged, together with poverty and inequality. Since the preexisting health-care infrastructure was shattered, the states also face the expenses of rebuilding hospitals, training medical personnel to replace those who had died or fled, and confronting the needs of a population that is further impoverished, undernourished, and stricken with infectious diseases. Since schools in the three countries were closed for a year, governments also face the expense of making good the educational deficit.

It is too early to predict what long-term results will follow with regard to such additional considerations as political stability, the development of civil society, and the vulnerability of the countries to civil strife and the ongoing challenge of infectious diseases. The renewed outbreak of Ebola in 2018 is a reminder that, especially in nations of such severe deprivation, continued challenges are inevitable. Indeed, in that context one of the bitterest ironies of the 2013–2016 crisis is that the expense of combatting the epidemic is estimated to be threefold the cost of setting up a functioning health infrastructure. Such an infrastructure perhaps could have prevented the outburst altogether while providing access to care for other afflictions. Emergency response to contain a conflagration already under way is expensive, inefficient, and inhumane.

Conclusion

Ebola painfully exposed the extent of global unpreparedness to face the challenge of epidemic disease despite the warning provided by SARS. But however heavy the burden of suffering in West Africa was, the world was fortunate that the calamity was not greater. By a consensus of informed opinion, Ebola reached the brink of spreading uncontrollably and internationally; it was on the verge of being transmitted across Africa and beyond, with incalculable consequences.

Such a degree of unpreparedness resulted from a combination of circumstances, which are still in effect today. One is the treatment of health as a commodity in the market rather than as a human right. Well before Ebola erupted, market decisions prevented West Africa from having tools to confront the emergency. Pharmaceutical companies prioritize treating the chronic diseases of industrialized nations, where profits are to be made, over the development of drugs and vaccines for the infectious diseases of the impoverished. As a result, tools to deal with diseases like Ebola lag far behind in the pipeline.

A further consequence of the perspective of health-for-profit was painfully evident in 2013–2016. This was the absence of functioning health-care systems accessible to everyone. Ebola circulated silently for months in West

Africa because no means of surveillance were in place. A public health infrastructure and guaranteed access to it are the essential means needed to sound the alarm, provide timely information, isolate infectious cases, and administer treatment. In Guinea, Liberia, and Sierra Leone, no sentinels had been posted, so Ebola, undetected for months, circulated freely.

Treating health as a commodity implies that decisions affecting the life and health of millions are placed in the hands of politicians whose power depends on generating development, trade, and profit. In theory, the nations of West Africa espoused the lofty goal of health for all as embodied in idealistic declarations such as the Millennium Development Goals of 2000, and they pledged themselves to creating health infrastructures at the Abuja conference of 2001. Those objectives were dear to spokespeople for public health, and to medical and humane interests. For political leaders, however, the very different principles enunciated by the World Bank, the International Monetary Fund, and the G8 countries—economic growth, privatization, and unfettered markets rather than public spending—took precedence. In practice, therefore, public health was abandoned. The siren song of military expenditure further completed the diversion of resources away from the construction of a robust health-care infrastructure, leaving West Africa perilously vulnerable.

Finally, Ebola was able to cause an epidemic because of the prevailing illusion that national borders matter in a globalized medical environment. As epidemic disease erupted in the "distant" Mano River Basin, the developed world slumbered in the peaceful belief that disease in Africa was a humanitarian issue at most, not one that raised the dismaying prospect that lives everywhere were directly at stake. But epidemic diseases are an ineluctable part of the human condition, and modernity, with its vast population, teeming cities, and rapid means of transport between them, guarantees that the infectious diseases that afflict one country have the potential to affect all. The public health disaster of West Africa was founded on the failure to make decisions regarding health from the perspective of the sustainable welfare of the human species as a whole rather than the unsustainable interests of individual nations. To survive the challenge of epidemic disease, humanity must adopt an internationalist perspective that acknowledges our inescapable interconnectedness.

The analysis here leads to a disconcerting corollary. The ongoing assault on tropical forests in Central and West Africa explains the fact that, since the emergence of Ebola in 1976, outbreaks of the disease have become more numerous and larger in scale. Nor is there any indication of a halt to the progression. Indeed, at the time of this book's conclusion in the fall of 2018, the Congo was experiencing yet another outbreak that was rapidly becoming the most extensive in the history of the country. This upsurge began on August 1, 2018—this time in the northeast province of Kivu where the country's borders intersect with those of Rwanda, Burundi, and Uganda.

A hopeful factor in the emergency anti-Ebola response is the availability of a trial vaccine that offers promise and is being administered to health-care personnel and people at serious risk. Unfortunately, however, the development of such a potentially useful tool is outweighed by powerful negative considerations in addition to the continuing advance of deforestation and the unpreparedness of Congo for a humanitarian emergency. One of these is the presence in Kivu Province of a million refugees from civil disorder. This large population is mobile, highly susceptible to infectious disease, and far beyond the surveillance of a fragile and crumbling health-care system. A further discouraging factor is the fact that Kivu is a war zone torn by strife among rival militias that make the attempt to provide medical care dangerous and largely impracticable. Indeed, the CDC has found it necessary to withdraw its emergency response personnel because they have come under fire and their safety cannot be guaranteed. For these reasons, the virologist Robert Redfield, the director of the CDC at the end of 2018, has warned that he fears two possible consequences that cannot be excluded. One is that, by escaping all control, Ebola for the first time may establish its presence as an endemic disease in Central Africa—with consequences that are unknowable. Redfield's second anxiety is that the epidemic may spread beyond the Congo, with serious international repercussions. It seems likely that the human experience with Ebola virus disease is far from over.

For these reasons, the experience of Ebola clearly indicates three initial steps that urgently need to be taken to prepare for the inevitable—and possibly far greater—next health challenge, whether from Ebola virus or from a different microbe. The first is the establishment of functioning health-care systems everywhere. As former CDC director William Foege argues, public health is the protection of the health of all, and it therefore implies measures of social justice. Second, it is essential to ensure direction and coordination from an internationalist perspective through a well-funded, competently staffed, and ever-vigilant World Health Organization. The West African epidemic revealed that neither measure has yet been implemented and that, in their absence, the world runs a severe risk of tragic and avoidable suffering.

Lastly, the relationship between the global international system and public health cannot be ignored. An economic system that neglects what economists euphemistically call "negative externalities" will ultimately exact a heavy cost in terms of public health. Chief among these externalities are the negative effects of certain models of development on the relationship between human beings and their natural and societal environments. The establishment of oil palm monocropping and chaotic, unplanned urbanization in West and Central Africa are just two examples among many. Epidemic diseases are not random events. As we have seen throughout this book, they spread along fault lines marked by environmental degradation, overpopulation, and poverty. If we wish to avoid catastrophic epidemics, it will therefore be imperative to make economic decisions that give due consideration to the public health vulnerabilities that result and to hold the people who make those decisions accountable for the foreseeable health consequences that follow. In the ancient but pertinent wisdom, *salus populi suprema lex esto*—public health must be the highest law—and it must override the laws of the marketplace.

Notes

CHAPTER 2
Humoral Medicine

1. Homer, *The Iliad*, Book I, trans. Samuel Butler, http://classics.mit.edu/Homer/iliad.1.i.html, accessed September 20, 2017.
2. "The Rev. Jerry Falwell," *Guardian*, May 17, 2007, https://www.theguardian.com/media/2007/may/17/broadcasting.guardianobituaries.
3. "Luther's Table Talk," Bartleby.com, https://www.bartleby.com/library/prose/3311.html, accessed August 16, 2018.
4. Hippocrates, "On the Sacred Disease," trans. Francis Adams, http://classics.mit.edu/Hippocrates/sacred.html, accessed September 17, 2017.
5. Charles-Edward Amory Winslow, *The Conquest of Epidemic Disease: A Chapter in the History of Ideas* (Princeton: Princeton University Press, 1943), 55–56.
6. Vivian Nutton, "Healers and the Healing Act in Classical Greece," *European Review* 7, no. 1 (February 1999): 31.
7. Quoted in Vivian Nutton, "The Fortunes of Galen," in R. J. Hankinson, ed., *The Cambridge Companion to Galen* (Cambridge: Cambridge University Press, 2008), 361.
8. Ibid., 355.

CHAPTER 3
Overview of the Three Plague Pandemics

1. Procopius, Medieval Sourcebook: Procopius: The Plague, 542, "History of the Wars, II.xxi–xxxiii" (scanned from *History of the Wars,* trans. H. B. Dewing, Loeb Library of the Greek and Roman Classics [1914]), https://sourcebooks.fordham.edu/source/542procopius-plague.asp, accessed September 20, 2017.
2. William Chester Jordan, *The Great Famine: Northern Europe in the Early Fourteenth Century* (Princeton: Princeton University Press, 1997), 24.
3. Per Lagerås, *Environment, Society and the Black Death: An Interdisciplinary Approach to the Late-Medieval Crisis in Sweden* (Oxford: Oxbow, 2016), 8.
4. Ibid., 7.

CHAPTER 4
Plague as a Disease

1. Procopius, Medieval Sourcebook: Procopius: The Plague, 542, "History of the Wars, II.xxi–xxxiii" (scanned from *History of the Wars,* trans. H. B. Dewing, Loeb Library of the Greek and Roman Classics [1914]), https://sourcebooks.fordham.edu/source/542procopius-plague.asp, accessed September 20, 2017.

2. Quoted in Andrew Cunningham and Ole Peter Grell, *The Four Horsemen of the Apocalypse: Religion, War, Famine and Death in Reformation Europe* (Cambridge: Cambridge University Press, 2000), 283.
3. See, for instance, this description in 1903 in Giles F. Goldsbrough, ed., *British Homeopathic Society* 11 (London, 1903), 256; and also in 2012, Theresa J. Ochoa and Miguel O'Ryan, "Etiologic Agents of Infectious Diseases," in *Principles and Practice of Pediatric Infectious Diseases,* 4th ed. (Elsevier, 2012) (see ScienceDirect, "Bubo," https://www.sciencedirect.com/topics/medicine-and-dentistry/bubo, accessed August 17, 2018).
4. Jane L. Stevens Crawshaw, *Plague Hospitals: Public Health for the City in Early Modern Venice* (Farnham: Ashgate, 2012), 143.
5. Rodrigo J. Gonzalez and Virginia L. Miller, "A Deadly Path: Bacterial Spread during Bubonic Plague," *Trends in Microbiology* 24, no. 4 (April 2016): 239–241, https://doi.org/10.1016/j.tim.2016.01.010.
6. Rachel C. Abbott and Tonie E. Rocke, *Plague,* U.S. Geological Survey Circular 1372, National Wildlife Health Center 2012, p. 7, https://pubs.usgs.gov/circ/1372/, last modified November 23, 2016.
7. Michael of Piazza quoted in Susan Scott and Christopher J. Duncan, *Return of the Black Death: The World's Greatest Serial Killer* (Chichester: Wiley, 2004), 14–15.
8. Roger D. Pechous, Vijay Sivaraman, Nikolas M. Stasulli, and William E. Goldman, "Pneumonic Plague: The Darker Side of Yersinia pestis," *Trends in Microbiology* 24, no. 3 (March 2016): 194, 196.
9. M. Drancourt, "Finally Plague Is Plague," *Clinical Microbiology and Infection* 18, no. 2 (February 2012): 105.
10. Giovanni Boccaccio, Medieval Sourcebook: Boccaccio: The Decameron—Introduction (scanned from *The Decameron,* trans. M. Rigg [London, 1921]), https://sourcebooks.fordham.edu/source/boccacio2.asp, accessed August 18, 2018.

CHAPTER 5
Responses to Plague

1. Daniel Defoe, *Journal of the Plague Year* (Cambridge: Chadwyck-Healey, 1996), 111–112.
2. Prayer from late-medieval primers quoted in Rosemary Horrox, ed., *The Black Death* (Manchester: Manchester University Press, 1994), 125.

CHAPTER 6
Smallpox before Edward Jenner

1. Quoted in Michael B. A. Oldstone, *Viruses, Plagues, and History* (Oxford: Oxford University Press, 2000), 8.
2. Donald R. J. Hopkins, *Princes and Peasants: Smallpox in History* (Chicago: University of Chicago Press, 1983), 3.
3. Quoted in C. W. Dixon, *Smallpox* (London: J. & A. Churchill, 1962), 8–11.

CHAPTER 7
The Historical Impact of Smallpox

1. Thomas Babington Macaulay, *The Complete Works of Lord Macaulay*, vol. 8 (Boston: Houghton, Mifflin, 1900), 272.
2. Charles Dickens, *Bleak House* (London: Bradbury and Evans, 1953), 354.
3. William Makepeace Thackeray, *The Works of William Makepeace Thackeray*, vol. 14: *Henry Esmond* (New York: George D. Sproul, 1914), 91.
4. Ibid., 103.
5. Edward Jenner, *On the Origin of the Vaccine Inoculation* (London: D. N. Shury, 1801), 8.
6. Quoted in Sam Kean, "Pox in the City: From Cows to Controversy, the Smallpox Vaccine Triumphs," *Humanities* 34, no. 1 (2013), https://www.neh.gov/humanities/2013/januaryfebruary/feature/pox-in-the-city.
7. United States Congress, Committee on Appropriations, Subcommittee on Departments of Labor, Health and Human Services, Education, and Related Agencies, *Global Eradication of Polio and Measles*, S. Hrg. 105-883, Special Hearing, United States Senate, One Hundred Fifth Congress, Second Session (Washington, DC: US Government Printing Office, 1999), 2.

CHAPTER 8
War and Disease

1. "The Haitian Declaration of Independence: 1804," Duke Office of News & Communications, https://today.duke.edu/showcase/haitideclaration/declarationstext.html, accessed August 21, 2018.
2. *Life and Correspondence of Robert Southey*, vol. 2, 1850, quoted in Flávia Florentino Varella, "New Races, New Diseases: The Possibility of Colonization through Racial Mixing in *History of Brazil (1810–1819)* by Robert Southey," *História, Ciencias, Saúde-Manguinhos* 23, suppl. 1 (2016), http://www.scielo.br/scielo.php?pid=S0104-59702016000900015&script=sci_arttext&tlng=en.
3. Robin Blackburn, "Haiti, Slavery, and the Age of the Democratic Revolution," *William and Mary Quarterly* 63, no. 4 (2006): 647–648.
4. Quoted in Philippe R. Girard, "Caribbean Genocide: Racial War in Haiti, 1802–4," *Patterns of Prejudice* 39, no. 2 (2005): 144.
5. Laurent Dubois, *Avengers of the New World: The Story of the Haitian Revolution* (Cambridge, MA: Harvard University Press, 2004), 113.
6. "Decree of the National Convention of 4 February 1794, Abolishing Slavery in All the Colonies," Liberty, Equality, Fraternity, https://chnm.gmu.edu/revolution/d/291/, accessed August 21, 2018.
7. Quoted in Girard, "Caribbean Genocide," 145–146.
8. Quoted in Philippe R. Girard, "Napoléon Bonaparte and the Emancipation Issue in Saint-Domingue, 1799–1803," *French Historical Studies* 32, no. 4 (Fall 2009): 604.
9. John S. Marr and John T. Cathey, "The 1802 Saint-Domingue Yellow Fever Epidemic and the Louisiana Purchase," *Journal of Public Health Management Practice* 19, no. 1 (2013): 79.
10. "History of Haiti, 1492–1805: General Leclerc in Saint-Domingue," https://library.brown.edu/haitihistory/9.html, last updated October 27, 2015.

11. Gilbert, *Histoire médicale de l'armée française en l'an dix; ou mémoire sur la fièvre jaune* (Paris: Guilleminet, 1803), 55.
12. Quoted in Girard, "Napoléon Bonaparte," 614.
13. Philippe R. Girard, *The Slaves Who Defeated Napoleon: Toussaint Louverture and the Haitian War of Independence* (Tuscaloosa: University of Alabama Press, 2011), 165.
14. Quoted in Girard, "Napoléon Bonaparte," 615.
15. Quoted in Girard, *The Slaves Who Defeated Napoleon*, 272.

CHAPTER 9
War and Disease

1. Eugene Tarle, *Napoleon's Invasion of Russia, 1812* (New York: Oxford University Press, 1942), 3.
2. Quoted in ibid., 54.
3. Quoted in ibid., 46–47.
4. Quoted in J. Christopher Herold, ed., *The Mind of Napoleon: A Selection of His Written and Spoken Words* (New York: Columbia University Press, 1955), 270.
5. Philippe de Ségur, *History of the Expedition to Russia Undertaken by the Emperor Napoleon in the Year 1812*, vol. 1 (London, 1840), 135.
6. Raymond A. P. J. de Fezensac, *A Journal of the Russian Campaign of 1812*, trans. W. Knollys (London, 1852), 38.
7. Ségur, *History of the Expedition*, 258.
8. Carl von Clausewitz, *The Campaign of 1812 in Russia* (London: Greenhill, 1992), 11–12.
9. Fezensac, *Journal*, 39.
10. Ségur, *History of the Expedition*, 258.
11. Ibid., 233.
12. Dominique Jean Larrey, *Memoir of Baron Larrey* (London, 1861), 120.
13. George Ballingall, *Practical Observations on Fever, Dysentery, and Liver Complaints as They Occur among the European Troops in India* (Edinburgh, 1823), 49.
14. Ségur, *History of the Expedition*, 195.
15. Ibid., 184.
16. Stephan Talty, *The Illustrious Dead: The Terrifying Story of How Typhus Killed Napoleon's Greatest Army* (New York: Crown, 2009), 156.
17. Leo Tolstoy, *The Physiology of War: Napoleon and the Russian Campaign*, trans. Huntington Smith (New York, 1888), 41–43.
18. Ségur, *History of the Expedition*, 339.
19. Fezensac, *Journal*, 53.
20. Tarle, *Napoleon's Invasion*, 201.
21. Tolstoy, *Physiology of War*, 56–57.
22. Ségur, *History of the Expedition*, 79.
23. Jean Baptiste François Bourgogne, *Memoirs of Sergeant Bourgogne (1812–1813)*, trans. Paul Cottin and Maurice Henault (London: Constable, 1996), 56–57.
24. Tolstoy, *Physiology of War*, 84.
25. Larrey, *Memoir*, 135.
26. Ségur, *History of the Expedition*, 231.

27. Talty, *Illustrious Dead,* 205.
28. Quoted in Adam Zamoyski, *1812: Napoleon's Fatal March on Moscow* (London: Harper Collins, 2004), 51.
29. Larrey, *Memoir,* 167.
30. D. Campbell, *Observations on the Typhus, or Low Contagious Fever, and on the Means of Preventing the Production and Communication of This Disease* (Lancaster, 1785), 35.
31. Quoted in Talty, *Illustrious Dead,* 167.
32. Rudolf Carl Virchow, *On Famine Fever and Some of the Other Cognate Forms of Typhus* (London, 1868), 9.
33. Fezensac, *Journal,* 88, 126.
34. Ibid., 148–149.
35. Bourgogne, *Memoirs,* 77.
36. Charles Esdaile, *Napoleon's Wars: An International History, 1803–1815* (London: Allen Lane, 2007), 13–14.

CHAPTER 10
The Paris School of Medicine

1. Quoted in Asbury Somerville, "Thomas Sydenham as Epidemiologist," *Canadian Public Health Journal* 24, no. 2 (February 1933), 81.
2. Quoted in Charles-Edward Amory Winslow, *The Conquest of Epidemic Disease: A Chapter in the History of Ideas* (Princeton: Princeton University Press, 1943), 166.
3. George Weisz, "Reconstructing Paris Medicine: Essay Review," *Bulletin of the History of Medicine* 75, no. 1 (2001): 114.
4. Abstract of "Inaugural Lecture at the Paris School of Medicine by M. Gubler, Professor of Therapeutics," *Lancet* 93, no. 2382 (1869): 564–565.
5. Eugène Sue, *The Mysteries of Paris,* vol. 3 (London, 1846), 291–292.

CHAPTER 11
The Sanitary Movement

1. Edwin Chadwick, *Report on the Sanitary Condition of the Labouring Population of Great Britain,* ed. M. W. Flinn (Edinburgh: Edinburgh University Press, 1965; 1st ed. 1842), 210.
2. Thomas Southwood Smith, *Treatise on Fever* (Philadelphia, 1831), 205, 212.
3. Ibid., 206.
4. Chadwick, *Sanitary Report,* 80.
5. Ibid., 81.
6. Ibid., 84–85.
7. Ibid., 266–267.
8. Ibid., 268.
9. Quoted in Socrates Litsios, "Charles Dickens and the Movement for Sanitary Reform," *Perspectives in Biology and Medicine* 46, no. 2 (Spring 2003): 189.

CHAPTER 12
The Germ Theory of Disease

1. John Snow, *On the Mode of Communication of Cholera* (1855), available at UCLA Department of Epidemiology, Fielding School of Public Health, "The Pathology of Cholera Indicates the Manner in Which It Is Communicated," http://www.ph.ucla.edu/epi/snow/snowbook.html.
2. Ibid., "Instances of the Communication of Cholera through the Medium of Polluted Water in the Neighborhood of Broad Street, Golden Square," http://www.ph.ucla.edu/epi/snow/snowbook2.html.
3. Quoted in Emily C. Parke, "Flies from Meat and Wasps from Trees: Reevaluating Francesco Redi's Spontaneous Generation Experiments," in *Studies in History and Philosophy of Science, Part C: Studies in History and Philosophy of Biological and Biomedical Sciences* 45 (March 2014): 35.
4. Quoted in Robert Gaynes, *Germ Theory: Medical Pioneers in Infectious Diseases* (Washington, DC: ASM, 2011), 155.
5. Quoted in Nancy Tomes, *The Gospel of Germs: Men, Women, and the Microbe in American Life* (Cambridge, MA: Harvard University Press, 1998), 26–27.
6. Thomas Schlich, "Asepsis and Bacteriology: The Realignment of Surgery and Laboratory Science," *Medical History* 56, no. 3 (July 2012), 308–334. Available at https://www.ncbi.nlm.nih.gov/pmc/articles/PMC3426977/.
7. Quoted in Tomes, *Gospel of Germs,* 184.

CHAPTER 13
Cholera

1. A. J. Wall, *Asiatic Cholera: Its History, Pathology and Modern Treatment* (London, 1893), 39.
2. Frank Snowden, *Naples in the Time of Cholera* (Cambridge: Cambridge University Press, 1995), 17.
3. Mark Twain, *Innocents Abroad* (Hartford, CT, 1869), 316.
4. Axel Munthe, *Letters from a Mourning City,* trans. Maude Valerie White (London: J. Murray, 1887), 35.
5. "The Sanitary Condition of Naples," *London Times,* September 27, 1884.
6. Report on the budget of 1881, July 14, 1881, *Atti del consiglio comunale di Napoli,* 1881, 371.
7. Giuseppe Somma, *Relazione sanitaria sui casi di colera avvenuti in sezione Porto durante la epidemia dell'anno 1884* (Naples, 1884), 4; "Plague Scenes in Naples," *New York Times,* September 14, 1884.
8. "Un mois à Naples pendant l'épidémie cholérique de 1884," *Gazette hebdomadaire des sciences médicales de Montpellier* 7 (1885): 125.
9. "Southern Italy," *London Times,* September 4, 1884.
10. Quoted in Roger Atwood, "Cholera Strikes 1500 a Day in Peru," *Los Angeles Times,* March 24, 1991, p. A8.
11. Quoted in "Peruvian Cholera Epidemic Spreading through Lima Slums," *Globe and Mail,* February 15, 1991, p. A12.
12. Nathaniel C. Nash, "Cholera Brings Frenzy and Improvisation to Model Lima Hospital: Amid Poverty, a Disease Is Growing Fast," *New York Times,* February 17, 1991, p. 3.

13. United States Congress, House Committee on Foreign Affairs, Subcommittee on Western Hemisphere Affairs, *The Cholera Epidemic in Latin America,* Hearing before the Subcommittee on Western Hemisphere Affairs of the Committee on Foreign Affairs, House of Representatives, One Hundred Second Congress, First Session, May 1, 1991 (Washington, DC: US Government Printing Office, 1991).
14. Quoted in Nathaniel C. Nash, "Fujimori in the Time of Cholera: Peru's Free Fall," *New York Times*, February 24, 1991, p. E2.
15. Quoted in Atwood, "Cholera Strikes 1500 a Day in Peru."
16. "Fact Sheet: Cholera," World Health Organization, February 1, 2018, http://www.who.int/en/news-room/fact-sheets/detail/cholera.
17. UN Offices for the Coordination of Humanitarian Affairs, "Haiti: Cholera Figures (as of 27 April 2018)," April 27, 2018, ReliefWeb, https://reliefweb.int/report/haiti/haiti-cholera-figures-27-april-2018.
18. J. Glenn Morris, Jr., "Cholera—Modern Pandemic Disease of Ancient Lineage," Emerging Infectious Diseases 17, no. 11 (November 2011): 2099–2104, https://wwwnc.cdc.gov/eid/article/17/11/11-1109_article.

CHAPTER 14
Tuberculosis in the Romantic Era of Consumption

1. Maurice Fishbert, *Pulmonary Tuberculosis*, 3rd ed. (Philadelphia: Lea & Febiger, 1922), 68.
2. John Bunyan, *The Life and Death of Mr. Badman* (London, 1808).
3. Fishbert, *Pulmonary Tuberculosis*, 72, 92.
4. Ibid., 397.
5. Charles L. Minor, "Symptomatology of Pulmonary Tuberculosis," in Arnold C. Klebs, ed., *Tuberculosis* (London: D. Appleton, 1909), 172.
6. Francis Pottenger, *The Diagnosis and Treatment of Pulmonary Tuberculosis* (New York: William Wood, 1908), 77.
7. Addison P. Dutcher, *Pulmonary Tuberculosis: Its Pathology, Nature, Symptoms, Diagnosis, Prognosis, Causes, Hygiene, and Medical Treatment* (Philadelphia, 1875), 168.
8. Fishbert, *Pulmonary Tuberculosis,* 523.
9. Dutcher, *Pulmonary Tuberculosis,* 293.
10. John Keats, "La Belle Dame sans Merci: A Ballad," available at https://www.poetryfoundation.org/poems/44475/la-belle-dame-sans-merci-a-ballad, accessed August 10, 2018.
11. Carolyn A. Day, *Consumptive Chic: A History of Beauty, Fashion, and Disease* (London: Bloomsbury, 2017), 86.
12. Ibid., 108.
13. Harriet Beecher Stowe, *Uncle Tom's Cabin, or Life among the Lowly* (New York: Penguin, 1981; 1st ed. 1852), 424.
14. Arthur C. Jacobson, *Tuberculosis and the Creative Mind* (Brooklyn, NY: Albert T. Huntington, 1909), 3, 5, 38.
15. Dutcher, *Pulmonary Tuberculosis,* 271.
16. Quoted in Charles S. Johnson, *The Negro in American Civilization* (New York: Holt, 1930), 16.

17. John Keats, "When I Have Fears that I May Cease to Be," available at https://www.poets.org/poetsorg/poem/when-i-have-fears-i-may-cease-be, accessed September 10, 2017.
18. Katherine Ott, *Fevered Lives: Tuberculosis in American Culture since 1870* (Cambridge, MA: Harvard University Press, 1996), 31.
19. Anton Chekov, *The Cherry Orchard*, in Anton Chekov, *Five Plays*, trans. Marina Brodskaya (Stanford: Stanford University Press, 2011), 236.
20. O. Amrein, "The Physiological Principles of the High Altitude Treatment and Their Importance in Tuberculosis," *Transactions of the British Congress on Tuberculosis for the Prevention of Consumption, 1901*, vol. 3 (London: William Clows and Sons, 1902), 72.

CHAPTER 15
Tuberculosis in the Unromantic Era of Contagion

1. André Gide, *The Immoralist*, trans. Richard Howard (New York: Alfred A. Knopf, 1970), 21–22, 24–25.
2. Quoted in Linda Bryder, *Below the Magic Mountain: A Social History of Tuberculosis in Twentieth-Century Britain* (New York: Oxford University Press, 1988), 20.
3. "Disease from Books," *New York Tribune*, February 5, 1906, p. 4.
4. "Vanity, Greed and Hygiene Combine to Banish the Beard," *Atlanta Constitution*, February 23, 1902, p. A4.
5. "Exclusion of Consumptives," *New York Tribune*, December 22, 1901, p. 8.
6. Quotations in the following discussion are from National Tuberculosis Association, *A Directory of Sanatoria, Hospitals, Day Camps and Preventoria for the Treatment of Tuberculosis in the United States*, 9th ed. (New York: Livingston, 1931).
7. Sigard Adolphus Knopf, *Pulmonary Tuberculosis* (Philadelphia, 1899), 35–36.
8. Ibid., 58.
9. Ibid., 213.
10. Charles Reinhardt and David Thomson, *A Handbook of the Open-Air Treatment* (London: John Bale, Sons & Danielsson, 1902), 19.
11. Francis M. Pottenger, *The Diagnosis and Treatment of Pulmonary Tuberculosis* (New York: William Wood, 1908), 216.
12. Thomas Spees Carrington, *Tuberculosis Hospital and Sanatorium Construction* (New York: National Association for the Study and Prevention of Tuberculosis, 1911), 14.
13. Pottenger, *Diagnosis and Treatment*, 216.
14. Knopf, *Pulmonary Tuberculosis*, 211.
15. F. Rufenacht Walters, *Sanatoria for Consumptives in Various Parts of the World* (London, 1899), 2.
16. A. E. Ellis, *The Rack* (Boston: Little Brown, 1958), 342.
17. Ibid., 142.
18. *Tuberculosis Dispensary Method and Procedure* (New York: Vail-Ballou, 1916), 10–11.
19. Quoted in Cynthia Anne Connolly, *Saving Sickly Children: The Tuberculosis Preventorium in American Life, 1909–1970* (New Brunswick: Rutgers University Press, 2008), 27.
20. Quoted in Jeanne E. Abrams, "'Spitting Is Dangerous, Indecent, and against the Law!': Legislating Health Behavior during the American Tuberculosis Crusade," *Journal of the History of Medicine and Allied Sciences* 68, no. 3 (July 2013): 425.

21. Tuberculosis Commission of the State of Maryland, *Report of 1902–1904* (Baltimore: Sun Job Printing Office, 1904), n.p.
22. Quoted in *Transactions of the British Congress on Tuberculosis for the Prevention of Consumption, 1901*, vol. 3 (London: William Clowes and Sons, 1902), 2–4.
23. Quoted in *Tuberculosis Dispensary*, 59.
24. Quoted in Hearing before the Subcommittee on Health and the Environment of the Committee on Energy and Commerce, House of Representatives, One Hundred Third Congress, First Session, "The Tuberculosis Epidemic," March 19, 1993 (3), Serial No. 103-36 (Washington, DC: US Government Printing Office, 1993), 1.
25. Ibid.
26. Quoted in Christian W. McMillen, *Discovering Tuberculosis: A Global History, 1900 to the Present* (New Haven: Yale University Press, 2015), 131.

CHAPTER 16
The Third Plague Pandemic

1. "Concerning Plague in Oporto," *Public Health Reports* 14, no. 39 (September 29, 1899): 1655–1656.
2. Frank G. Carpenter, "The Black Death," *Los Angeles Times*, July 15, 1894.
3. "The 'Black Death' in China," *The Interior* 25, no. 1266 (August 30, 1894): 1095.
4. Ibid.
5. *The Bubonic Plague* (Washington, DC: Government Printing Office, 1900), 10.
6. "The Black Death in China," *New York Tribune*, June 26, 1894, p. 6.
7. "Life in Hong Kong," *Austin Daily Statesman*, October 8, 1894, p. 7.
8. "Black Plague," *St. Louis Post-Dispatch*, July 29, 1894, p. 21.
9. "The Plague in Hong Kong," *British Medical Journal* 2, no. 1758 (September 8, 1894): 539–540.
10. Ibid.
11. Quoted in Christos Lynteris, "A 'Suitable Soil': Plague's Urban Breeding Grounds at the Dawn of the Third Pandemic," *Medical History* 61, no. 3 (July 2017): 348.
12. "Fighting the Black Plague," *Globe*, September 12, 1894, p. 6.
13. "Plague Haunts: Why the Poor Die," *Times of India*, May 1, 1903, p. 4.
14. "British Give Up Fight on Plague," *Chicago Daily Tribune*, November 29, 1903, p. 15.
15. "Plague Commission in Bombay," *Times of India*, February 15, 1899, p. 3.
16. "The Bubonic Plague in India," *Chautauquan*, March 26, 1898, p. 6.
17. "The Report of the Indian Plague Commission," *British Medical Journal* 1, no. 2157 (May 3, 1902): 1093.
18. Ibid., 1094.
19. "A Floating Population: Novel Plague Specific," *Times of India*, June 4, 1903, p. 5.
20. "India's Fearful Famine," *New York Times*, July 1, 1905, p. 5.
21. "India Like a Volcano: Widespread and Threatening Discontent," *New York Tribune*, July 3, 1897, p. 7.
22. "The Recent Riots in Bombay," *Times of India*, June 8, 1898, p. 5.

CHAPTER 17
Malaria and Sardinia

A version of this chapter appeared in Frank M. Snowden and Richard Bucala, eds., *The Global Challenge of Malaria: Past Lessons and Future Prospects.* It is used here by permission of the publisher. Copyright © 2014 World Scientific Publishing Co. Pte. Ltd.

1. Giovanni Verga, "Malaria," in *Little Novels of Sicily,* trans. D. H. Lawrence (New York: Grove Press, 1953), 73–74.
2. W. L. Hackett, *Malaria in Europe: An Ecological Study* (Oxford: Oxford University Press, 1937), 15–16, 108; "atomic bomb": quoted in Margaret Humphreys, *Malaria: Poverty, Race, and Public Health in the United States* (Baltimore: Johns Hopkins University Press, 2001), 147.
3. John Logan, "Estimates 1949, Malaria," October 25, 1948, Rockefeller Archive Center, Record Group 1.2, Series 700 Europe, box 12, folder 101, "Rockefeller Foundation Health Commission—Typus, Malaria, 1944" (February–October).
4. Eugenia Tognotti, *Per una storia della malaria in Italia: Il caso della Sardegna,* 2nd ed. (Milan: Franco Angeli, 2008), 23.
5. Relazione dell'Ufficio Centrale composto dei senatori Pantaleoni, Moleschott, Verga e Torelli, "Bonificamento delle regioni di malaria lungo le ferrovie d'Italia," *Atti parlamentari: Senato del Regno, sessione del 1880-81-92, documenti, n. 19-A,* Appendix 13 (my translation).
6. Quoted in Tognotti, *Per una storia,* 230–231.
7. "Mosquito Eradication and Malaria Control," excerpt from Trustees' Confidential Report, January 1, 1954, Rockefeller Archive Center, Record Group 1.2, Series 700 Europe, box 12, folder 101, p. 17.
8. Letter of John A. Logan, Sardinia *Anopheles* Eradication Project, August 1948, Rockefeller Archive Center, Record Group 1.2, Series 700 Europe, box 13, folder 113.
9. Letter of Paul Russell to Alberto Missiroli, November 3, 1949, Rockefeller Archive Center, Record Group 1.2, Series 700 Europe, box 14, folder 116.
10. Silvio Sirigu, "Press Digest: UNRRA Assistance to Sardinia," from *Il Nuovo Giornale d'Italia,* December 12, 1946, United Nations Archive, United Nations Relief and Rehabilitation Administration, 1943–1949, PAG-4/3.0.14.0.0.2:1.
11. B. Fantini, "*Unum facere et alterum non omittere:* Antimalarial Strategies in Italy, 1880–1930)," *Parassitologia* 40, nos. 1–2 (1998): 100.

CHAPTER 18
Polio and the Problem of Eradication

1. David Oshinsky, *Polio: An American Story* (New York: Oxford University Press, 2005), 53.
2. Paul De Kruif, "Polio Must Go," *Ladies' Home Journal* 52, no. 7 (July 1, 1935): 22.
3. Jonas E. Salk, "Considerations in the Preparation and Use of Poliomyelitis Virus Vaccine," *Journal of the American Medical Association* 158 (August 6, 1955): 1239–1240.
4. De Kruif, "Polio Must Go," 22.
5. Salk, "Considerations," 1239.

6. "Polio Victory May Spell End of All Virus Diseases," *New York Herald Tribune*, April 17, 1955, p. A2.
7. Bonnie Angelo, "Salk, Sabin Debate How to Fight Polio," *Newsday*, March 18, 1961, p. 7.
8. Alexander Langmuir, "Epidemiological Considerations," US Department of Health, Education, and Welfare, "Symposium on Present Status of Poliomyelitis and Poliomyelitis Immunization," Washington, DC, November 30, 1960, Albert Sabin Archives, Series 7, box 7.5, folder 10.
9. Paul A. Offit, "The Cutter Incident, 50 Years Later," *New England Journal of Medicine* 352 (April 7, 2005): 1411–1412.
10. Quoted in Alison Day, "'An American Tragedy': The Cutter Incident and Its Implications for the Salk Polio Vaccine in New Zealand, 1955–1960," *Health and History* 11, no. 2 (2009): 46.
11. Zaffran quoted in Leslie Roberts, "Alarming Polio Outbreak Spreads in Congo, Threatening Global Eradication Efforts," *Science*, July 2, 2018, http://www.sciencemag.org/news/2018/07/polio-outbreaks-congo-threaten-global-eradication.

CHAPTER 19
HIV/AIDS

1. Suzanne Daley, "AIDS in South Africa: A President Misapprehends a Killer," *New York Times*, May 14, 2000, p. WK4.
2. "Parliamentary Speeches of Mr. P. W. Botha," National Archives, United Kingdom, FCO 45/2369/73, p. 2.
3. "New Figures Show Staggering Rate of Urbanisation in SA," *Rand Daily Mail*, May 26, 2015.
4. Jeremy Seekings, *Policy, Politics and Poverty in South Africa* (London: Palgrave Macmillan, 2015), 2.
5. Greg Nicolson, "South Africa: Where 12 Million Live in Poverty," *Daily Maverick*, February 3, 2015, http://www.dailymaverick.co.za/article/2015-02-03-south-africa-where-12-million-live-in-extreme-poverty/#.V5zS7I5ErVo.
6. Seekings, *Policy, Politics*, 7.
7. Jason M. Breslow, "Nelson Mandela's Mixed Legacy on HIV/AIDS," *Frontline*, December 6, 2013, http://www.pbs.org/wgbh/frontline/article/nelson-mandelas-mixed-legacy-on-hivaids/.
8. Peter Duesberg, Claus Koehnlein, and David Rasnick, "The Chemical Bases of the Various AIDS Epidemics: Recreational Drugs, Anti-Viral Chemotherapy and Malnutrition," *Journal of Biosciences* 28, no. 4 (June 2003): 383.
9. Quoted in Declan Walsh, "Beetroot and Spinach the Cure for AIDS, Say Some in S Africa," *Irish Times*, March 12, 2004, https://www.irishtimes.com/news/beetroot-and-spinach-the-cure-for-aids-say-some-in-s-africa-1.1135185.
10. The letter is available at Frontline, "Thabo Mbeki's Letter," April 3, 2000, https://www.pbs.org/wgbh/pages/frontline/aids/docs/mbeki.html.
11. Makgoba: Chris McGreal, "How Mbeki Stoked South Africa's Aids Catastrophe," *Guardian*, June 11, 2001, https://www.theguardian.com/world/2001/jun/12/aids.chrismcgreal; Kaunda: André Picard, "AIDS Summit Convenes at Ground Zero," *Globe and Mail*, July 8, 2000, p. A2.

12. Quoted in André Picard, "AIDS Deniers Should Be Jailed: Head of AIDS Body Slams Fringe Movement," *Globe and Mail*, May 1, 2000, p. A3.
13. "Mandela's Only Surviving Son Dies of AIDS," *Irish Times*, January 26, 2005, p. 10.
14. UNAIDS, *Report on the Global AIDS Epidemic 2008*, August 2008, http://www.unaids.org/sites/default/files/media_asset/jc1510_2008globalreport_en_0.pdf.
15. Quoted in Celia W. Dugger, "Study Cites Toll of AIDS Policy in South Africa," *New York Times*, November 25, 2008, https://www.nytimes.com/2008/11/26/world/africa/26aids.html.

CHAPTER 20
HIV/AIDS

I thank Dr. Margaret Snowden for her contributions to this chapter, which is built on the lecture she delivered to the original course at Yale University in 2010.

1. Karl Marx and Frederick Engels, *Manifesto of the Communist Party* (1848), available at https://www.marxists.org/archive/marx/works/download/pdf/Manifesto.pdf, p. 16, accessed June 10, 2016.
2. Ibid.
3. Randy Shilts, *And the Band Played On: Politics, People, and the AIDS Epidemic* (New York: St. Martin's, 1987), 15.
4. Quoted in Marlene Cimons, "Ban on Explicit AIDS Education Materials to End," *Los Angeles Times*, December 15, 1991, http://articles.latimes.com/1991-12-15/news/mn-993_1_aids-educational-materials.
5. *The Federal Response to the AIDS Epidemic: Information and Public Education*, Hearing before a Subcommittee of the Committee on Government Operations, House of Representatives, One Hundredth Congress, First Session, March 16, 1987 (Washington, DC: US Government Printing Office, 1987), 18–19.
6. Ibid., 4.
7. Ibid., 2.
8. Ibid., 16, 19.
9. *Hearing before the Human Resources and Intergovernmental Relations Subcommittee of the Committee on Government Operations of the House of Representatives, One Hundred Third Congress, Second Session*, "AIDS and HIV Infection in the African-American Community," September 16, 1994 (Washington, DC: US Government Printing Office, 1995), 14.
10. Centers for Disease Control and Prevention, "HIV/AIDS among African Americans," 2003, http://permanent.access.gpo.gov/lps63544/afam.pdf.
11. Paul Denning and Elizabeth NiNenno, "Communities in Crisis: Is There a Generalized HIV Epidemic in Impoverished Urban Areas of the United States?," Centers for Disease Control and Prevention, last updated August 28, 2017, https://www.cdc.gov/hiv/group/poverty.html.
12. NAACP, "Criminal Justice Fact Sheet," http://www.naacp.org/pages/criminal-justice-fact-sheet, accessed August 16, 2016.
13. Centers for Disease Control and Prevention, "HIV/AIDS among African Americans."
14. *Hearing before the Human Resources and Intergovernmental Relations Subcommittee*, 75.

15. Ibid., 76, 80.
16. Jon Cohen, "The Sunshine State's Dark Cloud: New Efforts Aim to Curb Florida's Startlingly High HIV Infection Rate," *Science* 360, no. 6394 (June 15, 2018): 1176–1179, http://science.sciencemag.org/content/360/6394/1176.
17. Ibid.

CHAPTER 21
Emerging and Reemerging Diseases

This chapter is largely based on my earlier publication, "Emerging and Reemerging Diseases: A Historical Perspective," *Immunological Reviews* 225 (2008): 9–26. Copyright © 2008 The Author.

1. Aidan Cockburn, *The Evolution and Eradication of Infectious Diseases* (Baltimore: Johns Hopkins University Press, 1963), 133.
2. Ibid., 150.
3. A. Cockburn, ed., *Infectious Diseases: Their Evolution and Eradication* (Springfield, IL: C. C. Thomas, 1967), xi–xiii.
4. Frank Macfarlane Burnet, *Natural History of Infectious Disease,* 4th rev. ed. (Cambridge: Cambridge University Press, 1972; 1st ed. 1953), 1.
5. Ibid., 263.
6. Robert G. Petersdorf, "An Approach to Infectious Diseases," in *Harrison's Principles of Internal Medicine,* 7th rev. ed. (New York: McGraw-Hill, 1974), 722.
7. US Department of Health, Education, and Welfare, *Healthy People: The Surgeon General's Report on Health Promotion and Disease Prevention, 1979* (Washington, DC: US Public Health Service, 1979).
8. The White House, Office of Science and Technology Policy, "Fact Sheet: Addressing the Threat of Emerging Infectious Diseases," June 12, 1996, http://fas.org/irp/offdocs/pdd_ntsc7.htm.
9. United States Congress, Senate Committee on Labor and Human Resources, *Emerging Infections: A Significant Threat to the Nation's Health* (Washington, DC: US Government Printing Office, 1996), 3.
10. J. Brooke, "Feeding on 19th Century Conditions, Cholera Spreads in Latin America," *New York Times,* April 21, 1991, sec. 4, p. 2.
11. L. K. Altman, "A 30-year Respite Ends: Cases of Plague Reported in India's Largest Cities," *New York Times,* October 2, 1994, sec. 4, p. 2.
12. C. J. Peters and J. W. LeDuc, "An Introduction to Ebola: The Virus and the Disease," *Journal of Infectious Diseases* 179, suppl. 1 (1999): x.
13. "Author Richard Preston Discusses the Deadly Outbreak of the Ebola Virus in Zaire," CBS News transcripts, May 15, 1995, *Journal of Infectious Diseases* 179, suppl. 1 (1999): 1.
14. J. R. Davis and J. Lederberg, eds., *Public Health Systems and Emerging Infections: Assessing the Capabilities of the Public and Private Sectors* (Washington, DC: National Academy Press, 2000), 1.
15. Institute of Medicine, *Emerging Infections: Microbial Threats to Health in the United States* (Washington, DC: National Academy Press, 1992), 2.
16. Centers for Disease Control and Prevention, *Addressing Emerging Infectious Disease Threats: A Prevention Strategy for the United States* (Atlanta: CDC, 1994), 3.

17. United States Congress, *Emerging Infections*, 30.
18. Joshua Lederberg, "Infectious Disease as an Evolutionary Paradigm," speech given at the National Conference on Emerging Foodborne Pathogens," Alexandria, VA, March 24–26, 1997; published in *Emerging Infectious Diseases* 3, no. 4 (1997), https://wwwnc.cdc.gov/eid/article/3/4/97-0402_article.
19. R. J. Rubin, C. A. Harrington, A. Poon, K. Dietrich, J. A. Greene, and A. Molduddin, "The Economic Impact of *Staphylococcus aureus* Infection in New York City Hospitals," *Emerging Infectious Diseases* 5, no. 1 (1999), 9.
20. B. J. Marshall, "Helicobacter Connections," Nobel Lecture, December 8, 2005, http://nobelprize.org/nobel_prizes/medicine/laureates/2005/marshall-lecture.html.
21. Kate E. Jones, Nikkita G. Patel, Marc A. Levy, Adam Storeygard, Deborah Balk, John L. Gittleman, and Peter Daszak, "Global Trends in Emerging Infectious Diseases," *Nature* 451 (2008): 990–993.
22. United States Department of Defense, *Addressing Emerging Infectious Disease Threats: A Strategic Plan for the Department of Defense* (Washington, DC: US Government Printing Office, 1998), 1.

CHAPTER 22
Dress Rehearsals for the Twenty-First Century

1. Central Intelligence Agency, "The Global Infectious Disease Threat and Its Implications for the United States," NIE 99-17D, January 2000, http://permanent.access.gpo.gov/websites/www.cia.gov/www.cia.gov/cia/reports/nie/report/nie99-17d.html.
2. Jennifer Brown and Peter Chalk, *The Global Threat of New and Reemerging Infectious Diseases: Reconciling U.S. National Security and Public Health Policy* (Santa Monica: RAND Corporation, 2003).
3. J. Lederberg "Infectious Disease—A Threat to Global Health and Security," *Journal of the American Medical Association* 275, no. 5 (1996): 417–419.
4. The White House, Office of Science and Technology Policy, "Fact Sheet: Addressing the Threat of Emerging Infectious Diseases," June 12, 1996, http:/www.fas.org/irp/offdocs/pdd_ntsc7.htm.
5. United Nations, "Declaration of Commitment on HIV/AIDS: Global Crisis—Global Action," UN Special Session on HIV/AIDS, June 25–27, 2001, http://un.org/ga/aids/conference.html.
6. P. Caulford "SARS: Aftermath of an Outbreak," *Lancet* 362, suppl. 1 (2003): 2.
7. *The Ebola Epidemic: The Keys to Success for the International Response*, Hearing before the Subcommittee on African Affairs of the Committee on Foreign Relations, US Senate, December 10, 2014, S. Hrg. 113-625, p. 13, https://www.foreign.senate.gov/imo/media/doc/121014_Transcript_The%20Ebola%20Epidemic%20the%20Keys%20to%20Success%20for%20the%20International%20Response.pdf.
8. Derek Byerlee, Walter P. Falcon, and Rosamond L. Naylor, *The Many Dimensions of the Tropical Oil Revolution* (Oxford: Oxford University Press, 2016), 2.
9. James Grundvig, 'The Ebola Bats: How Deforestation Unleashed the Deadly Outbreak," *Epoch Times*, October 23, 2014, p. A17.
10. Ibid.
11. Maria Cristina Rulli, Monia Santini, David T. S. Hayman, and Paolo D'Odorico, "The Nexus between Forest Fragmentation in Africa and Ebola Virus Disease

Outbreaks, *Scientific Reports* 7, article no. 71613, February 14, 2017, https://doi.org/10.1038/srep41613.

12. Quoted in Stephan Gregory Bullard, *A Day-by-Day Chronicle of the 2013–2016 Ebola Outbreak* (Cham: Springer International Publishing AG, 2018), 32.
13. United Nations Development Programme, *UN Human Development Report, 2016: Human Development for Everyone*, Table 6, "Multidimensional Poverty Index: Developing Countries," p. 218, http://hdr.undp.org/sites/default/files/2016_human_development_report.pdf.
14. Adam Nossiter, "Epidemic Worsening, Sierra Leone Expands Quarantine Restrictions," *New York Times*, September 26, 2014, p. A10.
15. Alison Healy, "Cost of Treating Ebola Three Times What It Would Cost to Build a Health Service," *Irish Times*, March 26, 2015, p. 11.
16. Christopher Logue, "Everyone Has Underestimated This Outbreak: Ebola Is Not Going Away," *Irish Times*, September 16, 2014, p. B6.
17. Quoted in Adam Nossiter, "Ebola Reaches Guinean Capital, Stirring Fears," *New York Times*, April 2, 2014, p. A4.
18. MSF, "Ebola: Pushed to the Limit and Beyond," March 23, 2015, available at https://www.msf.org/ebola-pushed-limit-and-beyond.
19. Quoted in Nana Boakye-Yiadom, "UN Seeks to Calm Ebola Fears in West Africa," *Globe and Mail*, July 3, 2014, p. A6.
20. Lisa O'Carroll, "West Blamed for 'Almost Zero' Response to Ebola Outbreak Crisis in West Africa," *Irish Times*, August 20, 2014, 10.
21. Editorial Board, "A Painfully Slow Ebola Response," *New York Times*, August 16, 2014, p. A18.
22. Borneo Post online, "Ebola Outbreak under Control, Says Guinea President," May 1, 2014, http://www.theborneopost.com/2014/05/01/ebola-outbreak-under-control-says-guinea-president/.
23. David Quammen, "Ebola Is Not the Next Pandemic," *New York Times*, April 10, 2014, p. A25.
24. Quoted in Kelly Grant, "Canadian Doctor Describes Heartbreaking Scene of Ebola Outbreak," *Globe and Mail*, August 20, 2014, last updated May 12, 2018, https://www.theglobeandmail.com/life/health-and-fitness/health/canadian-doctor-describes-heart-breaking-scenes-of-ebola-outbreak/article20148033/.
25. Andrew Siddons, "U.S. and Global Efforts to Contain Ebola Draw Criticism at Congressional Hearing," *New York Times*, August 8, 2014, p. A11.
26. "Ebola Demands Urgent US Action," *Washington Post*, September 5, 2014, p. A20.
27. Adam Nossiter, "Ebola Epidemic Worsening: Sierra Leone Expands Quarantine Restrictions," *New York Times*, September 26, 2014, p. A10.
28. David Lewis and Emma Farge, "Liberia Shuts Schools, Considers Quarantine to Curb Ebola," Reuters, July 30, 2014, https://www.reuters.com/article/us-health-ebola-liberia-idUSKBN0FZ22H20140730.
29. Stephen Douglas, "In Sierra Leone, We've Stopped Shaking Hands," *Globe and Mail*, August 5, 2014, p. A9.
30. Quoted in Bullard, *Day-by-Day Chronicle*, 82.

Selected Bibliography

GENERAL

Ackernecht, Erwin H., *History and Geography of the Most Important Diseases* (New York: Hafner, 1965).
Bynum, William F., *Science and the Practice of Medicine in the Nineteenth Century* (Cambridge: Cambridge University Press, 1994).
Creighton, Charles, *A History of Epidemics in Britain* (Cambridge: Cambridge University Press, 1891–1894).
Crosby, Alfred W., *The Columbian Exchange: The Biological and Cultural Consequences of 1492* (Westport, CT: Greenwood, 1972).
Diamond, Jared, *Guns, Germs, and Steel: The Fate of Human Societies* (New York: Norton, 1997).
Ewald, Paul W., *Evolution of Infectious Disease* (New York: Oxford University Press, 1994).
Farmer, Paul, *Infections and Inequalities: The Modern Plagues* (Berkeley: University of California Press, 2001).
Foucault, Michel, *The Birth of the Clinic: An Archaeology of Medical Perception*, trans. A. M. Sheridan Smith (New York: Pantheon, 1973).
———, *Discipline and Punish: The Birth of the Prison*, trans. Alan Sheridan (New York: Vintage, 1995).
Garrett, Laurie, *The Coming Plague: Newly Emerging Diseases in a World Out of Balance* (New York: Penguin, 1994).
Harkness, Deborah E., *The Jewel House: Elizabethan London and the Scientific Revolution* (New Haven: Yale University Press, 2007).
Harrison, Mark, *Climates and Constitutions: Health, Race, Environment and British Imperialism in India, 1600–1850* (New Delhi: Oxford University Press, 1999).
Hays, J. N., *The Burdens of Disease: Epidemics and Human Response in Western History* (New Brunswick: Rutgers University Press, 2009).
Keshavjee, Salmaan, *Blind Spot: How Neoliberalism Infiltrated Global Health* (Oakland: University of California Press, 2014).
Krieger, Nancy, *Epidemiology and the People's Health: Theory and Practice* (New York: Oxford University Press, 2011).
Magner, Lois N., *A History of Infectious Diseases and the Microbial World* (Westport: Praeger, 2009).
McNeill, William H., *Plagues and Peoples* (New York: Anchor, 1998; 1st ed. 1976).
Miller, Arthur, *The Crucible* (New York: Penguin, 1996).

Nelson, Kenrad E., Carolyn Williams, and Neil Graham, *Infectious Disease Epidemiology: Theory and Practice* (Boston: Jones and Bartlett, 2005).

Pati, Bisamoy, and Mark Harrison, *Health, Medicine, and Empire: Perspectives on Colonial India* (Hyderabad: Orient Longman, 2001).

Ranger, Terence, and Paul Slack, eds., *Epidemics and Ideas* (Cambridge: Cambridge University Press, 1992).

Rosenberg, Charles E., *Explaining Epidemics and Other Studies in the History of Medicine* (Cambridge: Cambridge University Press, 1993).

Watts, Sheldon J., *Epidemics and History: Disease, Power and Imperialism* (New Haven: Yale University Press, 1997).

Winslow, Charles-Edward Amory, *The Conquest of Epidemic Disease: A Chapter in the History of Ideas* (Princeton: Princeton University Press, 1943).

Zinsser, Hans, *Rats, Lice and History* (Boston: Little, Brown, 1935).

HUMORALISM AND ANCIENT MEDICINE

Bliquez, Lawrence J., *The Tools of Asclepius: Surgical Instruments in Greek and Roman Times* (Leiden: Brill, 2015).

Cavanaugh, T. A., *Hippocrates' Oath and Asclepius' Snake: The Birth of a Medical Profession* (New York: Oxford University Press, 2018).

Edelstein, Ludwig, *Ancient Medicine: Selected Papers* (Baltimore: Johns Hopkins University Press, 1967).

Eijk, Philip J. van der, *Hippocrates in Context: Papers Read at the XIth International Hippocrates Colloquium, University of Newcastle upon Tyne, 27–31 August 2002* (Leiden: Brill, 2005).

Galen, *Selected Works* (Oxford: Oxford University Press, 1997).

Grmek, Mirko D., ed., *Western Medical Thought from Antiquity to the Middle Ages* (Cambridge, MA: Harvard University Press, 1939).

Hankinson, R. J., ed., *The Cambridge Companion to Galen* (Cambridge: Cambridge University Press, 2008).

Hart, Gerald D., *Asclepius, the God of Medicine* (London: Royal Society of Medicine Press, 2000).

Hippocrates, *The Medical Works of Hippocrates* (Oxford: Blackwell, 1950).

Horstmanshoff, Manfred, and Cornelius van Tilburg, *Hippocrates and Medical Education: Selected Papers Presented at the XIIth International Hippocrates Colloquium, University of Leiden, 24–26 August 2005* (Leiden: Brill, 2010).

Jouanna, Jacques, *Hippocrates* (Baltimore: Johns Hopkins University Press, 1999).

King, Helen, *Hippocrates' Woman: Reading the Female Body in Ancient Greece* (London: Routledge, 1998).

Langholf, Volker, *Medical Theories in Hippocrates' Early Texts and the "Epidemics"* (Berlin: Walter de Gruyter, 1990).

Levine, Edwin Burton, *Hippocrates* (New York: Twayne, 1971).

Lloyd, Geoffrey Ernest Richard, *In the Grip of Disease: Studies in the Greek Imagination* (Oxford: Oxford University Press, 2003).

———, *Magic, Reason, and Experience: Studies in the Origin and Development of Greek Science* (Cambridge: Cambridge University Press, 1979).
———, *Principles and Practices in Ancient Greek and Chinese Science* (Aldershot: Ashgate, 2006).
Mitchell-Boyask, Robin, *Plague and the Athenian Imagination: Drama, History and the Cult of Asclepius* (Cambridge: Cambridge University Press, 2008).
Nutton, Vivian, *Ancient Medicine* (Milton Park: Routledge, 2013).
———, "The Fatal Embrace: Galen and the History of Ancient Medicine," *Science in Context* 18, no. 1 (March 2005): 111–121.
———, ed., *Galen: Problems and Prospects* (London: Wellcome Institute for the History of Medicine, 1981).
———, "Healers and the Healing Act in Classical Greece," *European Review* 7, no. 1 (February 1999): 27–35.
Oldstone, Michael B. A., *Viruses, Plagues, and History* (Oxford: Oxford University Press, 2000.)
Schiefsky, Mark John, *Hippocrates on Ancient Medicine* (Leiden: Brill, 2005).
Shakespeare, William, *The Taming of the Shrew* (Guilford: Saland, 2011).
Smith, W. D., *The Hippocratic Tradition* (Ithaca, NY: Cornell University Press, 1979).
Temkin, Owsei, *Galenism: Rise and Decline of a Medical Philosophy* (Ithaca: Cornell University Press, 1973).
———, *Hippocrates in a World of Pagans and Christians* (Baltimore: Johns Hopkins University Press, 1991).
———, *Views on Epilepsy in the Hippocratic Period* (Baltimore: Johns Hopkins University Press, 1933).
Thucydides, *The Peloponnesian War* (Oxford: Oxford University Press, 2009).
Tuplin, C. J., and T. E. Rihll, eds., *Science and Mathematics in Ancient Greek Culture* (Oxford: Oxford University Press, 2002).
Wear, Andrew, ed., *Medicine in Society: Historical Essays* (Cambridge: Cambridge University Press, 1992).

PLAGUE

Advisory Committee Appointed by the Secretary of State for India, the Royal Society, and the Lister Institute, "Reports on Plague Investigations in India," *Journal of Hygiene* 11, Plague Suppl. 1, Sixth Report on Plague Investigations in India (December 1911): 1, 7–206.
———, "Reports on Plague Investigations in India," *Journal of Hygiene* 6, no. 4, Reports on Plague Investigations in India (September 1906): 421–536.
———, "Reports on Plague Investigations in India Issued by the Secretary of State for India, the Royal Society, and the Lister Institute," *Journal of Hygiene* 10, no. 3, Reports on Plague Investigations in India (November 1910): 313–568.
Alexander, John T., *Bubonic Plague in Early Modern Russia: Public Health and Urban Disaster* (Baltimore: Johns Hopkins University Press, 1980).

Archaeologica Medica XLVI, "How Our Forefathers Fought the Plague," *British Medical Journal* 2, no. 1969 (September 24, 1898): 903–908.
Ariès, Philippe, *The Hour of Our Death*, trans. Helen Weaver (New York: Knopf, 1981).
———, *Western Attitudes toward Death from the Middle Ages to the Present*, trans. Patricia M. Ranum (Baltimore: Johns Hopkins University Press, 1974).
Arnold, David, *Colonizing the Body: State Medicine and Epidemic Disease in Nineteenth-Century India* (Berkeley: University of California Press, 1993).
Bannerman, W. B., "The Spread of Plague in India," *Journal of Hygiene* 6, no. 2 (April 1906): 179–211.
Barker, Sheila, "Poussin, Plague and Early Modern Medicine," *Art Bulletin* 86, no. 4 (December 2004): 659–689.
Benedictow, O. J., "Morbidity in Historical Plague Epidemics," *Population Studies* 41, no. 3 (November 1987): 401–431.
Bertrand, J. B., *A Historical Relation of the Plague at Marseilles in the Year 1720*, trans. Anne Plumptre (Farnborough: Gregg, 1973; 1st ed. 1721).
Biraben, Jean Noel, *Les hommes et la peste en France et dans les pays européens et méditerranéens*, 2 vols. (Paris: Mouton, 1975).
Blue, Rupert, "Anti-Plague Measures in San Francisco, California, U.S.A.," *Journal of Hygiene* 9, no. 1 (April 1909): 1–8.
Boccaccio, *The Decameron*, trans. M. Rigg (London: David Campbell, 1921). Also available at Medieval Sourcebook: Boccaccio: The Decameron, https://sourcebooks.fordham.edu/source/boccacio2.asp, accessed August 22, 2018.
Boeckl, Christine M., "Giorgio Vassari's *San Rocco Altarpiece*: Tradition and Innovation in Plague Iconography," *Artibus et Historiae* 22, no. 43 (2001): 29–40.
Boelter, W. R., *The Rat Problem* (London: Bale and Danielsson, 1909).
Bonser, W., "Medical Folklore of Venice and Rome," *Folklore* 67, no. 1 (March 1956): 1–15.
Butler, Thomas, "Yersinia Infections: Centennial of the Discovery of the Plague Bacillus," *Clinical Infectious Diseases* 19, no. 4 (October 1994): 655–661.
Calmette, Albert, "The Plague at Oporto," *North American Review* 171, no. 524 (July 1900): 104–111.
Calvi, Giulia, *Histories of a Plague Year: The Social and the Imaginary in Baroque Florence* (Berkeley: University of California Press, 1989).
Camus, Albert, *The Plague*, trans. Stuart Gilbert (New York: Knopf, 1948).
Cantor, Norman F., *In the Wake of the Plague: The Black Death and the World It Made* (New York: Free Press, 2001).
Carmichael, Ann G., *Plague and the Poor in Renaissance Florence* (Cambridge: Cambridge University Press, 1986).
Catanach, I. J., "The 'Globalization' of Disease? India and the Plague," *Journal of World History* 12, no. 1 (Spring 2001): 131–153.
Centers for Disease Control and Prevention, "Human Plague—United States, 1993–1994," *Morbidity and Mortality Weekly Report* 43, no. 13 (April 8, 1994): 242–246.
———, "Plague—United States, 1980," *Morbidity and Mortality Weekly Report* 29, no. 31 (August 1980): 371–372, 377.

———,"Recommendation of the Public Health Service Advisory Committee on Immunization Practices: Plague Vaccine," *Morbidity and Mortality Weekly Report* 27, no. 29 (July 21, 1978): 255–258.

Chase, Marilyn, *Barbary Plague: The Black Death in Victorian San Francisco* (New York: Random House, 2003).

Cipolla, Carlo M., *Cristofano and the Plague: A Study in the History of Public Health in the Age of Galileo* (Berkeley: University of California Press, 1973).

———, *Faith, Reason, and the Plague in Seventeenth-Century Tuscany* (New York: Norton, 1979).

———, *Fighting the Plague in Seventeenth-Century Italy* (Madison: University of Wisconsin Press, 1981).

Cohn, Samuel Kline, *The Black Death Transformed: Disease and Culture in Early Renaissance Europe* (London: Arnold, 2002).

Condon, J. K., *A History of the Progress of Plague in the Bombay Presidency from June 1896 to June 1899* (Bombay: Education Society's Steam Press, 1900).

Crawford, R. H. P., *Plague and Pestilence in Literature and Art* (Oxford: Clarendon, 1914).

Crawshaw, Jane L. Stevens, *Plague Hospitals: Public Health for the City in Early Modern Venice* (Farnham: Ashgate, 2012).

Creel, Richard H., "Outbreak and Suppression of Plague in Porto Rico: An Account of the Course of the Epidemic and the Measures Employed for Its Suppression by the United States Public Health Service," *Public Health Reports (1896–1970)* 28, no. 22 (May 30, 1913): 1050–1070.

Defoe, Daniel, *Journal of the Plague Year* (Cambridge: Chadwyck-Healey, 1996).

Dols, Michael W., *The Black Death in the Middle East* (Princeton: Princeton University Press, 1977).

Drancourt, Michel, "Finally Plague Is Plague," *Clinical Microbiology and Infection* 18, no. 2 (February 2012): 105–106.

Drancourt, Michel, Gérard Aboudharam, Michel Signoli, Olivier Dutour, and Didier Raoult, "Detection of 400-year-old *Yersinia pestis* DNA in Human Dental Pulp: An Approach to the Diagnosis of Ancient Septicemia," *Proceedings of the National Academy of Sciences of the United States of America* 95, no. 21 (October 13, 1998): 12637–12640.

Echenberg, Myron J., "Pestis Redux: The Initial Years of the Third Bubonic Plague Pandemic, 1894–1901," *Journal of World History* 13, no. 2 (Fall 2002): 429–449.

———, *Plague Ports: The Global Urban Impact of Bubonic Plague, 1894–1901* (New York: New York University Press, 2007).

Ell, Stephen R., "Three Days in October of 1630: Detailed Examination of Mortality during an Early Modern Plague Epidemic in Venice," *Reviews of Infectious Diseases* 11, no. 1 (January–February 1989): 128–139.

Gilman, Ernest B., *Plague Writing in Early Modern England* (Chicago: University of Chicago Press, 2009).

Gonzalez, Rodrigo J., and Virginia L. Miller, "A Deadly Path: Bacterial Spread during Bubonic Plague," *Trends in Microbiology* 24, no. 4 (April 2016): 239–241.

Gregory of Tours, *History of the Franks*, trans. L. Thorpe (Baltimore: Penguin, 1974).
Herlihy, David, *The Black Death and the Transformation of the West* (Cambridge, MA: Harvard University Press, 1997).
Hopkins, Andrew, "Plans and Planning for S. Maria della Salute, Venice," *Art Bulletin* 79, no. 3 (September 1997): 440–465.
Jones, Colin, "Plague and Its Metaphors in Early Modern France," *Representations* 53 (Winter 1996): 97–127.
Kidambi, Prashant, "Housing the Poor in a Colonial City: The Bombay Improvement Trust, 1898–1918," *Studies in History* 17 (2001): 57–79.
Kinyoun, J. J., Walter Wyman, and Brian Dolan, "Plague in San Francisco (1900)," *Public Health Reports (1974–)* 121, suppl. 1, Historical Collection, 1878–2005 (2006): 16–37.
Klein, Ira, "Death in India, 1871–1921," *Journal of Asian Studies* 32, no. 4 (August 1973): 639–659.
———, "Development and Death: Bombay City, 1870–1914," *Modern Asian Studies* 20, no. 4 (1986): 725–754.
———, "Plague, Policy and Popular Unrest in British India," *Modern Asian Studies* 22, no. 4 (1988): 723–755.
Lantz, David E., *The Brown Rat in the United States* (Washington, DC: US Government Printing Office, 1909).
Ledingham, J. C. G., "Reports on Plague Investigations in India," *Journal of Hygiene* 7, no. 3, Reports on Plague Investigations in India (July 1907): 323–476.
Link, Vernon B., "Plague on the High Seas," *Public Health Reports (1896–1970)* 66, no. 45 (November 9, 1951): 1466–1472.
Little, Lester K., ed., *Plague and the End of Antiquity: The Pandemic of 541–750* (Cambridge: Cambridge University Press, 2007).
Lynteris, Christos, "A 'Suitable Soil': Plague's Urban Breeding Grounds at the Dawn of the Third Pandemic," *Medical History* 61, no. 3 (July 2017): 343–357.
Maddicott, J. R., "The Plague in Seventh-Century England," *Past & Present* 156 (August 1997): 7–54.
Manzoni, Alessandro, *The Betrothed*, trans. Bruce Penman (Harmondsworth: Penguin, 1972).
———, *The Column of Infamy*, trans. Kenelm Foster and Jane Grigson (London: Oxford University Press, 1964).
Marshall, Louise, "Manipulating the Sacred: Image and Plague in Renaissance Italy," *Renaissance Quarterly* 47, no. 3 (Autumn 1994): 485–532.
McAlpin, Michelle Burge, "Changing Impact of Crop Failures in Western India, 1870–1920," *Journal of Economic History* 39, no. 1 (March 1979): 143–157.
Meiss, Millard, *Painting in Florence and Siena after the Black Death* (Princeton: Princeton University Press, 1951).
Meyer, K. F., Dan C. Cavanaugh, Peter J. Bartelloni, and John D. Marshall, Jr., "Plague Immunization: I. Past and Present Trends," *Journal of Infectious Diseases* 129, suppl. (May 1974): S13—S18.

Moote, A. Lloyd, and Dorothy C. Moote, *The Great Plague: The Story of London's Most Deadly Year* (Baltimore: Johns Hopkins University Press, 2004).

National Institutes of Health, US National Library of Medicine, "Plague," MedlinePlus, http://www.nlm.nih.gov/medlineplus/ency/article/000596.htm, last updated August 14, 2018.

Newman, Kira L. S., "Shutt Up: Bubonic Plague and Quarantine in Early Modern England," *Journal of Social History* 45, no. 3 (Spring 2012): 809–834.

"Observations in the Plague Districts in India," *Public Health Reports (1896–1970)* 15, no. 6 (February 9, 1900): 267–271.

Orent, Wendy, *Plague: The Mysterious Past and Terrifying Future of the World's Most Dangerous Disease* (New York: Free Press, 2004).

Palmer, Darwin L., Alexander L. Kisch, Ralph C. Williams, Jr., and William P. Reed, "Clinical Features of Plague in the United States: The 1969–1970 Epidemic," *Journal of Infectious Diseases* 124, no. 4 (October 1971): 367–371.

Pechous, R. D., V. Sivaraman, N. M. Stasulli, and W. E. Goldman, "Pneumonic Plague: The Darker Side of Yersinia pestis," *Trends in Microbiology* 24, no. 3 (March 2016): 194–196.

Pepys, Samuel, *The Diary of Samuel Pepys*, ed. Robert Latham and William Matthews, 10 vols. (Berkeley: University of California Press, 2000).

Petro, Anthony M., *After the Wrath of God: AIDS, Sexuality, and American Religion* (Oxford: Oxford University Press, 2015).

"The Plague: Special Report on the Plague in Glasgow," *British Medical Journal* 2, no. 2071 (September 8, 1900): 683–688.

Pollitzer, Robert, *Plague* (Geneva: World Health Organization, 1954).

"The Present Pandemic of Plague," *Public Health Reports (1896–1970)* 40, no. 2 (January 9, 1925): 51–54.

The Rat and Its Relation to Public Health (Washington, DC: US Government Printing Office, 1910).

Risse, Guenter B., *Plague, Fear, and Politics in San Francisco's Chinatown* (Baltimore: Johns Hopkins University Press, 2012).

Ruthenberg, Gunther E., "The Austrian Sanitary Cordon, and the Control of the Bubonic Plague, 1710–1871," *Journal of the History of Medicine and the Allied Sciences* 28, no. 1 (January 1973): 15–23.

Scasciamacchia, S., L. Serrecchia, L. Giangrossi, G. Garofolo, A. Balestrucci, G. Sammartino et al., "Plague Epidemic in the Kingdom of Naples, 1656–1658," *Emerging Infectious Diseases* 18, no. 1 (January 2012), http://dx.doi.org/10.3201/eid1801.110597.

Shakespeare, William, *Romeo and Juliet* (London: Bloomsbury Arden Shakespeare, 2017).

Shrewsbury, John Findlay Drew, *A History of Bubonic Plague in the British Isles* (Cambridge: Cambridge University Press, 1970).

Slack, Paul, *The Impact of Plague on Tudor and Stuart England* (Oxford: Clarendon, 1985).

Steel, D., "Plague Writing: From Boccaccio to Camus," *Journal of European Studies* 11 (1981): 88–110.

Taylor, Jeremy, *Holy Living and Holy Dying: A Contemporary Version by Marvin D. Hinten* (Nashville, TN: National Baptist Publishing Board, 1990).
Twigg, G., *The Black Death: A Biological Reappraisal* (New York: Schocken, 1985).
Velimirovic, Boris, and Helga Velimirovic, "Plague in Vienna," *Reviews of Infectious Diseases* 11, no. 5 (September–October 1989): 808–826.
Verjbitski, D. T., W. B. Bannerman, and R. T. Kapadia, "Reports on Plague Investigations in India," *Journal of Hygiene* 8, no. 2, Reports on Plague Investigations in India (May 1908): 161–308.
Vincent, Catherine, "Discipline du corps et de l'esprit chez les Flagellants au Moyen Age," *Revue Historique* 302, no. 3 (July–September 2000): 593–614.
Wheeler, Margaret M., "Nursing of Tropical Diseases: Plague," *American Journal of Nursing* 16, no. 6 (March 1916): 489–495.
Wyman, Walter, *The Bubonic Plague* (Washington, DC: US Government Printing Office, 1900).
Ziegler, Philip, *The Black Death* (New York: Harper & Row, 1969).

SMALLPOX

Artenstein, Andrew W., "Bifurcated Vaccination Needle," *Vaccine* 32, no. 7 (February 7, 2014): 895.
Basu, Rabindra Nath, *The Eradication of Smallpox from India* (New Delhi: World Health Organization, 1979).
Bazin, H., *The Eradication of Smallpox: Edward Jenner and the First and Only Eradication of a Human Infectious Disease* (San Diego: Academic Press, 2000).
Carrell, Jennifer Lee, *The Speckled Monster: A Historical Tale of Battling Smallpox* (New York: Dutton, 2003).
Dickens, Charles, *Bleak House* (London: Bradbury and Evans, 1953).
Dimsdale, Thomas, *The Present Method of Inoculating of the Small-pox. To Which Are Added Some Experiments, Instituted with a View to Discover the Effects of a Similar Treatment in the Natural Small-pox* (Dublin, 1774).
Fenn, Elizabeth Anne, *Pox Americana: The Great Smallpox Epidemic of 1775–1782* (New York: Hill and Wang, 2001).
Fielding, Henry, *The Adventures of Joseph Andrews* (London, 1857).
———, *The History of Tom Jones, a Foundling* (Oxford: Clarendon, 1974).
Foege, William H., *House on Fire: The Fight to Eradicate Smallpox* (Berkeley: University of California Press, 2011).
Franklin, Benjamin, *Some Account of the Success of Inoculation for the Small-pox in England and America. Together with Plain Instructions, by Which Any Person May Perform the Operation, and Conduct the Patient through the Distemper* (London, 1759).
Glynn, Ian, *The Life and Death of Smallpox* (London: Profile, 2004).
Henderson, Donald A., *Smallpox: The Death of a Disease* (Amherst, NY: Prometheus, 2009).
Herberden, William, *Plain Instructions for Inoculation in the Small-pox; by Which Any Person May Be Enabled to Perform the Operation, and Conduct the Patient through the Distemper* (London, 1769).

Hopkins, Donald R., *The Greatest Killer: Smallpox in History* (Chicago: University of Chicago Press, 2002).
———, *Princes and Peasants: Smallpox in History* (Chicago: University of Chicago Press, 1983).
James, Sydney Price, *Smallpox and Vaccination in British India* (Calcutta: Thacker, Spink, 1909).
Jenner, Edward, *An Inquiry into the Causes and Effects of the Variolae Vaccinae, A Disease Discovered in Some of the Western Counties of England, Particularly Gloucestershire, and Known by the Name of the Cow Pox* (Springfield, MA, 1802; 1st ed. 1799).
———, *On the Origin of the Vaccine Inoculation* (London, 1801).
Koplow, David A., *Smallpox: The Fight to Eradicate a Global Scourge* (Berkeley: University of California Press, 2003).
Langrish, Browne, *Plain Directions in Regard to the Small-pox* (London, 1759).
Mann, Charles C., *1491: New Revelations of the Americas before Columbus* (New York: Knopf, 2005).
———, *1493: Uncovering the New World Columbus Created* (New York: Knopf, 2011).
Ogden, Horace G., *CDC and the Smallpox Crusade* (Atlanta: US Dept. of Health and Human Services, 1987).
Reinhardt, Bob H., *The End of a Global Pox: America and the Eradication of Smallpox in the Cold War Era* (Chapel Hill: University of North Carolina Press, 2015).
Rogers, Leonard, *Small-pox and Climate in India: Forecasting of Epidemics* (London: HMSO, 1926).
Rowbotham, Arnold Horrex, *The "Philosophes" and the Propaganda for Inoculation of Smallpox in Eighteenth-Century France* (Berkeley: University of California Press, 1935).
Rush, Benjamin, *The New Method of Inoculating for the Small-pox* (Philadelphia, 1792).
Schrick, Livia, Clarissa R. Damaso, José Esparza, and Andreas Nitsche, "An Early American Smallpox Vaccine Based on Horsepox," *New England Journal of Medicine* 377 (2017): 1491–1492.
Shuttleton, David E., *Smallpox and the Literary Imagination, 1660–1820* (Cambridge: Cambridge University Press, 2007).
Thakeray, William Makepeace, *The History of Henry Esmond* (New York: Harper, 1950).
Thomson, Adam, *A Discourse on the Preparation of the Body for the Small-pox; And the Manner of Receiving the Infection* (Philadelphia, 1750).
Waterhouse, Benjamin, *A Prospect of Exterminating the Small-pox; Being the History of the Variolae vaccinae, or Kine-Pox, Commonly Called the Cow-Pox, as it Appeared in England; with an Account of a Series of Inoculations Performed for the Kine-pox, in Massachusetts* (Cambridge, MA, 1800).
Winslow, Ola Elizabeth, *A Destroying Angel: The Conquest of Smallpox in Colonial Boston* (Boston: Houghton Mifflin, 1974).
World Health Organization, *The Global Eradication of Smallpox: Final Report of the Global Commission for the Certification of Smallpox Eradication* (Geneva: World Health Organization, 1979).

———, *Handbook for Smallpox Eradication Programmes in Endemic Areas* (Geneva: World Health Organization, 1967).
———, *Smallpox and Its Eradication* (Geneva: World Health Organization, 1988).

NAPOLEON: HAITI AND YELLOW FEVER

Blackburn, Robin, "Haiti, Slavery, and the Age of the Democratic Revolution," *William and Mary Quarterly* 63, no. 4 (October 2006): 643–674.
Dubois, Laurent, *Avengers of the New World: The Story of the Haitian Revolution* (Cambridge, MA: Belknap, 2005).
Dunn, Richard S., *Sugar and Slaves* (Chapel Hill: University of North Carolina Press, 1972).
Geggus, David Patrick, *Haitian Revolutionary Studies* (Bloomington: Indiana University Press, 2002).
Gilbert, Nicolas Pierre, *Histoire médicale de l'armée française, à Saint-Domingue, en l'an dix: ou mémoire sur la fièvre jaune, avec un apperçu de la topographie médicale de cette colonie* (Paris, 1803).
Girard, Philippe R., "Caribbean Genocide: Racial War in Haiti, 1802–4," *Patterns of Prejudice* 39, no. 2 (2005): 138–161.
———, "Napoléon Bonaparte and the Emancipation Issue in Saint-Domingue, 1799–1803," *French Historical Studies* 32, no. 4 (September 2009): 587–618.
———, *The Slaves Who Defeated Napoleon: Toussaint Louverture and the Haitian War of Independence* (Tuscaloosa: University of Alabama Press, 2011).
Herold, J. Christopher, ed., *The Mind of Napoleon: A Selection of His Written and Spoken Words* (New York: Columbia University Press, 1955).
James, Cyril Lionel Robert, *Black Jacobins: Toussaint L'Ouverture and the San Domingo Revolution* (New York: Vintage, 1963).
Kastor, Peter J., *Nation's Crucible: The Louisiana Purchase and the Creation of America* (New Haven: Yale University Press, 2004).
Kastor, Peter J., and François Weil, eds., *Empires of the Imagination: Transatlantic Histories and the Louisiana Purchase* (Charlottesville: University of Virginia Press, 2009).
Lee, Debbi, *Slavery and the Romantic Imagination* (Philadelphia: University of Pennsylvania Press, 2002).
Leroy-Dupré, Louis Alexandre Hippolyte, ed., *Memoir of Baron Larrey, Surgeon-in-chief of the Grande Armée* (London, 1862).
Marr, John S., and Cathey, John T., "The 1802 Saint-Domingue Yellow Fever Epidemic and the Louisiana Purchase, *Journal of Public Health Management Practice* 19, no. 1 (January–February 2013): 77–82.
———, "Yellow Fever, Asia and the East African Slave Trade, *Transactions of the Royal Society of Tropical Medicine and Hygiene* 108, no. 5 (May 1, 2014): 252–257.
McNeill, John Robert, *Mosquito Empires: Ecology and War in the Greater Caribbean, 1620–1914* (New York: Cambridge University Press, 2010).
Rush, Benjamin, *An Account of the Bilious Remitting Yellow Fever as It Appeared in the City of Philadelphia, in the Year 1793* (Philadelphia, 1794).

Scott, James, *Weapons of the Weak: Everyday Forms of Peasant Resistance* (New Haven: Yale University Press, 1985).
Sutherland, Donald G., *Chouans: The Social Origins of Popular Counter-Revolution in Upper Brittany, 1770–1796* (Oxford: Oxford University Press, 1982).
Teelock, Vijaya, *Bitter Sugar: Sugar and Slavery in Nineteenth-Century Mauritius* (Moka, Mauritius: Mahatma Gandhi Institute, 1998).
Tilly, Charles, *The Vendée* (Cambridge, MA: Harvard University Press, 1976).

NAPOLEON: TYPHUS, DYSENTERY, AND THE RUSSIAN CAMPAIGN

Alekseeva, Galina, "Emerson and Tolstoy's Appraisals of Napoleon," *Tolstoy Studies Journal* 24 (2012): 59–65.
Armstrong, John, *Practical Illustrations of Typhus Fever, of the Common Continued Fever, and of Inflammatory Diseases, &c.* (London, 1819).
Austin, Paul Britten, *1812: Napoleon in Moscow* (South Yorkshire: Frontline, 2012).
Ballingall, George, *Practical Observations on Fever, Dysentery, and Liver Complaints as They Occur among the European Troops in India* (Edinburgh, 1823).
Bell, David Avrom, *The First Total War: Napoleon's Europe and the Birth of Warfare as We Know It* (Boston: Houghton Mifflin, 2007).
Bourgogne, Jean Baptiste François, *Memoirs of Sergeant Bourgogne (1812–1813)*, trans. Paul Cottin and Maurice Henault (London: Constable, 1996).
Burne, John, *A Practical Treatise on the Typhus or Adynamic Fever* (London, 1828).
Campbell, D., *Observations on the Typhus, or Low Contagious Fever, and on the Means of Preventing the Production and Communication of This Disease* (Lancaster, 1785).
Cirillo, Vincent J., "'More Fatal than Powder and Shot': Dysentery in the U.S. Army during the Mexican War, 1846–48," *Perspectives in Biology and Medicine* 52, no. 3 (Summer 2009): 400–413.
Clausewitz, Carl von, *The Campaign of 1812 in Russia* (London: Greenhill, 1992).
———, *On War*, trans. J. J. Graham (New York: Barnes and Noble, 1968).
Collins, Christopher H., and Kennedy, David A., "Gaol and Ship Fevers," *Perspectives in Public Health* 129, no. 4 (July 2009): 163–164.
Esdaile, Charles J., "De-Constructing the French Wars: Napoleon as Anti-Strategist," *Journal of Strategic Studies* 31 (2008): 4, 515–552.
———, *The French Wars, 1792–1815* (London: Routledge, 2001).
———, *Napoleon's Wars: An International History, 1803–1815* (London: Allen Lane, 2007).
Fezensac, Raymond A. P. J. de, *A Journal of the Russian Campaign of 1812*, trans. W. Knollys (London, 1852).
Foord, Edward, *Napoleon's Russian Campaign of 1812* (Boston: Little, Brown, 1915).
Hildenbrand, Johann Val de, *A Treatise on the Nature, Cause, and Treatment of Contagious Typhus*, trans. S. D. Gross (New York, 1829).
Larrey, Dominique Jean, *Memoir of Baron Larrey* (London, 1861).
Maceroni, Francis, and Joachim Murat, *Memoirs of the Life and Adventures of Colonel Maceroni*, 2 vols. (London, 1837).

Palmer, Alonzo B., *Diarrhoea and Dysentery: Modern Views of Their Pathology and Treatment* (Detroit, 1887).

Rose, Achilles, *Napoleon's Campaign in Russia Anno 1812: Medico-Historical* (New York: Published by the author, 1913).

Rothenberg, Gunther E., *The Art of Warfare in the Age of Napoleon* (Bloomington: Indiana University Press, 1978).

Ségur, Philippe de, *History of the Expedition to Russia Undertaken by the Emperor Napoleon in the Year 1812*, vol. 1 (London, 1840).

Talty, Stephan, *The Illustrious Dead: The Terrifying Story of How Typhus Killed Napoleon's Greatest Army* (New York: Crown, 2009).

Tarle, Eugene, *Napoleon's Invasion of Russia, 1812* (New York: Oxford University Press, 1942).

Tolstoy, Leo, *The Physiology of War: Napoleon and the Russian Campaign*, trans. Huntington Smith (New York, 1888).

———, *War and Peace*, trans. Orlando Figes (New York: Viking, 2006).

Virchow, Rudolf Carl, *On Famine Fever and Some of the Other Cognate Forms of Typhus* (London, 1868)

———, "Report on the Typhus Epidemic in Upper Silesia, 1848," *American Journal of Public Health* 96, no. 12 (December 2006): 2102–2105 (excerpt from R. C. Virchow, *Archiv für pathologische Anatomie und Physiologie und für klinische Medicin*, vol. 2 [Berlin, 1848]).

Voltaire, *History of Charles XII, King of Sweden* (Edinburgh, 1776).

Xavier, Nicolas, Hervé Granier, and Patrick Le Guen, "Shigellose ou dysenterie bacillaire," *Presse Médicale* 36, no. 11, pt. 2 (November 2007): 1606–1618.

Zamoyski, Adam, *1812: Napoleon's Fatal March on Moscow* (London: HarperCollins, 2004).

THE PARIS SCHOOL OF MEDICINE

Ackerknecht, Erwin H., *Medicine at the Paris Hospital, 1794–1848* (Baltimore: Johns Hopkins University Press, 1967).

———, "Recurrent Themes in Medical Thought," *Scientific Monthly* 69, no. 2 (August 1949): 80–83.

Cross, John, *Sketches of the Medical Schools of Paris, Including Remarks on the Hospital Practice, Lectures, Anatomical Schools, and Museums, and Exhibiting the Actual State of Medical Instruction in the French Metropolis* (London, 1815).

Foucault, Michel, *The Birth of the Clinic: An Archaeology of Medical Perception*, trans. A. M. Sheridan Smith (New York: Pantheon, 1973).

Hannaway, Caroline, and Ann La Berge, eds., *Constructing Paris Medicine* (Amsterdam: Rodopi, 1998).

Kervran, Roger, *Laennec: His Life and Times* (Oxford: Pergamon, 1960).

Locke, John, *Essay Concerning Human Understanding* (Oxford: Clarendon, 1924).

Paracelsus, Theophrastus, *Four Treatises of Theophrastus von Henheim, Called Paracelsus*, trans. Lilian Temkin, George Rosen, Gregory Zilboorg, and Henry E. Sigerist (Baltimore: Johns Hopkins University Press, 1941).

Shakespeare, William, *All's Well That Ends Well* (Raleigh, NC: Alex Catalogue, 2001).
Somerville, Asbury, "Thomas Sydenham as Epidemiologist," *Canadian Public Health Journal* 24, no. 2 (February 1933): 79–82.
Stensgaard, Richard K., "All's Well That Ends Well and the Gelenico-Paraceslian Controversy," *Renaissance Quarterly* 25, no. 2 (Summer 1972): 173–188.
Sue, Eugène, *Mysteries of Paris* (New York, 1887).
Sydenham, Thomas, *The Works of Thomas Sydenham*, 2 vols., trans. R. G. Latham (London, 1848–1850).
Temkin, Owsei, "The Philosophical Background of Magendie's Physiology," *Bulletin of the History of Medicine* 20, no. 1 (January 1946): 10–36.
Warner, John Harley, *Against the Spirit of System: The French Impulse in Nineteenth-Century Medicine* (Baltimore: Johns Hopkins University Press, 2003).

THE SANITARY MOVEMENT

Barnes, David S., *The Great Stink of Paris and the Nineteenth-Century Struggle against Filth and Germs* (Baltimore: Johns Hopkins University Press, 2006).
Chadwick, Edwin, *Public Health: An Address* (London, 1877).
———, *The Sanitary Condition of the Labouring Population of Great Britain*, ed. M. W. Flinn (Edinburgh: Edinburgh University Press, 1965; 1st ed. 1842).
Chevalier, Louis, *Laboring Classes and Dangerous Classes in Paris during the First Half of the Nineteenth Century*, trans. Frank Jellinek (New York: H. Fertig, 1973).
Cleere, Eileen, *The Sanitary Arts: Aesthetic Culture and the Victorian Cleanliness Campaigns* (Columbus: Ohio State University Press, 2014).
Dickens, Charles, *The Adventures of Oliver Twist* (London: Oxford University Press, 1949).
———, *Dombey and Son* (New York: Heritage, 1957).
———, *Martin Chuzzlewit* (Oxford: Oxford University Press, 2016).
Douglas, Mary, *Purity and Danger: An Analysis of the Concepts of Pollution and Taboo* (London: Routledge & K. Paul, 1966).
Eliot, George, *Middlemarch* (New York: Modern Library, 1984).
Engels, Friederich, *The Condition of the Working Class in England*, trans. Florence Kelly Wischnewetsky (New York, 1887).
Finer, Samuel Edward, *The Life and Times of Sir Edwin Chadwick* (London: Methuen, 1952).
Foucault, Michel, *Discipline and Punish: The Birth of the Prison*, trans. Alan Sheridan (New York: Vintage, 1979).
Frazer, W. A. *A History of English Public Health, 1834–1939* (London: Baillière, Tindall & Cox, 1950).
Gaskell, Elizabeth Cleghorn, *North and South* (London: Smith, Elder, 1907).
Goodlad, Lauren M. E., "'Is There a Pastor in the House?': Sanitary Reform, Professionalism, and Philanthropy in Dickens's Mid-Century Fiction," *Victorian Literature and Culture* 31, no. 2 (2003): 525–553.
Hamlin, Christopher, "Edwin Chadwick and the Engineers, 1842–1854: Systems and Antisystems in the Pipe-and-Brick Sewers War," *Technology and Culture* 33, no. 4 (1992): 680–709.

———, *Public Health and Social Justice in the Age of Chadwick* (Cambridge: Cambridge University Press, 1998).

Hanley, James Gerald, "All Actions Great and Small: English Sanitary Reform, 1840–1865," PhD diss., Yale University, 1998.

Hoy, Sue Ellen, *Chasing Dirt: The American Pursuit of Cleanliness* (New York: Oxford University Press, 1995).

La Berge, Ann F., "Edwin Chadwick and the French Connection," *Bulletin of the History of Medicine* 62, no. 1 (1988): 23–24.

Lewis, Richard Albert, *Edwin Chadwick and the Public Health Movement, 1832–1954* (London: Longmans, 1952).

Litsios, Socrates, "Charles Dickens and the Movement for Sanitary Reform," *Perspectives in Biology and Medicine* 46, no. 2 (Spring 2003): 183–199.

Mayhew, Henry, *London Labour and the London Poor* (London, 1865).

McKeown, Thomas, *The Modern Rise of Population* (London: Edward Arnold, 1976).

———, *The Role of Medicine: Dream, Mirage or Nemesis?* (Princeton: Princeton University Press, 1976).

Pinkney, David H., *Napoleon III and the Rebuilding of Paris* (Princeton: Princeton University Press, 1958).

Richardson, Benjamin Ward, *Hygeia: A City of Health* (London, 1876).

Rosen, George, *A History of Public Health* (Baltimore: Johns Hopkins University Press, 1993).

Ruskin, John, *Modern Painters*, 5 vols. (London, 1873).

Sivulka, Juliann, "From Domestic to Municipal Housekeeper: The Influence of the Sanitary Reform Movement on Changing Women's Roles in America, 1860–1920," *Journal of American Culture* 22, no. 4 (December 1999): 1–7.

Snowden, Frank, *Naples in the Time of Cholera, 1884–1911* (Cambridge: Cambridge University Press, 1995).

Southwood Smith, Thomas, *A Treatise on Fever* (Philadelphia, 1831).

Tomes, Nancy, *The Gospel of Germs: Men, Women, and the Microbe in American Life* (Cambridge, MA: Harvard University Press, 1988).

THE GERM THEORY OF DISEASE

Bertucci, Paola, *Artisanal Enlightenment: Science and the Mechanical Arts in Old Regime France* (New Haven: Yale University Press, 2017).

Brock, Thomas D., *Robert Koch: A Life in Medicine and Bacteriology* (Washington, DC: ASM, 1999).

Budd, William, *Typhoid Fever: Its Nature, Mode of Spreading, and Prevention* (London, 1873).

Clark, David P., *How Infectious Diseases Spread* (Upper Saddle River, NJ: FT Press Delivers, 2010).

Conant, James Bryant, *Pasteur's and Tyndall's Study of Spontaneous Generation* (Cambridge, MA: Harvard University Press, 1953).

Dobell, Clifford, *Antony van Leeuwenhoek and His "Little Animals"* (New York: Russell & Russell, 1958).

Dubos, René, *Pasteur and Modern Science* (Garden City, NY: Anchor, 1960).

———, *Pasteur's Study of Fermentation* (Cambridge, MA: Harvard University Press, 1952).

Cheyne, William Watson, *Lister and His Achievement* (London: Longmans, Green, 1925).

Gaynes, Robert P., *Germ Theory: Medical Pioneers in Infectious Diseases* (Washington, DC: ASM, 2011).

Geison, Gerald, *The Private Science of Louis Pasteur* (Princeton: Princeton University Press, 1995).

Guthrie, Douglas, *Lord Lister: His Life and Doctrine* (Edinburgh: Livingstone, 1949).

Harkness, Deborah, *The Jewel House: Elizabethan London and the Scientific Revolution* (New Haven: Yale University Press, 2007).

Kadar, Nicholas, "Rediscovering Ignaz Philipp Semmelweis (1818–1865)," *Journal of Obstetrics and Gynecology* (2018), https://doi.org/10.1016/j.ajog.2018.11.1084.

Knight, David C., *Robert Koch, Father of Bacteriology* (New York: F. Watts, 1961).

Koch, Robert, *Essays of Robert Koch*, trans. K. Codell Carter (New York: Greenwood, 1987).

Laporte, Dominique, *History of Shit* (Cambridge, MA: MIT Press, 2000).

Latour, Bruno, *The Pasteurization of France*, trans. Alan Sheridan and John Law (Cambridge, MA: Harvard University Press, 1988).

Lehoux, Daryn, *Creatures Born of Mud and Slime: The Wonder and Complexity of Spontaneous Generation* (Baltimore: Johns Hopkins University Press, 2017).

Long, Pamela O., *Artisan/Practitioners and the Rise of the New Sciences, 1400–1600* (Corvallis: Oregon State University Press, 2011).

Metchnikoff, Elie, *Founders of Modern Medicine: Pasteur, Koch, Lister* (Delanco, NJ: Gryphon, 2006).

Nakayama, Don K., "Antisepsis and Asepsis and How They Shaped Modern Surgery," *American Surgeon* 84, no. 6 (June 2018): 766–771.

Nuland, Sherwin B., *Doctors: The Biography of Medicine* (New York: Random House, 1988).

———, *The Doctors' Plague: Germs, Childbed Fever, and the Strange Story of Ignác Semmelweis* (New York: W. W. Norton, 2004).

Pasteur, Louis, *Germ Theory and Its Applications to Medicine and Surgery* (Hoboken, NJ: BiblioBytes, n.d.).

———, *Physiological Theory of Fermentation* (Hoboken, NJ: BiblioBytes, n.d.).

Radot, René Vallery, *Louis Pasteur: His Life and Labours*, trans. Lady Claud Hamilton (New York, 1885).

Ruestow, Edward G., *The Microscope in the Dutch Republic: The Shaping of Discovery* (Cambridge: Cambridge University Press, 1996).

Schlich, Thomas, "Asepsis and Bacteriology: A Realignment of Surgery and Laboratory Science," *Medical History* 56, no. 3 (July 2012): 308–334.

Semmelweis, Ignác, *The Etiology, the Concept, and the Prophylaxis of Childbed Fever*, trans. F. P. Murphy (Birmingham, AL: Classics of Medicine Library, 1981).

Smith, Pamela H., *The Body of the Artisan: Art and Experience in the Scientific Revolution* (Chicago: University of Chicago Press, 2004).

Tomes, Nancy, *The Gospel of Germs: Men, Women, and the Microbe in American Life* (Cambridge, MA: Harvard University Press, 1998).

CHOLERA

Andrews, Jason R., and Basu Sanjay, "Transmission Dynamics and Control of Cholera in Haiti: An Epidemic Model," *Lancet* 377, no. 9773 (April 2011): 1248–1255.

Belkin, Shimson, and Rita R. Colwell, eds., *Ocean and Health Pathogens in the Marine Environment* (New York: Springer, 2005).

Bilson, Geoffrey, *A Darkened House: Cholera in Nineteenth-Century Canada* (Toronto: University of Toronto, 1980).

Colwell, Rita R., "Global Climate and Infectious Disease: The Cholera Paradigm," *Science* 274, no. 5295 (December 20, 1996): 2025–2031.

Delaporte, François, *Disease and Civilization: The Cholera in Paris, 1832* (Cambridge, MA: MIT Press, 1986).

Durey, Michael, *The Return of the Plague: British Society and the Cholera, 1831–1832* (Dublin: Gill and Macmillan, 1979).

Echenberg, Myron, *Africa in the Time of Cholera: A History of Pandemics from 1817 to the Present* (Cambridge: Cambridge University Press, 2011).

Evans, Richard J., *Death in Hamburg: Society and Politics in the Cholera Years* (New York: Penguin, 2005).

———, "Epidemics and Revolutions: Cholera in Nineteenth-Century Europe," *Past and Present*, no. 120 (August 1988): 123–146.

Eyler, J. M., "The Changing Assessment of John Snow's and William Farr's Cholera Studies," *Soz Praventivmed* 46, no. 4 (2001): 225–232.

Fang, Xiaoping, "The Global Cholera Pandemic Reaches Chinese Villages: Population Mobility, Political Control, and Economic Incentives in Epidemic Prevention, 1962–1964," *Modern Asian Studies* 48, no. 3 (May 2014): 754–790.

Farmer, Paul, *Haiti after the Earthquake* (New York: Public Affairs, 2011).

Fazio, Eugenio, *L'epidemia colerica e le condizioni sanitarie di Napoli* (Naples, 1885).

Giono, Jean, *The Horseman on the Roof*, trans. Jonathan Griffin (New York: Knopf, 1954).

Hamlin, Christopher, *Cholera: The Biography* (Oxford: Oxford University Press, 2009).

Howard-Jones, Norman, "Cholera Therapy in the Nineteenth Century," *Journal of the History of Medicine* 17 (1972): 373–395.

Hu, Dalong, Bin Liu, Liang Feng, Peng Ding, Xi Guo, Min Wang, Boyang Cao, P. R. Reeves, and Lei Want, "Origins of the Current Seventh Cholera Pandemic," *Proceedings of the National Academy of Sciences of the United States of America* 113, no. 48 (2016): E7730–E7739.

Huq, A., S. A. Huq, D. J. Grimes, M. O'Brien, K. H. Chu, J. M. Capuzzo, and R. R. Colwell, "Colonization of the Gut of the Blue Crab (*Callinectes sapidus*) by *Vibrio cholerae*," *Applied Environmental Microbiology* 52 (1986): 586–588.

Ivers, Louise C., "Eliminating Cholera Transmission in Haiti," *New England Journal of Medicine* 376 (January 12, 2017): 101–103.

Jutla, Antarpreet, Rakibul Khan, and Rita Colwell, "Natural Disasters and Cholera Outbreaks: Current Understanding and Future Outlook," *Current Environmental Health Report* 4 (2017): 99–107.

Koch, Robert, *Professor Koch on the Bacteriological Diagnosis of Cholera, Water-Filtration and Cholera, and the Cholera in Germany during the Winter of 1892–93*, trans. George Duncan (Edinburgh, 1894).

Kudlick, Catherine Jean, *Cholera in Post-Revolutionary Paris: A Cultural History* (Berkeley: University of California Press, 1996).

Lam, Connie, Sophie Octavia, Peter Reeves, Lei Wang, and Ruiting Lan, "Evolution of the Seventh Cholera Pandemic and the Origin of the 1991 Epidemic in Latin America," *Emerging Infectious Diseases* 16, no. 7 (July 2010): 1130–1132.

Longmate, Norman, *King Cholera: The Biography of a Disease* (London: H. Hamilton, 1966).

McGrew, Roderick E., *Russia and the Cholera, 1823–1832* (Madison: University of Wisconsin Press, 1965).

Mekalanos, John, *Cholera: A Paradigm for Understanding Emergence, Virulence, and Temporal Patterns of Disease* (London: Henry Stewart Talks, 2009).

Morris, J. Glenn, Jr., "Cholera—Modern Pandemic Disease of Ancient Lineage," *Emerging Infectious Diseases* 17, no. 11 (November 2011): 2099–2104.

Morris, Robert John, *Cholera 1832: The Social Response to an Epidemic* (London: Croom Helm, 1976).

Munthe, Axel, *Letters from a Mourning City*, trans. Maude Valerie White (London, 1887).

Pelling, Margaret, *Cholera, Fever, and English Medicine, 1825–1865* (Oxford: Oxford University Press, 1978).

Pettenkofer, Max von, *Cholera: How to Prevent and Resist It*, trans. Thomas Whiteside Hine (London, 1875).

Piarroux, Renaud, Robert Barrais, Benoît Faucher, Rachel Haus, Martine Piarroux, Jean Gaudart, Roc Magloire, and Didier Raoult, "Understanding the Cholera Epidemic, Haiti," *Emerging Infectious Diseases* 17, no. 7 (July 2011): 1161–1168.

Pollitzer, R., *Cholera* (Geneva: World Health Organization, 1959).

Ramamurthy, T., *Epidemiological and Molecular Aspects of Cholera* (New York: Springer Science and Business, 2011).

Robbins, Anthony, "Lessons from Cholera in Haiti," *Journal of Public Health Policy* 35, no. 2 (May 2014): 135–136.

Rogers, Leonard, *Cholera and Its Treatment* (London: H. Frowde, Oxford University Press, 1911).

Rosenberg, Charles E., *The Cholera Years: The United States in 1832, 1849, and 1866* (Chicago: University of Chicago Press, 1987).

Seas, C., J. Miranda, A. I. Gil, R. Leon-Barua, J. Patz, A. Huq, R. R. Colwell, and R. B. Sack, "New Insights on the Emergence of Cholera in Latin America during 1991: The Peruvian Experience," *American Journal of Tropical Medicine and Hygiene* 62 (2000): 513–517.

Shakespeare, Edward O., *Report on Cholera in Europe and India* (Washington, DC, 1890).

Snow, John, *Snow on Cholera* (New York: The Commonwealth Fund; and London: Oxford University Press, 1936).
Snowden, Frank, *Naples in the Time of Cholera: 1884–1911* (Cambridge: Cambridge University Press, 1995).
Somma, Giuseppe, *Relazione sanitaria sui casi di colera avvenuti nella sezione di Porto durante l'epidemia dell'anno 1884* (Naples, 1884).
Twain, Mark, *Innocents Abroad* (Hartford, CT, 1869).
United States Congress, House Committee on Foreign Affairs, Subcommittee on Western Hemisphere Affairs, *The Cholera Epidemic in Latin America.* Hearing before the Subcommittee on Western Hemisphere Affairs of the Committee on Foreign Affairs, House of Representatives, One Hundred Second Congress, First Session, May 1, 1991 (Washington, DC: US Government Printing Office, 1991).
Van Heyningen, William Edward, *Cholera: The American Scientific Experience* (Boulder, CO: Westview, 1983).
Vezzulli, Luigi, Carla Pruzzo, Anwar Huq, and Rita R. Colwell, "Environmental Reservoirs of Vibrio cholerae and Their Role in Cholera," *Environmental Microbiology Reports* 2, no. 1 (2010): 27–35.
Wachsmuth, I. K., P. A. Blake, and Ø. Olsvik, eds., *Vibrio cholerae and Cholera: Molecular to Global Perspectives* (Washington, DC: ASM, 1994).
Wall, A. J., *Asiatic Cholera: Its History, Pathology and Modern Treatment* (London, 1893).
World Health Organization, *Guidelines for Cholera Control* (Geneva: World Health Organization, 1993).

TUBERCULOSIS

Abel, Emily K., Rima D. Apple, and Janet Golden, *Tuberculosis and the Politics of Exclusion: A History of Public Health and Migration to Los Angeles* (New Brunswick: Rutgers University Press, 2007).
Barnes, David S., *The Making of a Social Disease: Tuberculosis in Nineteenth-Century France* (Berkeley: University of California Press, 1995).
Bryder, Linda, *Below the Magic Mountain: A Social History of Tuberculosis in Twentieth-Century Britain* (Oxford: Oxford University Press, 1988).
Bulstrode, H. Timbrell, *Report on Sanatoria for Consumption and Certain Other Aspects of the Tuberculosis Question* (London: His Majesty's Stationery Office, 1908).
Bynum, Helen, *Spitting Blood: The History of Tuberculosis* (Oxford: Oxford University Press, 2012).
Carrington, Thomas Spees, *Tuberculosis Hospital and Sanatorium Construction* (New York: National Association for the Study and Prevention of Tuberculosis, 1911).
Chekov, Anton, *Five Plays,* trans. Marina Brodskaya (Stanford, CA: Stanford University Press, 2011).
Comstock, George W., "The International Tuberculosis Campaign: A Pioneering Venture in Mass Vaccination and Research," *Clinical Infectious Diseases* 19, no. 3 (September 1, 1994): 528–540.

Condrau, Flurin, and Michael Worboys, *Tuberculosis Then and Now: Perspectives on the History of an Infectious Disease* (Montreal: McGill–Queen's University Press, 2010).

Connolly, Cynthia Anne, *Saving Sickly Children: The Tuberculosis Preventorium in American Life, 1909–1970* (New Brunswick: Rutgers University Press, 2008).

Crowell, F. Elizabeth, *Tuberculosis Dispensary Method and Procedure* (New York: Vail-Ballou, 1916).

Day, Carolyn A., *Consumptive Chic: A History of Beauty, Fashion, and Disease* (London: Bloomsbury, 2017).

Dubos, René, and Jean Dubos, *The White Plague: Tuberculosis, Man, and Society* (Boston: Little, Brown, 1952).

Dutcher, Addison P., *Pulmonary Tuberculosis: Its Pathology, Nature, Symptoms, Diagnosis, Prognosis, Causes, Hygiene, and Medical Treatment* (Philadelphia, 1875).

Ellis, A. E., *The Rack* (Boston: Little Brown, 1958).

Fishbert, Maurice, *Pulmonary Tuberculosis*, 3rd ed. (Philadelphia: Lea & Febiger, 1922).

Gide, André, *The Immoralist*, trans. Richard Howard (New York: Alfred A. Knopf, 1970).

Goffman, Erving, *Asylums: Essays on the Social Situation of Mental Patients and Other Inmates* (Chicago: Aldine, 1961).

Hearing before the Subcommittee on Health and the Environment of the Committee on Energy and Commerce, House of Representatives, One Hundred Third Congress, First Session, "The Tuberculosis Epidemic," March 9, 1993 (3) (Washington, DC: US Government Printing Office, 1993).

Jacobson, Arthur C., *Tuberculosis and the Creative Mind* (Brooklyn, NY: Albert T. Huntington, 1909).

Johnson, Charles S., *The Negro in American Civilization* (New York: Holt, 1930).

Jones, Thomas Jesse, "Tuberculosis among the Negroes," *American Journal of the Medical Sciences* 132, no. 4 (October 1906): 592–600.

Knopf, Sigard Adolphus, *A History of the National Tuberculosis Association: The Anti-Tuberculosis Movement in the United States* (New York: National Tuberculosis Association, 1922).

———, *Pulmonary Tuberculosis* (Philadelphia, 1899).

Koch, Robert, "Die Atiologie der Tuberkulose," *Berliner Klinische Wochenschrift* 15 (1882): 221–230.

Laennec, René, *A Treatise of the Diseases of the Chest*, trans. John Forbes (London, 1821).

Lawlor, Clark, *Consumption and Literature: The Making of the Romantic Disease* (New York: Palgrave Macmillan, 2006).

Madkour, M. Monir, ed., *Tuberculosis* (Berlin: Springer-Verlag, 2004).

Mann, Thomas, *The Magic Mountain*, trans. H. T. Lowe-Porter (New York: Modern Library, 1992).

McMillen, Christian W., *Discovering Tuberculosis: A Global History, 1900 to the Present* (New Haven: Yale University Press, 2015).

Muthu, C., *Pulmonary Tuberculosis and Sanatorium Treatment: A Record of Ten Years' Observation and Work in Open-Air Sanatoria* (London: Baillière, Tindall and Cox, 1910).

National Tuberculosis Association, *A Directory of Sanatoria, Hospitals, Day Camps and Preventoria for the Treatment of Tuberculosis in the United States*, 9th ed. (New York: Livingston, 1931).

———, "Report of the Committee on Tuberculosis among Negroes" (New York: National Tuberculosis Association, 1937).

———, *Twenty-five Years of the National Tuberculosis Association* (New York: National Tuberculosis Association, 1929).

New York City Department of Health, *What You Should Know about Tuberculosis* (New York: J. W. Pratt, 1910).

Ott, Katherine, *Fevered Lives: Tuberculosis in American Culture since 1870* (Cambridge, MA: Harvard University Press 1996).

Pope, Alton S., "The Role of the Sanatorium in Tuberculosis Control," *Milbank Memorial Fund Quarterly* 16, no. 4 (October 1938): 327–337.

Pottenger, Francis M., *The Diagnosis and Treatment of Pulmonary Tuberculosis* (New York: William Wood, 1908).

Ransome, Arthur, *Researches on Tuberculosis* (London, 1898).

———, "Tuberculosis and Leprosy: A Parallel and a Prophecy," *Lancet* 148, no. 3802 (July 11, 1896): 99–104.

Reinhardt, Charles, and David Thomson, *A Handbook of the Open-Air Treatment* (London: John Bale, Sons & Danielsson, 1902).

Rothman, Sheila M., *Living in the Shadow of Death; Tuberculosis and the Social Experience of Illness in American History* (New York: Basic, 1994).

Sontag, Susan, *Illness as Metaphor* (New York: Vintage, 1979).

Stowe, Harriet Beecher, *Uncle Tom's Cabin, or Life among the Lowly* (New York: Penguin, 1981; 1st ed. 1852).

Trudeau, Edward Livingston, *An Autobiography* (Garden City, NY: Doubleday, Page, 1916).

Tuberculosis Commission of the State of Maryland, *Report of 1902–1904* (Baltimore: Sun Job Printing Office, 1904).

Vickery, Heather Styles, "'How Interesting He Looks in Dying': John Keats and Consumption," *Keats-Shelley Review* 32, no. 1 (2018): 58–63.

Villemin, Jean Antoine, *De la propagation de la phthisie* (Paris, 1869).

———, *Études sur la tuberculose* (Paris, 1868).

Walksman, Selman, *The Conquest of Tuberculosis* (Berkeley: University of California Press, 1964).

Walters, F. Rufenacht, *Sanatoria for Consumptives* (London: Swann Sonnenschein, 1902).

World Health Organization, *Global Tuberculosis Control: WHO Report 2010* (Geneva: World Health Organization, 2010).

———, *Global Tuberculosis Report 2015* (Geneva: World Health Organization, 2015).

MALARIA

Carson, Rachel, *Silent Spring* (Greenwich, CT: Fawcett, 1962).

Clyde, David F., *Malaria in Tanzania* (London: Oxford University Press, 1967).

Cueto, Marcos, *Cold War, Deadly Fevers: Malaria Eradication in Mexico, 1955–1975* (Washington, DC: Woodrow Wilson Center Press, 2007).

Desowitz, Robert S., *The Malaria Capers: Tales of Parasites and People* (New York: W. W. Norton, 1993).

Faid, M. A., "The Malaria Program: From Euphoria to Anarchy," *World Health Forum* 1 (1980): 8–22.

Farley, John A., "Mosquitoes or Malaria? Rockefeller Campaigns in the American South and Sardinia," *Parassitologia* 36 (1994): 165–173.

Hackett, Lewis Wendell, *Malaria in Europe: An Ecological Study* (London: Oxford University Press, 1937).

Harrison, Gordon, *Mosquitoes, Malaria, and Man: A History of the Hostilities since 1880* (New York: E. P. Dutton, 1978).

Humphreys, Margaret, *Malaria: Poverty, Race, and Public Health in the United States* (Baltimore: Johns Hopkins University Press, 2001).

Litsios, Socrates, *The Tomorrow of Malaria* (Karori: Pacific Press, 1996).

Logan, John A., *The Sardinian Project: An Experiment in the Eradication of an Indigenous Malarious Vector* (Baltimore: Johns Hopkins University Press, 1953).

MacDonald, George, *The Epidemiology and Control of Malaria* (London: Oxford University Press, 1957).

Packard, Randall M., *Making of a Tropical Disease: A Short History of Malaria* (Baltimore: Johns Hopkins University Press, 2007).

Pampana, Emilio J., *A Textbook of Malaria Eradication* (London: Oxford University Press, 1963).

Ross, Ronald, *Malarial Fever: Its Cause, Prevention and Treatment* (London: Longmans, Green, 1902).

Russell, Paul, *Man's Mastery of Malaria* (London: Oxford University Press, 1955).

Sallares, Robert, *Malaria and Rome: A History of Malaria in Ancient Italy* (Oxford: Oxford University Press, 2012).

Sherman, Irwin W., *Magic Bullets to Conquer Malaria from Quinine to Qinghaosu* (Washington, DC: ASM, 2011).

Slater, Leo B., *War and Disease: Biomedical Research on Malaria in the Twentieth Century* (New Brunswick: Rutgers University Press, 2009).

Snowden, Frank M., *The Conquest of Malaria: Italy, 1900–1962* (New Haven: Yale University Press, 2006).

Soper, Fred L., and D. Bruce Wilson, *Anopheles Gambiae in Brazil, 1930–1943* (New York: Rockefeller Foundation, 1949).

Tognotti, Eugenia, *La malaria in Sardegna: Per una storia del paludismo nel Mezzogiorno, 1880–1950* (Milan: F. Angeli, 1996).

Verga, Giovanni, *Little Novels of Sicily*, trans. D. H. Lawrence (New York: Grove Press, 1953).

Webb, James L. A., Jr., *Humanity's Burden: A Global History of Malaria* (Cambridge: Cambridge University Press, 2009).

POLIO

Aylward, R., "Eradicating Polio: Today's Challenges and Tomorrow's Legacy," *Annals of Tropical Medicine & Parasitology* 100, nos. 5/6 (2006): 1275–1277.

Aylward, R., and J. Linkins, "Polio Eradication: Mobilizing and Managing the Human Resources," *Bulletin of the World Health Organization* 83, no. 4 (2005): 268–273.

Aylward, R., and C. Maher, "Interrupting Poliovirus Transmission: New Solutions to an Old Problem," *Biologicals* 34, no. 2 (2006): 133–139.

Closser, Svea, *Chasing Polio in Pakistan: Why the World's Largest Public Health Initiative May Fail* (Nashville, TN: Vanderbilt University Press, 2010).

Flexner, Simon, *Nature, Manner of Conveyance and Means of Prevention of Infantile Paralysis* (New York: Rockefeller Institute for Medical Research, 1916).

"Global Poliomyelitis Eradication Initiative: Status Report," *Journal of Infectious Diseases* 175, suppl. 1 (February 1997).

Jacobs, Charlotte, *Jonas Salk: A Life* (New York: Oxford University Press, 2015).

National Foundation for Infantile Paralysis, *Infantile Paralysis: A Symposium Delivered at Vanderbilt University, April 1941* (New York: National Foundation for Infantile Paralysis, 1941).

New York Department of Health, *Monograph on the Epidemic of Poliomyelitis (Infantile Paralysis) in New York City in 1916* (New York: Department of Health, 1917).

Offit, Paul A., *The Cutter Incident: How America's First Polio Vaccine Led to the Growing Vaccine Crisis* (New Haven: Yale University Press, 2005).

Oshinsky, David M., *Polio: An American Story* (New Haven: Yale University Press, 2005).

Paul, John., *History of Poliomyelitis* (New Haven: Yale University Press, 1971).

Renne, Elisha P., *The Politics of Polio in Northern Nigeria* (Bloomington: Indiana University Press, 2010).

Roberts, Leslie, "Alarming Polio Outbreak Spreads in Congo, Threatening Global Eradication Efforts," *Science* (July 2, 2018), http://www.sciencemag.org/news/2018/07/polio-outbreaks-congo-threaten-global-eradication.

Rogers, Naomi, *Dirt and Disease: Polio before FDR* (New Brunswick: Rutgers University Press, 1992).

Sabin, Albert, "Eradication of Smallpox and Elimination of Poliomyelitis: Contrasts in Strategy," *Japanese Journal of Medical Science and Biology* 34, no. 2 (1981): 111–112.

———, "Field Studies with Live Poliovirus Vaccine and Their Significance for a Program of Ultimate Eradication of the Disease," *Academy of Medicine of New Jersey Bulletin* 6, no. 3 (1960): 168–183.

———, "Present Status of Field Trials with an Oral, Live Attenuated Poliovirus Vaccine," *JAMA* 171 (1959): 864–868.

Salk, Jonas E., "Considerations in the Preparation and Use of Poliomyelitis Virus Vaccine," *Journal of the American Medical Association* 158 (1955): 1239–1248.

———, *Poliomyelitis Vaccine in the Fall of 1955* (New York: National Foundation for Infantile Paralysis, 1956).

Seytre, Bernard, *The Death of a Disease: A History of the Eradication of Poliomyelitis* (New Brunswick: Rutgers University Press, 2005).

Shell, Marc, *Polio and Its Aftermath: The Paralysis of Culture* (Cambridge, MA: Harvard University Press, 2005).

Wilson, Daniel J., *Living with Polio: The Epidemic and Its Survivors* (Chicago: University of Chicago Press, 2005).

Wilson, James Leroy, *The Use of the Respirator in Poliomyelitis* (New York: National Foundation for Infantile Paralysis, 1940).

World Health Organization, Seventy-First World Health Assembly, "Eradication of Poliomyelitis: Report by the Director-General," March 20, 2018, http://apps.who.int/gb/ebwha/pdf_files/WHA71/A71_26-en.pdf.

HIV/AIDS

Antonio, Gene, *The AIDS Cover-Up? The Real and Alarming Facts about AIDS* (San Francisco: Ignatius Press, 1986).

Baxen, Jean, and Anders Breidlid, eds., *HIV/AIDS in Sub-Saharan Africa: Understanding the Implications of Culture and Context* (Claremont: UCT Press, 2013).

Berkowitz, Richard, *Stayin' Alive: The Invention of Safe Sex, a Personal History* (Boulder, CO: Westview, 2003).

Berkowitz, Richard, and Michael Callen, *How to Have Sex in an Epidemic: One Approach* (New York: News from the Front Publications, 1983).

Bishop, Kristina Monroe, "Anglo American Media Representations, Traditional Medicine, and HIV/AIDS in South Africa: From *Muti* Killings to Garlic Cures," *GeoJournal* 77 (2012): 571–581.

Bonnel, Rene, *Funding Mechanisms for Civil Society: The Experience of the AIDS Response* (Washington, DC: World Bank, 2012).

Buiten, Denise, and Kammila Naidoo, "Framing the Problem of Rape in South Africa: Gender, Race, Class and State Histories," *Current Sociology* 64, no. 4 (2016): 535–550.

Decoteau, Claire Laurier, *Ancestors and Antiretrovirals: The Bio-Politics of HIV/AIDS in Post-Apartheid South Africa* (Chicago: Chicago University Press, 2013).

Dosekun, Simidele, "'We Live in Fear, We Feel Very Unsafe': Imagining and Fearing Rape in South Africa," *Agenda: Empowering Women for Gender Equity*, no. 74 (2007): 89–99.

Duesberg, Peter, Claus Koehnlein, and David Rasnick, "The Chemical Bases of the Various AIDS Epidemics: Recreational Drugs, Anti-Viral Chemotherapy and Malnutrition," *Journal of Biosciences* 28, no. 4 (June 2003): 383–422.

"The Durban Declaration," *Nature* 406, no. 6791 (July 6, 2000): 15–16.

Farmer, Paul, *AIDS and Accusation: Haiti and the Geography of Blame* (Berkeley: University of California Press, 2006).

Fourie, Pieter, *The Political Management of HIV and AIDS in South Africa: One Burden Too Many?* (New York: Palgrave Macmillan, 2006).

Gevisser, Mark, *Thabo Mbeki: The Dream Deferred* (Johannesburg: Jonathan Balo, 2007).

Gqola, Pumla Dineo, *Rape: A South African Nightmare* (Auckland Park: MF Books Joburg, 2015).

Grmek, Mirko, *History of AIDS: Emergence and Origin of a Modern Pandemic* (Princeton: Princeton University Press, 1990).

Gumede, William Mervin, *Thabo Mbeki and the Battle for the Soul of the ANC* (London: Zed Books, 2007).

Holmes, King, *Disease Control Priorities: Major Infectious Diseases* (Washington, DC: World Bank, 2016).

Hunter, Susan, *Black Death: AIDS in Africa* (New York: Palgrave Macmillan, 2003).

Johnson, David K., *The Lavender Scare: The Cold War Persecution of Gays and Lesbians in the Federal Government* (Chicago: University of Chicago Press, 2004).

Karim, S. S. Abdool, and Q. Abdool Karim, *HIV/AIDS in South Africa* (Cambridge: Cambridge University Press, 2010).

Koop, C. Everett, *Understanding AIDS* (Rockville, MD: US Department of Health and Human Services, 1988).

Kramer, Larry, *The Normal Heart and the Destiny of Me* (New York: Grove, 2000).

———, *Reports from the Holocaust: The Story of an AIDS Activist* (New York: St. Martin's, 1994).

Larson, Jonathan, *Rent* (New York: Rob Weisbach Books, William Morrow, 1997).

McIntyre, James, and Glenda Gray, "Preventing Mother-to-Child Transmission of HIV: African Solutions for an African Crisis," *Southern African Journal of HIV Medicine* 1, no. 1 (July 25, 2000): 30–31.

Naidoo, Kammila, "Rape in South Africa—A Call to Action," *South African Medical Journal* 103, no. 4 (April 2013): 210–211.

Patton, Cindy, *Globalizing AIDS* (Minneapolis: University of Minnesota Press, 2002).

Pépin, Jacques, *Origins of AIDS* (Cambridge: Cambridge University Press, 2011).

Piot, Peter, *No Time to Lose: A Life in Pursuit of Deadly Viruses* (New York: W. W. Norton, 2012).

Powers, T., "Institutions, Power and Para-State Alliances: A Critical Reassessment of HIV/AIDS Politics in South Africa, 1999–2008," *Journal of Modern African Studies* 12, no. 4 (December 2013): 605–626.

Rohleder, Poul, *HIV/ADS in South Africa 25 Years On: Psychosocial Perspectives* (New York: Springer-Verlag, 2009).

Sangaramoorthy, Thurka, *Treating AIDS: Politics of Difference, Paradox of Prevention* (New Brunswick: Routledge, 2014).

Shilts, Randy, *And the Band Played On: Politics, People, and the AIDS Epidemic* (New York: St. Martin's, 1987).

Statistics South Africa, "Statistical Release P0302: Mid-Year Population Estimates, 2017" (Pretoria, South Africa, 2017).

UNAIDS, *Global AIDS Update 2016* (Geneva: Joint United Nations Programme on HIV/AIDS, 2016), http://www.unaids.org/sites/default/files/media_asset/global-AIDS-update-2016_en.pdf.

———, *Report on the Global AIDS Epidemic 2008* (Geneva: Joint United Nations Programme on HIV/AIDS, 2008), http://www.unaids.org/sites/default/files/media_asset/jc1510_2008globalreport_en_0.pdf.

———, *UNAIDS Data 2017* (Geneva: Joint United Nations Programme on HIV/AIDS, 2017), http://www.unaids.org/sites/default/files/media_asset/20170720_Data_book_2017_en.pdf.

Vale, Peter, and Georgina Barrett, "The Curious Career of an African Modernizer: South Africa's Thabo Mbeki," *Contemporary Politics* 15, no. 4 (December 2009): 445–460.

Verghese, Abraham, *My Own Country: A Doctor's Story* (New York: Vintage, 1994).

Weinel, Martin, "Primary Source Knowledge and Technical Decision-Making: Mbeki and the AZT Debate," *Studies in History and Philosophy of Science* 38, no. 4 (2007): 748–760.

Whiteside, Alan, *HIV/AIDS: A Very Short Introduction* (New York: Oxford University Press, 2008).

EMERGING DISEASES: SARS AND EBOLA

Adams, Lisa V., *Diseases of Poverty: Epidemiology, Infectious Diseases, and Modern Plagues* (Hanover, NH: Dartmouth College Press, 2015).

African Development Fund, Agriculture and Agro-Industry Department, "Republic of Guinea: Completion Report on Diecke Oil Palm and Rubber Project, Phase III, SOGUIPAH III," April 2008, https://www.afdb.org/fileadmin/uploads/afdb/Documents/Project-and-Operations/ADF-BD-IF-2008-123-EN-GUINEA-PCR-SOGUIPAHIII.PDF.

Atlim, George A., and Susan J. Elliott, "The Global Epidemiologic Transition," *Health Education & Behavior* 43, no. 1 suppl. (April 1, 2016): 37S–55S.

Badrun, Muhammad, *Milestone of Change: Developing a Nation through Oil Palm 'PIR'* (Jakarta: Directorate General of Estate Crops, 2011).

Barani, Achmad Mangga, *Palm Oil: A Gold Gift from Indonesia to the World* (Jakarta: Directorate General of Estate Crops, 2009).

Beltz, Lisa A., *Bats and Human Health: Ebola, SARS, Rabies, and Beyond* (Hoboken, NJ: John Wiley & Sons, 2018).

———, *Emerging Infectious Diseases: A Guide to Diseases, Causative Agents, and Surveillance* (San Francisco: Jossey-Bass, 2011).

Brown, J., and P. Chalk, *The Global Threat of New and Reemerging Infectious Diseases: Reconciling U.S. National Security and Public Health Policy* (Santa Monica: RAND, 2003).

Bullard, Stephan Gregory, *A Day-by-Day Chronicle of the 2013–2016 Ebola Outbreak* (Cham: Springer International, 2018).

Burnet, Frank Macfarlane, *Natural History of Infectious Diseases*, 4th rev. ed. (Cambridge: Cambridge University Press, 1972; 1st ed. 1953).

Centers for Disease Control and Prevention, *The Road to Zero: CDC's Response to the West African Ebola Epidemic, 2014–2015* (Atlanta: US Department of Health and Human Services, 2015).

Childs, James E., ed., *Wildlife and Emerging Zoonotic Diseases: The Biology, Circumstances, and Consequences of Cross-Species Transmission* (Heidelberg: Springer-Verlag, 2007).

Close, William T., *Ebola: A Documentary Novel of Its First Explosion* (New York: Ivy Books, 1995).

Cockburn, Aidan, ed., *The Evolution and Eradication of Infectious Diseases* (Baltimore: Johns Hopkins, 1963).

———, ed. *Infectious Diseases: Their Evolution and Eradication* (Springfield, IL: C. C. Thomas, 1967).

Corley, R. H. V., and P. B. H. Tinker, *The Oil Palm*, 5th ed. (Chichester: John Wiley, 2016).

Davis, J. R., and J. Lederberg, eds., *Public Health Systems and Emerging Infections: Assessing the Capabilities of the Public and Private Sectors* (Washington, DC: National Academy Press, 2000).

Evans, Nicholas G., and Tara C. Smith, eds., *Ebola's Message: Public Health and Medicine in the Twenty-First Century* (Cambridge, MA: MIT Press, 2016).

Fidler, David P., *SARS: Governance and the Globalization of Disease* (New York: Palgrave Macmillan, 2004).

Fong, I. W., *Antimicrobial Resistance and Implications for the Twenty-First Century* (Boston: Springer Science and Business Media, 2008).

———, *Emerging Zoonoses: A Worldwide Perspective* (Cham, Switzerland: Springer International, 2017).

Garrett, Laurie, *The Coming Plague: Newly Emerging Diseases in a World Out of Balance* (New York: Hyperion, 2000).

Green, Andrew, "Ebola Outbreak in the DR Congo: Lessons Learned," *Lancet* 391, no. 10135 (May 26, 2018): 2096, https://doi.org/10.1016/S0140-6736(18)31171-1.

Gross, Michael, "Preparing for the Next Ebola Epidemic," *Current Biology* 28, no. 2 (January 22, 2018): R51–R54.

Hinman, E. Harold, *World Eradication of Infectious Diseases* (Springfield, IL: C. C. Thomas, 1966).

Institute of Medicine, *Emerging Infections: Microbial Threats to Health in the United States* (Washington, DC: National Academy Press, 1992).

Knobler, Stacey, Adel Mahmoud, Stanley Lemon, Alison Mack, Laura Sivitz, and Katherine Oberholtzer, eds., *Learning from SARS: Preparing for the Next Disease Outbreak* (Washington, DC: National Academies Press, 2004).

Lo, Terence Q., Barbara J. Marston, Benjamin A. Dahl, and Kevin M. De Cock, "Ebola: Anatomy of an Epidemic," *Annual Review of Medicine* 68 (2017): 359–370.

Loh, Christine, *At the Epicentre: Hong Kong and the SARS Outbreak* (Baltimore: Project MUSE, 2012).

Maconachie, Roy, and Hilson, Gavin, "'The War Whose Bullets You Don't See': Diamond Digging, Resilience and Ebola in Sierra Leone," *Journal of Rural Studies* 61 (July 2018): 110–122, https://doi.org/10.1016/j.jrurstud.2018.03.009.

Malaysian Palm Oil Board, *Going for Liquid Gold: The Achievements of the Malaysian Palm Oil Board* (Kuala Lumpur: Ministry of Plantation Industries and Commodities, 2010).

McLean, Angela, Robert May, John Pattison, and Robin Weiss, eds., *SARS: A Case Study in Emerging Infections* (Oxford: Oxford University Press, 2005).

Médecins Sans Frontières, *Pushed to the Limit and Beyond: A Year into the Largest Ever Ebola Outbreak,* March 23, 2015, https://www.msf.org/ebola-pushed-limit-and-beyond.

Mehlhorn, Heinz, *Arthropods as Vectors of Emerging Diseases* (Berlin: Springer, 2012).

Mol, Hanneke, *The Politics of Palm Oil Harm: A Green Criminological Perspective* (Cham: Springer, 2017).

Monaghan, Karen, *SARS: Down but Still a Threat* (Washington, DC: National Intelligence Council, 2003).

Mooney, Graham, "Infectious Diseases and Epidemiologic Transition in Victorian Britain? Definitely," *Social History of Medicine* 12, no. 3 (December 1, 2007): 595–606.

Nohrstedt, Daniel, and Erik Baekkeskov, "Political Drivers of Epidemic Response: Foreign Healthcare Workers and the 2014 Ebola Outbreak," *Disasters* 42, no. 1 (January 2018): 412–461.

Olsson, Eva-Karin, *SARS from East to West* (Lanham, MD: Lexington Books, 2012).

Omran, Abdel R., "A Century of Epidemiologic Transition in the United States," *Preventive Medicine* 6, no. 1 (March 1977): 30–51.

———, "The Epidemiologic Transition: A Theory of the Epidemiology of Population Change," *Milbank Quarterly* 83, no. 4 (2005): 731–757.

———, "The Epidemiologic Transition Theory: A Preliminary Update," *Journal of Tropical Pediatrics* 29, no. 6 (December 1983): 305–316.

Preston, Richard, *Hot Zone* (New York: Kensington, 1992).

Qureshi, Adnan I., *Ebola Virus Disease* (London: Academic Press, 2016).

Rulli, Maria Cristina, Monia Santini, David T. S. Hayman, and Paolo D'Odorico, "The Nexus between Forest Fragmentation and Ebola Virus Disease Outbreaks," *Scientific Reports* 7, 41613, doi: 10.1038/srep41613 (2017).

Satcher, David, "Emerging Infections: Getting Ahead of the Curve," *Emerging Infectious Diseases* 1, no. 1 (January–March 1995): 1–6.

United Nations Development Programme, Human Development Reports, *2016 Human Development Report,* http://hdr.undp.org/en/2016-report.

United States Congress, Senate Committee on Health, Education, Labor and Pensions and Subcommittee on Labor, Health and Human Services, Education and Related Agencies of the Senate Committee on Appropriations, *Joint Hearing Examining Ebola in West Africa, Focusing on a Global Challenge and Public Health Threat*, September 2014 (Washington DC: US Government Printing Office, 2017).

United States Congress, Senate Committee on Labor and Human Resources, *Emerging Infections: A Significant Threat to the Nation's Health* (Washington, DC: US Government Printing Office, 1996).

United States Department of Defense, *Addressing Emerging Infectious Disease Threats: A Strategic Plan for the Department of Defense* (Washington, DC: US Government Printing Office, 1998).

Wallace, Robert G., and Rodrick Wallace, eds., *Neoliberal Ebola: Modeling Disease Emergence from Finance to Forest and Farm* (Cham, Switzerland: Springer International, 2016).

Washer, Peter, *Emerging Infectious Diseases and Society* (New York: Palgrave Macmillan, 2010).

World Bank, *The Economic Impact of the 2014 Epidemic: Short and Medium Estimates for West Africa* (Washington, DC: World Bank, 2014).

World Rainforest Movement, "Oil Palm and Rubber Plantations in Western and Central Africa: An Overview," WRM Briefing, December 15, 2008, https://wrm.org.uy/wp-content/uploads/2013/01/Western_Central_Africa.pdf.

Zuckerman, Molly, "The Evolution of Disease: Anthropological Perspectives on Epidemiologic Transitions," *Global Health Action* 7 (2014): 1–8.

Index

Page numbers in *italics* refer to illustrations.